TELECOMMUNICATIONS INTERNETWORKING

MCGRAW-HILL TELECOMMUNICATIONS

Ali *Digital Switching Systems*
Ash *Dynamic Routing in Telecommunications Networks*
Azzam/Ransom *Broadband Access Technologies*
Bartlett *Cable Communications*
Bates *Broadband Telecommunications Handbook*
Bayer *Computer Telephony Demystified*
Bedell *Cellular/PCS Management*
Clayton *McGraw-Hill Illustrated Telecom Dictionary*
Davis *ATM for Public Networks*
Gallagher *Mobile Telecommunications Networking with IS-41*
Harte *Cellular and PCS: The Big Picture*
Harte *CDMA IS-95*
Harte *GMS Superphones*
Heldman *Competitive Telecommunications*
Lacks *Fiber Optics Communications*
Lee *Mobile Cellular Telecommunications 2/e*
Lee *Mobile Communications Engineering 2/e*
Macario *Cellular Radio 2/e*
Muller *Desktop Encyclopedia of Telecommunications*
Muller *Desktop Encyclopedia of Voice and Data Networking*
Muller *Mobile Telecommunications Factbook*
Pattan *Satellite-Based Cellular Communications*
Pecar *Telecommunications Factbook 2/e*
Richharia *Satellite Communications Systems 2/e*
Roddy *Satellite Communications 2/e*
Rohde/Whitaker *Communications Receivers 2/e*
Russell *Signaling System #7 3/e*
Russell *Telecommunications Protocols 2/e*
Russell *Telecommunications Pocket Reference*
Shepard *Telecommunications Convergence*
Simon *Spread Spectrum Communications Handbook*
Smith *Cellular System Design and Optimization*
Smith *Practical Cellular and PCS Design*
Smith *Wireless Telecom FAQs*
Turin *Digital Transmission Systems*
Winch *Telecommunications Transmission System 2/e*

Telecommunications Internetworking

P.J. Louis

McGraw-Hill
New York San Francisco Washington, D.C.
Auckland Bogotá Caracas Lisbon London
Madrid Mexico City Milan Montreal New Delhi
San Juan Singapore Sydney Tokyo Toronto

McGraw-Hill

A Division of The ***McGraw-Hill*** *Companies*

1 2 3 4 5 6 7 8 9 0 DOC/DOC 9 0 9 8 7 6 5 4 3 2 1 0 9

ISBN 0-07-135654-1

The sponsoring editor for this book was Steve Chapman and the production supervisor was Pamela A. Pelton. It was set in Century Schoolbook by D & G Limited, LLC.

Printed and bound by R. R. Donnelley & Sons Company

This book is printed on recycled, acid-free paper containing a minimum of 50 percent recycled, de-inked fiber.

Dedicated to my darling wife Donna and our children Eric and Scott. Their patience, tolerance, and love are at the heart of my soul. This book would not have been possible without their love and support.

Also dedicated to my late father, Richard Louis, who urged me to do better than he had done, go farther than he had gone, be self-reliant, learn from my mistakes and the mistakes of others, respect history, and be decisive. My father's favorite bit of advice: "Stop talking about it and do it."

CONTENTS

PREFACE

In the last 100 years, the telecommunications industry has grown increasingly complex. As we enter the next century, telecommunications professionals face the unenviable task of maintaining a network that is actually a network of networks. This network of networks serves a subscriber base that provides service to millions of people across the globe and is comprised of traditional wireline telephony, cellular service, *Personal Communications Service* (PCS), paging, satellite, Internet, and cable television. The list of services is, in fact, larger than what I have shown. The list of network and service providers is growing every day, and some providers are combining services while others are simply entering markets that have been largely ignored until today.

The telecommunications industry can be described as one that is growing explosively under an umbrella of deregulation. The picture appears so complex that telecommunications professionals might believe that the task of understanding how to design and engineer the network is nearly impossible. Even the telecommunications business professional might view the network of networks as impossible to understand as a business segment to manage.

The telecommunications professional who is attempting to understand the changes in the industry can begin with reading this book. Understanding telecommunications networks is not an overwhelming task. The first concept to understand is that all telecommunications networks have one factor in common: networks move information from one place to another. Telecommunications networks can be designed and engineered by using a set of common practices. Furthermore, network interconnection plays a pivotal role in today's networks. Interconnection has both a technical and regulatory aspect and is pivotal in the convergence of network technologies. This technology convergence that we refer to plays a key role in the growth and evolution of the telecommunications industry.

This book offers a balanced view of the technical and marketing issues that are associated with networks and their interconnection and covers the latest in technologies and systems. The book also has been organized to assist technical and non-technical readers alike.

The reader will be taken on a tour of the network(s)—a tour that begins with telecommunications basics, winds its way through the network and its various incarnations, and eventually finishes with an understanding of how interconnected and intertwined telecommunications has become. The telecommunications manager and business professional will gain a solid and broad understanding of the network.

INTRODUCTION

A network is a collection of objects that are interconnected and that transfer information. A network can consist of people, telephones, televisions, radios, computers, etc. The network enables the transfer of information between people and devices.

The objects can be close or far apart. The information can be voice, data, e-mail, video, or any combination of these elements. The information can also be moved over a variety of physical transmission media by using a variety of protocols.

Networks are designed to meet current and future needs of customers in a cost-effective manner. Networks tend to be classified by business segment (e.g., local telephone service, long-distance service, cable television, cellular service, paging service, *Personal Communications Services* or PCS, and LMDS). Traditional distinctions between local and long-distance providers or cellular and PCS providers are diminishing. The driving force behind the growth and changes in the network in the United States is the Telecommunications Act of 1996, which attempted to stimulate competition among service providers by reducing regulatory barriers. The Act was signed into law on February 8, 1996. The Act's profound impact on the U.S. telecommunications industry can be characterized by mergers, consolidation, and diversification. Once considered unimaginable, *Local Exchange Carriers* (LECs) are offering Internet access and cable TV; long-distance service providers are entering the local exchange carrier business and cable TV business; and cable TV providers are offering subscribers wireline telephony. The local exchange carriers are today entering the long-distance market. The lines of business and technical distinction are blurring.

The merging of wireless, wireline telephony, satellite, cable TV, data, the Internet, and entertainment to form information networks is known as *convergence*. Convergence accounts for both the technical and business aspects of integration of technology and business. Convergence is an old concept that has been given new life by the regulatory and business constructs of today.

As noted, interconnection plays a pivotal role in convergence. An important step towards convergence is the change in the way that fixed network interconnection has been perceived. Service integration follows later.

Today, the telecommunications industry can be difficult to comprehend as technical, regulatory, and business aspects of the industry intertwine. Technical decisions can rarely be made without cost optimization and therefore without understanding the business framework in which the technology will be used. The new regulations (or lack of regulations) have had a direct impact on how networks are interconnected, which services can be offered, which services must be offered (e.g., local number portability), and which companies can provide services. This book will provide the reader with both a technical foundation and an appreciation for the business opportunities that are available. The reader will be encouraged to take a step back and to observe common aspects of different types of networks, technology, and services.

Through the use of common network design concepts and network signaling, we can lay the foundation of service integration. Network interconnection between providers of different types of communications services is the first step toward integration and convergence. Network interconnection today among service providers, however, is still based on a network architecture hierarchy in which one network plays a dominant role while the other network is subservient. The relationship between service providers is as much a technical one as it is a business relationship. Future networks are likely to communicate with each other as peer networks.

This book lays the foundation for understanding the various networks in such a way that this vague concept called a network is simpler to understand. The subjects of the chapters in this book are as follows:

1. Chapter 1, "What Is a Telecommunications Network?," provides a basic introduction to the concept of a network and its components.
2. Chapter 2, "The Telecommunications Hub—Creating Value," provides the reader with a foundation in network interconnection, network services, and service provider perceptions. I will describe how interconnection facilitates business relationships. The reader will also be given a view of the relationship between the convergence of technology and network interconnection (to some extent, they facilitate each other). Basic engineering concepts shall also be addressed. Product managers will be given a foundation to understand what network interconnection is and how it can be used to facilitate products and can be treated as a product itself.
3. Chapter 3, "Basic Network Technologies," addresses network architectural issues and network technologies.
4. Chapter 4, "Network Signaling and Its Applications," provides a foundation in network signaling and its applications in supporting subscriber services.

5. Chapter 5, "Applications," provides a more detailed view of the kinds of applications and services that can be derived from the interconnection of networks. The reader will be asked the question, "What can you leverage in your network as a product?"
6. Chapter 6, "Wireline Telephone Networks," describes the wireline telephone system and its relation to other networks and gives a brief history of the telephone system, from its beginning to the AT&T divestiture. This chapter will provide a description of the basic network elements. A description of the basic network configurations used is also addressed.
7. Chapter 7, "Wireless—Cellular and Personal Communications Services Networks," is an overview of the cellular system and the PCS system. A description of the basic network elements of both systems and the basic network configuration used by both are included.
8. Chapter 8, "Paging Systems," addresses the paging network. A description of the basic network elements of the paging network is included, and the network configurations that are used by the paging network are described.
9. Chapter 9, "Satellite Communications Systems," provides an overview of the satellite system. This chapter includes a description of the basic network elements of the satellite network.
10. Chapter 10, "Cable Television Networks," addresses current cable television networks and projected networks.
11. Chapter 11, "The Internet," includes a description of the basic network elements of the Internet.
12. Chapter 12, "The Economics of and Requirements for Becoming a Telecommunications Carrier," addresses network economics, covering topics such as usage-based pricing, flat-rate pricing, and savings versus expenditures. This chapter also addresses the future of the network.

The appendices in this book are as follows:

1. Appendix A—Common Terms and Definitions
2. Appendix B—Network Interconnection Document Summary
3. Appendix C—Telecommunications Act of 1996 Summary

ACKNOWLEDGMENTS

The late Harry E. Young; my friend, superior, and mentor. Upon his death, many had asked who would replace Harry Young as the telecommunications industry's wireless network interconnection expert. My answer was and still is: No One.

To Charles P. Eifinger (a 40 year veteran of the Bell System who retired in 1990), who mentored me early in my career and continues to do so 20 years later. "Make sure you look at the problem from at least two viewpoints; remember to turn the bit on and then turn the bit off. Sleep on it, get back to me with your answer, and be prepared to explain your answer."

To the old Bell System, which taught me to think and love telecommunications.

To all of my colleagues who have influenced me through our interactions and debates.

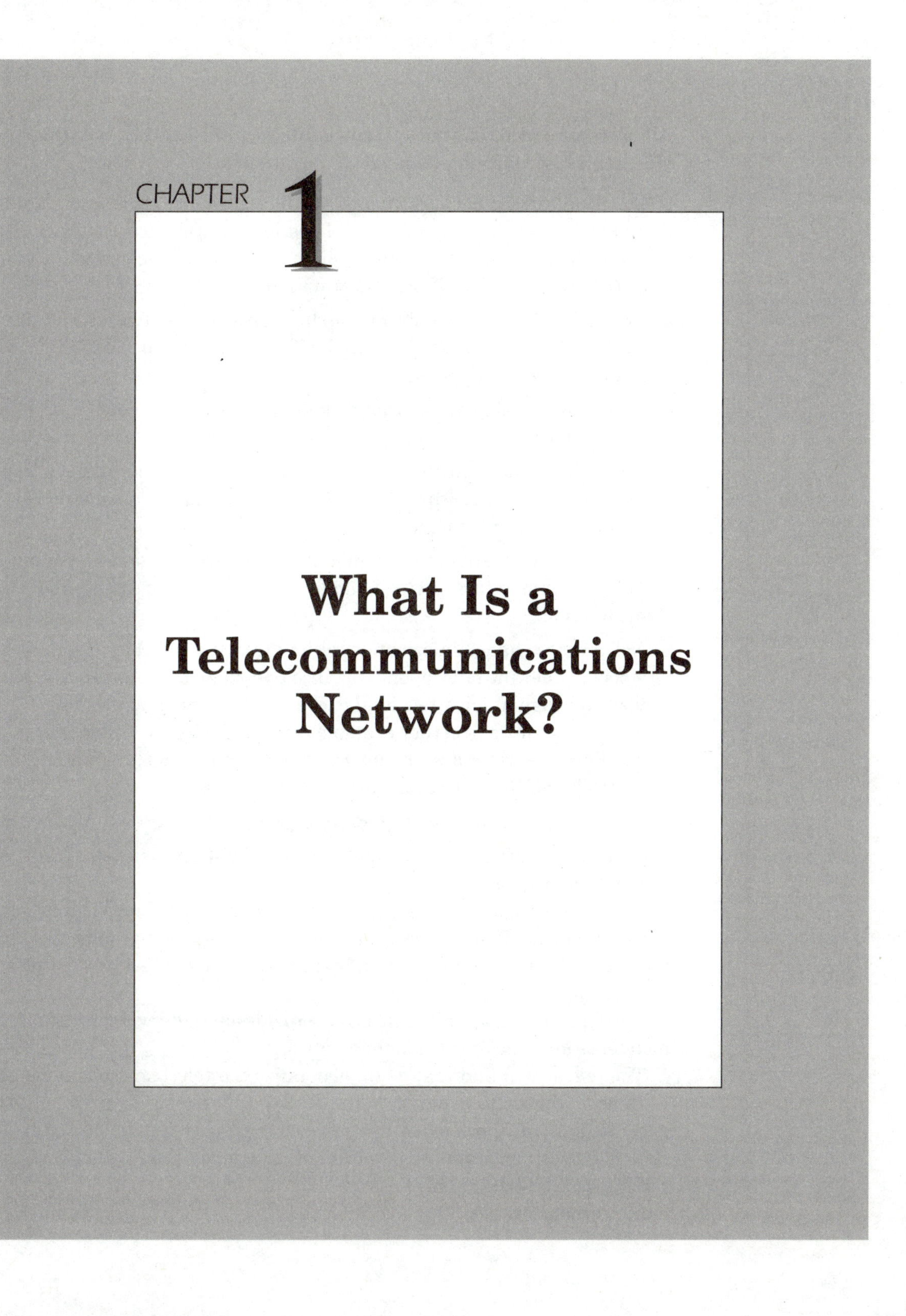

CHAPTER 1

What Is a Telecommunications Network?

All telecommunications networks, regardless of the services that they provide, share the following common characteristics:

1. Communications networks transport information (such as voice, data, and video) between the source and destination. Broadcast television and radio are examples of information from one source being distributed to many different destinations.
2. Networks concentrate information. In other words, networks contain entities that concentrate information from multiple information sources prior to transmission.
3. Networks contain entities that distribute information among multiple destinations.
4. Networks contain entities that simply carry information from one point in the network to another. These elements are called transmission facilities (or simply facilities).
5. Networks can modify the information to improve transmission quality. Simple examples are analog-to-digital conversion, line coding, and modulation.
6. Networks exchange information with other networks. For example, when a caller initiates a long-distance telephone call, the information flows through at least two local exchange networks and one long-distance network. Another example is when a wireless caller calls a wireline subscriber and the information flows through a wireless network and through a wireline network.

Local-Area Networks (LANs), *Wide-Area Networks* (WANs), *Virtual Private Networks* (VPNs), the Internet, and broadcast television are just some examples of networks.

The terms *call* and *information* are used interchangeably. Whether information takes the form of voice, video, or an electronic text/graphic file, all of these objects are classified as information. Networks carry information from one place to another.

The following diagram (Figure 1-1) provides a high-level view of a common telecommunications signaling network.

The reader will find that as we generalize the telecommunications network as an information network, the devices that enable information to be generated or transported can be described in generic functions. Figure 1-1 describes the more traditional telecommunications signaling network that carries voice and data. This network diagram represents a non-packet and non-Internet network. As I had indicated, the "network" carries informa-

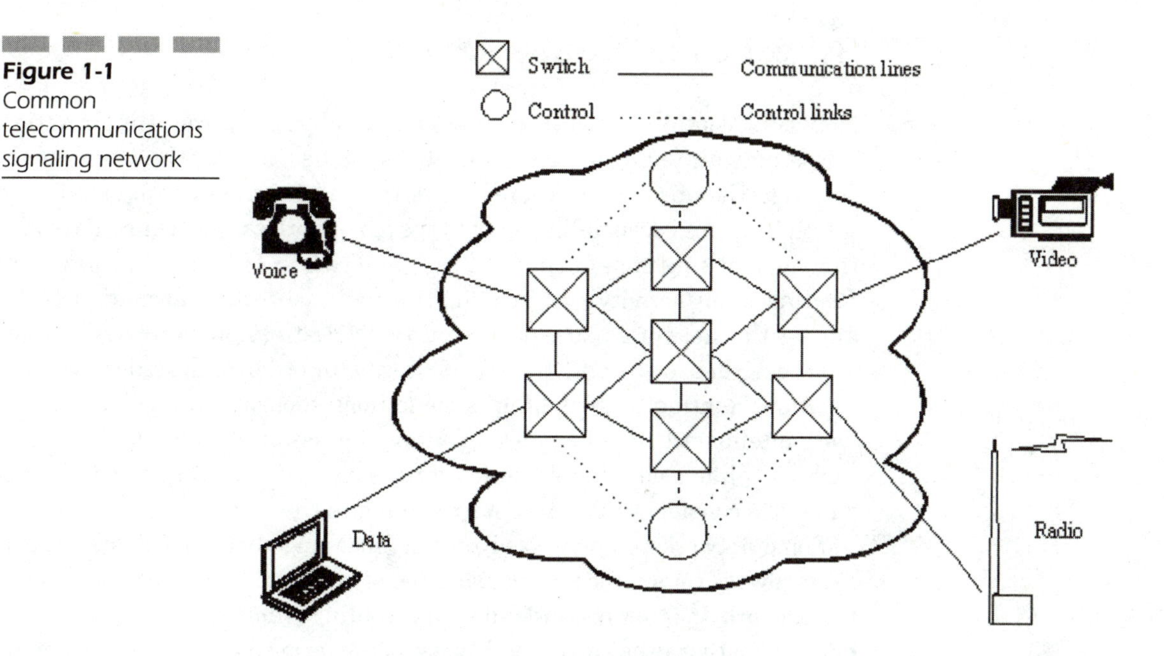

Figure 1-1 Common telecommunications signaling network

tion. All information networks, regardless of the type of network, use the same basic traffic engineering principles.

Network Components

Telecommunications networks are comprised of devices that tend to perform the same general functions, these devices or elements fall into the following general categories:

- Switches or routers
- Databases
- Transmission facilities
- Subscriber billing systems
- Customer care and provisioning
- Network management systems

These broad generalizations enable an individual to visualize and understand network architectures. By viewing the network as an entity that can be described as a set of functions, rather than simply boxes with names, the reader will see the network as a dynamic entity.

Switches or Routers

Networks contain network elements that physically route information. These network elements can be called switches, and in the world of the Internet, they are called routers. The routing function can be shared by multiple devices. Depending on the type of network that is being addressed, the element might be a *Signal Transfer Point* (STP), switch, router, etc. Routing is an activity or function. The router directs information traffic among the network nodes from source to destination. Industry laymen often use their own industry-specific terms in order to describe the same function. Routing or switching is sometimes separated by perceived network intelligence. In other words, routing implies intelligence and decision-making, while switching is merely a physical activity—a minor difference when one considers this concept from a high-level view.

Some networks employ fixed routing, meaning that a particular pair of users always communicates by using the same circuit(s), or path(s), through the network. In other networks, dynamic routing is employed, and therefore a different path is used each time. My use of the terms router and switch might appear to be confusing. My intent, however, is to broaden one's view of the routing or switching functions. The term *router* is typically used in the world of data networks, while *switch* is used in the world of traditional voice telephone networks. From a functional perspective, both routing and switching are the same. Information is processed and transmitted on a path or route.

Routing is an element of call control and will be explored further in the following sections.

Database

Multiple routing elements might be required for a single connection. Routers use routing tables to determine how and where to send a particular information signal. These routing tables are communications road maps that reside in a database. The database is a storage device for information and can be integrated within the switching device or can be external to the switching device, thereby serving multiple switching devices. The database might even store subscriber profile information, which can include billing preferences, subscriber features, restricted numbers, etc.

Transmission Facilities

Transmission facilities are used to interconnect the end user (subscriber) to the network, to interconnect one network to other networks, and to interconnect switching or concentrating transmission points within the network. Networks are often designed with duplication of transmission facilities between two points. This practice is called diversity routing and is used to improve reliability. Transmission facilities come in many different sizes and flavors—everything from 56Kbps to many gigabytes per second. Transmission facilities are primarily either metallic or glass, and these facilities carry information in either analog or digital format.

The information carried by the transmission facilities is either out of band or in band. An out-of-band information stream refers to a situation in which call control information is physically carried over one path, while the actual call or content information is sent over another path. In-band information refers to the situation in which call control and call/content information is physically carried over the same path in a serial fashion.

Network Management Systems

Service providers need some way to manage the various elements within their network. The management of the network elements can be done on site or remotely. All systems need to be managed. How that task is performed will vary from industry to industry and even within an industry. Network management practices and systems are as numerous as there are numbers of service providers.

Industry standardization or *Transmission Management Network* (TMN) standards notwithstanding, network management systems and practices are considered highly proprietary among service providers. An efficient network management system and set of practices might be the difference between a service provider's capability to show profit or the inability to show any profit at all. Network management systems enable a service provider to efficiently operate and control network-element functions and to provide visibility of alarm conditions.

The *Competitive Local Exchange Carrier* (CLEC) is a type of carrier that was created as a result of the Telecommunications Act of 1996. The opening of the local loop has enabled new entrants into the business of providing local exchange services. One of the challenges that these CLECs have been facing is in the area of network management.

Starting a CLEC forces one to focus on specific areas of need first. These areas of need include: a business plan, marketing, core products, sales, and revenue. Unfortunately, for many of the early CLECs, network management was the last aspect of the business on which to focus. CLECs usually do not require a sophisticated network management system early in a CLEC's operation; however, one should not expect to be in business for long unless a type of network-management system is in place.

Subscriber Billing Systems

Service providers of all types need to be paid for the services rendered. All networks need to record usage by the subscriber and bill the subscriber for the usage. Billing can be a function of usage volume, call volume, destination, or feature invocation. Service providers can offer service based on a variety of different parameters and subscriber activities. These parameters and subscriber activities might include calling party pays plans, toll-free numbers (such as 800, 888, and 877 numbers), credit card billing, geographic regional calling plans, time-sensitive promotional plans, and other types of billing arrangements.

Billing is an area of operations that is often overlooked. Too often, billing systems are treated as afterthoughts. In fact, the billing system is as important as the switching/routing capability. A billing system needs to do more than simply create a bill. A billing system needs to apply company billing rates to a call, state and local government surcharges, access charges, interconnect charges, and applicable discounts. These tasks are not simple.

Customer Care and Provisioning

Customer care and provisioning are activities and systems that might appear to be separate, but are in fact directly related to billing systems. Both maintain a record of the subscribers' names and addresses and types of features subscribed to and are used to support customer-support activities by customer-care personnel. These systems will be used by customer-care personnel to answer questions about a subscriber's bill and the type of service, to reflect payments, to remove or add new services, to request repair services, and most importantly to file customer complaints. Customers perceive all of these functions to be a single system. They are typically separate but interconnected data systems, however.

Network Signaling

Network signaling is the language through which we will communicate with other service providers and with network elements within your own network. The key component of network interconnection is the signaling. I want to focus on the basics of the two dominant signaling protocols used in North America for the mass telecommunications market: *Multi-Frequency* (MF) and *Signaling System 7* (SS7). Proceeding chapters will address up-and-coming signaling protocols in the communications industry.

Multifrequency Signaling

The MF and SS7 protocols are the primary signaling protocols used by the wireline and wireless service providers that provide services to the mass market. There are other service protocols; however, this book will address the major protocols that are in use today and that are gaining prominence in the marketplace.

Before MF signaling, the pulsing of digits of the called number conveyed address information. A single digit could require up to 10 pulses (interrupting electrical current flow). One second is needed for 8 to 12 pulses to be transmitted. When you include the pauses between digits (about 500 milliseconds), a single 10-digit telephone number could take several seconds to transmit. Few subscribers want to wait that long before their calls are completed. MF signaling uses tone pulsing in the voice frequency range. MF signaling (tone pulsing) transmits digits by combining two of six frequencies.

MF frequency pulsing of a 10-digit number takes about the same amount of time as it takes to transmit the 1 digit in older signaling systems. MF signaling utilizes tones in the 700 through 1,700 Hz range. When a MF tone is transmitted (representing a digit), two frequencies are sent simultaneously for each digit.

Table 1-1 illustrates the MF signaling code. Note that supplementary tones called *Key Pulsing Sender* (KP) and *Start Tone* (ST) (which stands for the completion of keying and the start of circuit operations) are also supported.

Table 1-1

MF signaling code

Digits	Frequencies (Hz)
1	700+900
2	700+1100
3	900+1100
4	700+1300
5	900+1300
6	1100+1300
7	700+1500
8	900+1500
9	1100+1500
0	1300+1500
KP	1100+1700
ST	1500+1700

MF pulsing existed before the electronic switch. MF was originally supported in the electro-mechanical telephone switching environment, and as a result, many of the terms and actions involved manual operations and electro-mechanical equipment. Let me go through an example of what used to be done during a MF pulsed call. Originally, a manually operated switchboard initiated MF pulsing. The operator would select an idle trunk, then press the KP button on his or her keyset to signal the distant sender or to register link equipment to connect to a MF receiver. The MF receiver resides in the terminating switch. The S lamp on the keyset would light when the terminating equipment was ready to receive the MF pulses. After the digits of the called number were keypulsed, the operator pressed the ST button, which would indicate the end of pulsing and would also disconnect the keyset from the operator's cord circuit.

The introduction of common-control landline switches (a.k.a., central offices) such as electronic switching and electro-mechanical switches (e.g., crossbars) required MF signals to be transmitted by MF outpulsing senders. The terminating switch (the called party's switch) contained equipment called MF receivers. These MF receivers were connected to the incoming sender or register-link circuit. As you might have gathered, MF pulsing is a high-speed activity. MF outpulsing only occurs during the period when a connection is being established.

Multifrequency Applications

Most service providers are converting their respective networks to networks that are based on *Common-Channel Signaling* (CCS). Therefore, I will not waste your time by listing any MF-based subscriber services. There are applications for MF signaling in current and future networks, however. These applications are not subscriber services; rather, they are network termination services. In other words, a service provider can provide call termination services to another carrier's network, assuming that the carrier on which the call is being terminated can support the activity. The added value is the potential for lower interconnection rates.

A typical MF information stream looks similar to a call by a subscriber to another party via an *Interexchange Carrier* (IC), as shown in Figure 1-2.

Figure 1-2 depicts a call involving an IC. Signaling sequences in the MF world are different for each type of call. The point of the diagram is to illustrate how serial-like the sequence is. In the world of CCS, the sequence consists of messages as opposed to tones.

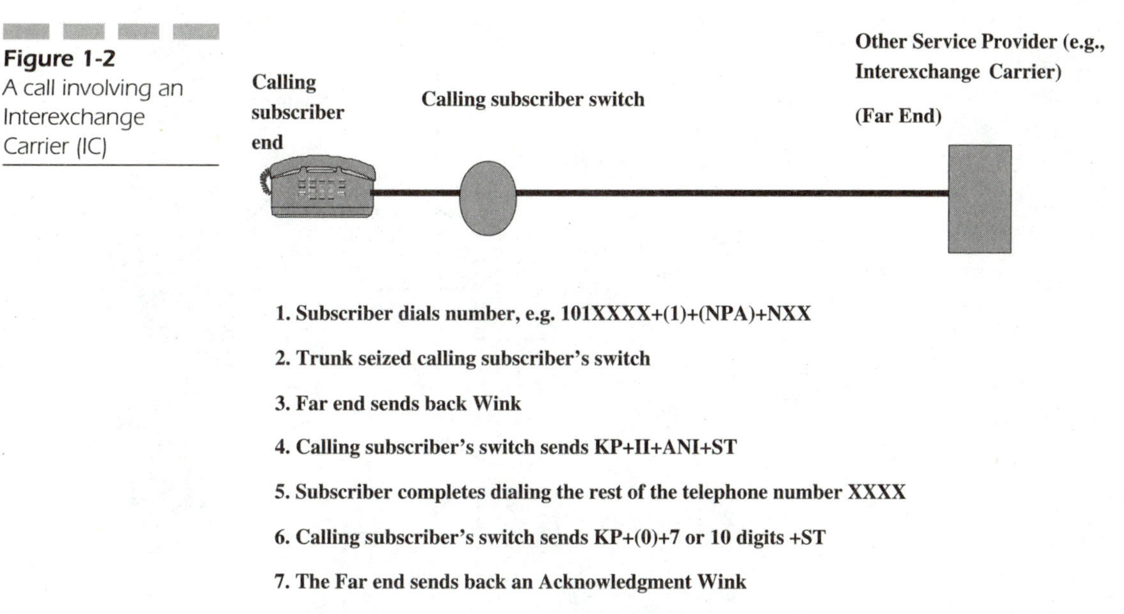

Figure 1-2 A call involving an Interexchange Carrier (IC)

Common-Channel Signaling (CCS)

CCS is a type of out-of-band signaling. The signaling used previously is circuit-associated/in-band signaling. While MF pulsing is tone based, CCS is actually a computer message-based signaling protocol. The MF signals used to set up facilities between the called and calling party occur on the same physical path as that of the conversation. In out-of-band signaling (or, in this case, CCS), the signaling that is associated with setting up the facility path for the conversation is different than the conversation path.

The following architecture is associated with CCS systems (refer to Figure 1-3).

CCS, specifically ANSI *Signaling System 7* (SS7), is the dominant form of CCS used in North America. SS7 is also being installed in all wireless systems. SS7 is the basis for all intelligent network platforms. Wireless systems (cellular and Personal Communications Services [PCS]) are installing enhanced service platforms, many of which require SS7 as the basic signaling protocol for their networks. SS7 is not the fastest network signaling protocol, but it is the most robust and mature out-of-band signaling protocol that exists to process calls—and it enables communication between network entities.

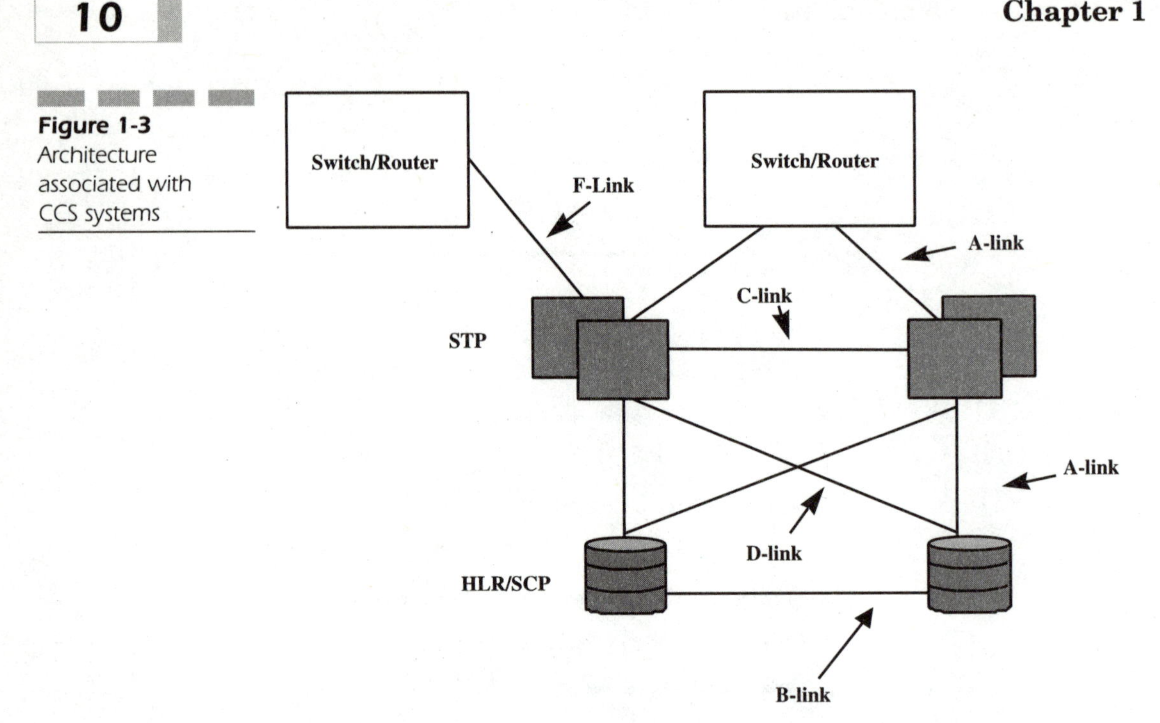

Figure 1-3 Architecture associated with CCS systems

CCS Network Architecture

The SS7 network architecture consists of a set of network elements that are generally referred to as nodes. These nodes are also called *Signaling Points* (SPs). A switch is a SP. These SPs are interconnected via point-to-point signaling links.

The following diagram (Figure 1-4) illustrates the basic SS7 architecture. The signaling links shown are as follows:

- Access link (A-link)—Used for Switch-STP signaling connections and STP-HLR
- Bridge link (B-link)—Used to connect STP-STP in other networks
- Cross link (C-link)—Used to connect diverse STPs of the same hierarchical level
- Diagonal link (D-link)—Used to connect STPs of different hierarchical levels or diverse STP-HLR/SCP combinations
- Extended link (E-link)—Links between switches and non-home STPs
- Fully Associated (F-link)—Optional links between switches

SS7 is the preferred network-signaling protocol—the service provider will interconnect with its resellers and interexchange carriers via this protocol. The service provider's SS7 network will need the capability to communicate the following minimum information between the service provider network and other SS7 networks: Point Codes (full point codes and cluster point codes), sub-system numbers, signaling link codes, timer values, the signaling link test, the congestion threshold, the routing priority, and global title translation.

The *Signal Transfer Point* (STP) relays messages at the network layer between switches. STPs are mated in pairs and are a crucial part of SS7 network design. STPs contain the routing templates that are used by carriers to route calls. These templates are called decision matrices.

The *Service Control Point* (SCP) is the database that contains the subscriber profiles. SCPs can also provide assistance in the routing of a call. *The Home Location Register* (HLR) is a wireless carrier SCP-type of database.

The SS7 protocol can be viewed as being composed of layers of functions/tasks. The protocol is similar to an architecture that was established by the *International Organization for Standardization* (ISO), called the *Open System Interconnection* (OSI) architecture.

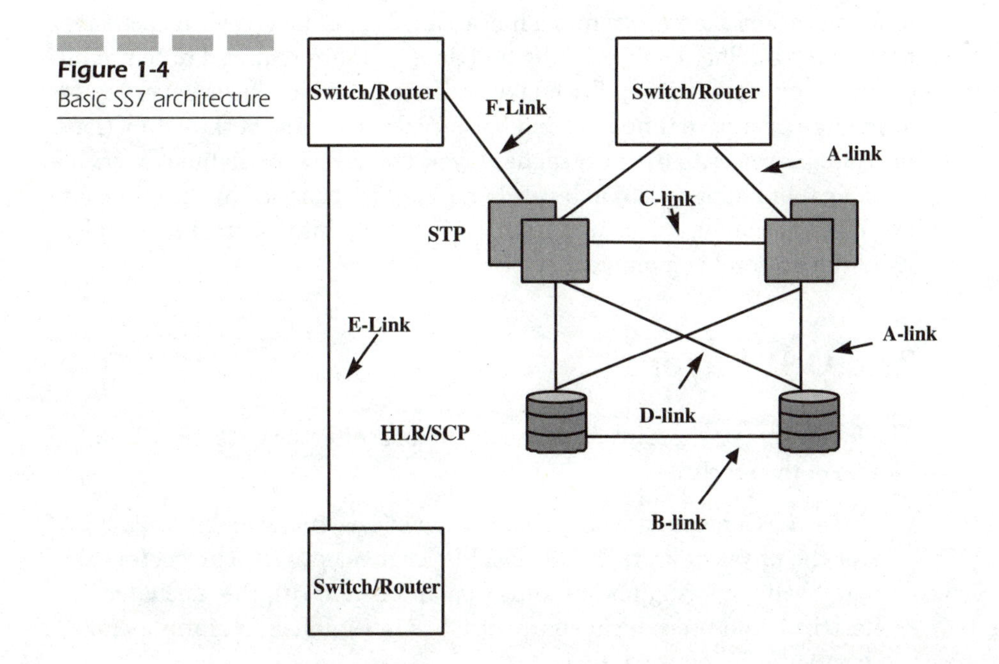

Figure 1-4
Basic SS7 architecture

When communication is desired among computers (switches are nothing more than computers) from different manufacturers/vendors, the software development effort can be difficult. Different vendors use different data formats and data exchange protocols that do not enable computers to communicate with one another. The OSI model is an engineering model that breaks everything down into simple and discrete tasks or layers.

The OSI model was actually created after SS7. The OSI model was developed by the ISO for use in a computing environment. The SS7 model is similar, because some of the concepts behind switching information and layering functions are similar in the telecommunications world.

The two models do not align perfectly because of the different times at which they were developed (and to some extent because of the perspective from which they were developed). You should note, however, that there are similarities in the models. Both support the transmission and application of information.

The OSI Model

The OSI model consists of seven layers (see Figure 1-5). The communication functions are broken down into a hierarchical set of layers, with each layer performing a related subset of the functions that are required to communicate with another system. Each layer relies on the next lower layer to perform more primitive functions and to conceal the details of those functions, providing services to the next higher layer. The layers are defined in such a manner that changes in one layer do not require changes in the other layers. By partitioning the communication functions into layers, the complexity of the protocol is manageable.

The OSI Layers

The following is a description of the layered architecture, starting from the bottom of the stack:

- *Physical* Concerned with transmission of unstructured bit streams over the physical link. The physical layer invokes such parameters as signal voltage swing and bit duration and deals with the mechanical, electrical, and procedural characteristics to establish, maintain, and deactivate the physical link.

- *Data Link* Provides for the reliable transfer of data across the physical link and sends blocks of data (frames) with the necessary synchronization, error control, and flow control
- *Network* Provides the upper layers with independence from the data transmission and switching technologies used to connect systems. The Network layer is responsible for establishing, maintaining, and terminating connections.
- *Transport* Provides reliable, transparent transfer of data between end points and provides end-to-end error recovery and flow control
- *Session* Provides the control structure for communication between applications and establishes, manages, and terminates connections (sessions) between cooperating applications
- *Presentation* Performs generally useful transformations on data to provide a standardized application interface and to provide common communications services. The Presentation layer also provides services such as encryption, text compression, and reformatting.
- *Application* Provides services to the users of the OSI environment and provides services for *File Transfer Protocol* (FTP), transaction server, network management, end-user services, etc.

The following diagram (Figure 1-5) illustrates the OSI model. Notice that the following model displays a stack of layers. The foundation layer is the

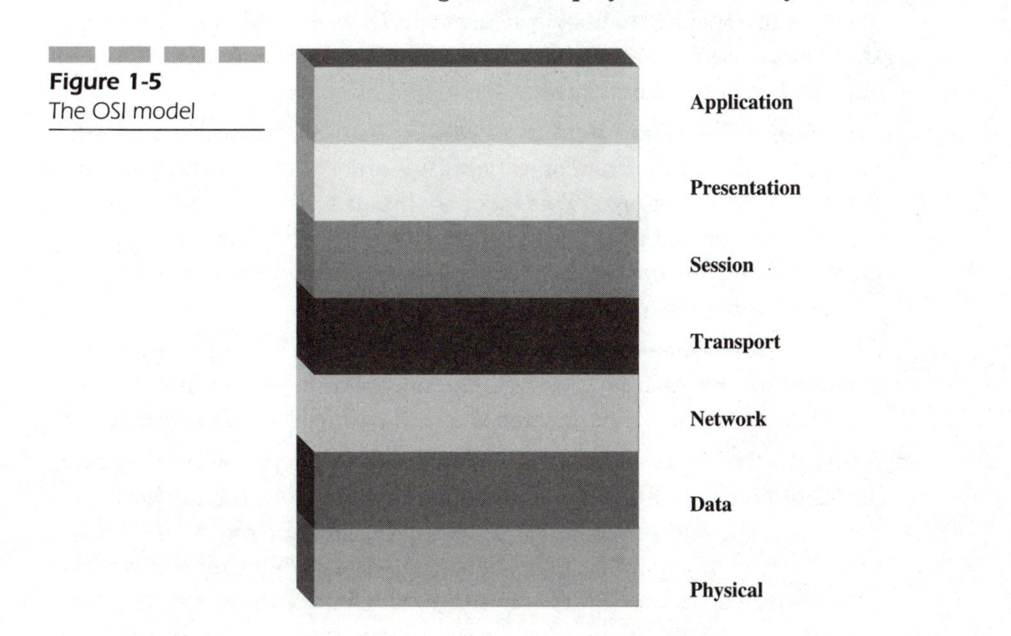

Figure 1-5
The OSI model

Physical layer. Every other layer is built on top of the Physical layer and on top of each other. You can see how these layers are interdependent.

The SS7 Layers

The SS7 protocol is similar to the OSI model in that the SS7 model supports the layering of functions. SS7 supports four functional layers, while the OSI model supports seven layers. The functions of the four SS7 layers are similar to the seven stacks of the OSI model. The OSI model presents a broad view of data transmission and applications, while the SS7 model has been specifically designed to meet the evolving needs of the telecommunications industry.

- *Physical* The Physical layer in this model is nearly the same. Unlike the OSI model, the SS7 model enables specific network interfaces to be identified. Today, the most frequently used interface is the DS0-level network interface. This interface represents a data rate of 56Kbps; therefore, all SS7 links today operate at the DS0 level. Industry practices in North America (as specified by Bellcore) support link rates at the DS0 level. The ANSI SS7 standard does not specify a specific interface.

  People who are conducting current industry work within North America are seeking to use *Asynchronous Transfer Mode* (ATM) for the transport of SS7 messages. Briefly, ATM will be implemented to support broad-band networks.
- *Data Link* The Data Link layer provides the SS7 protocol stack with error detection and correction capabilities and enables the sequential delivery of SS7 messages. This layer is similar to the data layer in the OSI model. The Data Link layer addresses only the transmission of data from one network node to another network node, and this layer is not responsible for routing.

  As the SS7 message is transmitted from one node to another, each node performs digit screening (review). The information gained from the dialed digits is used to determine the next node for the SS7 message.
- *Network* The Network layer provides for message review, routing, and distribution capabilities. The routing function enables the network to determine the address to which the message must be routed. The message review capability enables the network to determine to whom the message is addressed. The message distribution enables the network to identify the user part to which the message is addressed.

Within this layer, there are three network management functions: link management, route management, and traffic management. These capabilities are an integral part of the layer and enable the previously mentioned broader functions to work.

- *User Parts and Application Parts* This layer consists of several different components, including the following primary components:
  - *ISDN User Part* (ISUP)
  - *Transaction Capabilities Application Part* (TCAP)
  - *Operations, Maintenance, and Administration Part* (OMAP)
  - *Mobile Application Part* (MAP)

Figure 1-6 illustrates the SS7 model.

Layers 1, 2, and 3 correspond to Layers 1, 2, and part of Layer 3 in the OSI model. Layer 4 corresponds to Layers 3 through 7 of the OSI model. See Figure 1-7.

Figure 1-8 is an example of how a SS7 call is processed.

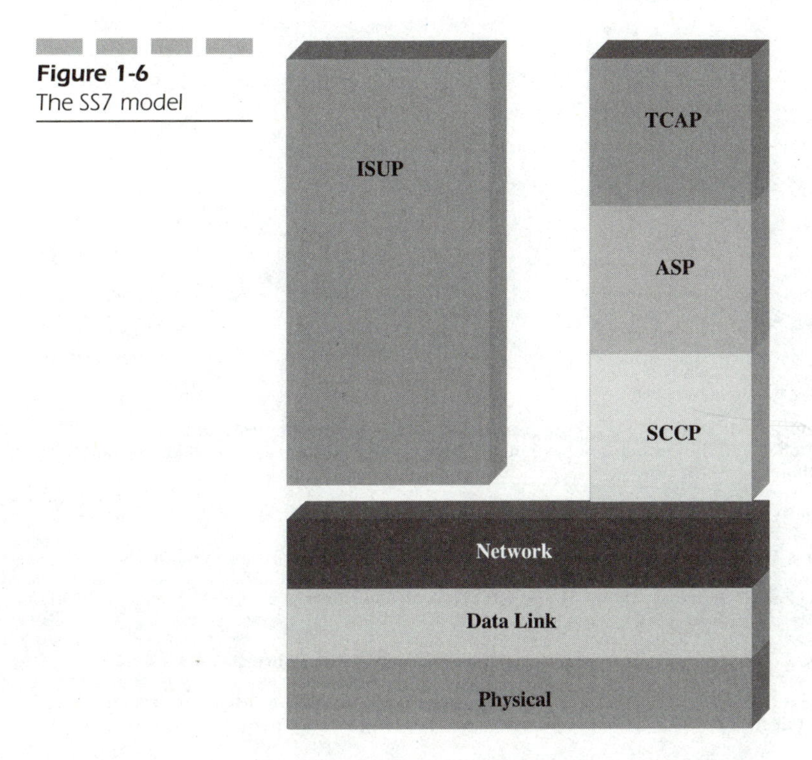

Figure 1-6
The SS7 model

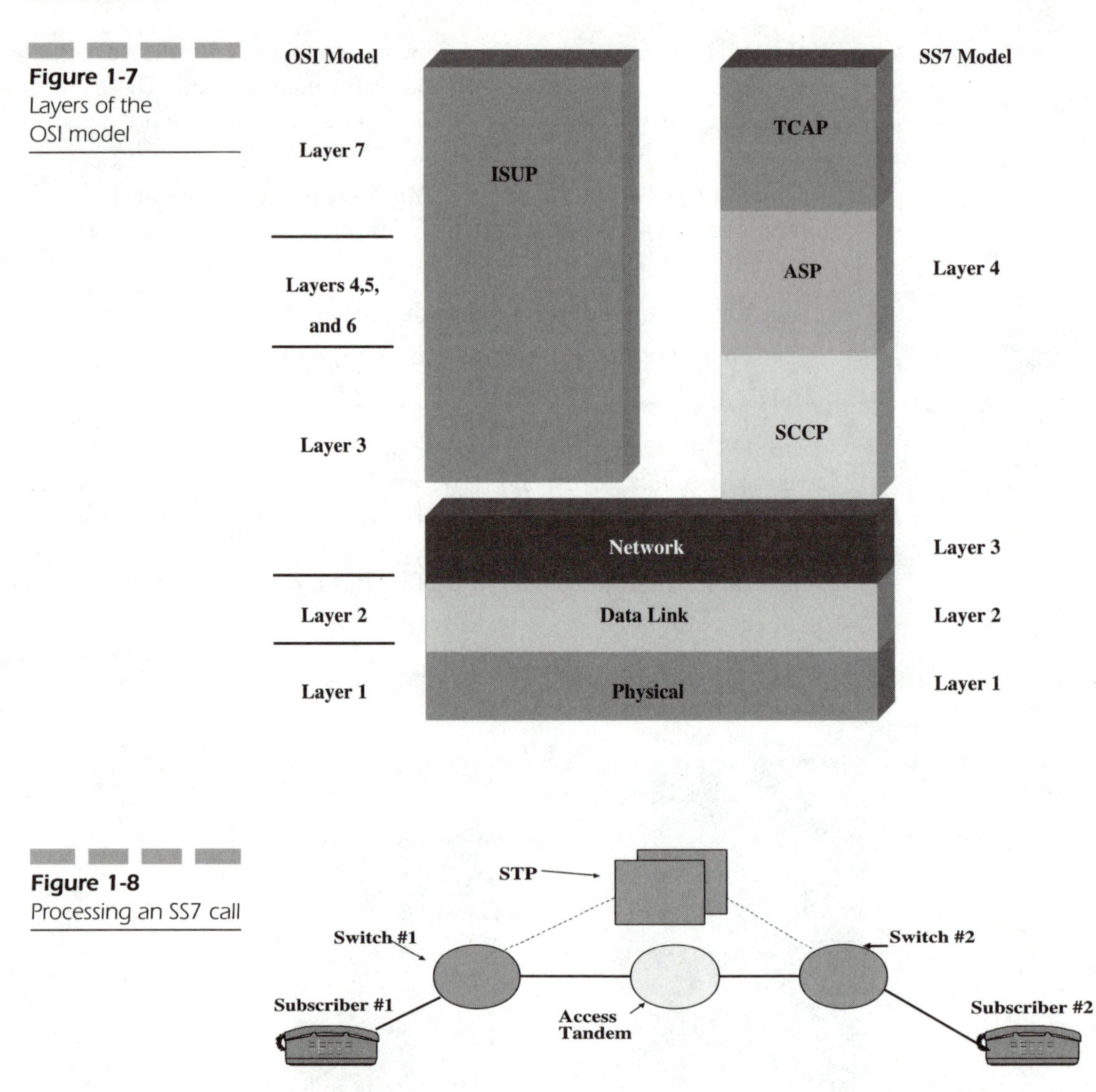

Figure 1-7 Layers of the OSI model

Figure 1-8 Processing an SS7 call

Subscriber # 1 dials number

Initial Address Message (IAM) is sent over the SS7 link—Switch #1 sends an IAM with the called number and the calling number potentially with COT to convey any other relevant information.

A Continuity Message (COT) is sent towards Subscriber # 2.

The Access Tandem accepts the IAM and COT and passes this information to the Far End.

When the Far End receives the IAM and COT, it sends an Address Complete Message (ACM) to the Access Tandem.

When the Access Tandem receives the ACM, it send an ACM to Subscriber # 1's switch.

When the called subscriber (Subscriber # 2) goes off-hook, an Answer Message (ANM) is sent from the Far End to the originating party.

Network Types

This book will focus on the current major commercial network types. I have classified them by business segment type, as follows:

- Wireline/landline telephone networks
- Cellular and PCS networks
- Paging networks
- Satellite networks
- Cable television networks
- The Internet

Some readers might wish to classify data networks as a separate network type. The fact is that the term *data networks* is nearly meaningless when one views this term in the context of business-segment type. Almost every industry segment supports data. The oldest version of data networking was computer-to-computer communication. Up until 15 years ago, computer-to-computer communication was restricted to mainframe computer-to-mainframe computer communication. Even then, however, there were many in the wireline telephony industry who claimed the banner of data networking as well. Today, the most popular use of computer data networks is the Internet, and the Internet is growing exponentially.

Brief descriptions of these network types appear in the following sections.

Wireline/Landline Telephone Networks

I have classified *Local Exchange Carriers* (LECs), *Competitive Local Exchange Carriers* CLECs), *Interexchange Carriers* (IXCs or ICs), and *Competitive-Access Providers* (CAPs) as wireline or landline carriers. This categorization appears to be a fairly loose view of the landline carrier. In reality, the terms *landline* and *wireline* typically refer to these previously mentioned carriers; however, the advances in technology have enabled even wireless carriers to enter the local services businesses.

The landline carriers comprise what is known as the *Public Switched Telephone Network* (PSTN). The term PSTN originally referred to AT&T, the 22 Baby Bells, and all of the landline-independent telephone companies. In 1984, at the time of the AT&T divestiture, even the long-distance carriers were not considered part of the PSTN. Over the years, with the walls of regulation tumbling down, the use of the term PSTN has been relaxed.

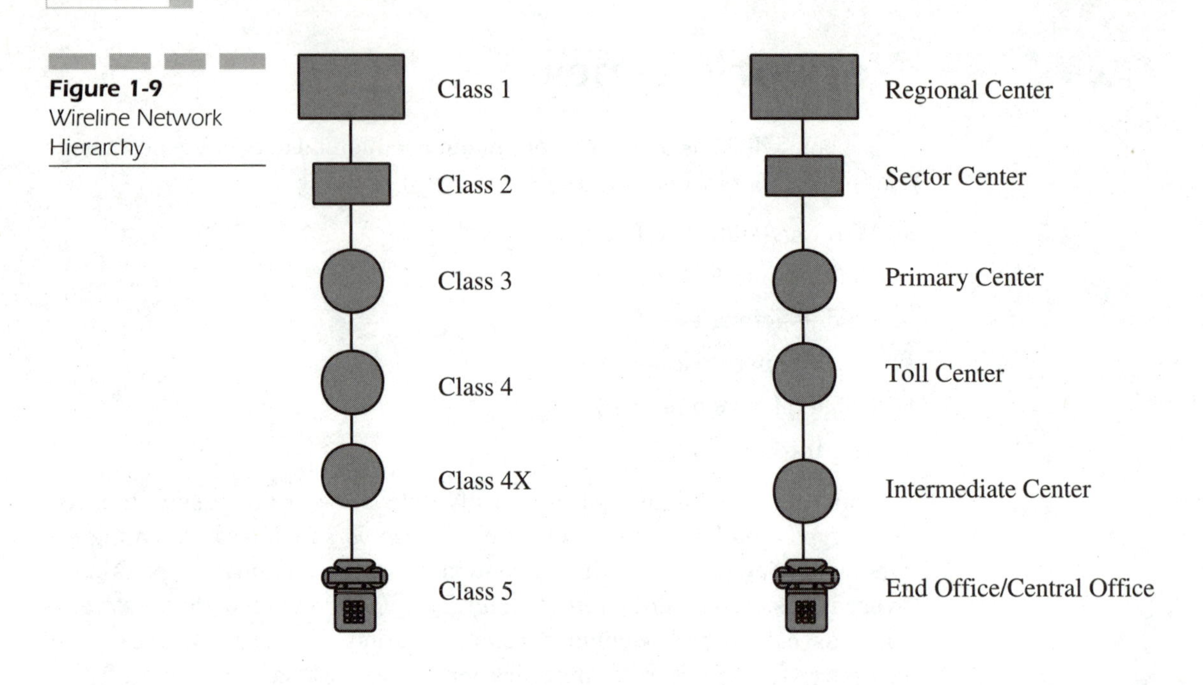

Figure 1-9 Wireline Network Hierarchy

In the context of this book, I will use PSTN to refer to the wireline/landline carriers. I will also use the term wireline, as opposed to landline. Figure 1-9 illustrates the hierarchical structure followed in the wireline network.

Cellular and *Personal Communications Services* (PCS) Networks

Cellular In 1946, two-way mobile radio service was introduced. Soon after its introduction, the disadvantages/weaknesses of two-way mobile service became apparent. From a customer's standpoint, there was an issue of competing for RF channels and interference. From an engineer's perspective, these issues were exactly the same. The technical challenges involved the questions, "How do you go about giving the subscriber a larger pool of RF channels so that they can make his or her calls?" and "How do you reduce interference between subscribers?" The quick and simple solution could have been to not worry about giving subscribers more RF channels and to physically separate the radio coverage areas to ensure that there would be no overlap. Fortunately, no engineer took that approach.

Instead, engineers from Bell Telephone Laboratories began to explore a concept that would reuse frequencies in small radio coverage areas. These coverage areas (called cells) would be linked by a switch that would enable calls to be made while moving. Time had to pass and computer/switching technology had to improve before mobile radio service became commercially viable. Availability was another issue, however, because of regulatory delays. Cellular became commercially available in the United States in 1983. Today, there are two basic radio technologies that are commercially available in the cellular world: analog cellular and digital cellular.

Figure 1-10 illustrates the basic Cellular and PCS network. The reader will find that the networks are nearly the same.

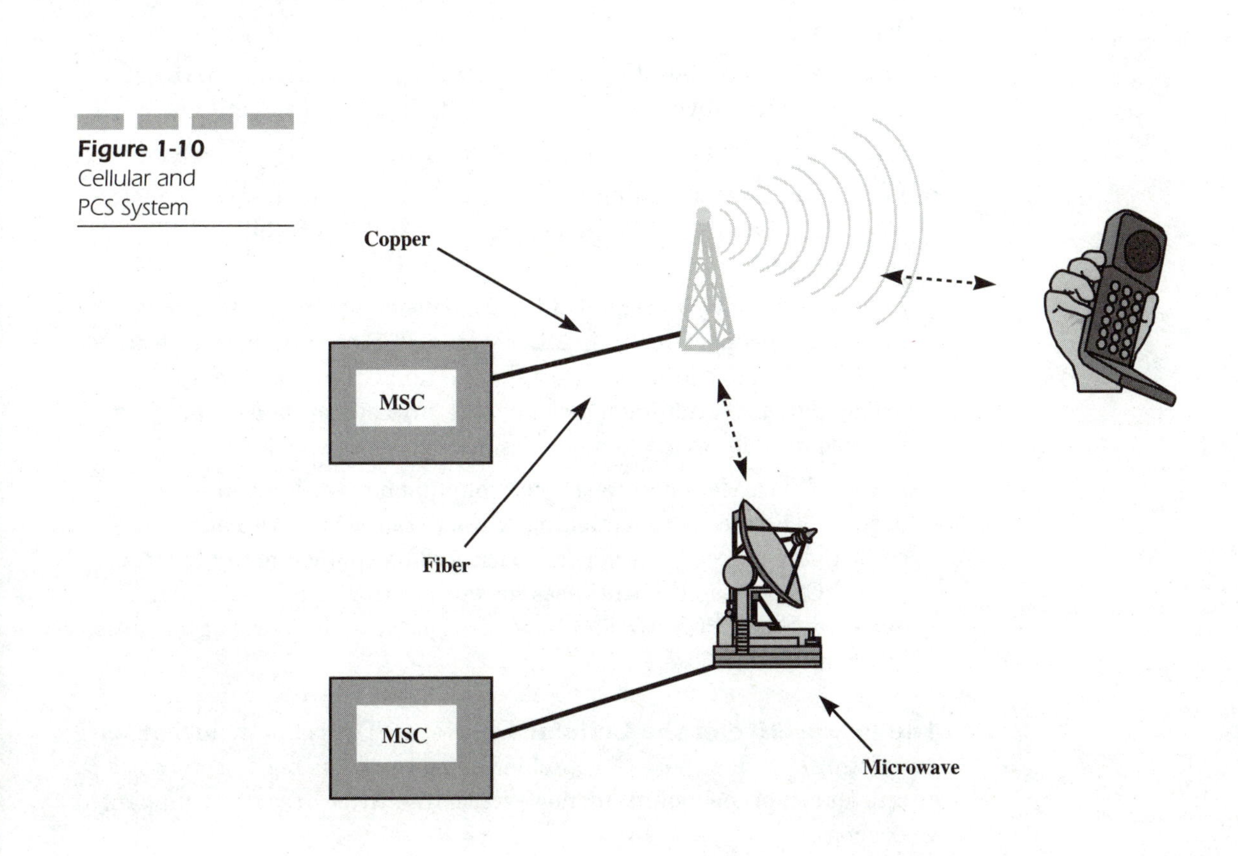

Figure 1-10 Cellular and PCS System

Personal Communications Services (PCS)

PCS was originally envisioned as a service that would enable people to access the PSTN, regardless of their physical location. Today, most people believe that PCS is digital cellular—something better than cellular or cellular without the old baggage (legacy equipment).

In the mid-1990s, many people envisioned PCS as having features that supported terminal personal and service mobility. PCS was supposed to combine many emerging intelligent network capabilities of the public networks (CCS, ISDN, and IN, for example) with sophisticated wireless access technologies and related radio network mobility control capabilities.

From a technical and concept standpoint, there is not much of a difference between cellular and PCS. The same engineering practices are employed, and the same traffic design techniques are used. I believe the biggest differences between PCS and cellular are as follows:

- PCS must contend with high capital costs for equipment and deployment.
- PCS has faced competition since day one. Cellular carriers in their infancy did not have much to worry about as far as competition.
- The competitors are both new and established.
- PCS carriers are installing the latest technology. Cellular carriers are also doing the same but still have to maintain the established equipment base.
- PCS carriers have a fraction of the time needed to make their networks commercially operational. In other words, PCS carriers have to work hard and fast to capture market share. The PCS carriers have the disadvantage of building a new network infrastructure as rapidly as possible in order to make money as quickly as possible.
- Many PCS carriers had to spend a lot of money to obtain their licenses. The recent reauctioning of the C-Block licenses and bankruptcy proceedings of some carriers has enabled many to either enter PCS or keep their licenses for far less than one would have expected. Some PCS carriers have also closed their doors for business; going out of business.

The Perspective of the Cellular Carriers The cellular carriers have an existing customer base. The problem with being the big dog on the block or king/queen of the mountain, however, is that there are people nipping at your shoes:

- Cellular carriers have the burden of existing infrastructure. PCS carriers are deploying new infrastructure (i.e., brand new "stuff").
- Cellular carriers have a need to upgrade existing equipment to compete with new carriers that are coming into the marketplace. Upgrading existing equipment costs money.
- PCS carriers have the burden of generating enormous expenditures in order to create a profit-making network.
- PCS carriers are dealing with a double-edged sword. Cellular carriers can make the same case (upgrading aging equipment while deploying new infrastructure to meet the needs of an expanding marketplace and new coverage areas). In the case of the cellular carrier, the double-edged sword is not as big or as nasty. The major differences between cellular and PCS have more to do with business and less to do with technology.

Figure 1-11 illustrates the technical and physical similarities between the cellular network and the PCS network. If the reader compares the two network types from an equipment provisioning and network operations perspective, the networks are the same.

Figure 1-11 Cellular and PCS System

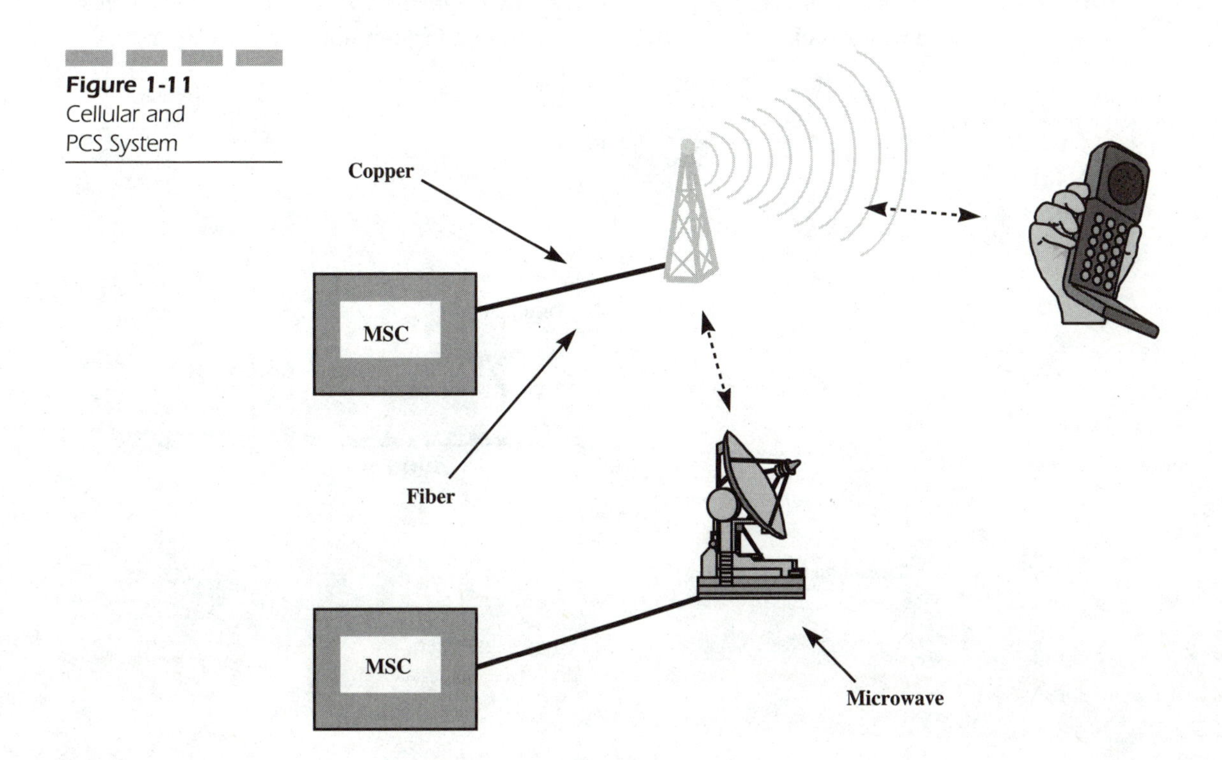

Paging Systems

Paging is usually considered to be the low end of mobile communications. Paging is less expensive than other mobile communications systems because it is a one-way system. The paging receiver alerts the user to the call but does not verify or respond in any way to the base station. Because the cost and bulk of a typical mobile transceiver is associated with the transmit portion (and this feature is missing from a paging receiver), a pager can therefore be small and cheap.

Compared to the cellular and PCS networks, paging networks need far fewer transmitters. The devices are primarily one-way communications devices, but two-way communication is envisioned as the desired primary mode for the industry. Currently, there is limited voice and limited one-way textual data capability. Figure 1-12 illustrates the paging network.

Satellites

Satellite communications are the result of work in the radio field with the objective of achieving the greatest coverage and capacity at the lowest cost. Satellite communications systems can be broken down into two parts:

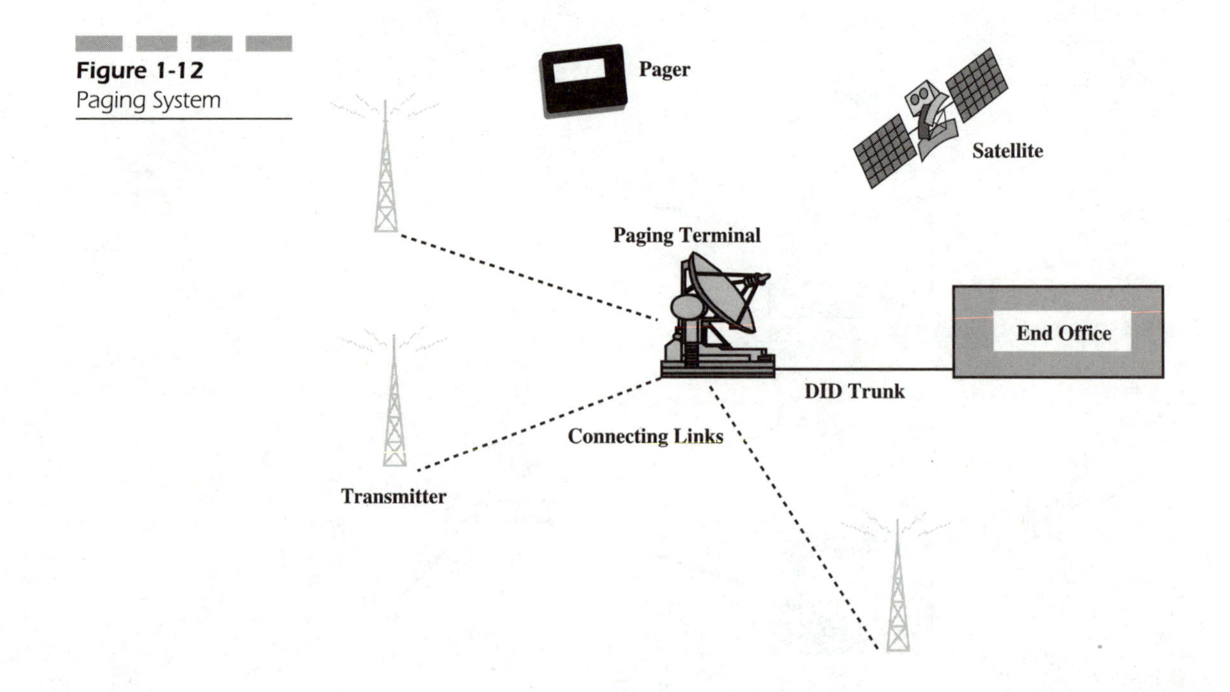

Figure 1-12 Paging System

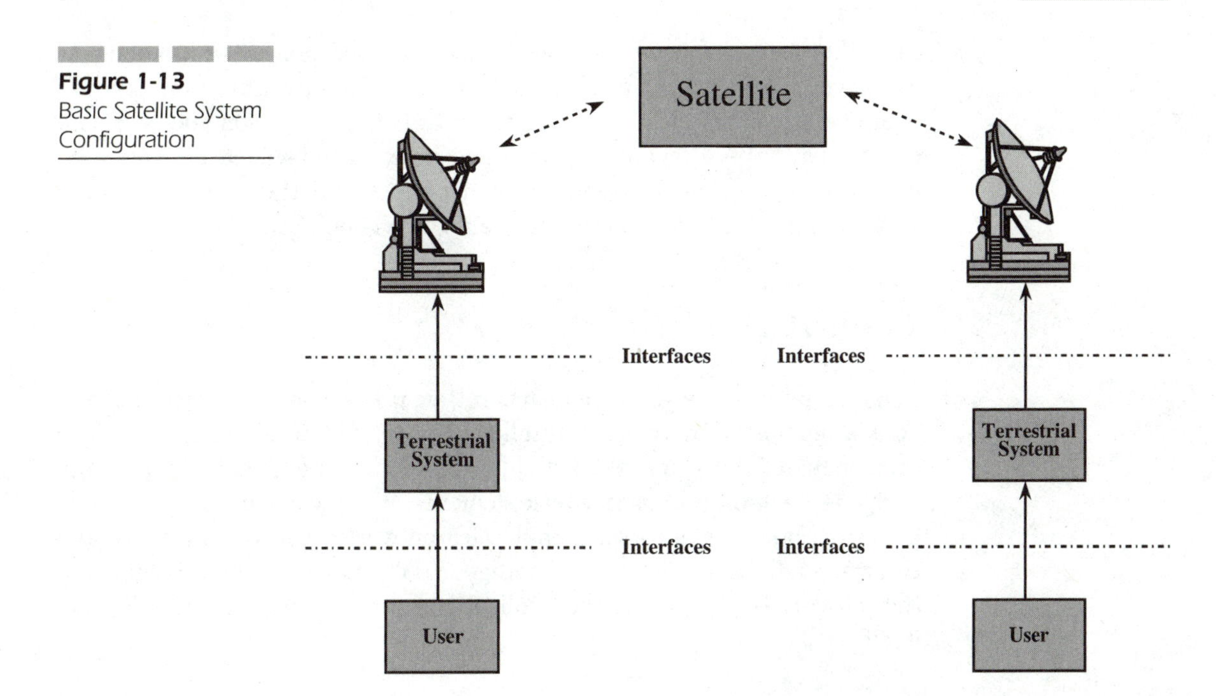

Figure 1-13 Basic Satellite System Configuration

1. *Space portion* Includes the satellite and the means on Earth that are necessary for launching
2. *Earth portion* Includes Earth transmission and receiving stations

Unlike Earth-based communications systems, satellite communications systems require support in a number of non-communication related areas—such as rocket launchers, power supply in outer space, orbital propulsion motors, etc. Figure 1–13 is an illustration of a satellite communications network.

Cable Television Networks

Cable television provides lots of different entertainment channels with clear pictures. CATV, the acronym, actually stands for Community-Access Television, but CATV has become associated with the term cable TV.

Cable television networks are currently designed to transmit multiple, conventional, analog television signals to multiple subscriber locations. This system is a one-way system for distributing the same set of signals to each subscriber location, but historically these systems have had limited capability for return transmissions from designated subscriber locations. As we move forward, that situation will change. The advent of new types of equipment that enable a television to act as both a television and as a computer terminal will eventually convert the cable television network into broad-band multimedia information networks and the Internet. Figure 1-14 is an illustration of a the basic cable television network.

The Internet

The Internet is not a single network; rather, it is a web of networks (or perhaps a network of networks). Intelligence does not reside within a single component of the Internet. Instead, intelligence is embedded within the components of individual network elements. When you compare the Internet with the older and traditional voice networks, you will find that the Internet is decentralized in its intelligence. The power of the Internet is in the software that supports the applications and is also within the protocol itself.

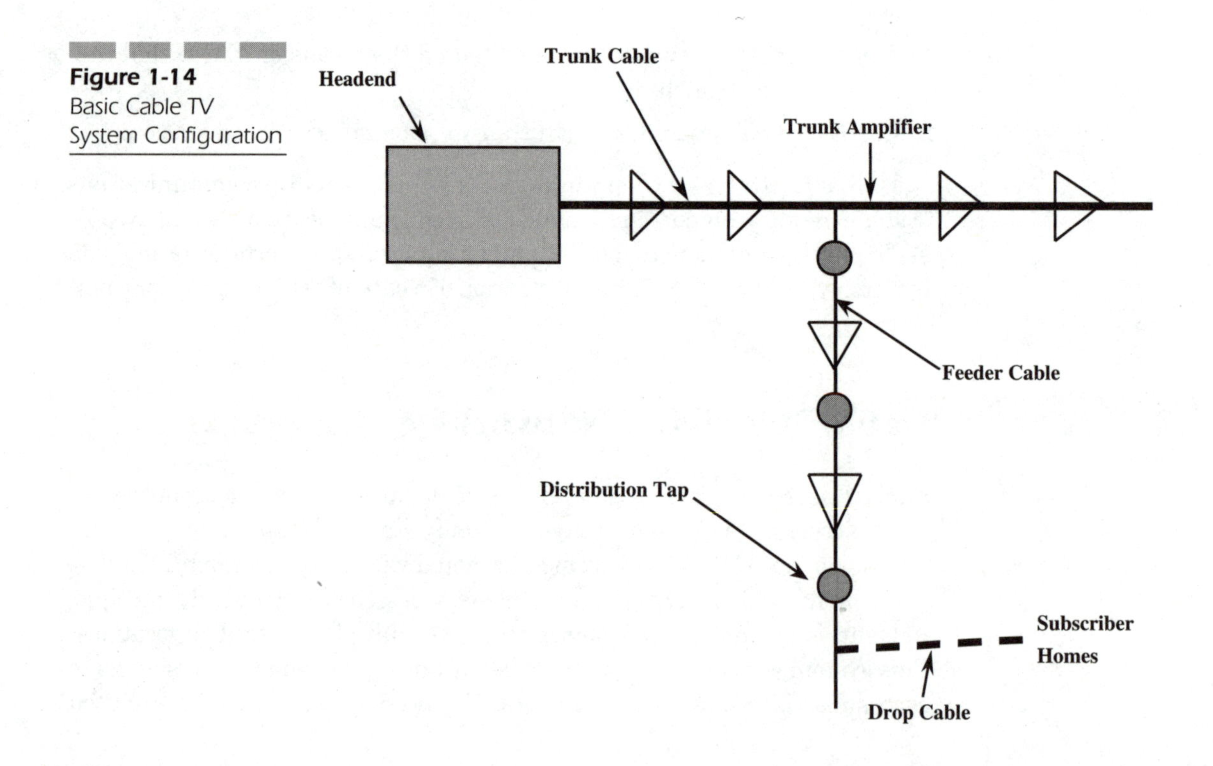

Figure 1-14 Basic Cable TV System Configuration

The difference between the Internet and the other networks that have been briefly described is that the Internet is not controlled by a single company. The Internet is highly popular and is heavily used by millions of individuals, and it carries voice, video, and data. Today, the Internet exists as an overlay network. Physically, the Internet needs the existing transmission network (primarily under the control of the Incumbent Local Exchange Carriers and Interexchange Carriers) to carry its information packets. The underlying switching/routing matrix is a separate set of network elements, however, that are typically owned and operated by the *Internet Service Provider* (ISP).

The Internet has become a catalyst for convergence. The following graphic (Figure 1-15) illustrates the nature of the Internet.

Interconnection Types

Network interconnection involves the interconnecting of two or more networks. Telecommunications interconnection has a long history, starting with the PSTN and how it interconnected to other types of networks.

Figure 1-15
The Internet

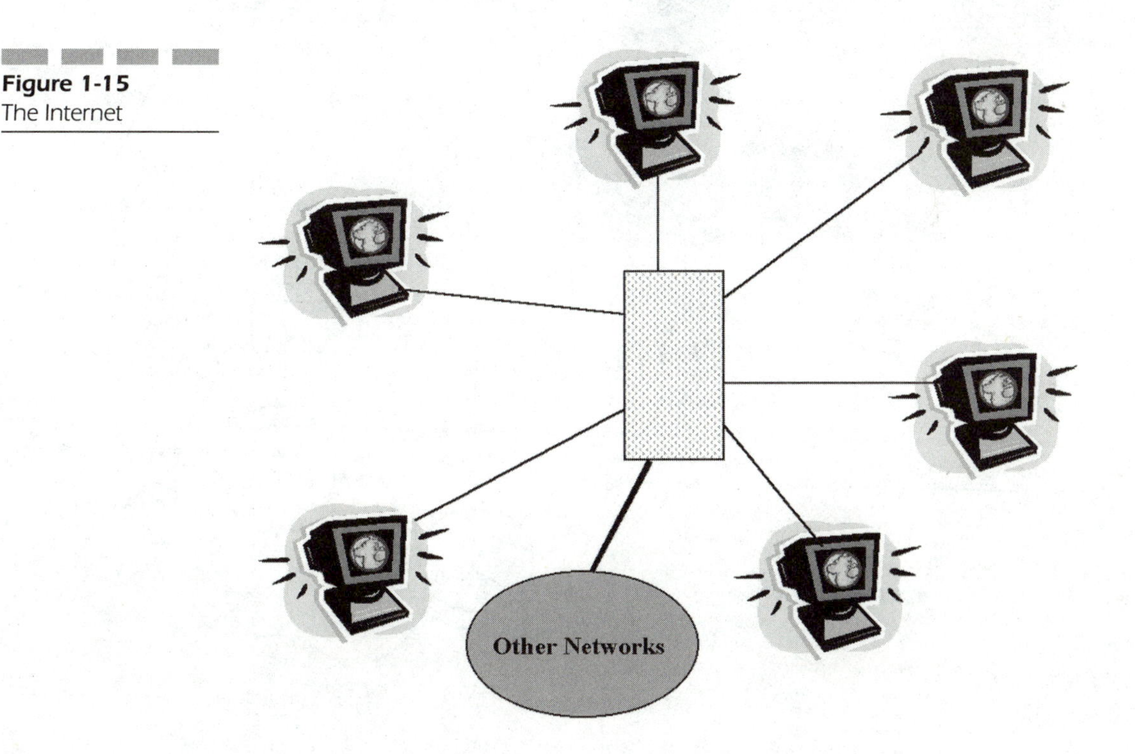

There are two elements of interconnection: physical and protocol. The physical refers to the type of network transmission facility that is being used. There are various grades of transmission facility, and each type is capable of transmitting information at various speeds. The protocol refers to the network language that is being used to communicate information between two or more dissimilar and similar networks.

Interconnecting is the first step towards convergence and thereby creating new values from the network. A broader view of interconnection would include the business relationship between the service providers. In many cases, the business issues define and dominate the relationship. Bear in mind, however, that interconnection is an infrastructure product that can be leveraged by service providers to support the provisioning of other services or access to other services (Access Services) and is the foundation for the resale of infrastructure services.

The concept of network interconnect is shown in Figure 1-16. Network interconnect is applicable to all types of networks.

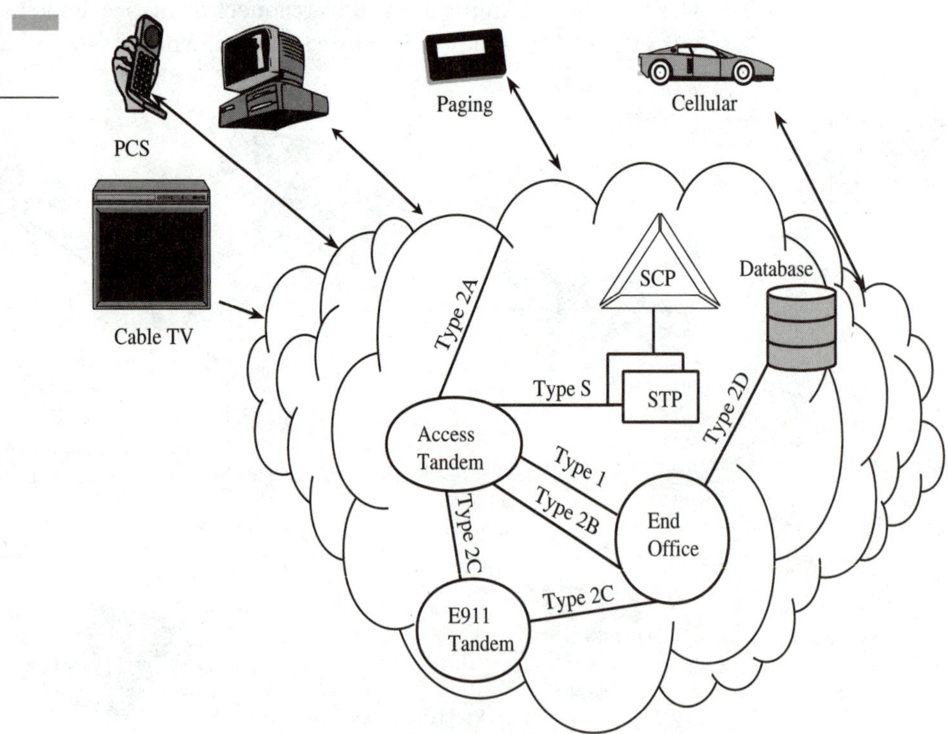

Figure 1-16 Interconnection

Network Intelligence

Several years ago, the term *Intelligent Network* (IN) was used in almost every manner of technical discussion that was taking place. Many pundits claimed that the intelligent network would have the capability to address all of the service provider's market needs and the subscribers' needs. These words became meaningless after a period of time.

The fact is that the intelligent network has been evolving for years. The IN is conceived to be a network that supports access and control of information. The SS7 network became the foundation for the intelligent network. To many people, SS7 represents the first step towards network intelligence—where the logic was no longer dictated by the hard-wiring of network terminals but was instead dictated by software. IN involves communication and interaction between network elements that reside outside the carrier's core/central switching equipment. Essentially, service logic and portions of the switching logic reside in external network elements.

SS7 brought the following features to network evolution:

- Physically separating databases that could support multiple switches and respond to queries from the switches
- Out-of-band signaling; separate paths for the voice and control signals, which led to flexibility in routing
- Flexible routing

The SS7 network screens the digits that are dialed, conducts a series of database queries, determines whether the calling party has any special restrictions or requirements, and establishes control-signal connectivity to the network. This network's returned responses also instruct the signaling node to either set up voice trunks or not. This rudimentary network intelligence was light years ahead of the MF network. The mere fact that the telecommunications network (initially, the PSTN) could make queries was revolutionary. The SS7 network's first application was support for the original toll-free number, the 800 number. Caller ID was the next major application, followed by versions of existing services such as call forwarding. SS7 has become so commonplace that one can easily overlook its historical and operational significance.

IN enables users and carriers to configure new services more easily. While SS7 services rely on specific parameters to create services, the IN relies on interpreting a parameter. The interpretation aspect is the intelligence of the IN. The interpretation leads to an event, and the parameter in this case is called a trigger.

The term *trigger* is normally used when speaking of intelligent networks. A trigger is "something" that causes "something" to occur in the network. That "something" might be an event, a type of call, a time of day, a subscriber location, a day of the week, a calling location, a piece of terminal equipment, etc. A trigger causes the network to take some type of action, and triggers are used to support services.

For example, a calling subscriber decides to call another subscriber on a Monday. The called subscriber had the service provider set up the network to route all calls to his home in Montana. The same called subscriber has his profile set up so that any call received on a Tuesday is sent to his home in California. These types of triggers are called terminating triggers. Most triggers are terminating triggers, and calls are given special routing treatment at the terminating location. Originating triggers would include the location of the originating party or the type of terminal equipment that originated the call. An example of an originating trigger would be a subscriber making a computer-based telephony call. The network identifies that the terminal equipment is a computer. The subscriber's service profile has been set up to so that all called computer users recorded on a specific list are alerted to the subscriber's location. Furthermore, the subscriber's service profile has been set up so that all calls from that specific terminal, at a specific time of day, are billed to a specific account. The trigger was the terminal type.

The IN was envisioned to eventually include the following key capabilities:

- A user-friendly service-creation environment for quick deployment of services
- Custom tailoring of services

The IN has been evolving for years to what it is today: the *Intelligent Network* (IN). The IN will continue to evolve over the years. Intelligent networking will be addressed in greater depth later.

Network Elements

The following list represents network elements within the SS7 network. These elements are also the functional forerunners of the current IN network elements. The following elements support the SS7 infrastructure:

- *Service Switching Point* (SSP)
- *Signal Transfer Point* (STP)
- *Service Control Point* (SCP)

Service Switching Point **(SSP)** The SSP is the LEC's end office (or the *Mobile Switching Center* [MSC]) in the wireless carrier's network. The SSP provides the functionality of switch communication with the rest of the SS7 network via the use of messages and data packets used by the SS7 network.

Signal Transfer Point **(STP)** The STP is interconnected to the SSP and acts similar to a traffic cop for messages in the SS7 network. Another way of viewing the STP is that it acts similarly to a router in this network. In the MF world, the nearest analogous switching element would be the tandem. The tandem is used to interconnect network areas. The STP also interconnects network areas.

Service Control Point **(SCP)** The SCP is the database that is used by the SSPs to store customer information, such as subscriber service profiles, special routing instructions, special service numbers, calling card numbers, and any other customer information that can be stored. In a broader sense, the SCP is nothing more than a database that can be used in a variety of specific capacities. The SCP, in effect, can be broadened or narrowed in its mission.

The following are examples of the SCP:

- *Line Information Databases* (LIDB) Provide subscriber information about special billing applications, such as calling cards or toll-free number service
- *Home Location Register* (HLR) Maintains registration information as it relates to the wireless subscriber's current location
- *Visitor Location Register* (VLR) Maintains registration information as it relates to the presence of roaming subscribers in a wireless network

The intelligent network differs in that it not only requires modifications to all of the previously mentioned network elements but also requires additional network elements. Specifically, the following objects support the intelligent network:

- *Intelligent Peripheral* (IP)
- *Service Node* (SN)

Intelligent Peripheral **(IP)** An IP is a device that has no switching capability, responds to queries, and reacts to specific instructions. The IP also supports a specific application, such as voice mail. In the case of this version of voice mail, however, the IP responds to triggers such as time of day or day of the week and provides the appropriate message based on those parameters.

***Service Node* (SN)** A SN is an "IP plus database capabilities plus switching capabilities." The SN is not responsible for switching calls but is capable of directing the processing in such a fashion that it essentially initiates an IN trigger, rather than responding to one.

Network Services

Network services is a general category. These services can be sold to subscribers, carriers, and corporate users. Network services are the product of the carriers.

There are two major classifications for network services:

- Bearer service
- Teleservice

Bearer Service Bearer services define the transmission capacity and functions that are required from the network. The functionality provided by the user terminals is not considered in bearer services. Terminals simply provide the capability to transfer the information.

Teleservice Teleservices provide the service for the user and include terminal devices (in addition to the actual service). Typical services, such as call forwarding, are teleservices.

Operational Support Systems and Services

Operational support has historically been described as the hind end and least-interesting aspect of the traditional voice telecommunications industry (and of the newer types of telecommunications businesses). The reason for this poor reputation is because operational support is not sexy—because it is not perceived to be the power behind the provisioning of services to the subscriber.

The fact is that *Operational Support Systems* (OSSs) are a key component to service provisioning. Without OSSs, you cannot create a customer bill, run your network, address customer complaints, maintain the network, and fix network troubles. Some people might be right when they say that OSSs are not the sexy part of the business, but these systems are essential to running the business.

My view of an OSS is an integrated one. OSSs can be viewed as a set of systems that can be integrated into a single platform. Unfortunately for those folks in network operations, most support systems are not integrated. On the other hand, the industry is rapidly moving towards an integrated approach. The drivers for consolidating OSSs are as follows:

- Operations-cost reductions
- Reduced workforce
- Growing number of disparate systems that provide enhanced services to the subscriber and enhanced network capabilities. These complex systems need to be managed efficiently.
- Centralization of functions, partly as a result of cost reductions and partly because of the complexity of systems

The following diagram (Figure 1-17) illustrates my view of integrated OSSs. The key behind this concept is an information-mediation device that

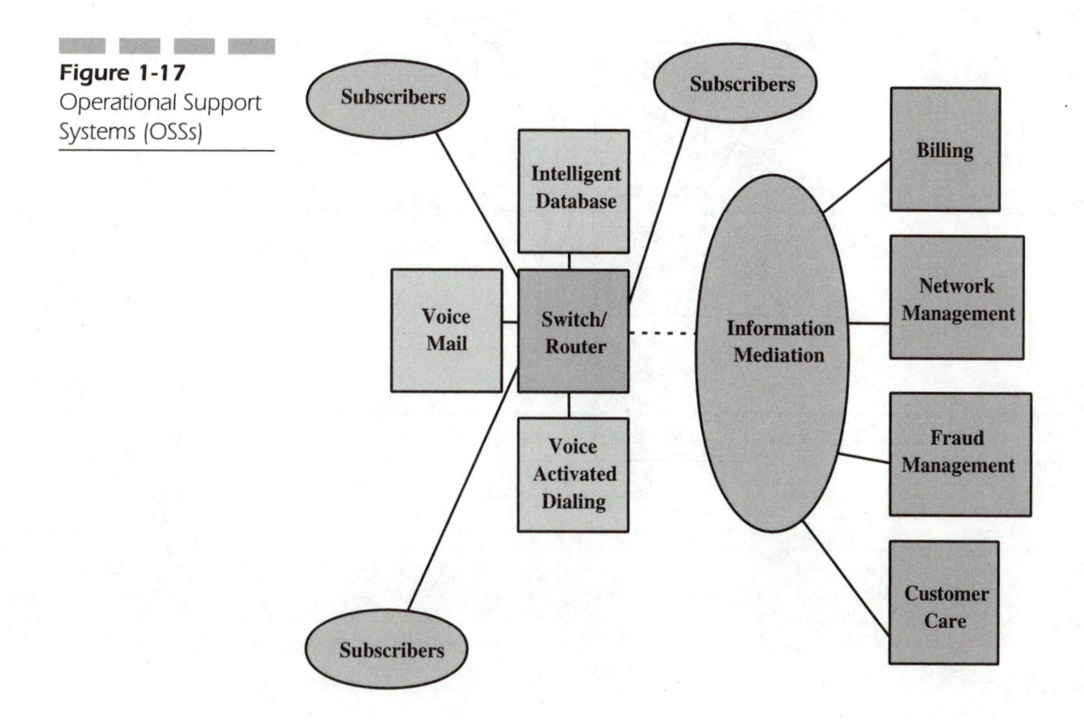

Figure 1-17 Operational Support Systems (OSSs)

can pull together the various disparate information sources and distribute this information to appropriate downstream applications. This topic alone warrants a book; however, for our purposes, a general description should be sufficient.

Operational support systems and services include a number of functions, which include the following:

- Information mediation
- Call-detail records and subscriber billing
- Network management
- Customer care

Figure 1-18 is a high level illustration of the *Operational Support System* (OSS). The systems all have one common function; support the running of the network and the business of the service provider. Up until recently, OSSs were treated as discrete systems. However, today these systems are viewed as a whole and the need for expertise in the design and operation of these systems is paramount today.

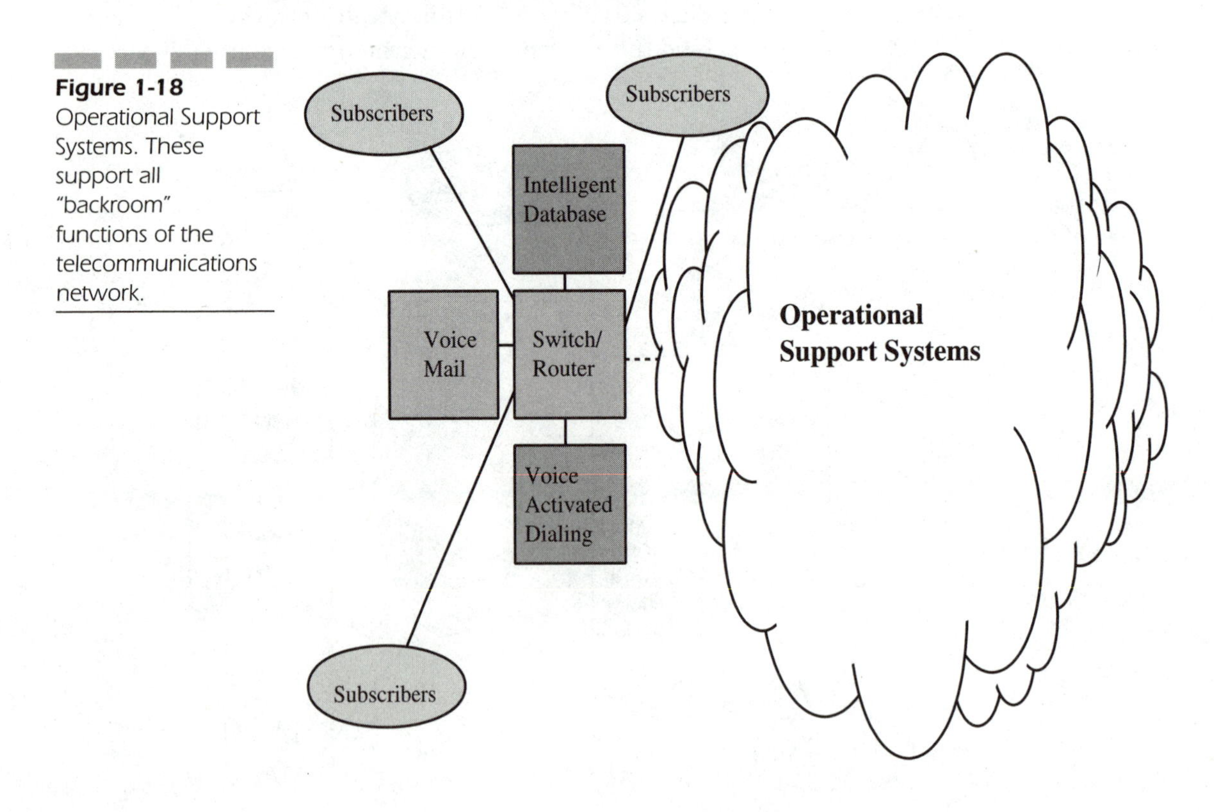

Figure 1-18 Operational Support Systems. These support all "backroom" functions of the telecommunications network.

The OSS network is literally a separate network that manages network elements, monitors the commercial side of the network, and is responsible for downstream internal applications.

As indicated previously, OSS is comprised of a number of functions that provide mission-critical support to the operations of the service provider.

Information Mediation

The information mediation function shown in the previous diagrams illustrates how the service provider must collect and process information from the commercial network. The mediation function collects, filters, and distributes data to the appropriate information-management applications with the service provider's organization.

Information mediation is a business in its own right. Collecting and processing data from multiple network elements for downstream processing by other types of network-management systems is no small task. Many network elements utilize a nearly standard protocol for internal element communication. Unfortunately, the standard language actually might be one of several languages—all with a slight proprietary twist.

The mediation function has become increasingly important within the last three years. Without this mediation capability, service providers will find it extremely difficult to contain and control costs. The mediation function's most obvious benefit is in the area of labor expense. Instead of having four network managers manning 20 different terminals for 20 different systems, the service provider can reduce its workforce by at least 50 percent and the number of different monitoring systems by a significant percentage.

Call-Detail Records and Subscriber Billing

Call detail is the raw information used by a billing system to create a subscriber bill. This information, called call detail, is essentially information that is generated by the network elements. Call detail usually includes the following types of information:

- Date the call or transaction is made
- Time of day the call or transaction is initiated
- Time of day the call or transaction is ended
- Who made the call or transaction

- Who was being called
- Type of terminal device or telephone line from which the call is made
- Use of calling or credit cards in order to initiate and pay for the call
- Is the call or transaction being billed to a third party?
- What features are turned on?
- When is a featured turned off?
- When is a feature turned on by the subscriber?
- When is a feature turned off by the subscriber?

The identity of the calling party can be displayed as a name, a terminal number, or a telephone number. The service provider, for its own routing and internal operations purposes, will identify the calling party with some type of number: a telephone number, a calling card number, a credit card number, etc. The identity of the called party can be displayed as a name, a terminal number, or a telephone number. The service provider will, for its own routing and internal purposes, identify the called party with some type of number: telephone number, calling card number, credit card number, etc. The type of terminal device can be a coin-operated telephone, a credit-card telephone, a wireless handset, or a home landline telephone with a calling card.

Call detail information can be generated by any network element that meters data. Most network elements generate some type of information that describes a call (voice or data). The issue usually involves the question, "Where can the service provider go within its network to collect this data?" This question represents the heart of call detail record collection: aggregating call detail data in an intelligent and meaningful manner.

The capability to collect data becomes complicated when the service provider's network is comprised of multiple network equipment from different vendors. The task of collecting meaningful call detail becomes complicated. The different vendor equipment might use different operating protocols for internal operations or even for communicating with external network elements. This situation makes it nearly impossible for a network manager to rapidly address troubles when they arise. When all of the communications between devices requires multiple computer terminal access, however, then repairing network troubles becomes a challenge. We will provide more information about network management systems in a later section.

Call detail record collection typically refers only to information regarding the call. This information is used to create bills. I have a more expanded view of call detail records, however. I define call detail to be any information that is metered by a network element. This view is fairly broad, but you will find that most of the raw information that is used to create a bill can be used for other purposes, as well. These other functions would require downstream application systems to process the data for these other purposes. Remember that the call detail being recorded is in a raw and unprocessed format. In other words, the information has not yet been processed by a computer system for a specific purpose.

Call detail can be used to support the following functions:

- Billing (previously explained)
- *Fraud management* The information used for billing can be used to identify fraudulent usage of a telephone. This task would require a software application that would, for example, calculate the times that multiple calls were made and the location from which they were made. The speed algorithm used by the service provider would tell the carrier whether or not the customer could have been able to travel from point A to point B in five minutes to make the call. In other words, if I made a call in New York and then five minutes later made a call in California, it would obviously be impossible for me to fly from New York to California in five minutes. Another example would be making calls from a city from which I have traditionally never called.
- *Customer care* Customer care systems would use call detail to address customer complaints, to issue repair orders, and to sometimes communicate information to the carrier's billing/accounting organization.

Network Management

The use of the term *network management* in this context is a little misleading. As I had indicated, the OSS is a set of internal operational functions that are needed to manage the network and customer base. Network management is an abstract term used to describe the management of the network and the customer base, whereas OSS is a more modern term that is used to describe more discrete activities or functions within network management. In the days prior to the *Telecommunications Management Network* (TMN) framework, OSSs were a collection of functions that changed in definition from person to person. As a matter of fact, the term was just gaining popularity. Prior to the mid-1980s, a network operations manager was (and still is) considered to be part miracle worker, part workhorse, part wielder of spit and baling wire, and one of the first few people to hear the customer shout in protest over a service problem.

Prior to TMN, the network operations center was a center of multiple systems that supported multiple-vendor equipment. You could walk into a network operations center and find monitoring equipment for every type of vendor system. Furthermore, each vendor system used different network management protocols (and usually proprietary ones, at that). Many engineers had largely ignored network operations. Fortunately for the industry, as the complexity of operating the network elements increased and became more software oriented, the need and capability to manage disparate network elements from a central point/terminal became necessary and easier.

TMN is a way of framing, organizing, and standardizing the whole matter of OSS. Network management today is comprised of the following components:

- Network-element management
- Network systems management (the entire collection of elements)
- Service management

Network-Element Management

Network-element management involves the capability to monitor and manage specific network elements. A network element can be defined as a database, the switch, a router, a voice mail recording system, or an adjunct system.

Typically, the network operations manager wants to know the following information:

- Maintenance schedules
- Routing table update schedules
- Number of subscribers being served by the system
- Number of roaming subscribers served by the system
- Services being provided to the subscribers
- Performance objectives for the network element
- Transaction processing time thresholds
- Types of diagnostic tools that are available
- Operating procedures documentation
- Failure levels/indications
- Alarm conditions indications

Network Systems Management

Network systems management involves the way one manages the entire collection of elements and all of the functions within the business that affect the physical network.

Network operations managers would ask questions similar to the ones that they would ask in relation to the network element. These questions would involve the following considerations:

- Service objectives for the system
- Network monitoring software tools
- Network diagnostic tools
- Anticipated traffic load on the system
- Overall traffic profile of the system on an hourly and daily basis
- Types of subscribers that are being served
- Level of network system control (can the entire system be rebooted, or is an involved, element-by-element process necessary?)
- Can traffic be rerouted?
- Types of alarm indications
- Types of network failure indications
- Overall network operating procedures documentation
- Any disaster recovery plans

Service Management

Service management links the network systems management function with the customer care function, which includes performance metrics and customer satisfaction metrics. Service management requires direct involvement with the customer and involves every mechanism and process that is required for delivery of services to the customer (i.e., operational support systems, sales, and marketing).

This area of network management includes the following:

- Establishing and following service-level agreements
- Market surveys
- Customer satisfaction surveys

- Competitive analyses
- Operations Cost analyses
- Customer complaints resolution
- Management of operations cost

Customer Care

Customer care is the one carrier operating area that cannot be totally mechanized/automated. True, there are automated systems to handle customer queries; however, when a subscriber wishes to speak with a representative of the carrier, the representative is usually someone who is in the customer care department. The customer care person must be as highly trained to handle customer queries, customer anger, and customer requests as a modern network manager needs to be in order to operate the network.

Automation in customer care enables a carrier to answer a variety of standard queries from the subscriber, but automation cannot address the customer who simply wants to speak to a person or wants to shout at someone. As silly as this idea might sound, think of a retail department store or grocery store. They all have a customer service desk. Now, think of how you would feel if the customer service representative did not treat you (the customer) with respect and simple common courtesy.

The customer care system is one of the few points of customer-carrier contact. The automated customer system needs to be pleasant to listen to, fast, accurate, and easy to use. The customer care representative needs to reflect the personality and face of the carrier, and this person also needs to be pleasant, fast, accurate, and personable.

SUMMARY

The first step in understanding the telecommunications network and business is to understand the fundamentals of the network and the business. Chapter 1 lets you (the reader) take the first step.

The Telecommunications Hub—Creating Value

The telecommunications hub I refer to is a traffic concept that is used in vehicular and air transport traffic management. Traffic hubs serve as central points of distribution. In the case of air transport, airlines designate certain cities to serve as principal bases of operations, repair, and takeoff/landing areas. In some cities, vehicles are brought into and out of the cities via a network of roadways that manage and distribute traffic as efficiently as possible.

The telecommunications hub is a traffic-management tool employed to manage the distribution of information. Although images of a wheel are invoked when the word *hub* is used, the hub does not need to look like a wheel. The following diagrams (Figures 2-1 through 2-3) illustrate how information traffic can be distributed.

Each one of the aforementioned diagrams represents a generic traffic configuration. We will explore configurations types in more detail in the next chapter. What is important to understand is that the telecommunications hub is a central place or focal point for information dissemination/ traffic flow. The hub has value beyond traffic-management efficiencies or even call-processing efficiencies. The telecommunications hub serves as a point of network convergence.

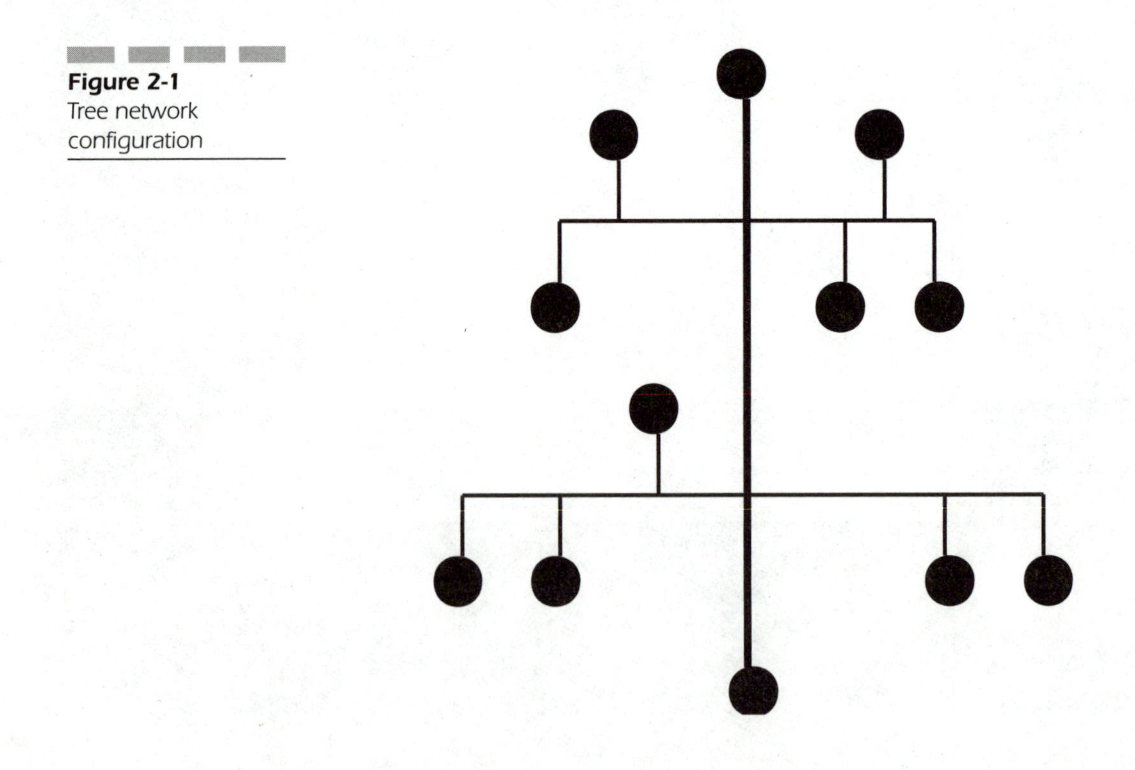

Figure 2-1 Tree network configuration

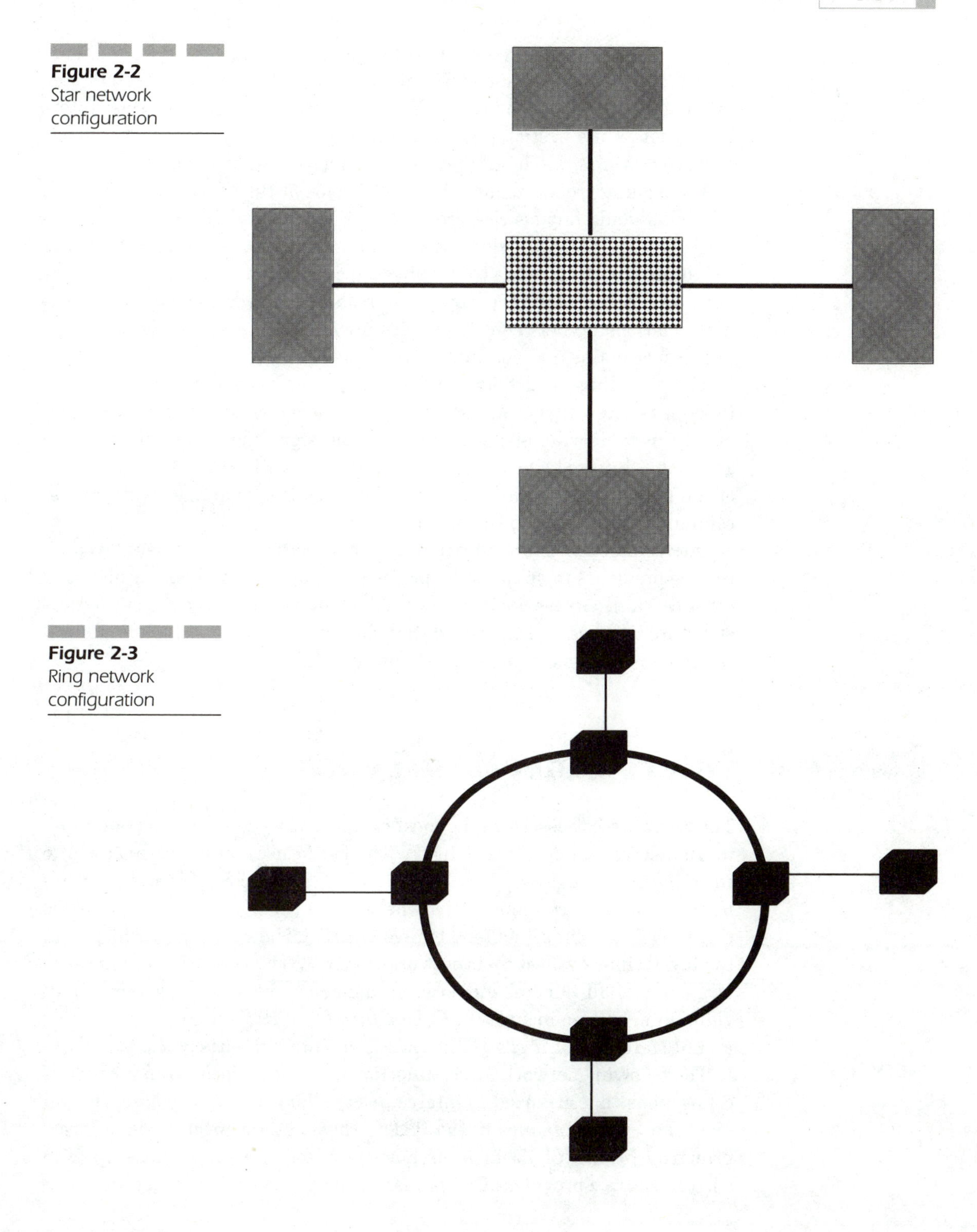

Figure 2-2
Star network configuration

Figure 2-3
Ring network configuration

Dissimilar networks (such as cable television, wireless, wireline, etc.) need to interconnect somehow. If interconnection does not occur, call flow or information flow from one network to another network cannot occur. Interconnection is the first step towards convergence—convergence of technology, services, and telecommunications business relationships.

The business relationships that I refer to lie at the heart of convergence. Why else would any service provider seek to converge his or her network with another service provider unless there was a business reason for doing so? Interconnection refers to the physical connection between two or more networks. A much broader view of interconnection would include the business relationship between the service providers. In many cases, the business issues define and dominate the relationship.

The signaling used between interconnected networks dictates which types of services can be supported between the networks (more information about this topic will appear later). Note that signaling is basically the language that is used by networks to communicate with each other. The limits of the language or the interpretation between the languages restricts the communication between the networks.

Interconnection is an infrastructure product that can be leveraged by service providers to support the provisioning of other services or access to other services (Access Services) and is the foundation for the resale of infrastructure services and the foundation for how networks and telecommunications business segments relate to one another.

Interconnection Types

The following list describes the most commonly used interconnection types in the telecommunications industry. The list focuses on interconnection to the *Public Switched Telephone Network* (PSTN). The PSTN is still the dominant commercial communications network in North America. Discussions regarding interconnection tend to revolve around how service providers (not the PSTN) interconnect their networks to the PSTN. The PSTN refers to the large and small (incumbent) local exchange carrier networks. Some individuals might even include the *Competitive Local Exchange Carrier* (CLEC) and *Interexchange Carrier* (IC) networks as integral parts of the PSTN.

The following network interconnections are the de facto, standard interconnections that are used to interconnect cellular carriers, paging companies, and PCS companies to the PSTN. These interconnect types are used primarily by the LECs in their business/interconnect relationship with wireless service providers. The names have as much a regulatory and business meaning as they do a technical meaning. Not all *Local Exchange Carriers* (LECs) support all of the following interconnection types:

- Dial-line
- *Direct Inward Dialing* (DID)
- Type 1
- Type 1 with ISDN BRI
- Type 1 with ISDN PRI
- Type 2A
- Type 2A with SS7
- Type 2B with SS7
- Type 2C
- Type 2D
- Type 2 D with SS7
- Type S
- Private-line

A detailed description of these interconnection types can be found in Appendix B. The following brief summaries follow.

Dial-Line The dial-line connection is a two-wire, line-side connection from a LEC end office. This connection is similar to the connections that are used for business and residential lines. This connection can be used on a bidirectional communications basis. Dial-line connections enable the MSC to access any valid telephone number, including NXX codes within the LEC's serving territory as well as codes that are accessible through an IC carrier's network (including international calls).

Dial-lines enable the MSC access (or other carrier's switch access) to any valid telephone number in the LEC's network.

The dial-line permits the mobile user to access N11 codes (for example, 911) and Service Access Codes (such as 700, 800, 888, 877, etc.).

Direct Inward Dialing (DID) The *Direct Inward Dialing* (DID) connection is a trunk-side, end-office connection. The DID connection is a two-wire circuit limited to one-way incoming service (LEC to wireless carrier). This connection gives the LEC the perception that the MSC is a *Private Branch Exchange* (PBX). In some cases, they are also considered to be Type 1 Connections. The DID connection literally places the terminating switch in a subservient role.

Type 1 The Type 1 connection is a trunk-side connection to an end office. The end office uses a trunk-side signaling protocol in conjunction with a feature known as *Trunk With Line Treatment* (TWLT). The TWLT feature enables the end office to combine some line-side and trunk-side features. TWLT enables the LEC to provide billing and IC presubscription. Note that presubscription is a process in which the subscriber selects his or her primary long-distance carrer, then subscriber dials 1 followed by the 10-digit directory number and is automatically connected to his or her long-distance carrier of choice.

The service provider will have access to numbers that reside in the interconnected LEC end office, including NXX codes within the LEC's territory as well as codes that are accessible through an interchange carrier's network (including international calls). Type 1 connections also permits the mobile user to reach Directory Assistance, N11 codes (such as 911) and Service Access Codes (such as 700, 800, 888, 877, 900, etc.).

Type 1 with ISDN The *Integrated Services Digital Network* (ISDN) connection is a ANSI-standard ISDN line between the LEC end office and the wireless carrier's MSC. The ISDN connection should be capable of providing connectivity to the PSTN that will support ISDN *Primary Rate Interface* (PRI) and ISDN *Basic Rate Interface* (BRI). The ISDN connection is a variation of the Type 1 connection.

ISDN BRI provides two bearer channels (each 64 kb/sec) and a data channel (16 kb/sec), while ISDN PRI provides 23 bearer channels and one data channel.

Type 2A The Type 2A connection is a trunk-side connection to the LEC's access tandem. This connection enables the wireless carrier's MSC to interface with the access tandem as if it were an LEC end office. This connection enables the wireless carrier's subscribers to obtain presubscription. The service provider will have access to any set of numbers within the LEC network. The Type 2A comes in two flavors: a MF version and a SS7 version.

Type 2B The Type 2B connection is similar to the high-usage trunk groups that are established by the LEC for its own internal routing purposes. In the case of wireless carrier interconnection, the Type 2B should be used in conjunction with the Type 2A. When a Type 2B is used, the first choice of routing is through a Type 2B with overflow through the Type 2A.

Type 2C The Type 2C connection is intended to support interconnection to a public safety agency via an LEC E911 tandem or local tandem. This connection enables a wireless carrier to route calls through the PSTN to the *Public Safety Answering Point* (PSAP). This connection supports a limited capability to transport location coordinate information to the PSAP via the LEC.

Type 2D The Type 2D connection is intended to support interconnection to the LEC's operator service position. This connection enables the LEC to obtain *Automatic Number Identification* (ANI) information about the calling wireless subscriber in order to create a billing record.

Type S The Type S connection is not a voice path connection. The Type S is a SS7 signaling link from the wireless carrier to the LEC. The Type S supports call setup via the *ISDN User Part* (ISUP) portion of the SS7 signaling protocol and TCAP querying.

The Type S is used in conjunction with the Type 2A, 2B, and 2D interconnections.

Private-Line The private-line connection is often used to connect the wireless carrier switch to a cell site or to connect two cell sites. This connection can be two-wire or four-wire analog, DS−1, or DS−3 circuits.

The previous typical descriptions are written from the perspective of a wireline LEC and represent interconnection from one type of carrier's perspective. Both in-band and out-of-band signaling are represented. Standards bodies such as the *Telecommunications Industry Association* (TIA) have expanded the view of interconnection to include the wireless perspective. The following diagram (Figure 2-4) is a pictorial representation of PSTN interconnection.

Importance of Interconnection

If you were the recipient of these previous interconnection offerings, you would have to consider the utility of each interconnection type in your business plan. If you are a LEC and are on the supplying end, then the decision is a no-brainer. You should leverage every network capability. In the case of the wireline telephone network, it usually means a lot of network power that you (as the recipient of the interconnect) do not necessarily need. A point of information needs to be raised. The Telecommunications Act of 1996 has added a new term to the industry lexicon: *Incumbent Local Exchange Carrier* (ILEC). The term ILEC refers to the dominant LEC in any given market. I will be using the terms ILEC and LEC interchangeably.

Figure 2-4
PSTN interconnection

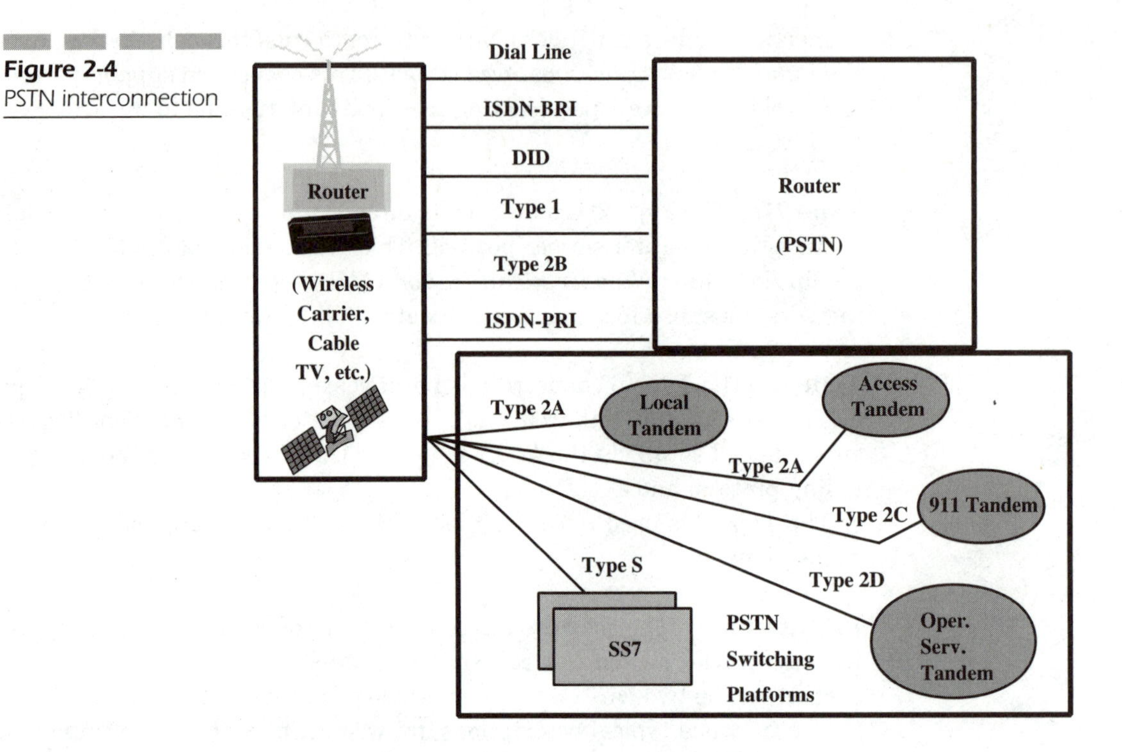

Let's assume that you are a network provider (a.k.a., a service provider with your own switches and infrastructure), and you want to interconnect with a wireline telephone company. You want to maximize the value of the interconnect that you are buying. You probably want the following features from your interconnect:

- *Visibility to other service providers.* In other words, you want to interconnect to the PSTN (the LEC) because your subscriber base wants to call subscribers of the LEC, and you want to give your subscribers the capability to call subscribers in other networks. The LECs today have the physical connectivity to scores of other service providers and therefore offer readily available access to these service providers.
- *Visibility for your own network.* Now everyone else can buy one transmission facility to the PSTN and reach Service Provider XYZ *and* your network.
- Reducing network operating costs and capital outlays

- Establishing a technical and business relationship with the LEC, which provides the capability to obtain services from the LEC. These services would include billing services, database dependent services, subscriber features, operator services, directory assistance, and switching and transmission facility provisioning.

The question facing many CLECs and other types of carriers is, "Which interconnect should I buy?". The short answer is that it depends on the type of CLEC or other carrier and who your customers are. The interconnect has to provide you with the capabilities that you require. These are good reasons for obtaining interconnect to the LEC/ILEC (or PSTN). As the number and size of new entrants grow, however, the question you should ask is, "Why can't a carrier other than the incumbent LEC provide the same interconnect?". This idea was first raised in the late 1980's.

You can take this new perspective one step further and come to the conclusion that another carrier other than the ILEC can provide vertical services and network support to other service providers. I once knew this concept as access services, wholesale services, or infrastructure services support. Today, some call it retail support, wholesale services, or infrastructure support. I have always preferred the term infrastructure support. The more popular term, however, is wholesale services.

Later in this chapter, we will delve further into the types of services that you can buy.

Interconnection between Different Network Providers

The following illustration (Figure 2-5) is what interconnection looks like from the perspective of the LEC (or PSTN). The reader who is primarily experienced in wireline telephony world will find that I am using terms such as router, as opposed to the word *switch*. The reason is simple. At a the most basic level of switching, all a switch does (whether the switch is a Class 5 end office or a Class 4 tandem) is route. By taking a broader view of semantics, we can better see the close relationship between network concepts (used by various service providers).

This figure illustrates the standardized interconnections that the LECs provide to the wireless carriers. These interconnections support access to

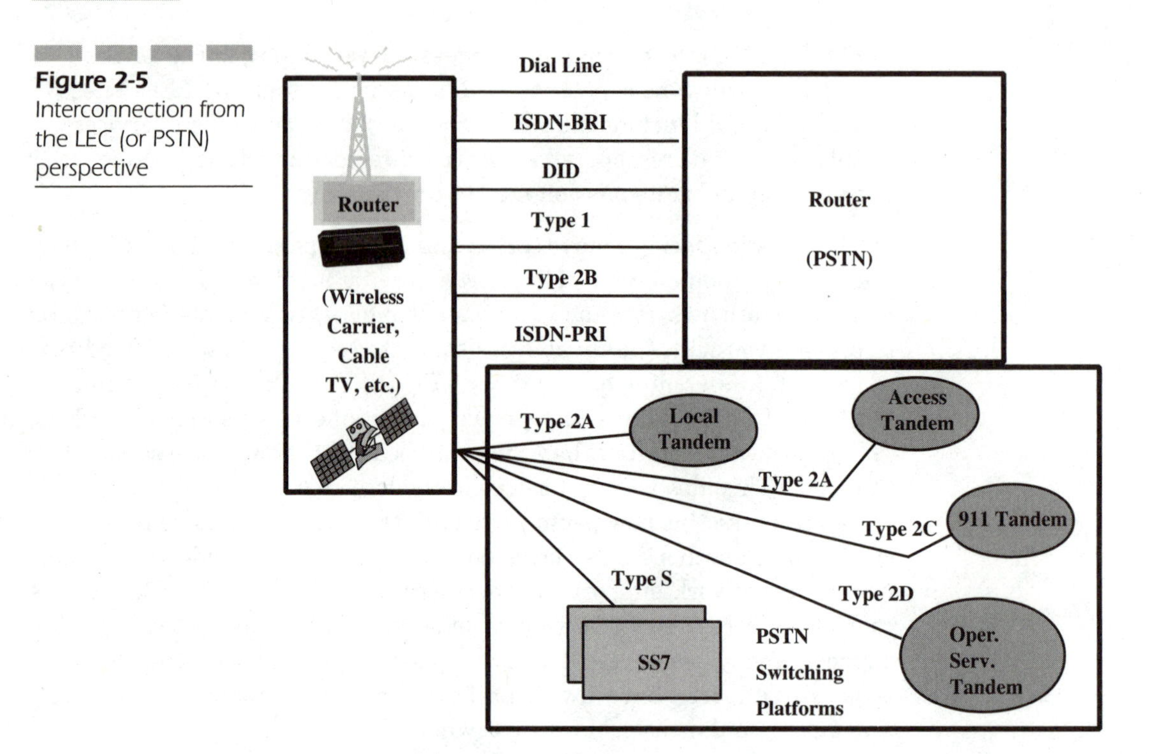

Figure 2-5 Interconnection from the LEC (or PSTN) perspective

and from the PSTN. Note that the PSTN also supports access to public safety agencies (via 911 tandems) and to operator services.

These interconnections are generally applicable to most other types of service providers. More importantly, the previous interconnection scenarios can be provided by other types of service providers—not just LECs. What I am alluding to is the fact that any service provider technically can become a provider of interconnect. When you look at these previous examples, you are really seeing the PSTN in the middle of all of the action—similar to the hub of a wheel. Can another type of carrier be at the center of the hub? We will examine the concept of wholesale later in this book. Refer to the following diagram (Figure 2-6).

Another way of viewing this concept can be seen in Figure 2-7 (for those who are used to the cloud types of diagrams). In these diagrams, a service provider of some unknown capability sits in the middle of the picture and is represented as a cloud, because you do not need to know what comprises the cloud (a.k.a., network).

The question that a CLEC, wireless carrier, or some other type of carrier should ask is, "Why can't I be the center of this telecommunications hub?". No technical reason exists as to why a CLEC, wireless carrier, or some other

Figure 2-6 PSTN Network Interconnection. Interconnection viewed from the perspective of the Incumbent LEC

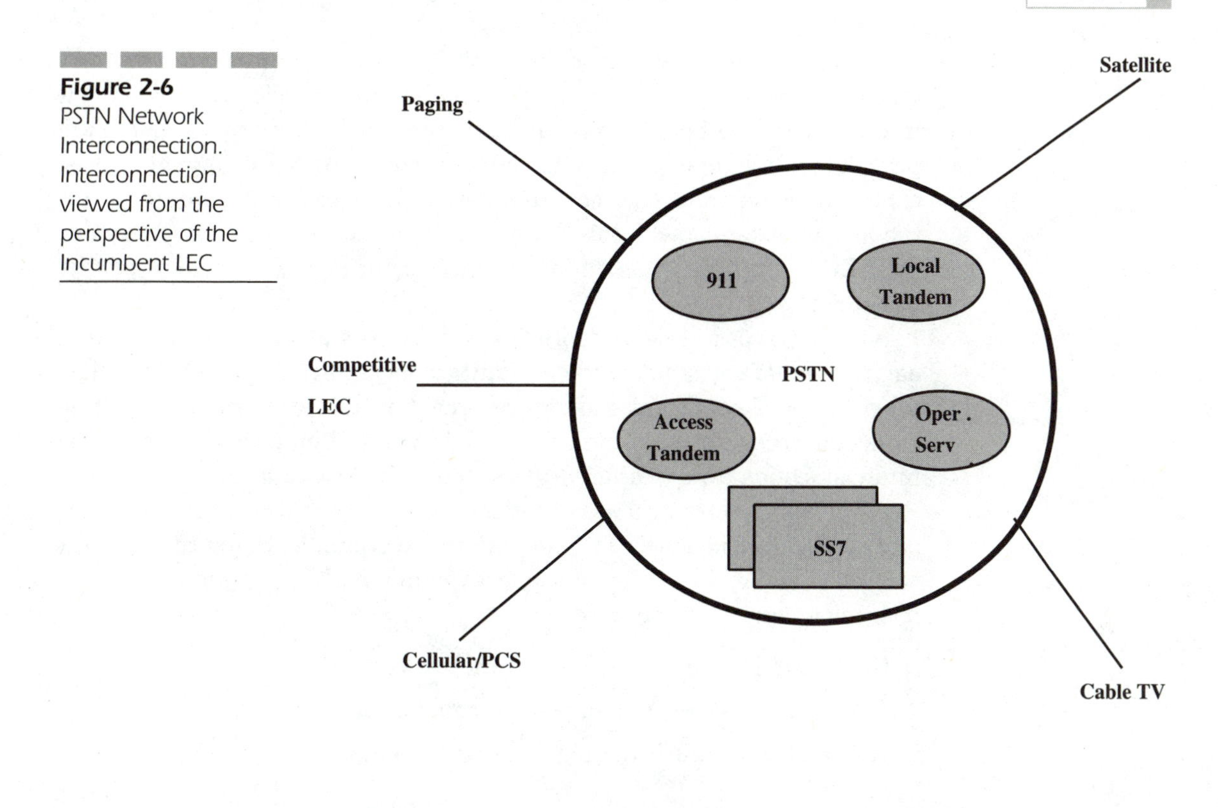

Figure 2-7 Another view of interconnected networks

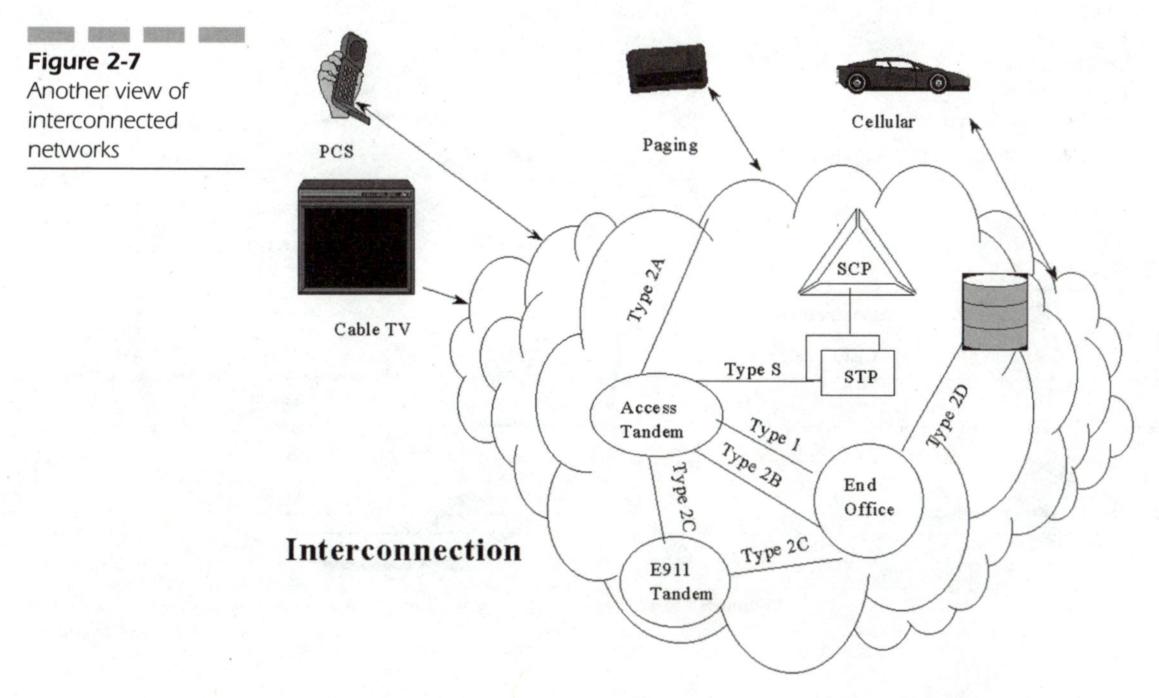

type of carrier cannot be at the center of the telecommunications hub. Technically, there is not any reason why any service provider cannot provide interconnect. The point to remember is that the LECs are the dominant providers of telecommunications services today. The ILEC has the largest capital investment in the local telecommunications market and typically the longest established local network in the local marketplace. Therefore, the LEC rightly serves as an interconnect point between different service providers.

As other types of service providers' networks grow, they will become aggregators of consumer telecommunications traffic. At some point, these other types of service providers are providing interconnect to other networks for other service providers. For interconnection purists, let's look at interconnection from a different perspective. Refer to Figure 2-8.

A key point to remember is that there are no technical impediments to any service provider being the provider in the middle. Being the hub, the core, the middle, the center, or the focal point (or whatever you wish to call it) brings many strengths:

- Call setup information transits your network.
- Information needed to create bills transits your network.
- Information traffic is aggregated on your network.

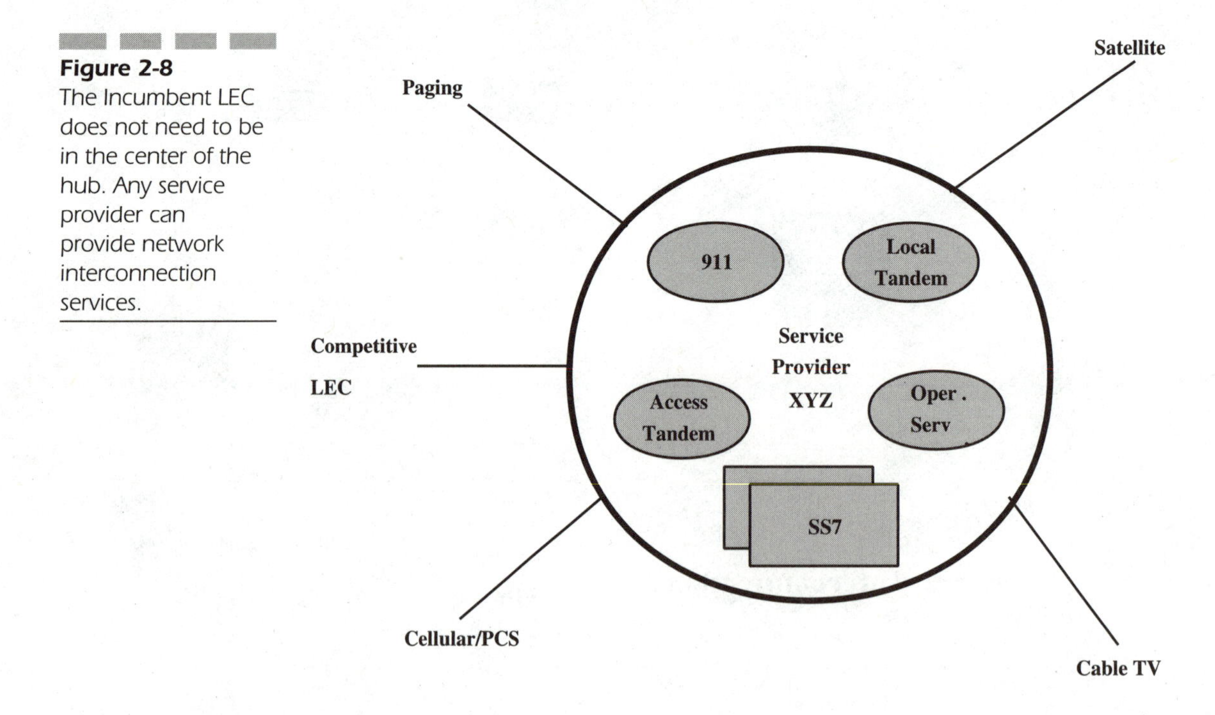

Figure 2-8
The Incumbent LEC does not need to be in the center of the hub. Any service provider can provide network interconnection services.

To those in the regulatory world, the word *interconnect* has a definite and distinct meaning. In fact, the term is defined in the Telecommunications Act of 1996 and is defined in state tariffs throughout the country. Interconnect is typically defined via a set of specific interconnection arrangements that are supported by the ILEC. Interconnect is a business/technical network relationship between the PSTN and other providers. Because the PSTN has been and still is the dominant local network provider, it makes sense that most people still view a Type 2A as a form of interconnect that the PSTN provides.

With new networks being built and new network providers coming online all the time, these other communications networks can provide a Type 2A, for example. A Type 2A is defined as follows: "The Type 2A connection is a trunk-side connection to the LEC's access tandem. This connection allows the MSC to interface with the access tandem as if it were a LEC end office. This interconnection enables the wireless carrier's subscribers to obtain presubscription. The Type 2A uses trunk-side signaling protocols; using E&M supervision with *multi-frequency* (MF) address pulsing. With the Type S signaling connection the PSTN can provide SS7 with the Type 2A."

Service Provider XYZ (as shown in Figure 2–8) can decide to call its Type 2A a Gateway Interface. A gateway switch is a switch that provides a single point of access out of and into Service Provider XYZ's network. A Type 2A essentially provides access out of and into the LEC's network. Service Provider XYZ, however, has decided to call the interface a Gateway Interface. What's the difference? The answer is none.

Service Provider Perceptions

The PSTN sees network interconnects such as the Type 2A as a service and as an enabler of business relations. In the current interconnect environment, the previously mentioned interconnect types are fairly inexpensive for a CLEC or other type of carrier to purchase. What the interconnect provides to the PSTN, however, is a chance to build a business relationship with the interconnected service provider. Think about this idea. Once you have established a technical and business relationship, why not simply build on that relationship and obtain additional services from the PSTN? Unless there is a business imperative to not work this closely with the LEC, no reason exists not to obtain additional services.

The original function of the interconnection between the PSTN and other service providers was to give the customer bases of the other service providers the capability to reach the much-larger PSTN customer base.

Times have changed since the inception of these basic interconnect types, and so have the uses of network interconnect.

Some network interconnect purists might say that cellular carriers do not have tandem switches; therefore, they cannot provide a Type 2A. A tandem is a switch that simply acts like a traffic cop directing traffic. Some large cellular carriers use *Mobile Switching Centers* (MSCs) as tandems—subtending traffic from multiple MSCs to the rest of the world. Other switching experts use the word *gateway* to describe these types of switches. Part of the problem that we will encounter in the communications industry as it evolves is that there is a legacy that many of us will carry with us. There is absolutely nothing wrong with this legacy, and this legacy has roots that date back to 1876.

Having an understanding of how the technology, practices, and industry evolved is valuable for any telecommunications engineer, because this knowledge gives perspective and design insight. The one problem that we will encounter, however, is a stubborn addiction or "stick-to-it-ness" to semantics. As we move forward, let's focus on the function of the network element and not the word. Examining a network and its network elements from a functional standpoint provides any network manager with a broader understanding of how the network should work.

If a cellular carrier were to provide a Type 2A, the carrier would provide the transmission path between its gateway MSC and the other provider. The Type 2A could be a T1, DS1, or even DS3 that supports not only MF but also SS7 networking. Note that SS7 networking support requires an additional facility to serve as the data link between the switch and the STP.

The following diagram (Figure 2-9) illustrates how the networks can be interconnected in an SS7 environment.

This figure can be made generic by removing the words MSC and ILEC Switch. The diagram now appears as follows (in Figure 2-10).

This service provider interconnection diagram can be applied to any service provider, including the PSTN. Bear this thought in mind as your view of interconnection expands.

Interconnection Products and Services

Telecommunications products are typically physical/tangible objects, such as subscriber calling services, bundled subscriber calling services, subscriber calling cards, and subscriber rate plans. Services would include customer care, billing plans (i.e., deferred payments, credit card payment, etc.), repair, and customer complaint. The reader might not agree with this

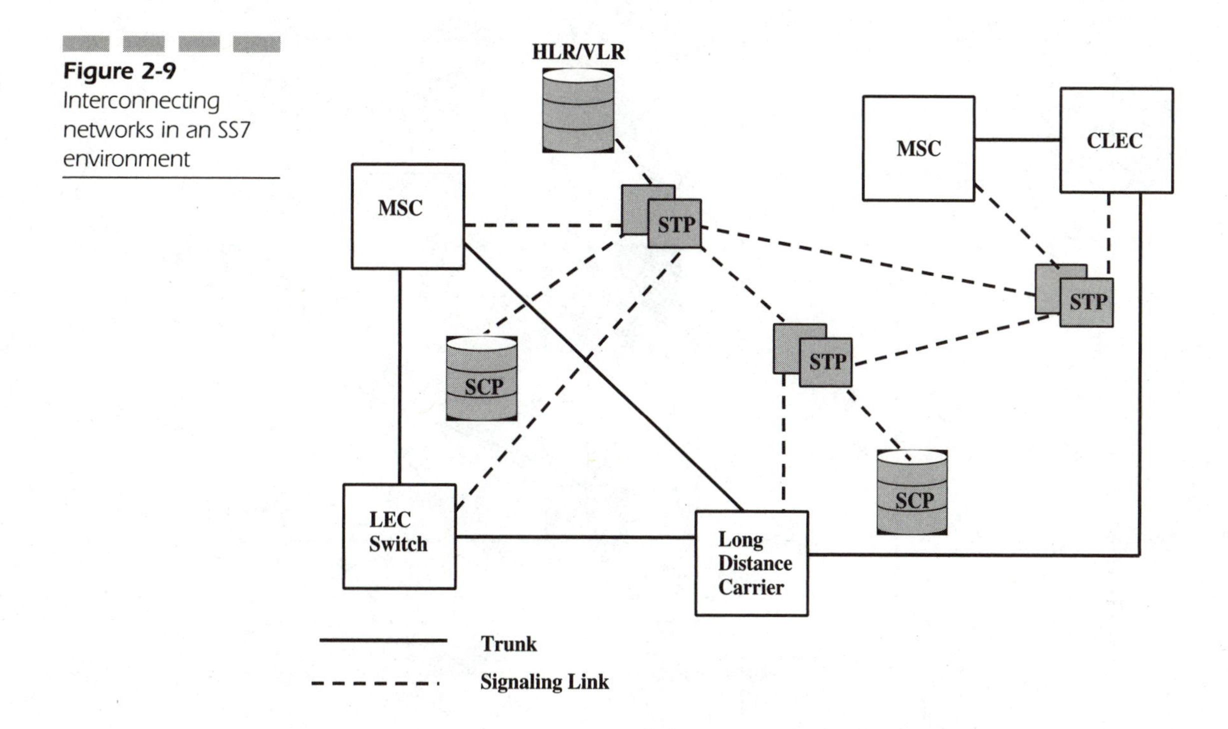

Figure 2-9 Interconnecting networks in an SS7 environment

breakdown, which is understandable. Many people interchange the words products and services, which there is no harm in doing. For the sake of clarity and understanding in this book, I will use products and services as I have described.

Products and services can be viewed as something broader than what I have just described. First, the customer does not have to be an individual or business. The customer can be another telecommunications service provider. In this case, the products would be operational support systems, radio infrastructure support, transmission facilities use, building space, utilities (i.e., water, power, and light), real estate access (possibly for antennas), database management, switching infrastructure support, directory assistance, etc. The service that a carrier would offer to another carrier would be repair, installation, network management, call detail recording, bill rendering, network maintenance, etc. We will explore this topic in more detail in the following chapters.

Interconnection can serve as an enabler of all of these services. The technical interconnection between the network does not enable these services; rather, it enables the business relationship that leads to these types of products and services.

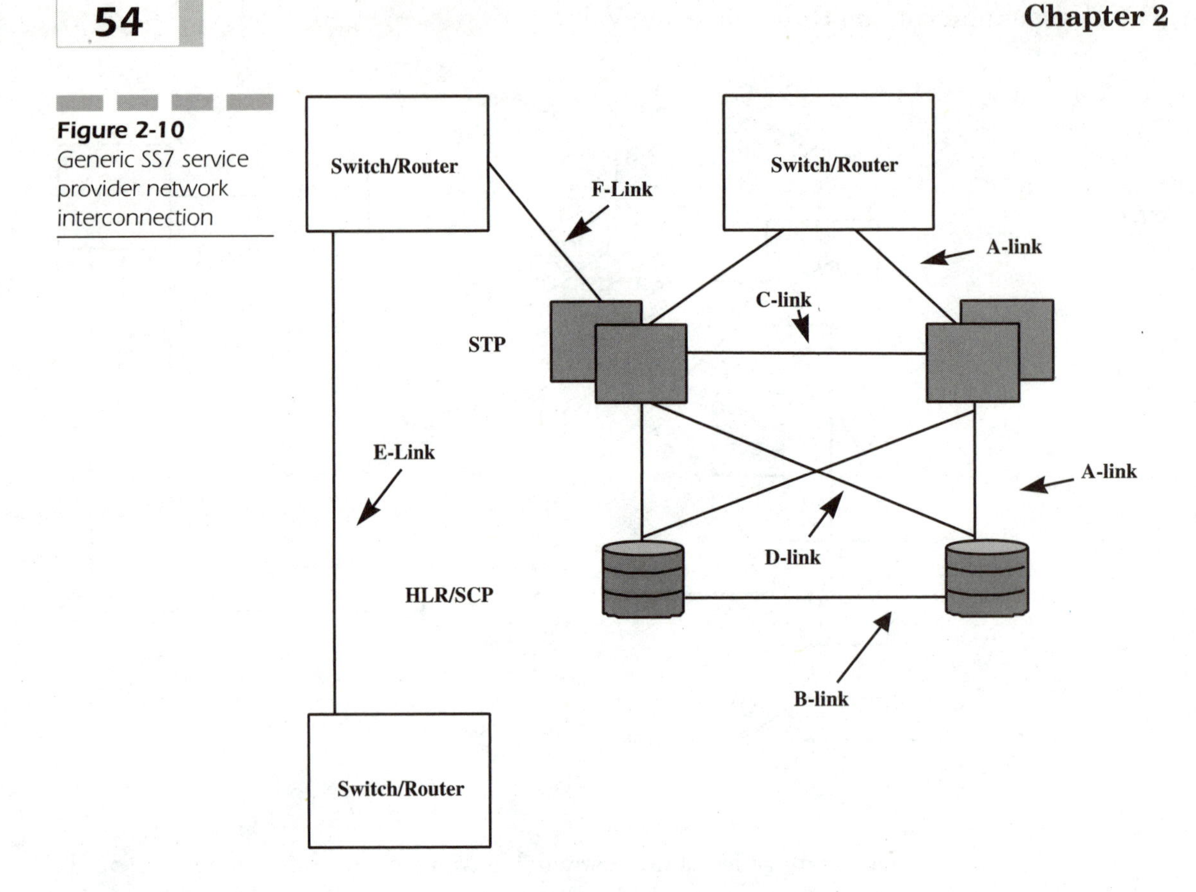

Figure 2-10 Generic SS7 service provider network interconnection

Before we take the next step toward understanding interconnection, we need to understand where most of this information resides. There are two industry documents that contain much of what we need to know from a purely technical perspective. One document is a standard, while the other is an industry practice.

The following paragraphs summarize the documents.

Specifications versus Standards

There are two documents that serve as the wireless industry's principal source of technical information for network interconnection. One document was produced by Bell Communications Research, Inc. (now known as Telcordia Technologies), and the other document was produced by the *Telecommunications Industry Association* (TIA). One should view these documents as tools for understanding how interconnect works as a product and as an enabler of products.

The Bellcore document is called *Generic Reference* (GR)−145-CORE, and the TIA document is called *Interim Standard* (IS)−93. The documents are informative and are quite helpful in understanding the technical aspects of interconnect. We will tackle the business aspects of interconnect after a brief summary of what these documents contain.

Work on TIA's IS−93 began in 1992 when it became clear that technology development had elevated the wireless switch from the technical functionality of a PBX to that of a switching system equal to that of a wireline telephone company switch. A number of wireless industry players had hoped to use IS−93 as a way of forcing the *Regional Bell Operating Companies* (RBOCs) into renegotiating their interconnection agreements. The most significant difference between IS−93 and TR-NPL−000145 Issues 1 and 2 was the acceptance of a bidirectional signaling relationship between the cellular carriers and the PSTN. By acknowledging that the SS7 signaling occurred in two directions, the RBOCs could interact with the cellular carrier as a co-carrier; therefore, the cellular carriers could force acceptance of mutual compensation. GR−145-CORE was written (by Bellcore) in response to IS−93 Revision A. Bellcore's GR is a proprietary product sold by Bellcore. Some industry players believe that the Bellcore document was funded by the owners of Bellcore in order to keep their existing interconnection agreements in place. Everyone has an opinion.

Both documents were written to support technical and business perspectives and objectives of the respective interest groups. GR−145-CORE supports the objectives of the RBOCs, and IS−93 supports the objectives of the wireless industry—specifically, the *Cellular Telecommunications Industry Association* (CTIA) and the TIA. IS−93 is intended to support the entire wireless community and is a standard. GR−145-CORE is a company proprietary document that is sold for profit. The intent of the documents is not relevant within the context of a technical discussion.

GR−145-CORE is services and implementation oriented. In other words, GR−145-CORE interconnection descriptions provide more information than just technical signaling protocols and parameter information. GR−145-CORE addresses interconnection from the perspective of how the LEC is or is not capable of providing information. For example, GR−145-CORE describes how the LEC provides Feature Group A, B, C, and D support. Essentially, GR−145-CORE is an implementation document for LEC interconnection, where the LEC is at the center of the interconnection view. GR−145-CORE also promotes Bellcore document products. TIA's IS−93 addresses the interconnection not only from a implementation perspective but also from the perspective of a wireless carrier wish list, where the wireless carrier is at

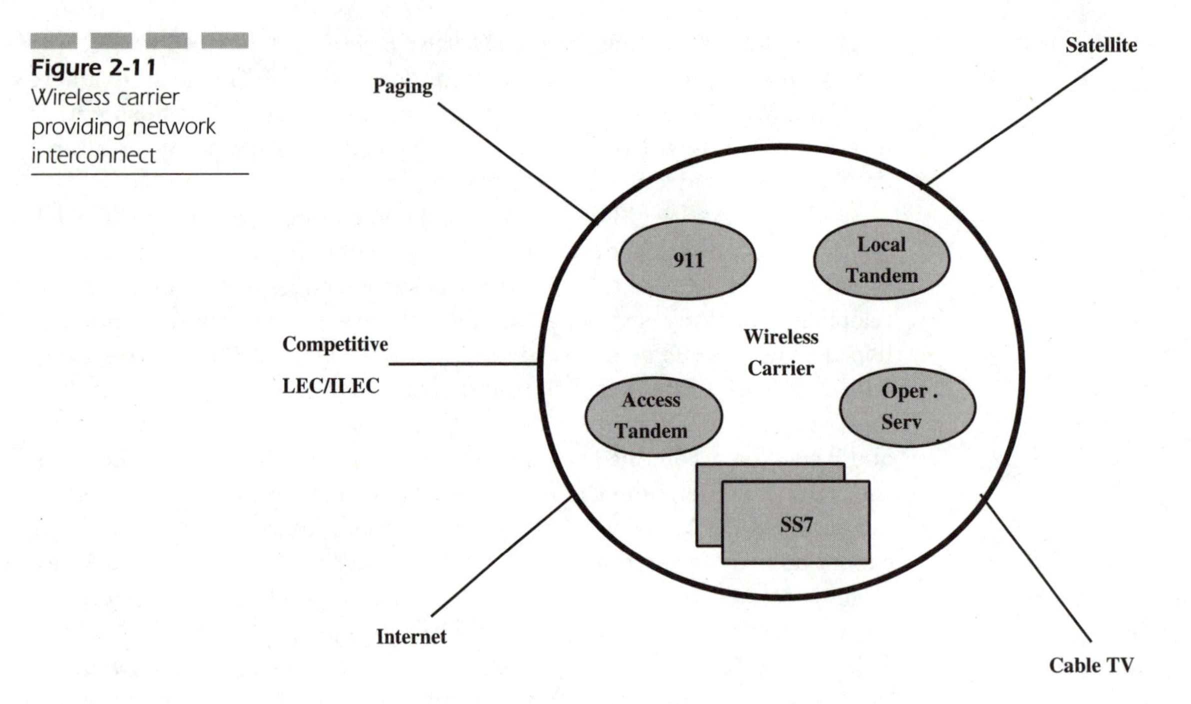

Figure 2-11 Wireless carrier providing network interconnect

the center of the interconnection view. The wireless carrier is assumed to be a network to which other carriers wish to interconnect.

Figure 2-11 illustrates what many wireless carrier proponents had supported: the wireless carrier providing network interconnect.

Figure 2-11 leads one to believe that interconnect might be easier said than done. The fact is that the complexity of interconnect is based on the type of information to be communicated between the carriers (or service providers). Furthermore, the type of information—whether voice or data (non-voice)—is determined by the business relationship.

Each type of interconnect is capable of supporting different types of calls between carriers, subscribers, etc. Appendix B details the capabilities of each interconnect type.

Given its long history, Bellcore's GR−145-CORE is still the predominantly quoted industry specification. The documents are officially titled as follows:

- EIA/TIA IS−93 Revision 0, December 1993, "Cellular Radio Telecommunications Ai-Di Interfaces Standard." Revision A was issued in September, 1998, which led to the approval of IS-93 as an *American Standards Institute* (ANSI) standard.

- GR−145-CORE Issue 1, March 1996, "Compatibility Information for Interconnection of a Wireless Services Provider and a Local Exchange Carrier Network." Since 1996, GR-145-CORE has been updated to reflect changing wireless carrier network technology.

Interface Types: Standards versus De Facto

GR−145-CORE focuses on 10 specific categories of interconnection, while IS−93 focuses on 13 specific categories of interconnection. The categories of both documents are largely the same from a protocol perspective.

GR−145-CORE supports the following categories:

1. Direct *Wireless Service Provider* (WSP) connection through a LEC end office using MF signaling, called Type 1
2. Direct WSP connection through a LEC end office using the ISDN protocol, also known as Type 1 with ISDN
3. Direct WSP connection with a LEC tandem office using MF signaling, called Type 2A
4. Direct WSP connection with a LEC tandem office using SS7 signaling, called Type 2A with SS7
5. Direct WSP connection with a LEC *Common Channel Signaling* (CCS) *Signaling Transfer Point* (STP), called Type S
6. Direct WSP connection with a specific LEC end office using MF signaling, called Type 2B
7. Direct WSP connection with a specific LEC end office using SS7 signaling, called Type 2B with SS7
8. Direct WSP connection with a LEC tandem office arranged for 911 emergency calls, called Type 2C
9. Direct WSP connection with a LEC tandem office arranged for LEC operator assisted calls or directory service using MF signaling, called Type 2D
10. Direct WSP connection with a LEC tandem office arranged for LEC operator-assisted calls or directory service using SS7 signaling, called Type 2D with SS7

TIA's IS−93 supports the following categories:

1. Trunk with Line Treatment using MF signaling, called *Point of Interface* (POI) -T1

2. General Trunk Access Signaling using MF signaling, called POI-T4
3. General Trunk Access Signaling using SS7 signaling, called POI-T5 and POI-S5
4. Direct Trunk Access Signaling using MF signaling, called POI-T6
5. Direct Trunk Access Signaling using SS7 signaling, called POI-T7 and POI-S7
6. Operator Services Access Signaling using MF signaling, called POI-T10
7. Operator Services Access Signaling using SS7 signaling, called POI-T11 and POI-S11
8. Call Management Features Signaling using MF signaling, called POI-T12
9. Call Management Features Signaling using SS7 signaling, called POI-T13 and POI-S13
10. Basic Signaling Transport using SS7 signaling, called POI-S14
11. Global Title Signaling Transport using SS7 signaling, called POI-S15
12. Cellular Nationwide Roaming Signaling using SS7 signaling, called POI-S16
13. TCAP Applications using SS7 signaling, called POI-S17

As you have noticed, based on the numbering sequence there are interfaces that are not fully described in the TIA document. The following are not fully supported in IS−93:

- Trunk with Line Treatment using ISDN-BRI signaling, called POI-T2
- Trunk with Line Treatment using ISDN-PRI signaling, called POI-T3
- Emergency Services Access Signaling using MF signaling, called POI-T8
- Emergency Services Access Signaling using SS7 signaling, called POI-T9 and POI-S9
- ISUP End-to-End Signaling using SS7 signaling, called POI-S18

The TIA acknowledges that these interfaces are not explicitly described in IS−93 but are recognized as legitimate interfaces that can be supported by the LEC. Details about these interfaces and the lack of substantive industry support can be found in Appendix B. Appendix B will also provide a more detailed comparison of both documents.

Although the interfaces/interconnect types have different names, they are functionally the same. The plethora of interfaces/interconnect types described in the two documents serve a purpose: connecting networks. The reader should bear this purpose in mind. Networks—whether they are wireless, wireline, data, Internet, satellite, paging, or some other type of communications network—can all be interconnected.

An interesting fact to note is that prior to SS7, all signaling was unidirectional. Call information such as *Automatic Number Identification* (ANI) information was sent from the party that was interconnected to the LEC. This unidirectional relationship to signaling dictated for years how carriers were perceived by the LEC. The advent of SS7-supported interconnects changed this perception. SS7 is a bidirectional signaling protocol. The bidirectional nature of the protocol meant that suddenly, carriers (at the time, wireless carriers) were now equal. To many in the wireless carrier community, this situation meant that the one-sided contracts and tariffs between themselves and the LEC could now be rewritten to support reciprocal and even mutual compensation. Nevertheless, it took the Telecommunications Act of 1996 to change the relationship between wireless and wireline carriers.

The business imperative is what drives the need to interconnect. The question the reader needs to address is, "How will interconnecting help the service provider's business?". The answers are that interconnecting will enable a service provider to establish business relationships and sell infrastructure capabilities. Interconnect enables network visibility between networks; therefore, subscribers of one service provider can communicate with subscribers of another service provider.

The most obvious example of infrastructure capabilities (services) is the network-based subscriber service. We will discuss network-based subscriber services in the following section. Note that there are a number of other capabilities that we will discuss in succeeding chapters.

Example of Infrastructure Services

Subscriber Services A subscriber service is a capability that telecommunications providers give to their customers. The subscriber service enhances the way that the subscriber makes calls, receives calls, and is billed for calls. In some cases, the subscriber service even limits whom you can call or where you can call. A network-based subscriber service is one that is provided by the communications service provider's infrastructure equipment, as opposed to customer-premise equipment. Centrex is an example of a network-based service. Centrex ia a trademarked Bell System product. Calling Number Identification Presentation (better known as Caller ID) is an example of a network-based service.

The types of services that the carrier provides will dictate the architecture of the carrier's network.

The previous discussion of interconnect types described one form of interconnect that might have the capability to support subscriber services (such as the ones described as follows). The form of interconnect that I refer to uses the network signaling type called SS7. This type of network signaling enables voice and call control signals to be carried on two separate transmission facilities. This feature is important, because it effectively gives a service provider greater control over the call and over the components of the call. This description is simplistic but is accurate enough for now. The key behind providing subscriber services to another service provider (let's call this provider Service Provider 2) is providing service in a manner that is invisible to the subscriber. To better understand this point, later chapters will describe how *Common Channel Signaling* (CCS) works. The information embedded within the signaling can be used to trigger database queries or to trigger events within the network.

SS7 is a form of CCS, and MF is a form of inband signaling. Inband signaling involves the transmission of both the voice (information content) and the call control information over the same physical transmission facility. The inband signaling is inherently less flexible from the perspective of using call control components for providing services.

Service providers seeking to provide subscriber services will probably end up supporting services such as the ones in the following list. Product managers have seen all of these services before, and almost any user of telecommunications services has probably seen all of these services as well:

- *Call Delivery* (CD)
- *Call Forwarding—Busy* (CFB)
- *Call Forwarding—Default* (CFD)
- *Call Forwarding—No Answer* (CFNA)
- *Call Forwarding—Unconditional* (CFU)
- *Call Transfer* (CT)
- *Call Waiting* (CW)
- *Cancel Call Waiting* (CCW)
- *Calling Number Identification Presentation* (CNIP)
- *Calling Number Identification Restriction* (CNIR)
- Calling Name Identification Presentation
- Calling Name Identification Restriction
- *Conference Calling* (CC)

- *Three-Way Calling* (3WC)
- *Do Not Disturb* (DND)
- *Flexible Alerting* (FA)
- *Message Waiting Notification* (MWN)
- *Mobile Access Hunting* (MAH)
- *Password Call Acceptance* (PCA)
- *Preferred Language* (PL)
- *Remote Call Forwarding* (RCF)
- *Remote Feature Control* (RFC)
- *Screen List Editing* (SLE)
- *Selective Call Acceptance* (SCA)
- *Selective Call Forwarding* (SCF)
- *Selective Call Rejection* (SCR)
- *Short Message Service—Point-to-Point* (SMS-PP)
- *Messaging Delivery Service* (MDS)
- Paging Message Service
- Voice mail
- Fax mail
- Computer-based telephony
- Data services: broad-band, narrow-band, bursty, etc.
- *Single Number Service* (SNS)

These services all require some degree of switching intelligence. The CNIP and CNIR services require CCS support. Features such as CNIP and CNIR require calling identification parameters that are found only in a CCS network. Other services require high-speed voice/video protocols. You should understand how network signaling affects the way in which we provide the services. In the end, this process does not just involve marketing. We have to know why and how the signaling works—or else the service provider might end up not meeting the customer's needs.

Some of these services can be provided without CCS. The question is, can you provide the same services by using inband signaling? The answer is not exactly. A subscriber service can be packaged from a marketing perspective in such a manner that a customer could not tell whether the service was being provided via inband or out-of-band signaling. Services such as CNIP, however, require out-of-band signaling. I must reiterate, therefore, that you must understand the general capabilities of the network in order to provide the appropriate services.

Other Types of Infrastructure Services These examples are subscriber related, and there are other types of infrastructure services that support other carriers. These services would be part of a carrier's wholesale service product line. The following list describes examples of infrastructure services that are provided to other carriers:

- Rooftop access
- Wireless tower access
- Utilities (water, power, and light)
- Building space
- Transmission facility leasing

The reseller is a carrier that purchases the subscriber services of another carrier. The reseller owns the customer/subscriber account; however, the reseller is dependent on the wholesaler to provide the service.

Points of View

New subscriber services are being developed at this moment. One might wonder how many different types of subscriber services can be developed.

The marketplace has reached a point in which new (and hopefully inventive) ways of accessing and using these services (possibly in concert with each other) are now desired. Once you have Caller ID, what else can you want? (Calling Name ID, Calling Name ID with Call Waiting, maybe two new types of Call Forwarding, etc.) The market has reached a point in its maturity that the industry must find new ways of leveraging synergies between different segments in order to continue growing. The industry's current flurry of mergers and acquisitions are examples of this point.

The provisioning of services to a subscriber base can be viewed from the perspective of a business model. In other words, the subscriber needs to understand the type of carrier. I am not referring to technology or to the telecommunications segment. The business model can be broken into three basic categories:

- Resale
- Wholesale
- Retail

The resale model refers to the purchase, usually at a discount, of subscriber services by one carrier from another carrier. The reseller essentially acts as an agent for the wholesaler. The wholesale model refers to the carrier that sells its subscriber services to another carrier.

The retail model is the one business model that most subscribers are used to seeing and understanding. Retailers create and sell their subscriber services directly to a subscriber. The following figure (Figure 2-12) presents a graphical representation of the business models that I have noted and describes their relationship to each other.

The role that a carrier plays within the larger network of networks has a great deal to do with how the networks are interconnected. Each one of the network interconnects that I have described is written from the perspective of the controlling carrier.

Evaluating New Services

Evaluating a new service requires the following information:

- An analysis of marketplace needs
- An analysis of current subscriber usage of subscriber services
- An analysis of subscriber service penetration rates
- A business case to support the new service
- Impact on the carrier's business plan

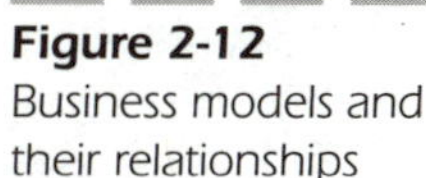

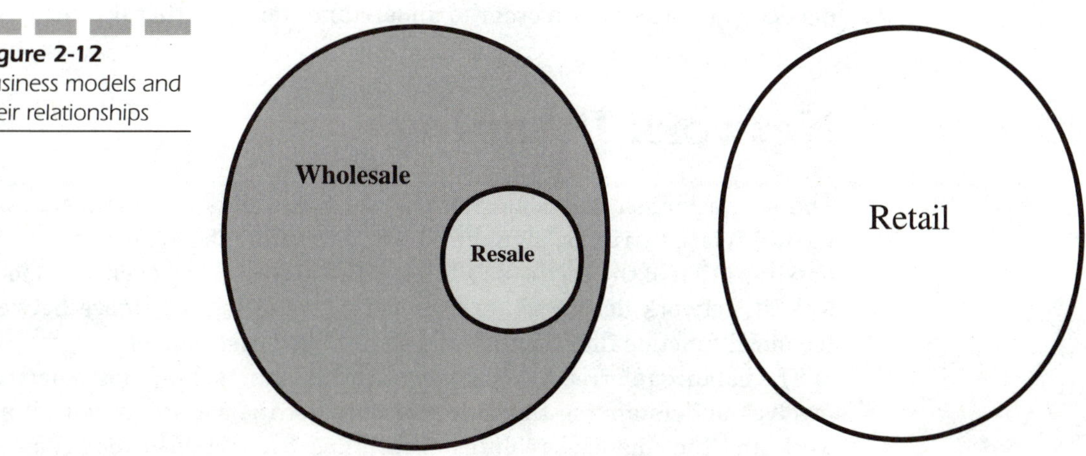

Figure 2-12 Business models and their relationships

Some years ago, when CCS meant only Hundred Call Seconds and not Custom Calling Services or Common Channel Signaling, telecommunications managers participated in the development of a survey of a set of services called *Custom Local Area Signaling Services* (CLASS). These CLASS features included Speed Calling, Calling Number Identification, Calling Number Identification Restriction, Call Forwarding, Call Waiting, and Cancel Call Waiting. The surveys told the telecommunication marketing managers that subscribers wanted the services. The technology vision supported the survey and basically stated that the carrier was positioning the network and subscriber base for the future with a network and a set of features that would enable the subscriber to create services on his or her own and to improve network performance.

This type of strategic planning was technology based and was often done in the early 1980s. Today, subscriber services development and deployment is based on a business case that supports revenue generation. Note that given the competitiveness of the marketplace, the individual subscriber service is not always supported by a revenue-based business case. Rather, many services are deployed as part of a bundle of services. Effectively, the value (of the service) to the subscriber is not in the individual service, but rather in the set of services. Each individual service in a set of services normally supports the value of another. Some key questions to ask yourself when evaluating a new service are as follows:

1. Is it useful?
2. Is it easy to use?
3. Are the benefits of the service readily apparent to the user?

There are so many network technologies that purport to be "The Future." You have to be careful not to be lulled into a sense of security. You have to dig deep into a service in order to understand the value that it brings.

Network Planning

The single largest component of the network tends to be transmission facility-related (trunks, signaling links, etc.). You should not ignore the maximization of the core capabilities of the network, however. You should look at network design as an exercise in achieving a balance between technical functionality (facility and switch) and cost control.

The balance referred to is called network planning. Planning a network involves understanding the technical needs of the service provider's network and the financial realities of fulfilling the technical requirements.

Network interconnection requires a service provider to carefully plan the design and implementation of its network. There is a cost to interconnecting networks, which involves (but is not limited to) network signaling protocols, the cost of transmission facilities, routing plans, and the cost of network interface units.

Network planning is essential to any service provider. While planning a network—whether the goal is to support Internet services, satellite services, or paging services—there are a set of common planning (design) activities required. A description of the planning involved should give the reader a better understanding of just how common or synergistic many networks are.

Some basic tools are needed to design a voice network. These tools are applicable to all network types and are not software; rather, they are conceptual in nature:

1. *Architectural Switching plan* This plan would include handoff methodology, anchor switching design, switching selection points, and a routing plan; need efficient routing of calls or data packets based on network efficiency or business arrangements. Route diversity is a major concern for many service providers, but there is a dollar cost involved.
2. *Transmission plan* Types of facilities used and transmission design requirements
3. Addressing plan, numbering plan, point codes, etc.
4. Network signaling plan (more detail will be introduced in a later chapter)
5. Subscriber services plan (more detail will be introduced in a later chapter)
6. Subscriber billing/customer support plan
7. Network management and operational support systems plan

Architectural Switching Plan

How do you reduce the number of links that are used to connect the user to the network while reducing the length of those links and maximizing the use of the network's other resources? The answer is by using a combination of three network configurations: Tree, Ring, and Star.

The Tree, Ring, and Star network configurations represent basic user-network connections/topologies that serves as the first step towards maximizing network resources.

Tree Architecture The Tree architecture looks like a tree. The characteristics of the tree architecture are the same as that of a typical LAN, as some landline telephone company networks (party lines), or as some cellular carrier networks. Since its inception, nearly all cable television systems have been deployed by using the Tree configuration. The Tree architecture is the most efficient way of distributing the same set of communications signals to multiple terminals. Figure 2-13 is an illustration of the Tree network configuration.

Ring Architecture The Ring architecture loops traffic (voice or data) so that it returns to its original starting point. One example of the Ring architecture is the Token Ring and the Self-Healing Ring. Service providers (telephone companies and wireless carriers) commonly use the Self-Healing Ring.

The Self-Healing Ring is a ring architectural configuration in which the same messages are transmitted simultaneously in opposite directions on parallel rings. In case of a service interruption on the ring, the messages are automatically transferred to the other ring. This architecture provides a high degree of redundancy. At each port, node, or drop on the ring, messages can be transferred to a Tree, Star, or another Ring network.

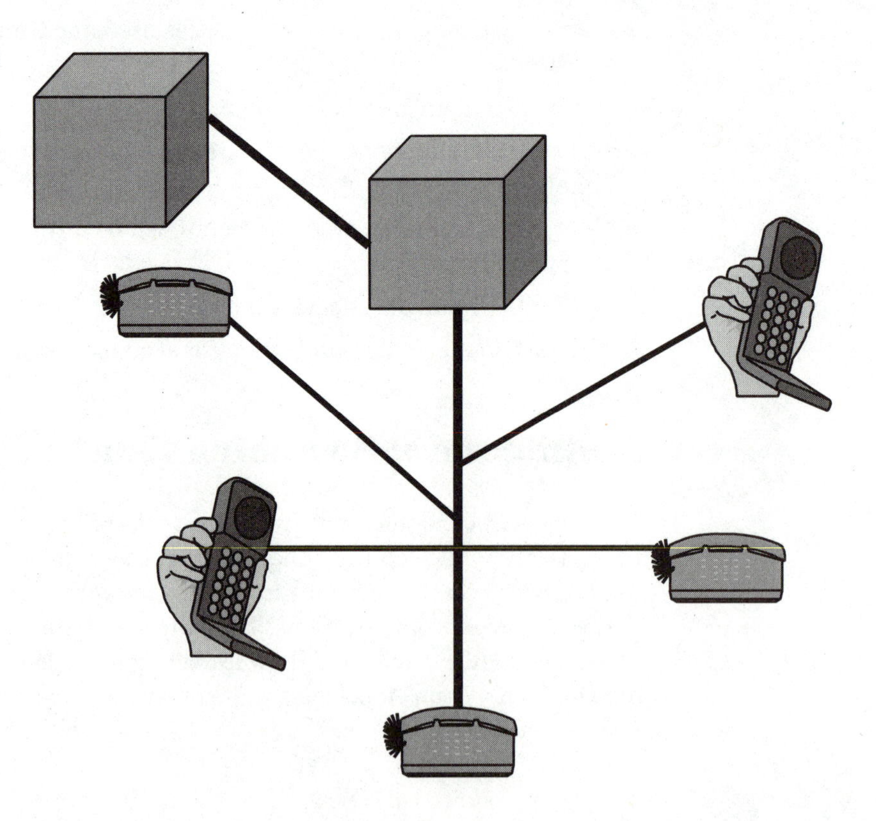

Figure 2-13
Tree configuration

Ring networks for cable television, wireline telephone companies, or wireless carriers normally utilize fiber optic facilities for digital transmission. The Self-Healing Ring is expensive, because you are supporting at least two networks that are carrying the same messaging for the sake of redundancy. The expense is worth the disaster recovery capabilities that are inherent within this architecture, however. Figure 2-14 is a graphic of the basic Ring network configuration.

Star Architecture The Star architecture is the most typical wireline telephone company configuration. The Star topology enables separate transmission paths to be established to each subscriber. Each path can be designed to carry one message or different messages. The Star network configuration is illustrated in Figure 2-15.

While Tree and Star architectures are common in the cable television and telephone industry, the Ring architecture is being increasingly used. The need to recover from network failures and damage is important and is becoming an increasingly difficult goal to meet as networks grow in complexity. The architecture of a network plays an integral role in maintaining network health.

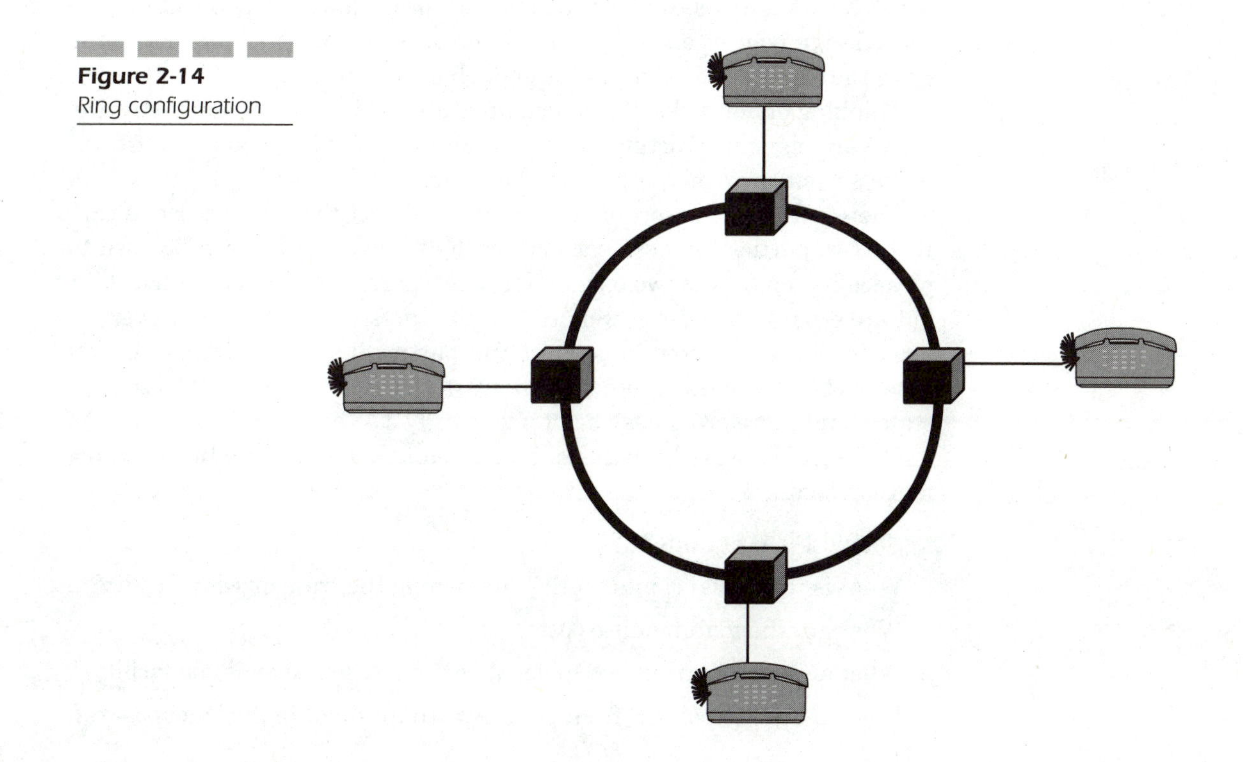

Figure 2-14
Ring configuration

Figure 2-15
Star configuration

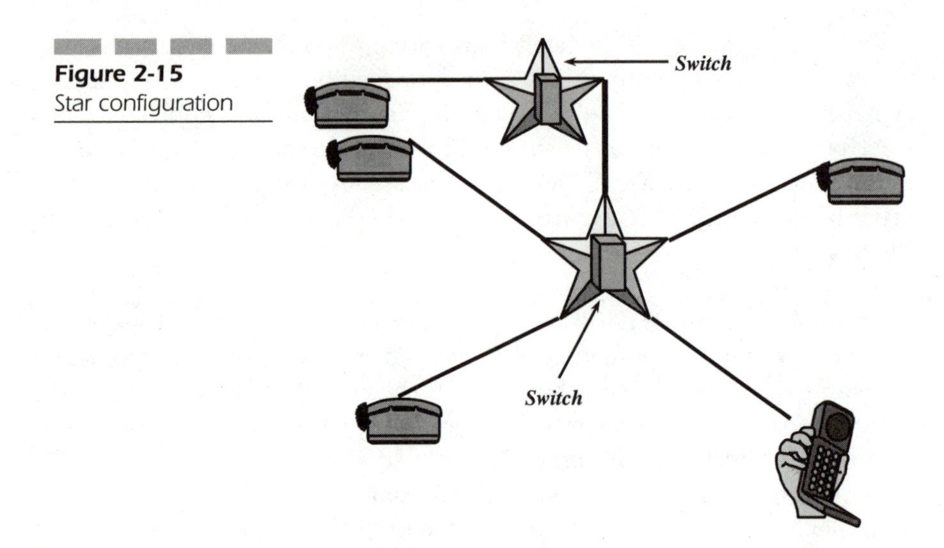

The question of how much redundancy and how much network protection to design into a network is a difficult question to answer. There is a crossover between network cost and network disaster recovery. The point at which a carrier reaches too much network protection is dependent on network design policies and corporate financial objectives. In other words, it is up to the carrier to decide what is enough or not enough.

Thinking of networks like vehicular roadways is a useful mental device. There are signs instructing drivers when to stop, yield, make a left turn, make a right turn, merge, go, and slow down.

There will be some sort of box that will connect the subscriber to a call. If the two parties are near each other, the same black box can be used to physically connect the two parties. If the two parties are not near each other but are separated by several miles, towns, states, or countries, however, we need to figure out how to connect the parties. Laying facilities directly between two switches or routers that are (for example) separated by several states (and therefore miles) is not necessarily cost effective.

One should always keep the following points in mind in relation to transmission facilities:

- Should I lease or buy?
- What is the initial capital outlay for owning the transmission facility?
- What are the maintenance costs?
- What are the recurring fees (if you do not own the transmission facility)?
- How do I get to point B from point A with minimal financial exposure?

At this point, you might be asking the question, "What is a technical book doing asking a business-related question?". The answer is that finance is one of the most critical components of designing a network. If you are not thinking in terms of dollars and cents, then you are headed for failure.

The direct connect approach leads to connecting each party to every switch and every customer. If you took this approach, you would find yourself buried under switching equipment and network facilities. Therefore, all networks are designed so that switches concentrate the users in local networks and route the calls via specific paths.

This type of approach to network design enables the service provider to establish high-usage communications paths. As you can see, an objective of a service provider is to maximize equipment usage in order to gain the greatest cost benefit from the equipment. In effect, the service provider establishes a grade of service that assumes an acceptable level of call blockage (incomplete call origination). What is considered acceptable and what is not is based on telecommunications traffic policy.

All of this information leads to a network hierarchical switching design methodology (alternate and dynamic routing). This design approach enables us to design an efficient network.

Routing Techniques

Assuming that the service provider has decided on a particular network topology, the type of routing plan used will require a routing plan. The routing plan dictates how calls shall be routed through the network. The routing plan is a decision-tree matrix (i.e., the call is routed based on a number of what-if conditions or dynamic traffic conditions). There are two routing techniques:

- Alternate routing is applied to voice traffic by providing a first choice (high-usage trunk route) and one or more alternate routes if the first route is unavailable.
- Dynamic routing refers to the updating of routing patterns on a real-time or short-time interval on the basis of traffic statistics collected by a network management system.

Routing requires a database that will store subscriber service profiles, network routing information, and algorithms.

Alternate Routing Alternate routing provides the opportunity to minimize the cost per unit of carried traffic. The traffic load is allocated to high usage and final routes in the most economical manner. This type of routing also permits the meshing of traffic streams that have differing busy hours or seasons. Alternate routing requires you to predetermine the alternate routes, which is also known as hierarchical routing.

In the following figure (Figure 2-16), you will see a one-way high-usage trunk from switch A to switch B with an alternate (final) route via a gateway/tandem switch. In general, the direct or high-usage route is shorter and less expensive than the alternate route. Because each segment of the alternate route is used by other calls or information packs, however, a number of traffic items can be combined for improved efficiency on that route. The challenge then, is to minimize the cost of carrying the offered load.

More on this subject will appear in Chapter 3, "Basic Network Technologies."

Dynamic Routing Dynamic routing is a traffic routing method in which one or more central switching controllers determine near real-time routes for a switching network. The near real-time decision is based on the state of network congestion measured as trunk group busy/idle status and switch congestion. The choice of traffic routes is not predetermined.

In a network that uses dynamic routing, traffic can be more effectively distributed over the network trunk groups and switches than traffic that is routed by using the alternate routing method. Dynamic routing enables a network to select routes that provide lower blocking than the hierarchical fixed-route network can achieve. Dynamic routing can reduce the level of demand servicing and can react more easily to relieve problems that are associated with forecast errors. Furthermore, dynamic routing enables networks to carry significantly more traffic (if a switch or fiber route fails) than networks that are dependent on hierarchical (alternate) routing.

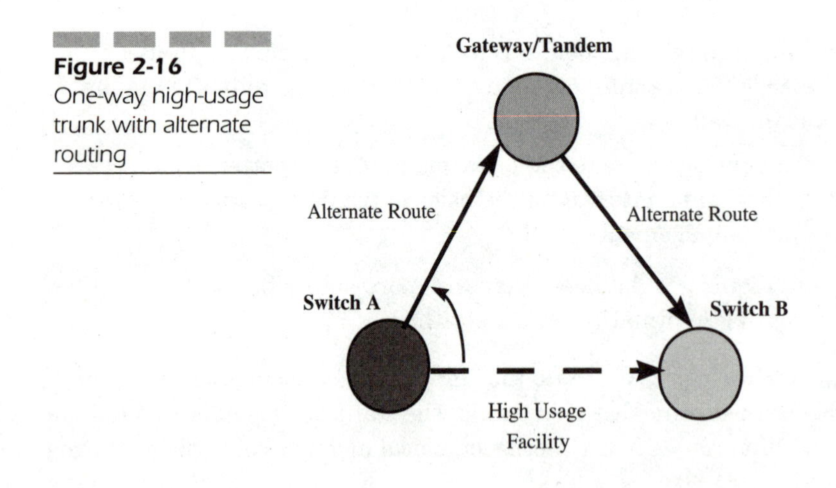

Figure 2-16 One-way high-usage trunk with alternate routing

Packet-switch networks and the Internet are superb examples of networks that are analogous to the dynamic routing that I have described. In fact, both packet switching and Internet communications by nature inherently behave (route) in a dynamic manner. Rather than establishing routing tables (i.e., fixed/pre-determined instructions) to find the route with the lowest blocking probability, the packet switch and Internet router simply seek connectivity. This connectivity can be achieved through another party's router or through a packet-switch network and is more spontaneous than traditional telecommunications. One can determine the route that a dynamically routed call takes in a traditional voice telecommunications network. In the case of a packet-switch network and the Internet, however, the precise path cannot be traced. In a packet and Internet environment, the information is literally thrown out into the networked environment, and based on the addressing information in the information packet, the call/information packet finds its way or connects to the other network. Dynamic routing can be illustrated by the following diagram in Figure 2-17.

This diagram does not easily lend itself to the fixed, hard-and-fast hierarchical routing rules and illustrates how we can combine all of the basic configurations (Tree, Ring, and Star) to support a group of network elements.

The basic nodal equation used by electrical engineers for circuit analysis (and by all traffic engineers) confirms that the ideal number of network connections is 10. This formula is $N(N-1)/2$. You should note that the boxes in

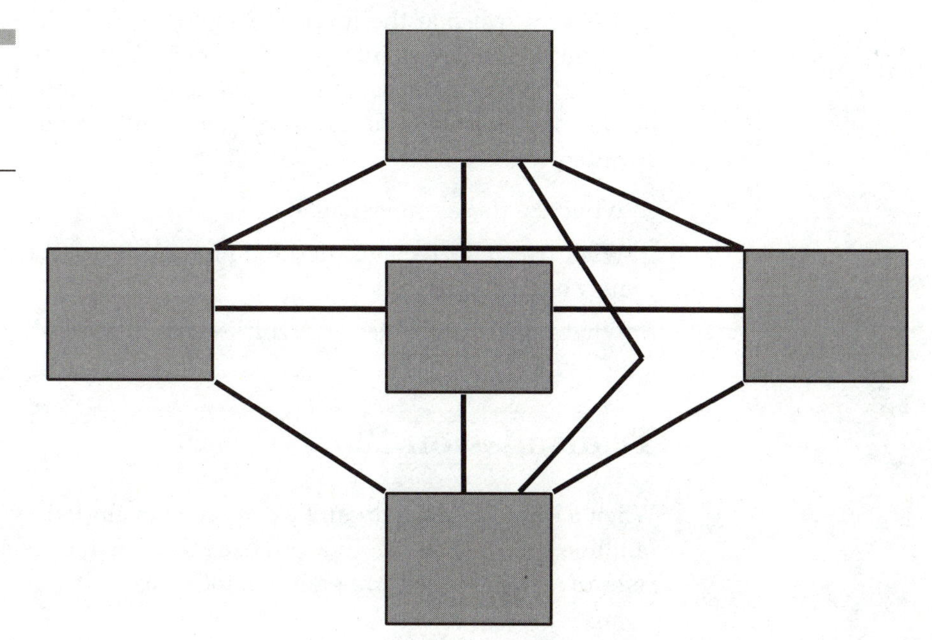

Figure 2-17 Information can be routed along many different paths.

Figure 2-17 do not represent any particular network element. In fact, it does not matter what the network element is. What is important to focus on is that the figure represents how information flows from one point to another point.

Routing Relationships The routing schema that I have described is not only applicable to specific network elements within a given network, but it also applies to the relationship of one network to another. To some extent, one would think that there must be some type of order between one network versus another. In the traditional voice telecommunications world, when given the post-divestiture and Pre-Telecommunications Act of 1996 environments, networks were related to one another in the following fashion:

- *Local Exchange Carrier* (LEC) networks serve the local loop.
- LEC networks carry calls within their own network to their own subscribers.
- When a LEC subscriber crosses the regulatory *Local Access Transport Area* (LATA) boundary, the call must be transported by an IC network.

The call flow has the appearance of climbing a ladder from one floor to another floor.

In the case of non-voice telecommunications networks (i.e., data or the Internet), there is a type of network information flow. Currently, *Internet Service Providers* (ISPs) need to connect to a LEC in order to transport their call/information from one point to another point outside the LEC's jurisdiction—or even inside the jurisdiction of the LEC. Essentially, there is an order about the flow of information from an ISP to another ISP.

Once the routing schema has been determined, the following questions need to be asked about the network and the number of connections that are required:

- What are these connections?
- What transport technologies and protocols are out there to support the network configuration?
- What type of transmission facilities should be used?

Transmission Plan

When a call is made, data and video are transmitted—whether wireless or landline based—and facilities are used to transmit the call. The facilities I refer to are also called transmission facilities.

Transmission facilities can be broken into the following basic network types:

- Metallic
- Fiber optic (glass)
- Microwave

The major differentiator between facilities types is speed (sometimes referred to as bandwidth). Speed refers to the speed at which information is transmitted. Time is money, and the higher the information speed supported by the facility, the more money that is made and saved. Transmission facilities' sole function is to serve as the distribution media for information.

Today, the most common transmission facility is the T1. The word T1 refers to a specific type of telephone transmission equipment that is used to support analog voice transmission. The T1 facility is a metallic facility that offers two-way connectivity at an information speed rate of 96000 Hz (96,000 bits per second). T1 facilities support analog voice inputs that have been multiplexed into 24 channels. Before the T1, voice was transmitted over solid copper wire—one strand of wire for every conversation. Several decades ago, transmission facilities of this type resulted in towns and cities being overwhelmed with overhead wires and telephone poles. *Frequency Division Multiplexing* (FDM) enabled service providers to take the single copper wire and make use of its full bandwidth. In other words, multiple voice conversations were transmitted over specific frequency ranges. Conversation 1 was transmitted over frequency range 0–4000 Hz, while conversation two was transmitted over frequency range 4000–8000 Hz (and so on).

Analog amplification and multiplexing creates one problem, however: noise. The T1 facility was susceptible to nearby ambient electrical signals that were created by electrical motors, electrical power lines, other communications facilities, and any other source of electromagnetic radiation. These unwanted signals are called noise.

The noise problem was fixed/addressed when the telecommunications industry developed digital transmission. Digital transmission addresses the problem caused by analog transmission—that the signal can assume almost any value. On the other hand, a digital signal can only assume two values: zero or one. At this time, a brief description of analog and digital is in order.

Analog versus Digital Analog can be defined as an electromagnetic or audio signal that can assume any value. Digital can only assume discrete

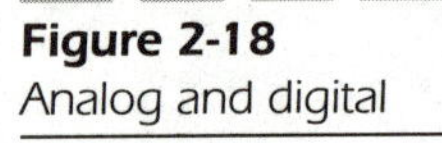

Figure 2-18
Analog and digital

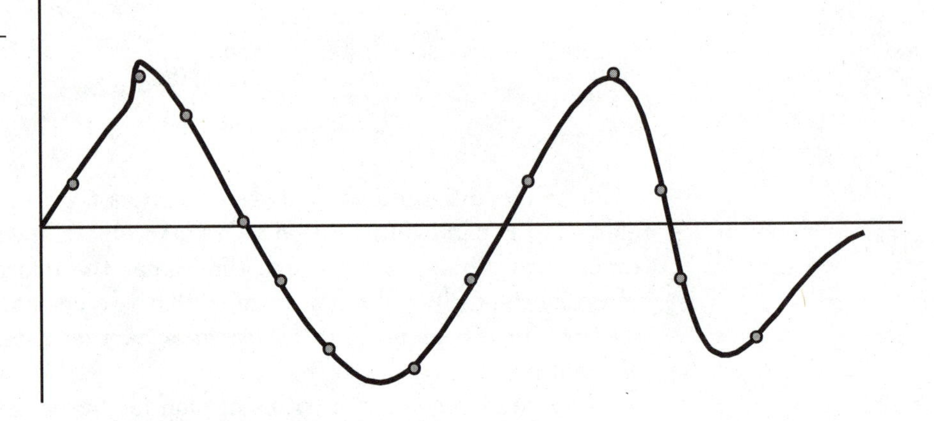

values within the signal. A voice, a sound, or an electromagnetic signal can be pictorially drawn as a wave (refer to Figure 2-18).

The analog waveform can assume any shape. The shape can be irregular (as in Figure 2-18), or it can be smooth and symmetric on both sides of the axes. The dots represent discrete values or points on the waveform. These points represent a digital sample of the waveform. The waveform is sampled by using some technique, most likely *Pulse Code Modulation* (PCM). PCM works as follows:

1. The analog signal is sampled or looked at by the system (modulator) at a rate of 8,000 times per second. The modulator sends out each sample throughout the network, and each sample has a discrete value that can be placed on top of the original analog waveform. When you assemble all of the pieces, the result looks similar to a connect-the-dots picture of the original waveform.
2. The value or height of the signal sample is then converted to a binary (digital) value by an encoding device. The result is an eight-bit binary word. The whole process results in the analog signal being converted into a digital signal of 64,000 discrete bits per second.

A single voice requires 300–3300 Hz bandwidth. Taking the previously mentioned information into account, we find that when the voice is converted to a digital equivalent, the bandwidth required is 64000 bits per second. When you implement this bandwidth over a T1, you can still place 24 channels on the T1. In the digital environment, each channel's data rate is 64000 bits per second.

Now, we digress for a moment. The older term for *bits per second* (bps) was baud rate. Baud is the number of times per second that the carrier signal shifts value. For example, a 1,200 bit-per-second modem is said to be running at 300 baud. This modem would be moving four bits per baud ($4 \times 300 = 1{,}200$ bits per second).

At 24 channels, a T1 is capable of a total data rate of 1,536,000 bps plus 8,000 bps for control and synchronization. The maximum number of simultaneous conversations that can be carried over a T1 facility is 24. This rate is standard in North America, Japan, and Australia. In Europe and in other *International Telecommunications Union* (ITU) countries, 30 voice channels and two signaling channels are multiplexed at a rate of 2,048,000 bps. The T1 is used to support transmission of information between switch locations.

Today, the term T1 has become an overused term to refer to the digital transmission rate of DS−1. The term T1 is often interchanged with the term DS−1.

The current digital transmission rates are as follows:

Signal Rate	Bits per Second	Number of Channels	Analog Equivalent
DS−0	64,000 bps	1	none
DS−1	1,544,000 bps	24	T1
DS−1C	3,152,000 bps	48	T1C
DS−2	6,312,000 bps	96	T2
DS−3	44,736,000 bps	672	T3
DS−4	274,176,000 bps	4032	T4

Synchronous Optical Network (SONET) is a high-speed transmission protocol format that is much more than just a souped-up DS−3-level format. SONET is an optical format that is capable of supporting speeds ranging from 150 Mbps to several gigabits per second. SONET is meant to address network access by customer-premise equipment and network interconnection between providers.

Metallic wire (usually copper) tends to be the typical medium for DS−0 through DS−2 transmission facilities, whereas fiber optics and even microwave tend to be the typical media for DS−3, DS−4, and higher.

The provisioning of facilities for transmission is a business for many service providers who have excess transmission capacity. These service providers can be in the cable television business, landline telephone business, cellular business, PCS business, or even in the power utility business. The fact is, any company that has a communications transmission network can provide facility support to another service provider.

Transmission facilities provisioning is a business opportunity for the service provider. Such providers came into being years ago. Today, however, any service provider can provision transmission facilities if desired. If you have the extra bandwidth, then use it.

Addressing/Numbering Plans

A numbering plan or destination address describes the location of the originator and receiver of the call.

This book will address two types of numbering plans: the public telephone numbering plan and the Internet address schema. The Internet providers and public telecommunications network providers currently use these two destination address plans.

Public Telecommunications Service Providers Numbering Plans

For the purposes of this chapter, I define service providers as follows:

- Landline telephony companies (PSTN, cable television, and the Internet)
- Wireless service providers (cellular, personal communications service providers, paging, and satellite)

These service providers provide service to the mass consumer market. Cable TV fits into this category if you include the fact that there is work underway to develop and market cable TV-based telephone service. The Internet currently provides telephone service for many users.

Landline (Wireline) Telephony and Wireless Services The first numbering plan originated in the landline telephony world (the PSTN). The numbering plan is simply the telephone number of the subscriber. In North America, the telephone number is a 10-digit number. The format of this 10-digit number is as follows:

- *Numbering Plan Area* (NPA) + *Number Exchange* (*NXX*) + four-digit station number (*XXXX*)
- The NPA is the three-digit area code. Its format is *NXX*, where *N* can be any digit two through nine, and *X* can be any digit zero through nine.
- The Number Exchange is a three-digit code that is typically associated with the service provider's switch. Its format is *NXX*, where *N* can be any digit two through nine, and *X* can be any digit zero through nine.

In North America, the NPA defines a specific geographical area. Given the growing shortage of numbers, a new numbering format has been adopted called Interchangeable NPAs. The Interchangeable NPA format involves interchanging an NPA and an *NXX*. In other words, in the past, the NPA-*NXX* combination of 301–555 would not be reused as 555–301, because this new number would violate past numbering format rules stating that the first digit could be any number from two through nine. The second digit could only be a zero or a one. The third digit can be any number from two through nine. Those rules made sense at the time; however, given the massive growth of telecommunications services and products requiring telephone numbers, a new format needed to be adopted. Note that these services referred to in this chapter range from office facsimile machines, home facsimile machines, new service providers requiring numbers, etc. The new numbering format is the one that I described previously.

The landline telephony world also supports the following additional numbering formats:

- N11, which includes 411, 611, and 911
- *Services Access Codes* (SACs) The SAC is an NPA in the format N00 and NNN. This format would include 500, 600, 700, 900, and the well-known 800 number (toll-free number) groups. The new numbers 888 and 877 have also been activated to support toll-free access. These codes are non-geographic in nature and have been reserved for ICs, LECs, and wireless carriers. These codes are used to support mass-media/market services that all carriers can offer, such as 800 toll-free, 888 toll-free, 877 toll-free, 500 personal numbers, 900 market services, etc.
- *Vertical Services Codes / Feature Codes* These codes are used to activate or deactivate numbers. The codes are in the format *XX or *XX0. The features that are typically supported here are Call Forwarding, Call Back, etc.

These numbering/dialing patterns are supported by all public telecommunications carriers (including wireless carriers). Given the growth of the telecommunications business, it will not be long before the 10-digit numbering plan is replaced by a numbering plan that supports a 15-digit numbering plan.

Local number portability is a recent *Federal Communications Commission* (FCC) mandate that we will describe later in this book. You might consider local number portability a new type of numbering plan. Local number portability follows the same basic principles of establishing telephone numbers, however.

Internet Addressing The Internet uses an addressing scheme that appears to the layman as a fairly flexible identification plan. A subscriber can select a name or number that has meaning only to the subscriber or that simply identifies the subscriber. The only limitation is that the Internet address cannot already be used by someone else. The Internet is a whole new type of network that does not quite seem to fit the traditional telecommunications business or network model. The Internet is the new major player on the block, however. We will discuss the Internet in more detail later in this book.

The Internet uses *Transmission Control Protocol/Internet Protocol* (TCP/IP) for transmission. TCP/IP is a simple data protocol. The mapping of mobile station addresses to an Internet server can be approached in the same manner that we approach protocol conversion (e.g., X.25 to SS7 or GSM MAP to IS−41 MAP). The Internet address is a 4-tuple composed of a source *Internet Protocol* (IP) address, source TCP port, destination IP address, and destination TCP port.

If an Internet user moves, his or her IP address changes—and therefore, the associated TCP connection changes. The TCP connection is hard wired and fixed, and the IP address serves as both a routing directive and an endpoint identifier.

Internet sites also have also alpha names, known as *domain names*. Domain names have two or more parts, which are separated by periods. The part on the left is a specific pointer to a computer, and the part on the right is a pointer that points in some general direction. A given computer or set of computers might have more than one domain name, but a given domain name points to only one computer.

Some different domain names are as follows:

- `jacko.net` is a domain name that points to a computer.
- `bigdog.jacko.net` is a domain name of computer A.
- `bigcat.jacko.net` is also a domain name of the same computer A.

All of these domain names can refer to the same computer, but each domain name can refer to no more than one computer.

Network Signaling Plan

A service provider communicates with other service providers and with network elements in its own network through a specific type of network signaling. Network signaling refers to the language (protocol) that the network uses to speak to other networks (and to networks within its own framework). Network signaling plans identify the signaling protocols that will be supported by the service provider.

There are several different network signaling protocols in use today. Each signaling protocol is optimized to function in specific environments for specific reasons. The following represents the more popular protocols in use today:

- *Signaling System 7* (SS7)
- *Asynchronous Transfer Mode* (ATM)
- Frame Relay
- TCP/IP
- *Multi-Frequency* (MF)

More information about these protocols will appear in the next chapter.

Network Management Plan

The Network Management Plan refers to the systems that are used to manage the service provider's network. The network management systems would include the following:

- Network Element Management
- Network Systems Management
- Service Management

Planning these systems requires the coordination of functions between each of the systems listed. If the plan is executed properly, the result will be a well-run *Network Operations Center* (NOC). In general, the network operations manager is concerned with information and system tools that enable him or her to run/manage the network. Managing the network involves scheduling of activities and resources, managing network traffic load, routine maintenance, trouble reporting, and diagnostics. A generic description of network management would be network maintenance and health.

The following sections summarize Network Element Management, Network Systems Management, and Service Management. All of the following information is applicable when managing any network.

Network Element Management As noted in Chapter 1, "What Is a Telecommunications Network?", network-element management involves the capability to monitor and manage specific network elements. A network element can be defined as a database, a switch, a router, a voice mail recording system, or an adjunct system.

- Maintenance schedules
  - Maintenance schedules refer to the scheduling of activities concerned with the provisioning of high-quality subscriber service and network stability.
- Routing table update schedules
  - Routing table update schedules refer to the scheduling of updates to the switching system's routing tables. The term *routing table* has its historical roots in the traditional public telecommunications world. The fact is that every telecommunications system, whether Internet, satellite, or even paging, maintains a database that describes the switch/router queries for instructions on how, when, and where to route a call or a packet of information.
- Number of subscribers being served by the system
  - Knowing the number of subscribers should be a given. Without knowing how many customers are being served, the network manager cannot properly manage the traffic load on the network. Traffic-load management entails moving telecommunications traffic from one part of the network to different parts of the network and even blocking new traffic to ensure that existing traffic is routed properly.
- Number of roaming subscribers being served by the system
  - Roaming is associated with wireless carrier (cellular and PCS) subscribers. The network manager needs to know the number of visitors to the network. This piece of information is used to determine network database impacts and traffic load impacts.
- Services being provided to the subscribers
  - Network managers need a complete picture of the services being made available to the subscriber base. These managers are not concerned about who specifically has what subscriber service; rather, they are concerned about the services that are available to the population of subscribers as a whole. In some instances, specific hardware or software might need to be available in the network in order to provide a service.
- Performance objectives for the network element
  - Measuring the performance of a network element is way of measuring network health. The network manager needs to have an understanding of how each element in the network is expected to behave. Using these benchmarks, the network manager can address malfunctions or anomalous conditions.

- Transaction processing time thresholds
  - Processing time of any transaction is another indicator of network health.
- Types of diagnostic tools that are available
- Operating procedures documentation
  - Typically, this item is the last one that is addressed by most people in any type of business. Documentation of procedures is often treated as more of an annoyance than as a requirement or necessity. Nevertheless, this process (and the documentation of the process) is absolutely critical if one is expected to ensure that activities are handled in the same manner each time the activities are performed. Replication of an event is necessary if a network manager is to ensure that he or she can identify network trouble.
- Failure levels/indications
  - Network elements will fail at some point in the life of the device, and this failure needs to be reported to the network manager. The network operations centers need to have the capability to reflect the failures in some manner, whether they are audible or visual alarms.
- Alarm conditions indications
  - As I had indicated, a failure needs to be announced in some manner —as well as anomalous conditions. The alarm indications need to reflected via visual or audible means. More importantly, the alarm indication should assist the network manager with resolving the alarm condition. In other words, the indication must report an meaningful event. Within the context of network management, the event must be a real network trouble or unusual occurrence.

Network Systems Management Network systems management involves the way that one manages the entire collection of elements and all of the functions within the business that affect the physical network. Network systems management involves total management of the network and its components.

Network operations managers seek to have the same questions and issues addressed about the whole network as they would about the individual network element. These questions/issues are as follows:

- Service objectives for the system
- Network monitoring software tools
- Network diagnostic tools
- Anticipated traffic load on the system

- Overall traffic profile of the system on a hourly and daily basis
- Types of subscribers being served
- Level of network system control (in other words, can the entire system be rebooted, or is a complex element-by-element process involved?)
- Can traffic be rerouted?
- Types of alarm indications
- Types of network failure indications
- Overall network operating procedures documentation
- Any disaster-recovery plans

Managing a network is similar to taking care of one's own physical health. There are medical doctors who specialize in specific areas of medicine and in specific parts of the body. There are also doctors (even today) who are generalists and look at the person as a whole. Managing a network requires the network manager be a generalist. The network manager needs to be able to understand each network component while simultaneously understanding how each network element relates to each other. The network manager needs to be able to assess the health of the network from a micro and from a macro level.

Service Management Service management links the network system's management function with the customer care function, including performance and customer satisfaction metrics. Service management requires direct involvement with the customer and involves every mechanism and process required for delivery of services to the customer (i.e., operational support systems, sales, and marketing).

Service management links the network to the financial/revenue portion of the telecommunications service business. As indicated in Chapter 1, this area of network management includes the following:

- Establishing and following service-level agreements
  - These agreements govern the relationship between the service provider and the subscriber.
- Market surveys
  - Market surveys are an obvious necessity—without which a service provider cannot meet the needs of the marketplace.
- Customer satisfaction surveys
  - Customer satisfaction surveys are a must; otherwise, the service provider places itself in the unenviable position of waiting for

customer complaints in order to determine the perceived quality of service. Although many of these questions involve subjective concerns, one will discover that the business of providing service to any customer (whether or not it is a telecommunications service) is a subjective matter. Furthermore, at this point, the customer is always right—because the customer pays the provider's income. The surveys will address issues and questions such as the following:

  - Timeliness in resolving customer complaints. Often, the surveys will ask the customer to check off specific time intervals listed.
  - Pleasantness of the operators, if any
  - Were customer orders for new features handled in an expeditious manner?
  - Were the customer service representatives pleasant and helpful?
  - Have you suffered any service outages?
  - Were all billing questions resolved satisfactorily?
  - If services were bundled, did the customer see the value in the service offerings? If not, then maybe the customer believes that he or she was forced into purchasing a feature that was undesirable.

- Competitive analyses
- Service providers' competitive analyses not only address the types of services that the competitors are providing but also the way in which the customer perceives the provided service. Specifically, the customer pays attention to questions and issues such as the following:
  - Static on the line
  - Intelligibility of the parties' voices
  - Can the data be transmitted without impact on quality? Does the data received have errors that can be attributed to noisy transmission lines?
  - Customer complaint handling between the two competitors is examined. The service provider that is perceived to be the most customer oriented can easily win over a customer who is provided with the same type of service based on customer handling alone.
  - Cost of the competitor's service
  - Does the service cost impact the customer's decision to use that particular provider?
- Operations cost analyses

- Operations analyses of this type address the cost of running the network. Often, the manner in which a competitor runs his or her network provides important information about how to run the business effectively and describes what shortcomings exist in the competitor's business. Such information can be used to drive up the costs of the competitor's network. Operations costs can mean the difference between meeting financial margins and operating in the red ink column.

- Customer complaints resolution
  - The customer complaint resolution process is without a doubt a key facet of the business. CLECs and other carriers can purchase/lease the use of facilities (switches, transmission, etc.). Customers expect to speak to people who can answer their questions and listen to their complaints, however. Every carrier needs a customer complaint resolution process. This process is, in many ways, the most critical. A service provider can live with system troubles as long as these problems are eventually fixed, but it is impossible to operate even for a day without a proper customer-complaint bureau. Customers typically cancel their service if they are angered once.
- Management of operations cost
  - Managing costs is essential to every business. In the case of capital-intensive service providers (carriers that own and operate their own equipment), it is imperative that every minute of the day, every item of spare parts inventory, every procedural step, and every tool should be managed in the most cost-effective manner possible.

Subscriber Billing/Customer Support Plan

The customer is involved throughout all aspects of a service provider's business. The customer is at the heart of the product offerings of the carrier, the managing of the network, and even the network architecture of the network. Most providers (not all) create plans that address subscriber billing/customer support. The customer is so vital to any business that how the customer is treated is singled out as a major planning effort. This type of plan does not exactly involve marketing to the customer. The Subscriber Billing/Customer Support Plan addresses the following topics:

- Types of billing plans
- Customer payment options

- Customer bill formats
- Customer billing periods
- Customer service representatives' training
- Customer interaction
- Customer complaint resolution process
- *Customer support* This area is not just limited to Internet companies. Many would be surprised to know that a customer can ask even the traditional wireline LECs a technical question and have this topic addressed.
- Installation and repair departmental coordination

The subscriber billing/customer support plan details the way in which the public views the carrier. This perspective goes beyond any sophisticated advertising campaign. Customers speak to operators, to customer service representatives, and even to the field technicians. These employees represent the face of the carrier and can wreck the reputation and image of the carrier more than any poorly designed intelligent network service ever could.

The following diagram (Figure 2-19) illustrates the relationship of customer support with the overall business.

SUMMARY

The public telecommunications network is a network of networks, no matter which type of subscriber access (wireless, wireline, Internet, etc.) is supported. What the reader has discovered is that public networks (and even

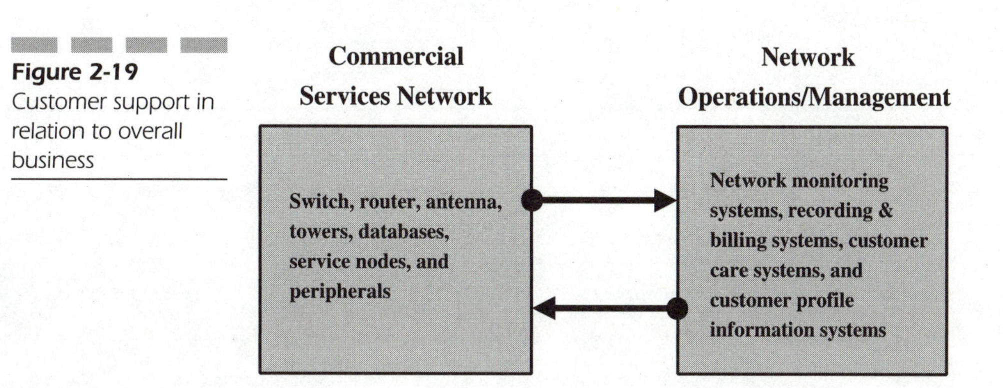

Figure 2-19 Customer support in relation to overall business

some private networks) need to be interconnected in order to have subscribers of one network (a service provider) reach subscribers of another network (a service provider). The tools needed to plan a network are actually the same for all service provider types. Knowing and understanding the simplicity and similarities of the networks should ease the concern of any network manager who is building and operating his or her own telecommunications carrier. Until recently, the LEC was the heart of all public telecommunications. The changing regulatory and market environments have enabled many other types of carriers to view interconnection and to recognize their place within the network of networks.

The fact is that any carrier can be at the center of the "network of networks." What matters is your perception and how you act on that perception.

Basic Network Technologies

The purpose of this chapter is to describe the various and popular network transport technologies (transmission media) and network configurations that are available today. We will also address the wireline carriers' local loop.

This chapter will further provide the reader with more detailed information about basic engineering concepts. These concepts will assist you with understanding how common network engineering concepts can be applied across different sectors of the telecommunications business. What the following chapter will illustrate is how closely related these networks are and how they are capable of supporting each other.

One of the goals of the network provider (voice or data) is to reduce cost through the allocation of network resources. In other words, the provider wants to operate a network that has been engineered to function economically and that meets the needs of the customer.

The fundamental rules governing traffic engineering are as follows:

1. Reduce the number of links (trunks) that are needed to connect all users of the network.
2. Reduce the length of the links (lines and trunks) that is needed to connect users to the network.
3. Maximize the usage of network facility resources in a given period of time by increasing the number of different connections (trunks) to enable users to communicate with each other or with other resources.

Understanding how a network is designed will enable the reader to understand the reasoning behind a network's design. For example, two wireless service providers provide service to two different markets. The markets, however, have exactly the same type of subscriber base and provide the same types of services—yet their networks are as different as night and day. Why?

The answer could be either of the following:

1. Simply economic
2. The result of a difference in business plan; i.e., the service provider might be emphasizing data rather than voice

Before planning any network, the service provider should know "what business one is supposed to be in." The type of telecommunications segment in which one operates will dictate the technologies employed and the type of network configuration that is optimal for the segment.

Network Structures

The following diagram (Figure 3-1) is a generic depiction of a collection of telecommunications nodes. The nodes do not represent any particular device(s) or network element(s). The purpose of this figure is to demonstrate how a network can be configured. One can configure a network in any manner desired; however, configuration characteristics and network efficiency drive a carrier to choose one configuration or a combination of configurations.

The following configurations were briefly discussed in Chapter 2, "The Telecommunications Hub-Creating Value." The following sections will enable us to explore a little further the benefits of one type of network configuration over another (and even combinations of the configurations).

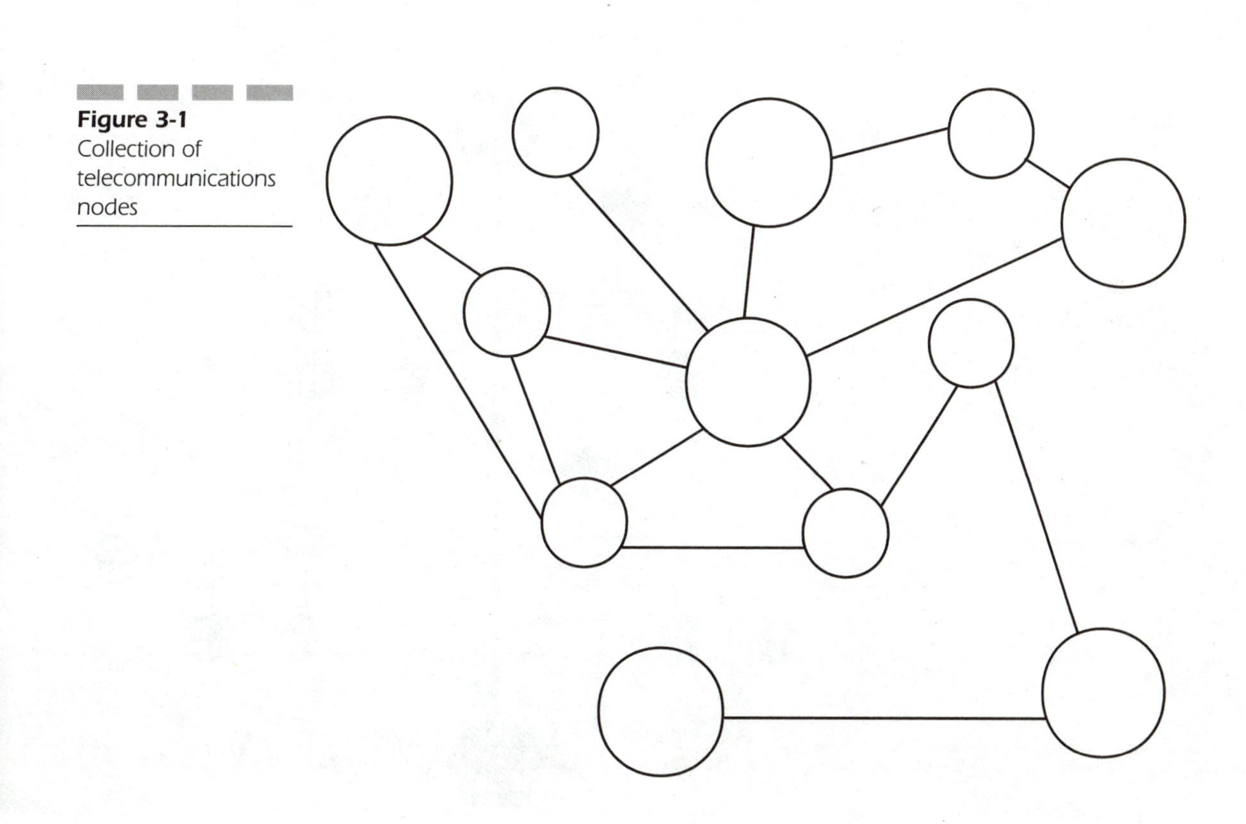

Figure 3-1 Collection of telecommunications nodes

Tree Architecture

In plain terms, the Tree architecture looks similar to a tree. The characteristics of the tree architecture are the same as that of a typical LAN in some landline telephone company networks (party lines), in cellular carrier networks, and in cable television systems.

As indicated in Chapter 2, the Tree architecture is the most efficient way of distributing the same set of communications signals to multiple terminals. The configuration's network efficiency is supposed to lie in its capability to reach multiple terminals. A Ring and Star configuration can distribute the same signals to multiple terminals just as well as the Tree configuration. The difference, however, is that the Tree configuration does not provide support for individual terminal devices to establish their own paths and does not even support disaster recovery. The Tree network configuration is meant to distribute signals and not necessarily to support terminal interaction with other terminals or with other central network nodes.

The Figure 3-2 diagram is a depiction of the Tree network configuration.

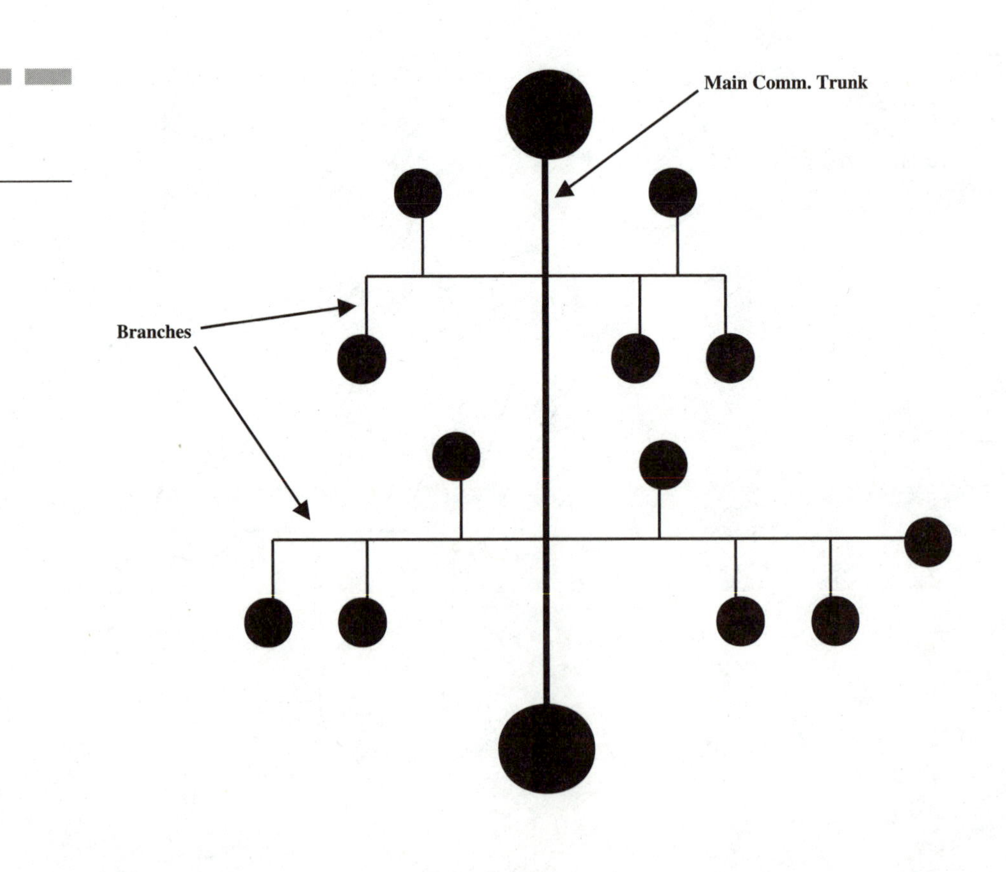

Figure 3-2 Tree network configuration

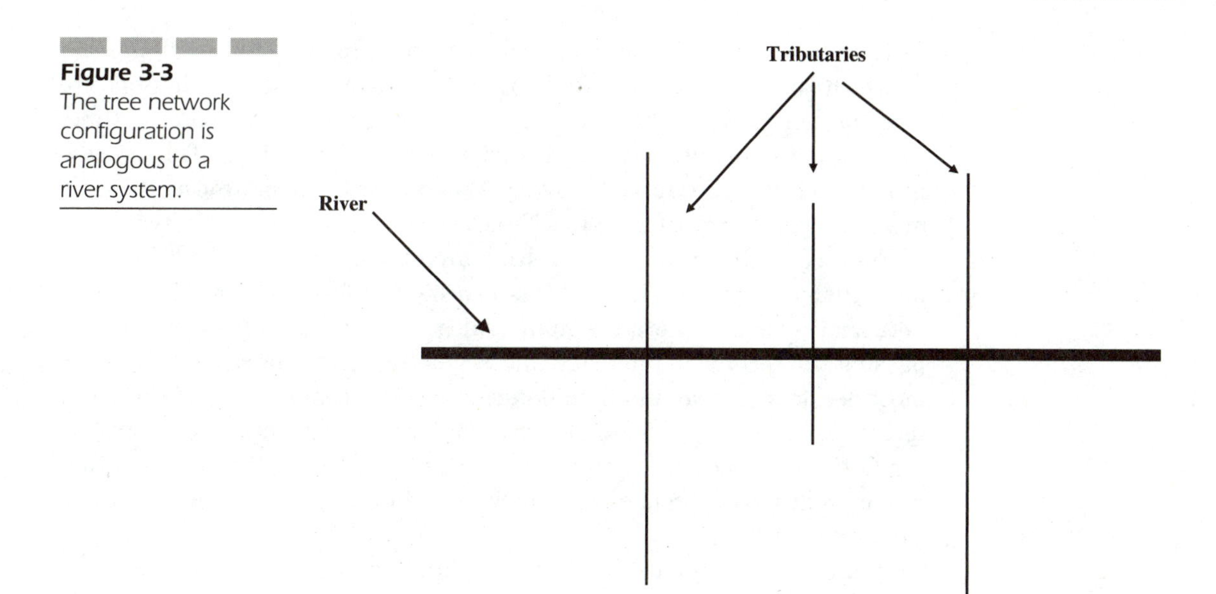

Figure 3-3 The tree network configuration is analogous to a river system.

In Figure 3-3, I have simplified the architecture even further. What we are looking at is similar to a river that feeds a bunch of tributaries. The river feeds the tributaries (and not the other way around). The Tree architecture is meant to distribute information to terminals and to not necessarily have the terminals interact with the principal source of information in any significant manner.

Ring Architecture

As noted in Chapter 2, the Ring architecture loops traffic so that it returns to its original starting point. Two examples of the Ring architecture are the Token Ring and the Self-healing Ring. Service providers (telephone companies and wireless carriers) commonly use the Self-healing Ring.

The Ring architecture is fairly new to traditional telecommunications and has only been deployed in various communities in the last 15 years. The Ring architecture is still not dominant in this segment of telecommunications, however. In the world of traditional data (i.e., computer-to-computer communication), the Ring architecture has been deployed far more heavily for about 20 years. Data integrity (bit error rates of 1×10^{-9} and smaller) has been a greater concern in this traditional world than in the voice telecommunications world. Financial transactions require a high degree of precision, while the human ear is far more forgiving when it encounters voice transactions.

To reiterate, the Self-healing Ring is a Ring architectural configuration in which the same messages are transmitted simultaneously in opposite directions on parallel Rings. In case of a service interruption on the Ring, the messages are automatically transferred to the other Ring. This process provides a high degree of redundancy. At each port, node, or drop on the ring, messages can be transferred to a Tree, a Star, or another Ring network.

Figure 3-4 illustrates how the Ring architecture appears and the manner in which various architectures can be combined. The reader should note that although there are many rules to designing a network-encompassing the technical, financial, and regulatory-one should first and foremost decide on a goal and then determine which combination of rules and technologies can be executed. Many carriers are not happy with specific regulations, so they lobby to have the rules changed. Other carriers find unique solutions to problems in which straight-forward application of rules can find no solutions.

Ring networks for cable television, landline telephone companies, or wireless carriers normally utilize fiber optic facilities for digital transmission. The Self-healing Ring is expensive, because you are supporting at least two networks that are carrying the same messaging for the sake of

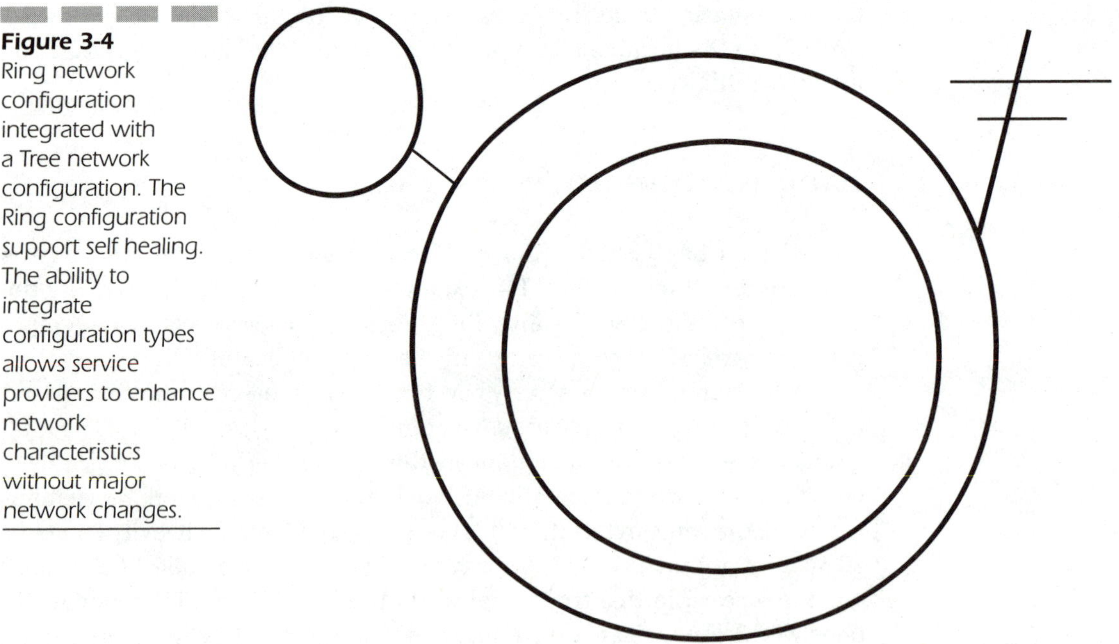

Figure 3-4 Ring network configuration integrated with a Tree network configuration. The Ring configuration support self healing. The ability to integrate configuration types allows service providers to enhance network characteristics without major network changes.

redundancy. The expense is worth the disaster recovery capabilities that are inherent within this architecture, however.

Disaster recovery is often an overlooked matter. Many people consider disaster recovery to be unnecessary. To those individuals who have lived through network overloads and network crashes, however, disaster recovery is an absolute necessity. New carriers (such as CLECs) often consider redundancy an unnecessary expense, given the fact that they are probably in a growth mode. One should understand, however, that a network is a "thing"—and that all things at some point fail, either due to poor maintenance or just simple wear and tear. If your network fails, you should remember the following points:

- The customer will suffer an outage of service.
- Customer-critical data might not be lost, but it will be late in getting to its destination.
- Loss of service damages customer confidence in the carrier.
- Loss of service means a potential impact on the business of the customer.
  - Loss of service could include a loss of *Automatic Teller Machines* (ATMs), voice service, Internet access, cable television programming, stock trading, broadcast television, intra-company customer communications, etc. The list is exhaustive and essentially represents a line of dominoes: the corporate customer might lose advertising revenue, consumer fees/charges, and future business. The individual telecommunications consumer might not even give the carrier a second chance.
  - Lost business for the customer means lost business for the carrier.

Star Architecture

The Star architecture is the most typical wireline telephone company configuration. The topology enables separate transmission paths to be established for each subscriber and customer location. Each path can be designed to carry the same information or different information. To some extent, the different transmission paths support a form of disaster recovery. Figure 3-5 is a high level representation of how the Star configuration can be implemented in a service provider's network.

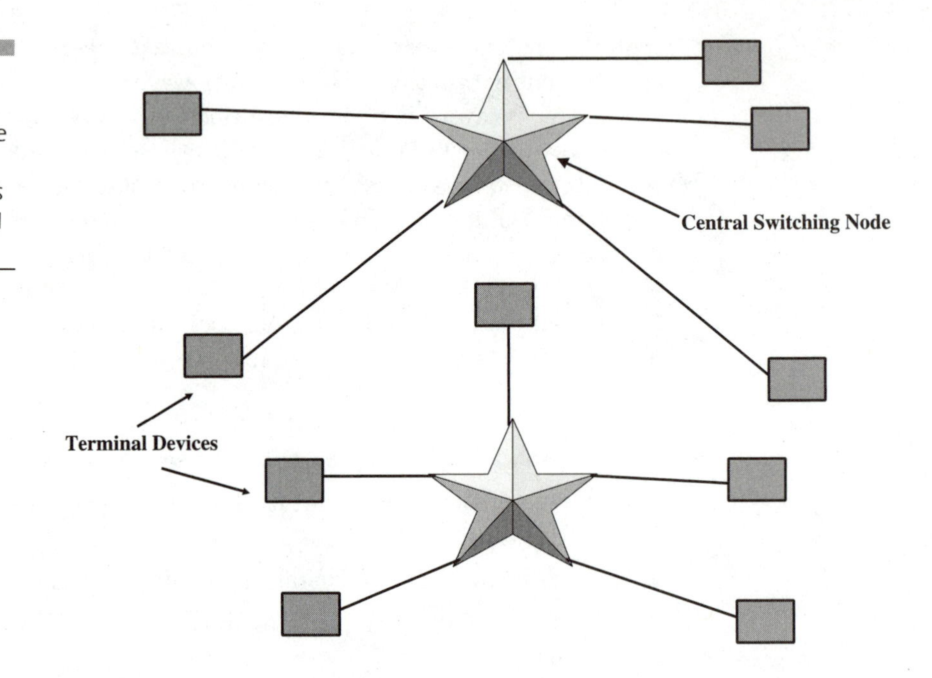

Figure 3-5
The Star network configuration can be implemented with the service provider's switch as the central node.

The following diagram (Figure 3-6) is another representation of the Star configuration.

I used the term *telecommunications hub* in Chapter 2. The term *hub* usually refers to the Star configuration. In comparison, the Star configuration supports distribution more cost effectively than either the Tree or Ring configurations.

The Tree network is subject to information/power losses as one travels farther and farther from the central node. Therefore, repeaters are needed to enhance the signal. The Star configuration does not inherently mean that one does not need repeaters. Carriers that deploy Star networks tend to establish an outer boundary around the central switching node, however, and as a result, repeaters are not normally used. The Tree network distributes traffic in a linear fashion, with one main feed into an area with distribution nodes off the main feed.

The Ring network configuration is a redundant architecture that supports disaster recovery scenarios far better than the Star configuration. The Ring architecture is simply expensive and is only cost effective when establishing redundancy on a grander scale, however (i.e., in large communities of interest such as Wall Street in New York City, or even in a small state where the overall needs of the business community warrant this redundancy). One does not deploy a Ring network in a neighborhood of 100 homes.

Figure 3-6
Star configuration

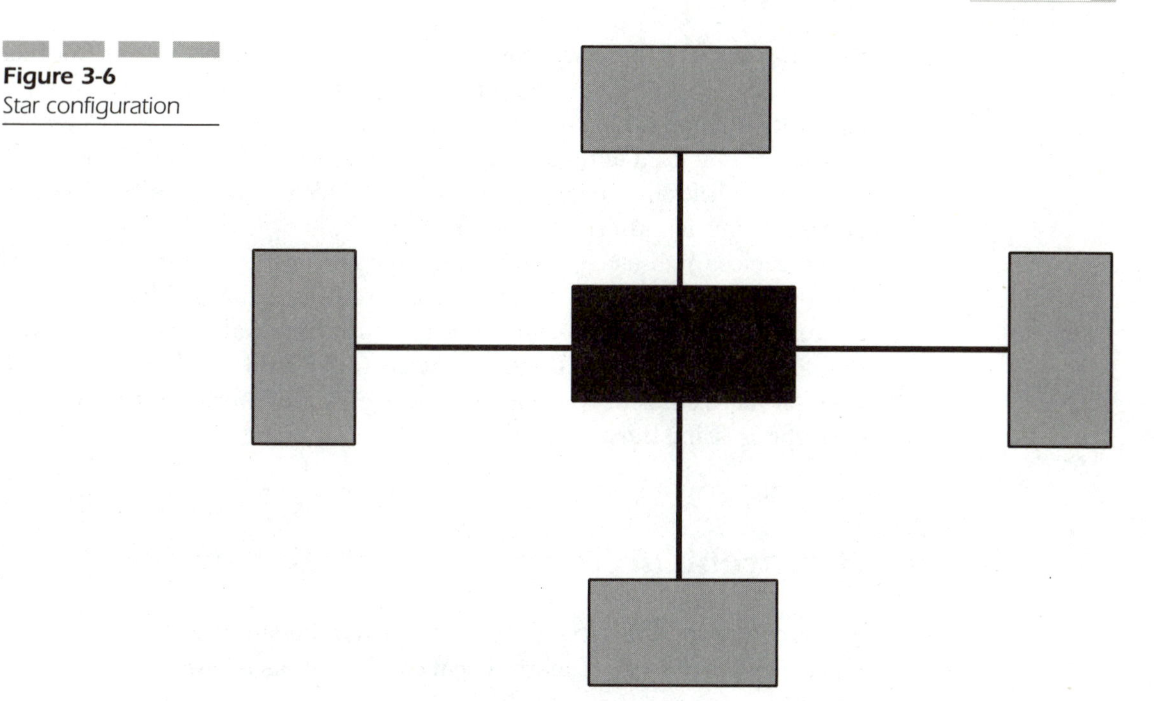

Determining when to deploy a Ring network involves economics. Essentially, all of the strengths of the Ring network are weaknesses of the Star configuration. The Star configuration's inherent strength is in its distribution characteristics.

By combining network configurations, one can play on the strengths of each configuration.

Routing Techniques

Routing is the action of directing information from one point to another. There are two principal forms of routing that are in use in the traditional telecommunications world: hierarchical routing (also known in today's world as alternate routing) and dynamic routing.

When you take a more expanded view of telecommunications and include the data world, however, there is a third type of routing: packet routing (i.e., packet switching of information).

Routing is not a transport technology; rather, it is a methodology of moving information around the transport structure that we will discuss later in this chapter. The network configuration is interrelated to the routing methodology that can be optimally executed. Given its importance, we will elaborate further on this topic.

The original form of routing that was first used in the old wireline voice telecommunications environment was hierarchical routing—also known as alternate routing. Alternate routing is applied to voice traffic by providing a first choice (high-usage trunk route) and one or more alternate routes if the first route is unavailable. This form of routing is hierarchical in nature because there is a predetermined order of routing that the switching system follows. Dynamic routing lends itself to flexibility and grew from the need for more efficient routing and management of the enormous volumes of traffic crossing the nation.

Hierarchical Routing/Alternate Routing

Hierarchical routing (alternate routing) has advantages and disadvantages. The advantages of hierarchical routing are as follows:

- Minimizing the cost per unit of traffic flow in a planned manner. This feature enables the carrier to closely manage traffic routing costs by deploying additional transmission media assets where necessary.
- Maximizing route usage during specific times of the day, days of the week, and weeks of the year. One should not leave capital assets sitting in the file underutilized.
- Maximizing the scheduling of network management and operations assets. Carriers that know the specific details of when and where traffic is being routed are capable of scheduling activities in areas of their network in which the traffic load is at its lowest. Remember, the carrier cannot shut down its network for maintenance; rather, it can only schedule maintenance during low periods of activity.

The disadvantages of hierarchical routing/alternate routing are as follows:

- Carriers need to monitor and analyze the traffic patterns in their network on a daily basis. This activity should take place daily; however, a carrier can save time and money if the carrier could reduce the number of times necessary to make network adjustments for the sake of traffic flow. Theoretically, people do not make drastically large changes or frequent changes in their traffic patterns. With the advent of the CLEC (wireless and wireline), however, a change in traffic patterns can mean a loss of customers or maybe even an influx of new customers.

- Routing plans must be accurate. This concept is not as unusual as it might seem. There have been instances since the divestiture of the old AT&T Bell System in which carriers have suffered outages due to severed transmission facilities and carriers overloading their networks. Despite the fact that extensive load planning was performed, not enough planning was done to ensure secondary and even tertiary routes. The disadvantage is that resources must be available and skill sets must be present in the planning organization.
- Time and money needs to be spent to ensure accurate planning.

As indicated in Chapter 2, the following figure (Figure 3-7) shows a one-way, high-usage transmission facility from switch X to switch Y, with an alternate (final) route via a switch that performs a traffic cop function called tandeming. In general, the direct or high-usage route is shorter and less expensive than the alternate route. Because each segment of the alternate route is used by other calls or information packets, a number of traffic items can be combined for improved efficiency on that route. The basic challenge then, is to minimize the cost of carrying the offered traffic load.

This example in Figure 3-7 is applicable to any telecommunications information traffic situation. The decision Tree basically looks similar to the following description:

Switch X needs to route information to Switch Y. Is the high-usage route available?

- If the high-usage route is available, then use it.
- If the high-usage route is not available, then use the alternate route.
- If the primary alternate route is not available, then use the secondary route. In most cases of voice telecommunications switching, there will be no secondary route, because we assume that blocking has been designed in the network. We will explore the concept of blocking later in this chapter. In the case of data transactions, companies that specialize in transporting financial data will probably have a secondary alternate route.
- A Switch C diagram was added to illustrate the point that in some cases, there might not be an alternate route. Information traffic

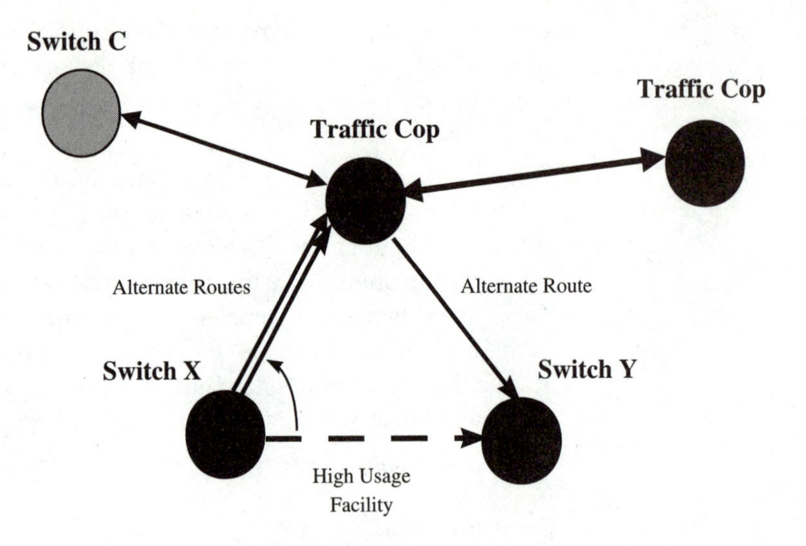

Figure 3-7 A one-way, high-usage transmission facility from switch X to switch Y

engineering and necessity will dictate whether or not there should be an alternate route.

The issue of alternate routes brings up the question, "When is enough enough?". We will delve into the issue of blocking and its roles after the following discussion of dynamic routing. The following section addresses the financial aspects of hierarchical routing.

A relationship exists between the number of trunks, the financial cost of the direct route, the financial cost of the alternate route, and the total dollar cost for serving the given load. As you will see, the high-usage facility cost is proportional to the number of high-usage trunks. If there are no high-usage trunks, all of the traffic must be sent on the alternate route, so the incremental alternate route money cost is high. If a high-usage group is available, the incremental cost of the alternate route is lower, because theoretically, less traffic is sent over the alternate route. The cost of alternate routing decreases the minute the first trunk is added to the high-usage trunk group. The flip side of this cost picture is that there is a point at which there will be too many high-usage trunks in the high-usage group. As more high-usage trunks are added to the high-usage group, each additional high-usage trunk will theoretically carry less traffic—while each alternate route trunk will continue to carry a significantly larger amount of traffic.

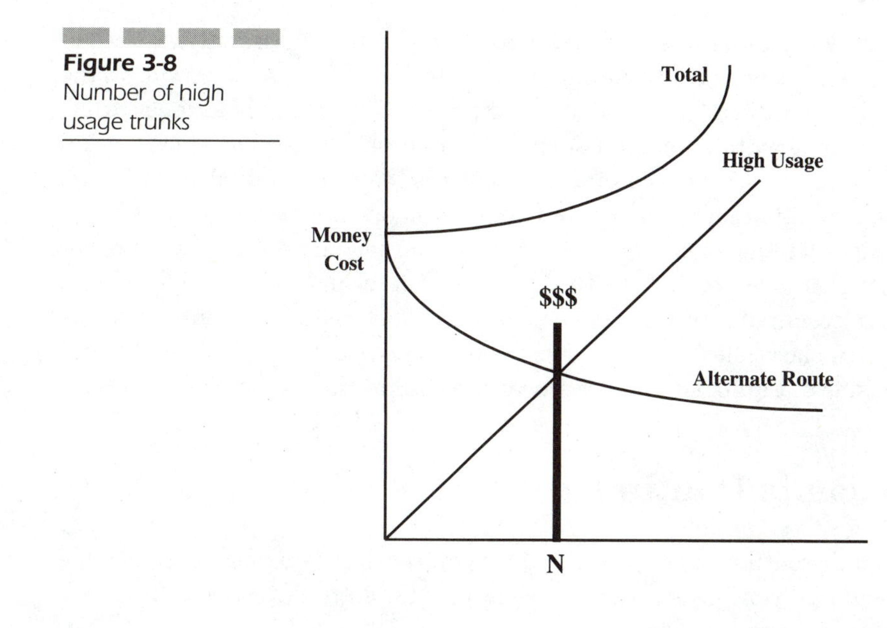

Figure 3-8
Number of high usage trunks

At some point, adding more high-usage trunks becomes cost prohibitive. This philosophical point is illustrated in Figure 3-8.

The point at which this threshold occurs is where the total cost is minimized. This point is designated as *N*.

A method commonly used to determine *N* is called *Economic Hundred Call Seconds* (ECCS) engineering. This method determines the maximum number of high-usage trunks for which the cost per *Hundred Call Seconds* (CCS) carried on the last trunk of the high-usage trunk group is less than or equal to the cost per CCS on an additional alternate-route trunk.

The following equation forms the basis of ECCS engineering:

$$\frac{ALT}{HU} = \frac{28}{ECCS}$$

ALT = cost of a path on the alternate route

HU = cost of a trunk on the high-usage route

28 = capacity in CCS added to the alternate route by the addition of a trunk

Blocking is when a call cannot be completed. Blocking is a traffic engineering concept that enables a service provider to establish a "grade of service objective" for its network. The number of calls that cannot be completed during a given time period is the "grade of service" provided by the carrier.

No carrier has the financial resources to build a network that can accommodate the completion of simultaneous calls from all subscribers within the switch's operational jurisdiction. Another way of looking at blocking is drawing an analogy to a line at a supermarket checkout line. The supermarket determines the optimum number of cashiers that are required at any given time making an assumption that the customers will be willing to wait to checkout for some period of time. A supermarket cannot hire a cashier for every customer; otherwise the supermarket would go out of business. The telecommunications service provider cannot install a circuit to process simultaneous calls from every subscriber; otherwise the carrier would go out of business. Equipment and space for the equipment costs money.

Dynamic Routing

Dynamic routing refers to the updating of routing patterns on a real-time or near real-time interval on the basis of traffic statistics collected by a network management system.

Dynamic routing is a traffic routing method in which one or more central controllers determine near real-time routes for a switching network based on the state of network congestion measured as trunk group busy/idle status and switch congestion. The choice of traffic routes is not predetermined. Dynamic routing schemas are used in situations where the complexity of routing the call is so high that applying the fixed rules and structure of hierarchical network configurations makes it difficult for a network manager to properly manage call flow. To some extent, there must be a degree of intelligence embedded within the network.

Figure 3-9 illustrates the complexity of routing information in a hierarchical-routed network versus routing information in a dynamic-routed network. Figure 3-9 is an illustration of a typical network.

Figure 3-10 is an illustration of an even more complex set of network relationships. Figure 3-10 is a pictorial description of the Internet. In today's World Wide Web of Internet sites, links are somewhat apparent between multiple sites—regardless of whether or not there is any relationship at all. Perform a simple word search by using any one of the plethora of search engines, and you will find hundreds of thousands of Web hits. All we had to do was type the search words `antique cars`. More information about this topic will appear in later chapters.

Figure 3-10 illustrates a number of networks that are interconnected in a variety of ways. Given the changing regulatory environment, the networks will be interconnected in this fashion (and will seem random, but actually they are not). Business relationships dictate how the networks of various

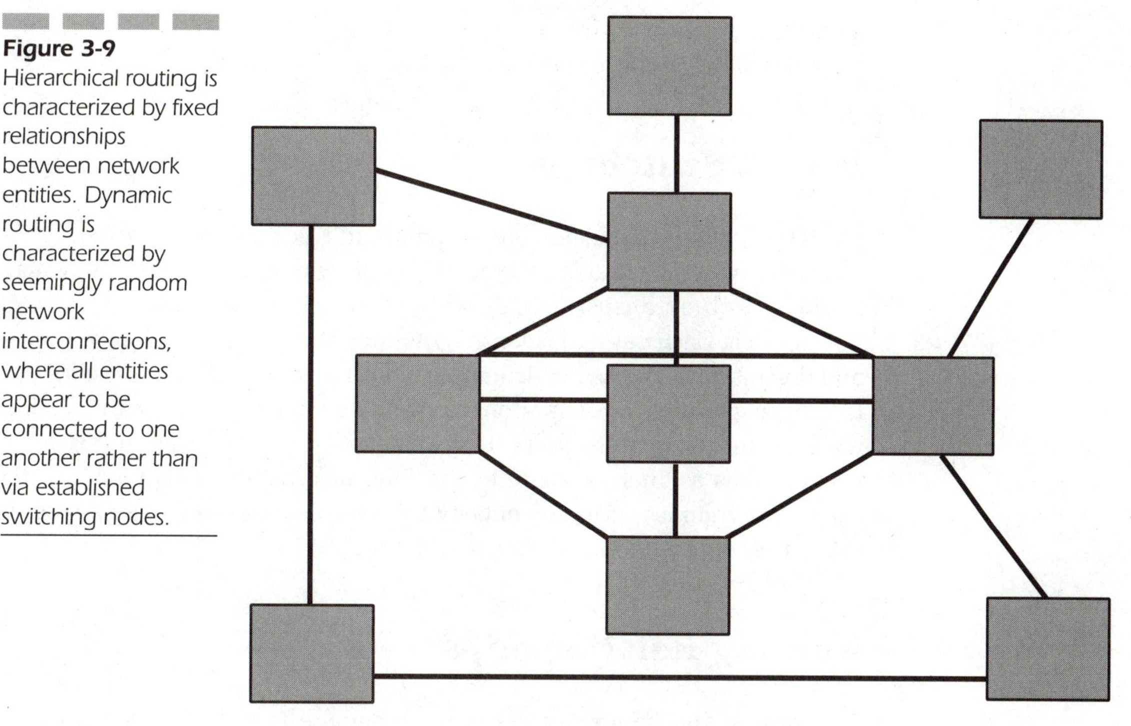

Figure 3-9 Hierarchical routing is characterized by fixed relationships between network entities. Dynamic routing is characterized by seemingly random network interconnections, where all entities appear to be connected to one another rather than via established switching nodes.

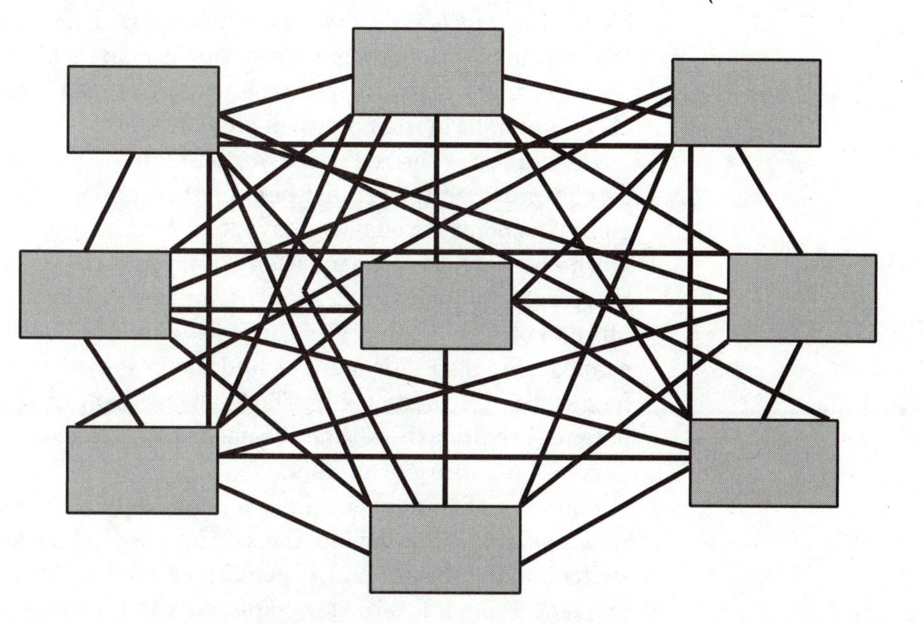

Figure 3-10 Internet network relationships appear to be weblike in nature. The Internet is a collection of networks that are all interconnected to one another. The Internet does not appear to maintain a rigid and hierarchical routing schema.

service providers are interconnected. Therefore, as stated previously, although the networks appear to be interconnected in a seemingly random fashion, the networks are in fact connected in a realistic fashion. Applying the rules of hierarchical routing is nearly impossible in this diagram.

Packet Switching

Packet switching represents one step beyond traditional voice telecommunications switching and is suitable for a data environment. Packet switching involves the transport of information in discrete packets or packages of information. Each packet is identification-tagged with its label, origination, and destination. The packet is launched into a network that is conditioned to support packets of information (rather than a continuous stream of information) and ultimately finds its way to the destination of choice. Each packet takes a different route to the final destination. This model is not exactly a dynamic routing schema, but it is fairly close—because the routes are not predetermined and change constantly.

Routing Relationships

As indicated in Chapter 2, the routing schema I have described is applicable to networks of different service providers. There should be a set of rules that define how the flow of information is managed between networks. I am not referring to signaling protocols; rather, I mean the management process of sending information between networks.

Regardless of the regulated or non-regulated nature of any information service provider, there must be a method of how information is sent from point A to point F. Today, we have ISPs that own the servers in a local area but do not own the transmission media that connects the servers. In fact, there are companies that simply transport information across the nation but do not have visibility to the consumer. Each service provider is interconnected in a specific manner, and the interconnection is totally governed by the business relationship. The following diagram (Figure 3-11) shows the potential relationships that are needed to transport information from one part of the country to another.

One way of circumventing this potentially complex set of management relationships is by having the carrier own and operate everything—all switching, databases, and transmission media. We will now digress for a moment. From a macro viewpoint, one can take this scenario even farther. If such a carrier existed, that carrier could install multi-functional switches that would support all forms of telecommunications. From a micro view-

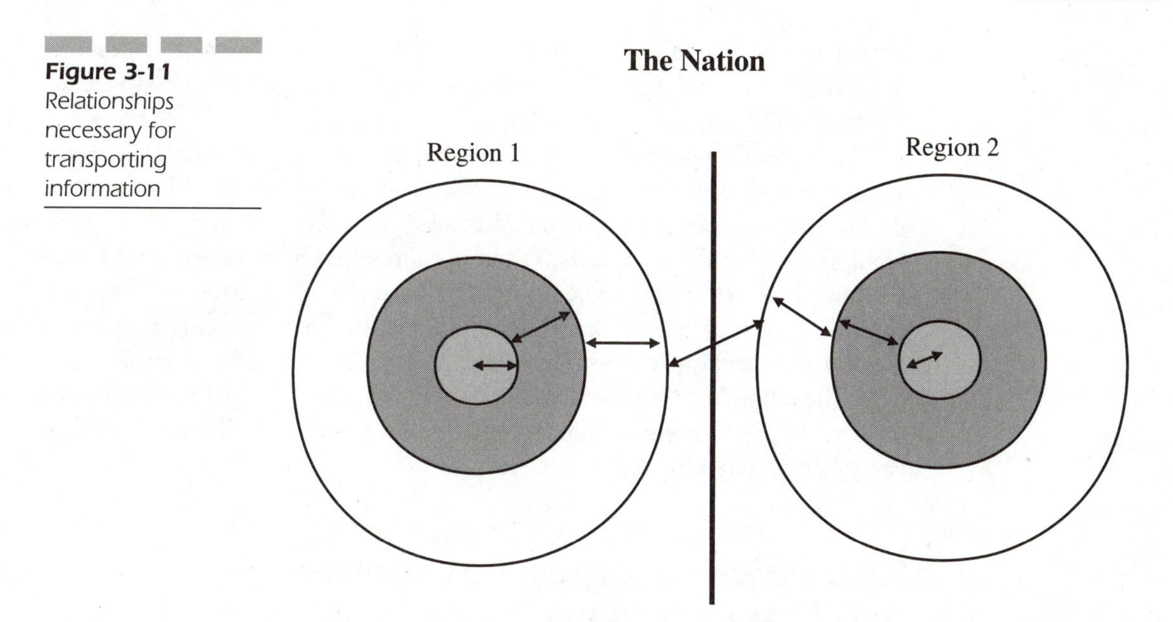

Figure 3-11 Relationships necessary for transporting information

point, one will see that technical design logic requires a degree of functional separation—i.e., "taking it one step at a time and looking at the individual pieces." Therefore, understanding relationships is still a necessity.

The next step in examining the network is reviewing what will be needed to physically connect the networks and transport the information. The reason for the concern is because each type of physical media has limitations in the amount of data that it can transmit in a given period of time. The speed of information transmission will have an impact on the decision-making process that is associated with network design and network interconnection.

Transmission Facilities

Transmission facilities are needed to connect networks. Transmission media (also known as outside plant) can be broken into the following basic types:

- Metallic
- Fiber optics (glass)
- Microwave

The reader should note that wireless carriers support a variety of radio technologies in different frequency bands. The wireless aspect of the wireless carrier industry is related to local device access to the subscriber and not to access between networks (or even between its own network elements). To some readers, this topic is more familiar if you refer to transmission facilities/media as backhaul.

The biggest concern of carriers (relating to transmission media) is speed. Time is money, and the higher the information speed supported by the facility, the more money that is made and saved. Each type of facility is capable of supporting different speeds. Transmission speed is governed by the controlling electronics that are employed at both the originating end and terminating end of a call. Each form of media has physical limitations associated with how much information it can transmit without errors, however. The following discussion will examine the types of facilities that can be used to haul volumes of information traffic between switching nodes or computing elements (i.e., data centers).

Transmission Types: Advantages and Disadvantages

Metallic wire Metallic wire (usually copper) tends to be the typical medium for DS–0 through DS–2 transmission facilities. Metallic transmission facilities have the following advantages:

- *Durability.* In the world of wireline telephony, the author is personally aware of situations in which a specific facility (physical "copper in the ground" or "on the poles") has been in use/operation for at least 50 years.
- *Easy to maintain.* In some communities, the LEC has existed for decades.
- *Easy to install.* Metallic transmission facilities can be mounted on a telephone pole or placed underground. Insulation can last for years without harm. Note that in the case of the telephone pole, I am aware of poles as old as 60 years that are still in use today.

There are disadvantages to using metallic transmission media, however:

- Because metallic wire is made of metal, it is subject to electromagnetic interference of all kinds, which can reduce the quality of the transmission. This interference ranges from radio transmissions to electrical sparking/noises that are caused by an electric toaster or even by an electric drill. Most of the interference can be overcome with better electrical grounding or even signal regenerators. There is a cost to these solutions, however.

- Metal ages, and eventually it will fail physically.
- Metallic wire is heavy to transport. The weight and size of metallic-insulated cable is a transportation logistics concern.
- This medium requires large storage areas in comparison to fiber optic cable.

Fiber Optics Fiber optics is also known as optical fiber. Some still refer to this medium as glass or simply as fiber. I will interchange the terms throughout the book. Fiber optics has many uses in transmission. Generally, fiber is used in applications that require the following functionalities:

- *Moving data at high speeds.* Optical transmission is classified as DS4 and higher signal levels. Optical rates are in their own category of formats:
- Optical Carrier (OC) -1, 3,12, 24, 48, and 192
  - OC-1 equates to a data speed of 51.840Mbps.
  - OC-3 equates to a data speed of 155.520Mbps.
  - OC-12 equates to a data speed of 622.08Mbps.
  - OC-24 equates to a data speed of 1244.16Mbps.
  - OC-48 equates to a data speed of 2488.32Mbps.
  - OC-192 equates to a data speed of 9953.28Mbps.
- Large bandwidth applications (e.g., video, Internet video, and audio applications)
- *Limited access locations.* Not all locations lend themselves to gouging out a large trench to lay outside plants.
- Transoceanic cable

  Fiber optics advantages are as follows:

- *Non-susceptibility to electromagnetic interference.* Glass is a non-ferrous/non-magnetic material.
- *Ground returns are eliminated.* All electrical systems need to be grounded, because grounding establishes a difference in electrical potential between electrical conductors. Light does not need to be grounded.
- *Small size.* One strand of fiber (the diameter of a human hair) can carry data at speeds in excess of 100 gigabits per second. Given the small size of fiber optic cable, carriers can lay fiber in places where laying (underground) metallic cable is obtrusive (for environmental reasons, property access issues, or because of limited room in existing

underground communication cable conduit).

- *Light weight.* Not counting the protective insulation, a fiber strand is as light as a human hair.
- Low maintenance requirements
- Glass does not age as fast as metal.

Fiber optics disadvantages are as follows:

- Expensive for low-traffic load needs
- Requires special electronics for the light source, light detection system, encoders, and decoders
- Susceptible to signal attenuation (loss). In comparison to metallic T1 lines, however, where repeaters need to be installed at one-mile intervals, fiber optic systems can operate with repeater distance intervals of up to 100 miles.
- *Signal dispersion.* The optical signal (light) needs to travel through the optical fiber without any reflection outside the fiber. In other words, the light pulses must remain within the structure of the fiber strand. As light travels through the fiber, it must be totally reflected within the fiber optic strand. This process can be affected by the quality of the fiber optic strands. Dispersion can be brought about by a number of factors. There are different forms of dispersion: material, modal, waveguide, and chromatic.
  - Material dispersion occurs because certain transmitted optical frequencies travel faster than others in the optical media. This situation results in some light packets arriving before others, which causes intersymbol interference at the terminating/receiving end.
  - Modal dispersion occurs when multiple modes (different transmissions) occur. Due to the difference in frequencies, some of the modes have different reflective properties, resulting in the interference of the transmissions with each other.
  - Waveguide dispersion occurs when the light wave penetrates the insulation (commonly known as the cladding) and suffers a change in the refractive index of the medium in which the light ray is traveling.
  - Chromatic dispersion is a combination of a material and waveguide dispersion.

Figure 3-12 illustrates a fiber optic system.

Figure 3-13 shows a cross-section view of a fiber optic cable. From a macro perspective, this cable is not unlike a metallic wire. Both possess

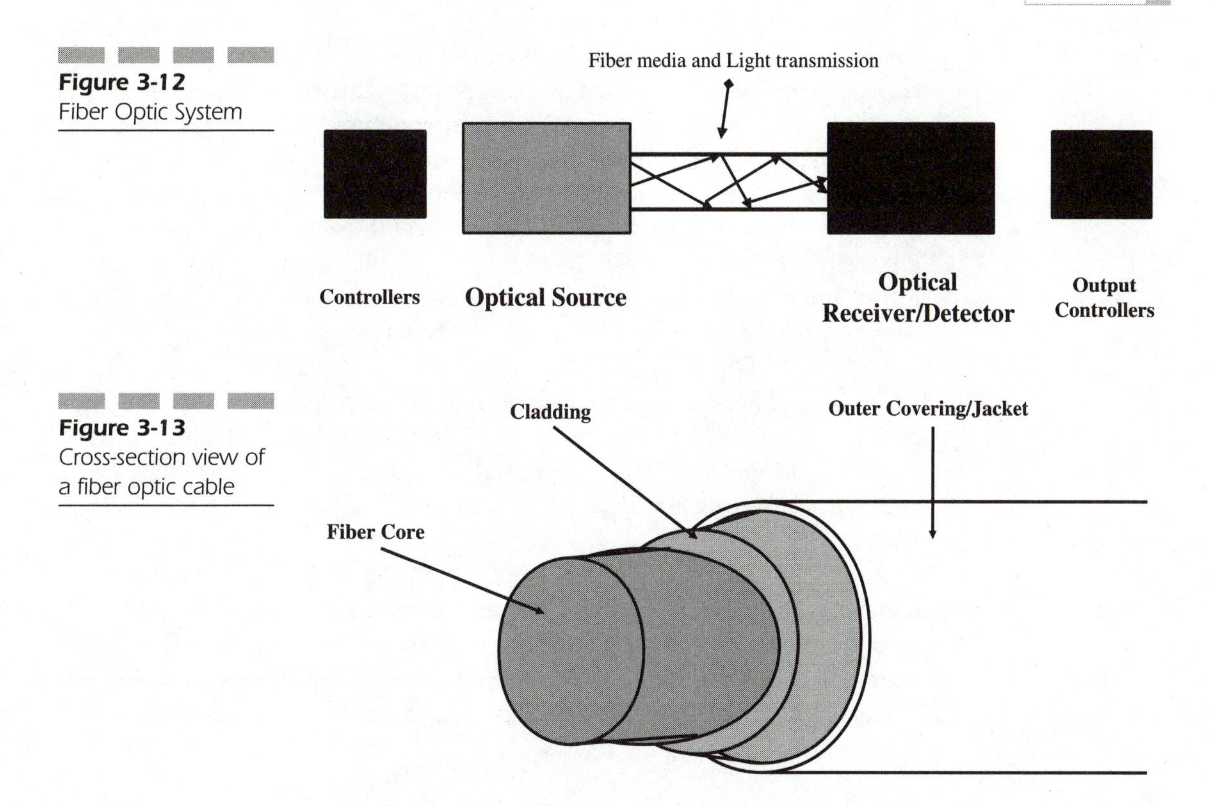

Figure 3-12
Fiber Optic System

Figure 3-13
Cross-section view of a fiber optic cable

insulation and some type of outer covering.

Different grades of fiber exist, ranging from multi-mode to single mode. Each grade has different bandwidth and physical properties. The core of a single-mode fiber has a diameter that is narrow enough to restrict light to one mode of travel. In other words, we have a one-way transmission facility. Multi-mode fiber facilities have a much wider diameter. Due to its wide diameter, multi-mode fiber is capable of allowing light to enter at different angles and to travel through the fiber along different criss-crossing paths.

Multi-mode fiber supports more than one mode, whereas single-mode supports one mode. Multi-mode has superior bandwidth properties in comparison to single-mode fiber.

Synchronous Optical Network (SONET)

Synchronous Optical Network (SONET) deserves some attention, because it is the North American digital format designed specifically for fiber optic transmission systems. The format was developed by Bellcore and the

American National Standards Institute (ANSI). SONET is a high-speed transmission-protocol format that defines the features and functionality of an optic transmission system based on synchronous multiplexing. This structure is much more than just a souped-up, DS−4-level, bit-rate format. SONET is an optical format that is capable of supporting speeds ranging from 150Mbps to several gigabits per second. SONET was designed to address both network access by customer-premise equipment and network interconnection between providers at high speeds. From an OSI model perspective, SONET is a Layer 1 protocol (Physical layer signaling).

Unlike T1, DS1, DS2, and DS3 formats (or even ETSI-based E1, E2, or E3), which support multiple signaling formats defined after the fact, SONET is a format in and of itself. SONET was designed specifically to take advantage of the full potential bandwidth of glass. SONET is a transport protocol, as opposed to a network control signaling protocol such as SS7.

A significant portion of the SONET signal structure (frame structures) is dedicated to network management and network maintenance. A SONET frame is 810 octets large. SONET was designed with supporting existing signaling formats in mind and has the capability and flexibility to support customer services and network-signaling formats (such as ATM, ISDN, X.25, and SS7).

Terrestrial Microwave Transmission Terrestrial microwave transmission consists of line-of-sight transmission through the atmosphere of focused microwave energy. Microwave antennas can be seen in all areas of the country and are typically mounted either on their own towers or on top of tall buildings. Figure 3-14 is a depiction of a terrestrial microwave system.

Microwave energy can be focused (FM or AM radio transmissions) into tight and narrow beams. Further, line of sight is required in order to ensure reception between antenna stations. The microwave energy can be absorbed by animate and inanimate objects, and the absorption will result in significant signal degradation. The microwave transmissions usually occur in 4GHz, 6GHz, and higher bands.

Line-of-sight refers to the transmission and reception visibility between microwave stations. The distance between stations is a function of tower height, terrain cover (i.e., buildings, trees, and water), topology (mountainous, flat, desert, water, etc.), the Earth's horizon, and atmospheric conditions. The typical distances between microwave stations is approximately 25 to 30 miles.

Figure 3-14
Terrestial microwave system

Microwave Transmission- Line of Sight

Figure 3-15 illustrates the line-of-sight concept.

Microwave transmission has a variety of telecommunications applications:

- Communication (e.g., military)
- Linking wireless carrier cell sites in a wireless carrier network (i.e., backhaul)
- Linking within wireline carrier networks between geographic areas. Rather than laying cable across an expanse of land at a considerable cost, you can use a microwave "shot" to cross "the divide." The cost I refer to can be unending bickering with landowners or nearly endless negotiations with the local boards of several different towns. Another cost is the cost of the physical cable, whether it is fiber or metallic.

Similar to fiber optics and metallic cable, microwave has its own advantages and disadvantages. The reader will note that these advantages and disadvantages are technical, operational, and financial in nature.

Microwave transmission advantages are as follows:

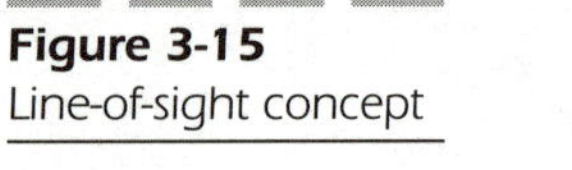

Figure 3-15
Line-of-sight concept

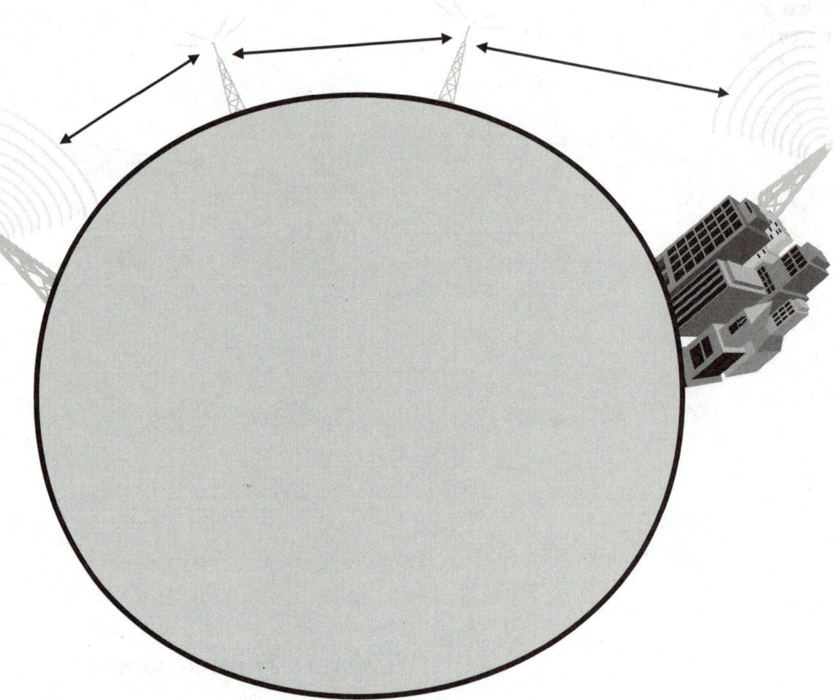

- Low-cost option when compared to fiber optics and metallic cable. This cost has two different facets: political and financial.
  - *Political* Local zoning boards can prevent the laying of physical transmission facilities for a variety of reasons, which would include damage to the environment and damage to public and private property.
  - *Financial* Unless the carrier can claim right-of-way, the carrier will be paying private landowners monthly or annual charges for the use of the land that the cable crosses. This situation can result in paying numerous high rents/leasing fees to numerous landowners—a potential real estate management problem that one can overcome by using microwave. Another cost is constructing/laying/burying cable in harsh environments. Harsh environments demand protection for the facility and safety of the facility construction crews.
- *Large bandwidth can support a variety of voice, video, and data applications* Microwave transmission is classified as a DS4 and higher signal level.

- *Can bridge harsh terrain where laying cable is nearly impossible* Harsh terrain can include mountains, deserts, canyons, valleys, dense forests, and bodies of water. Safety of the personnel would always be a concern. The fact is that to properly physically install cable of any kind, however, you might have to cross terrain that demands costly (from a financial perspective) construction and demolition methods. The money spent on performing the construction or demolition might be prohibitively high.
- Repeater spacings are on the order of 25 to 30 miles, as opposed to repeater spacings of one mile for a typical T1 span.

The disadvantages of microwave transmission are to some extent the flip side of the advantages just described:

- *High cost* The cost of deploying a microwave system can be described in two ways: political and financial.
  - *Political* Placing towers, especially those with the labels of cellular, PCS, radio, and microwave, can bring outcries from local community/town zoning boards and environmental groups. The towers (specifically, what the antennas are transmitting) will be tagged as unsightly, environmentally damaging, or health hazards.
  - *Financial* Some (but not all) landlords might decide to charge the carrier with incredibly high rent or leasing fees for the towers (and sometimes for just the antennas). Land, rooftops, and building facades are all used to support towers and antennas. Sometimes, the antennas are just installed on building facades. Regardless, these scenarios all involve real estate that can be rented or leased.
- Large bandwidth can support a variety of voice, video, and data applications. Unfortunately, the transmissions are not necessarily secure transmissions unless encoding is performed.
- *The transmissions are subject to the whims of weather.* The problem with using the Earth's atmosphere as a major transmission medium is the weather. The weather can be broken down as rain, fog, warm air, cold air, hot earth, cold earth, warm earth, cool earth, bodies of water, and any combination of these characteristics. The solutions are either higher towers or towers that are near each other.
- *Can bridge harsh terrain where laying cable is nearly impossible.* Harsh terrain can include mountains, deserts, canyons, valleys, dense forests, and bodies of water. Safety of the personnel would always be a

concern. The fact is, however, that properly and physically operating remote microwave stations requires environmentally controlled enclosures, electrical power, heating, ventilation, air conditioning, storage for tools, and access security for the station. If the terrain is harsh, there is a high probability that the ambient environment (i.e., weather) is also harsh.

Microwave transmission can also be viewed from a non-terrestrial perspective (i.e., satellite transmission). Satellite is growing as a medium to haul information around the globe—but not in any significant fashion for the public telecommunications systems in use. Satellite transmission will be addressed in later chapters as a technology and business opportunity.

Evaluating Transmission Facilities

Microwave and fiber optic and metallic cable all have advantages and disadvantages. The only way to evaluate the necessity to use any of these mediums or a combination of them is to perform a cost-benefit analysis. The analysis would seek to answer the following questions:

- What kind of data is the carrier transporting?
- What are the data transmission speed requirements?
- What level of security is required for the data?
- Where does the data need to be transmitted?
- Are there time-of-day and day-of-week transmission requirements?
- Are there any seasonal peaks/fluctuations in traffic volume?
- What signaling protocol will be used to transmit the data?
    - Certain protocols have speed limitations. If the data used is limited to a certain type of information traffic, the speed requirements might be high speed rather than lower speeds.
- Does the medium need to be fractionalized in any way?
    - If so, then how many channels need to be supported?
- What are the various cost components associated with providing a transmission facility that will support the needs of what has been answered up to this point?
    - Wireless engineers will also perform a path analysis, also known as a link budget. Link budgets address operating issues such as

transmitted power, antenna design, noise, and bit-error rates. These wireless design issues will also play a role in a wireless carrier's decision.

- What is the overall cost of the alternatives being examined? This total picture would include fees, rents, and leasing fees paid to landlords.

Each one of these questions has a technical component and a financial component. What often drives a carrier, especially a CLEC, are the financial budgets being approved (and not the perfect technical solution).

Transmission Business Opportunities

The provisioning of facilities for transmission is a business for many service providers that have excess transmission capacity. These service providers can be in the cable television business, landline telephone business, cellular business, PCS business, or even in the power utility business. As I have already noted, the fact is that any company that has a communications transmission network can provide facility support to another service provider. If you have the extra bandwidth, then use it.

Bandwidth is defined as capacity of a transmission facility or medium (hardware or radio spectrum). The capacity of transmission media is always a concern. The growth of new services, such as video-based services, requires lots of bandwidth. You can compress information only so much.

The transmission business can be seen not only as a provisioning business for those who have communications facilities, but also as a business for those who own real estate and especially those who possess right-of-way. Railroad companies, telephone companies, electric power utilities, natural gas supply companies, oil companies, and even towns all possess rights-of-way. The concept of right-of-way was legally established in this nation in the 19th century. The right-of-way enabled (and still enables) the aforementioned companies to construct their various facilities (railroads, streets, highways, oil pipelines, gas pipelines, telephone poles, and electric power transmission lines) without fear of legal reprisal from the community or even from the citizen (barring environmental and gross negligence issues). Effectively, the company is granted the right to install these components within defined tracts of land. Companies that possess real estate can either build their own transmission facilities (typically fiber loops) or towers and enter the transmission leasing business. You should remember that physical transmission facilities need to

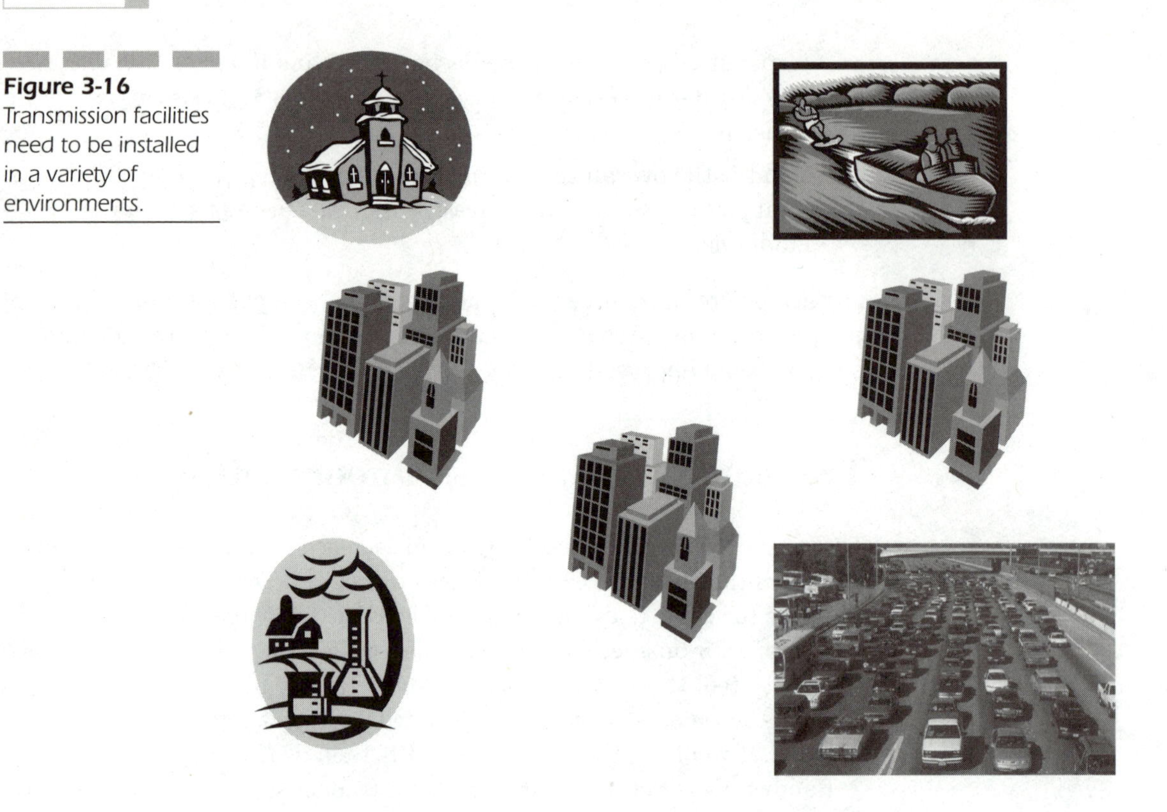

Figure 3-16 Transmission facilities need to be installed in a variety of environments.

placed, laid, buried, or hung. Figure 3-16 is a high level depiction of the kinds of environments that a transmission facility needs to be installed in.

Local Loop

The local loop is the portion of the physical telecommunications circuit between the LEC's central office (local switch, also known as a Class 5 switch) and the customer's premise. This circuit originally consisted of twisted-pair conductors (essentially, a continuous loop of wire). The original local loop was solid metal (copper) wire. Refer to Figure 3-17.

Despite the deployment of digital and high-speed access systems in the home and on the customer's premises, the majority of the wiring to the customer premise is still largely twisted pair. This fact should not be surprising when you realize how long the wireline telephone system has been in place (since 1876).

The greatest challenge for the CLECs has been penetrating the local loop. To some, penetrating the local loop might mean installing wire to the

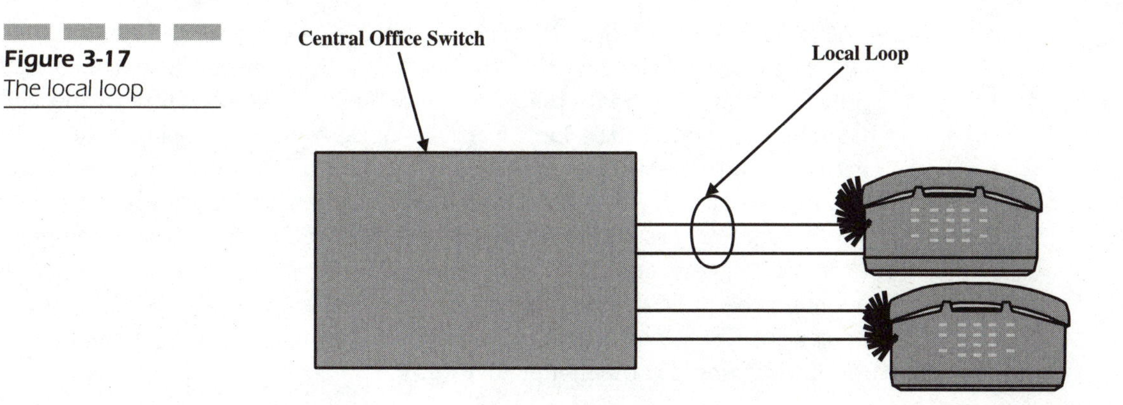

Figure 3-17
The local loop

premise or owning and maintaining the existing local wire. If a CLEC were to install its own local loop, it would require enormous expenditures on a scale far beyond the basic switching systems.

To many CLECs, penetrating the loop does not necessarily mean owning and maintaining the physical, outside plant; rather, it means owning the customer as the primary provider of basic voice and data telecommunications. The Telecommunications Act of 1996 and current state tariffs and regulations, however, require the ILEC to continue maintaining the wire in and around the neighborhoods and communities for regulated fixed fees/surcharges, which the CLEC can charge back to the subscriber. This method is how many CLECs can rightfully claim that they are using telco wire. CLECs, however, are facing challenges—creating profit margins in a business that requires huge capital and labor expenditures. Telco wire is an old term used to refer to "telephone company" owned transmission facilities. The answer is actually composed of multiple parts.

CLECs, to some degree, are basically all resellers of ILEC services and systems. CLECs might own and operate their own switches and building facilities, and they might own some of their own transmission facilities. CLECs might even obtain all of its backbone facilities (interswitch/intranetwork connectivity) from a party that is not an ILEC; however, this situation rarely occurs. The CLEC, however, will always have to work with the ILEC for local loop access. Given the critical nature of timely deployment and providing service to subscribers, the CLEC needs to find ways to create margin on its books in all facets of its business, including the local loop.

Some CLECs have begun deploying *x-Digital Subscriber Line* (xDSL) technology as a way of entering the home and business on a more cost-effective basis. The letter x in front of the term DSL represents a generic labeling for the family of DSL technologies.

Digital Subscriber Line Technology (DSL) is a loop transmission technology that was developed by the wireline LECs to optimize the so-called last mile to the home or small- to medium-size business. The last mile is another term used to address the local loop. The optimization of the last mile is, in actuality, the capability to provision wide-band multimedia video and data services to the home over the twisted pair of wires. Layer 1 and higher network signaling protocols/technologies such as SS7, ATM, and intelligent network triggers enhanced the way networks as a whole functioned—enabling the provisioning of new types of services. Essentially, xDSL technology is a LEC's way of getting more mileage from the existing twisted pair (or getting more for the money invested).

Reasons for Using xDSL As I had noted previously, xDSL technologies enable a carrier to optimize the use of its own local loop or a CLEC's use of the local loop. What is needed is an explanation of why the local loop has to be enhanced.

Much of the modernization of the network (or network of networks) has taken place at a network-to-network perspective and switch-to-switch perspective (i.e., backbone network signaling for large scale/regional/national networks). There have been little (if any) significant enhancements for the local loop. ISDN and X.25 signaling are examples of local loop enhancements in which data and bearer services would be brought directly to customer-premise terminals. Unfortunately, depending on your viewpoint, ISDN and X.25 were either ahead of their time or were developed in a total vacuum of market input and industry coordination. Regardless of one's defense of or opposition to ISDN and X.25, the results have been low levels of implementation and forced reduced pricing to stimulate interest in terminal equipment development. This situation is a tragedy when one considers that ISDN development began in 1976, and even today in 2000, penetration is low.

When the loop was first constructed, it was designed to carry only voice conversations. The human voice operates in the 300 to 3,300 Hz range. The loop was designed to carry information (the human voice) at frequencies from 0 Hz to 3,400 Hz. This narrow frequency band of operation enables modems to function quite well; however, the modems can transmit (with a high level of data integrity) at speeds up to 57,600 bps. Data transmission speeds that are any higher will result in either noise or useless received data. Given the history of the loop and 120 years of embedded outside plant, the question involves how to bring T1/DS1 and higher types of services to the home and to small/medium-size businesses without changing out the loop.

There have been attempts to bring fiber to the curb, fiber to the home, fiber to the neighborhood, fiber around the town, fiber around the county, etc. The installation of fiber directly to a home would mean that the LEC would have a wide-band information pipe directly to the wall jacks in the home. Discussion of signaling protocol was not even a priority issue at the time that these efforts were being executed, as far back as the mid-1980's. The goal was to get bandwidth into the home or to small/medium-sized businesses, because there was a fear that cable television would provide voice, video, and data services by using its coaxial cable transmission systems. Fortunately for the ILECs, cable television deregulation came in time to cause many cable TV systems to enter bankruptcy. With competition disappearing, the overwhelming need to deploy bandwidth to the home in a expeditious manner disappeared. Therefore, to some extent, the ILEC's incentive to develop xDSL technologies also disappeared.

What has changed within the last few years is the Telecommunications Act of 1996 and the resulting growing competition from the CLEC community. To reiterate, the CLECs are driving xDSL deployment.

The first step in developing and deploying DSL technologies is to understand the physical limitations of twisted pair wire and then to remove the 3,400 Hz restriction that the industry placed on the local loop. What the industry came to understand is that twisted pair wire can carry high-frequency transmissions. The simple solution would be to install larger gauge wire. Larger gauge wire, however, is more expensive and results in higher per meter (or foot) outside plant costs. Replacing the existing twisted pair would be unrealistic. Over a long period of time (on the order of several years), it is possible that the entire loop will be replaced; however, if past experience is a measure of what can occur, twisted pair will be around for decades.

In general, transmitting high-speed data over twisted pair will present some challenges. The transmission properties of twisted pair wire (bundled in a multi-pair bundle of wire) are affected by the following items:

- Wire gauge
- Type of insulation
- Twist lengths

Twisted pair wire is typically either 26 *American Wire Gauge* (AWG) or 24 AWG, which is extremely thin wire. Metallic wire is similar to any other medium in that it attenuates energy. The type of insulation can serve as a dielectric medium, thereby resulting in a capacitive effect. The twisting of wire can create an inductive effect. The capacitor is a basic electrical equipment component that stores energy. The dielectric material serves to support the storage of electrical energy. This effect can have a deleterious affect on the facility's ability to transmit information. The twisting of wire can create an inductive effect. Inductance is a basic electrical engineering concept in which the change in energy transmission rates can be slowed due to interacting electro-magnetic fields. Inductance can impede the transmission of information. High-frequency signals are used to support high-speed data transmission. At a given power level, the high frequencies will travel a specific distance over a given diameter of wire. At the same power level but at lower frequencies, the signal will travel farther. In other words, high-frequency energy will dissipate faster than lower-frequency signals. Increasing the gauge of the wire would mean larger diameter wire and would result in lower electromagnetic resistance. As I said, however, this option is too expensive to consider.

Various forms of DSL technology exist; hence the abbreviation xDSL. Each version can be characterized in the same manner. Because xDSL is a transport/access technology, it is completely neutral as it relates to signaling protocols. xDSL employs the following generic system components:

- Signal splitters
- Customer-premise modems (specific to each type of DSL) to multiplex and de-multiplex the channels over the twisted pair. These modems are essential to providing xDSL services to a home or business.
- Transceiver units at the service provider's switching facility
- Transceiver units at the subscriber's end

Figure 3-18 illustrates this general configuration.

xDSL requires changes to the outside plant but is far less costly than a total replacement of the local wire. There are several different flavors of digital subscriber-line technology:

- *High Data Rate Digital Subscriber Line* (HDSL)
- *Very High Data Rate Digital Subscriber Line* (VDSL)
- *ISDN Digital Subscriber Line* (IDSL)
- *Single Line Digital Subscriber Line* (S-DSL)
- *Symmetric Digital Subscriber Line* (SDSL)
- *Asymmetric Digital Subscriber Line* (ADSL)
- *Rate Adaptive Digital Subscriber Line* (RADSL)

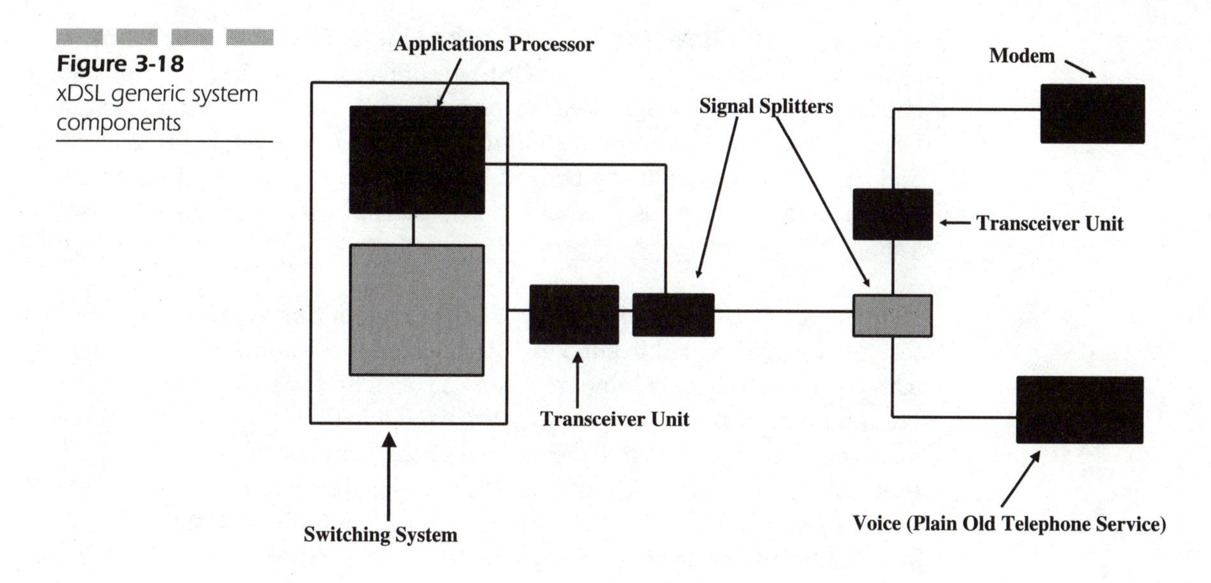

Figure 3-18 xDSL generic system components

Each one of these components was designed to meet certain perceived market needs. All versions of xDSL technology are vying for the position of de facto industry standard. At the time of this writing, all of these components are seeking some form of industry standardization from the ADSL Forum. The primary differences between the technology versions are the transmission distance between the switch and the customer premise and data rates that each supports. The distance issue is not a minor issue. The distance for which each technology can transmit directly impacts the carrier's (or CLEC's) cost structure. The data rate also has an impact on the cost structure of the CLEC; however, the reader should not take lightly the distance factor. The following subsections briefly describe the various versions of xDSL.

High Data Rate Digital Subscriber Line (HDSL) *High Data Rate Digital Subscriber Line* (HDSL) technology was the first of the xDSL technologies developed. HDSL was created in the 1991–1992 time frame. The goal of HDSL was to deliver T1 services/capabilities directly to the home over two pairs of twisted wire (four wires).

HDSL supports the splitting of the T1 bit rate (1.544 Mbps) over the two pairs of wires. Splitting the total bit rate across the four wires versus two wires enables the carrier to transmit lower frequencies across a single wire (in the four-wire configuration) versus higher frequencies on a single wire (in the two-wire configuration). See Table 3-1 for data rates.

Very High Data Rate Digital Subscriber Line (VDSL) *Very High Data Rate Digital Subscriber Line* (VDSL) technology splits the transmitted bandwidth across a single twisted pair. VDSL is an asymmetric form of DSL. There is an *Asymmetric Digital Subscriber Line* (ADSL) technology, and VDSL is a variation of ADSL. The application for VDSL is in the area of high bandwidth-intensive applications, such as high-definition video. See Table 3-1 for data rates.

ISDN Digital Subscriber Line (IDSL) *ISDN Digital Subscriber Line* (IDSL) technology was created to support ISDN access in the home. IDSL is a dedicated service that only supports data (not video). IDSL does not support the full range of ISDN services that are available under ISDN. Like ISDN, however, IDSL requires a separate line to the home or business. From an execution perspective, IDSL utilizes the same modem and transceiver units as the other forms of xDSL but also uses standard ISDN terminal adapters at the subscriber's premise. See Table 3-1 for data rates.

Table 3-1

XDSL Capabilities

xDSL Comparison Table

XDSL Technology	Data Transmission Rate and Rate Ranges	Directionality/ Mode	Loop Range Estimates: Distances Between Customer and Switch	Number of Wire Pairs
HDSL	1.544 Mbps (each pair will transport data at a rate of 784 kbps)	Two Way (Duplex)	10,000 to 12,000 feet	2
VDSL	1.544 Mbps to 2.3 Mbps 13 Mbps to 52 Mbps	Upstream Downstream	4500 feet	1
ISDL	128 kbps	Duplex	18,000 feet	1
S-DSL (HDSL2)	1.544 Mbps	Duplex	12,000 feet	1
SDSL	1.544 Mbps	Duplex	10,000 feet	1
ADSL	1.544 Mbps to 2 Mbps 1.544 Mbps to 6.1 Mbps 16 kbps to 640 kbps	Downstream Downstream Upstream	18,000 feet 12,000 feet 18,000 feet	1
RADSL	640 kbps to 7 Mbps 128 kbps to 1 MBps 640 kbps	Downstream Upstream Duplex	12,000 feet 18,000 feet 18,000 feet	1

Single Digital Subscriber Line (S-DSL) *Single Digital Subscriber Line* (S-DSL) is a variation of HDSL and is also known as HDSL2. S-DSL supports HDSL over a single twisted pair. Data rates are different for downstream (service provider to the subscriber) and the upstream (subscriber to the service provider), however. See Table 3-1 for data rates.

Symmetric Digital Subscriber Line (SDSL) *Symmetric Digital Subscriber Line* (SDSL) is also a form of HDSL. Data rates are symmetric for the downstream and upstream applications, however. According to the ADSL Forum, SDSL is a version of HDSL and *Plain Old Telephone Service* (POTS). See Table 3-1 for data rates.

Asymmetric Digital Subscriber Line (ADSL) At the time of this writing, *Asymmetric Digital Subscriber Line* (ADSL) is currently the more popular version of xDSL. Equipment to support ADSL is available and is being deployed widely not only by CLECs, but also by ILECs. ADSL appears to be the middle ground that I have alluded to between technology and cost-benefit. ADSL is capable of supporting video, voice, and data. True, the other versions of xDSL can support the same services; however, the difference is that the technology is available for the CLECs that are seeking to jump into business now (and not tomorrow). More often than not, the deciding factor in one technology's victory over another is simply availability. Because the technology is being deployed and is therefore generating profit for the manufacturer and for the service provider, the cost of the technology is in fact lower than other xDSL technologies.

The asymmetric nature of ADSL technology enables the service provider to run three separate channels over the same wires. One channel carries voice (i.e., POTS). The second channel carries signaling at a rate of 16 to 640 kbit/s to the network. The third channel is a higher-speed data connection that operates at a rate ranging from T1 (1.544 Mbps) to 9 Mbps. The second channel is used for uploading information from the subscriber to the network. The third channel is used for downloading information from the network to the user/subscriber. To a CLEC, you are getting the most out of the pair of wires; instead of two channels being supported, three channels are being supported. To an ILEC, you are also getting the most out of the pair of wires, and you are enabling the ILEC to rapidly meet the challenges presented by competition. More information about this topic will appear later in the book.

From an execution perspective, ADSL can be provided on a one-on-one basis. In other words, the necessary equipment upgrades can take place one subscriber at a time. Upgrades to a neighborhood or to a community can therefore take place in a controlled fashion.

ADSL deployment requires the following equipment to be installed:

- *ADSL Transceiver Unit Central Office* (ATU-C)
- *ADSL Transceiver Unit Remote* (ATU-R)
- Discrete Multi-Tone *(DMT) Modem*—If fiber to the home or business is available, then the DMT modem is not needed. A modem will still be needed at both the subscriber and service provider locations, however, to multiplex the channels to the customer.
- ADSL-supportive repeaters of the customer should be farther than 18,000 feet from the carrier's switching facility.

What is interesting to note is that both the CLECs and the ILECs see the same applications for ADSL. Given the current and future state of competition in the local loop market, the ILECs see ADSL as a way of finally converging voice, video, entertainment, data, and Internet access business opportunities by using the existing infrastructure (enhancements notwithstanding, the infrastructure remains largely unchanged). An interesting twist to the local loop scenario is that CLECs are also deploying the same technology directly to the subscriber by leasing dark/dry loops (no controlling electronics) from the ILECs and by deploying their own ATU-Cs in the ILEC central office switching facility. Collocation of CLECs in ILEC buildings is enabled under the Telecommunications Act of 1996. This procedure is analogous to inviting your own competitor to set up shop in your own storefront.

Rate Adaptive Digital Subscriber Line (RADSL) *Rate Adaptive Digital Subscriber Line* (RADSL) is a version of ADSL. The difference is that RADSL uses intelligent ADSL modems that are designed to sense the performance of the loop and to accordingly adjust transmission speeds. Effectively, RADSL supports both asymmetric and symmetric transmission of data.

Disadvantages of xDSL Despite the apparent advantages of xDSL technology, there are also disadvantages:

- To upgrade to xDSL, telecommunications operators need to install xDSL equipment at both ends of the loop that connects the subscriber and the central offices.
- A potential drawback for customers, however, is that equipment from different vendors is likely to be incompatible.
- Furthermore, there will be a need to make some modifications or changeouts of repeaters on the loop. This process leads to additional costs.
    - The cost occurs at both ends of the loop. Carriers need to answer the questions, "Who will pay for this?" and "How will the service be paid for?"

Advantages of xDSL The advantages, like the disadvantages, impact both the ILECs and the CLECs:

- xDSL technology can assist a wireline telephone company with keeping its subscriber base from jumping to another carrier, such as a cable TV company.
- Furthermore, most lines to homes and small businesses are still analog; therefore, the ILEC can continue gaining value from existing outside plants. xDSL provides a way of serving the subscribers with minimal impact on the service provider's network.
- CLECs might find technologies such as ADSL useful in starting up their networks quickly and in a cost-effective manner.
  - CLECs face high startup costs and short market windows. xDSL offers a way of providing a competitive service to the incumbent LECs. Unfortunately, xDSL should not be considered the answer to the CLEC startup's problems. There are so many other issues involved with starting up a company. xDSL enables a CLEC or other carrier to differentiate itself by providing high-speed data services to the subscriber.

SUMMARY

The network can be broken into specific functional categories:

- *Private* LANs, closed user groups, etc.
- *Local* Including towns, cities, and counties
- State
- *Regional/sector* Including multi-state network views and even national regions (East coast, Southeast coast, etc.)
- National
- *International* Addresses interconnectivity between nations

By looking at the network in a geographic, functional manner, one can begin understanding at a macro level how the network (the network of networks) functionally relates to itself and to other networks. The transmission of the information across the networks requires specific technology.

The transmission technologies used can be employed by any variety of carriers: wireless, wireline, Internet, data, and even satellite. These technologies and routing schemes can be used to interconnect different carriers for the purposes of completing calls/transactions/information delivery, for providing services to the subscriber, and for providing wholesale service to other carriers (more on this topic will appear in later chapters).

To a great extent, the network of networks can be described as similar to the way that a highway, street, and road system is set up. There are high-speed roadways that interconnect communities (a network backbone). There are streets that run through a town or city (interoffice, interswitch connectivity). There are also local roads leading to homes and businesses (the local loop).

Understanding the relationships between the various networks and the elements within a network is key to understanding how interconnection can be used to benefit a carrier. Figure 3-19 is a high level illustration of the network of networks.

The next step that needs to be taken is to look at the language that the networks will use to speak with each other.

Figure 3-19 The network of networks is comprised of multiple networks; interconnected to one another at various levels.

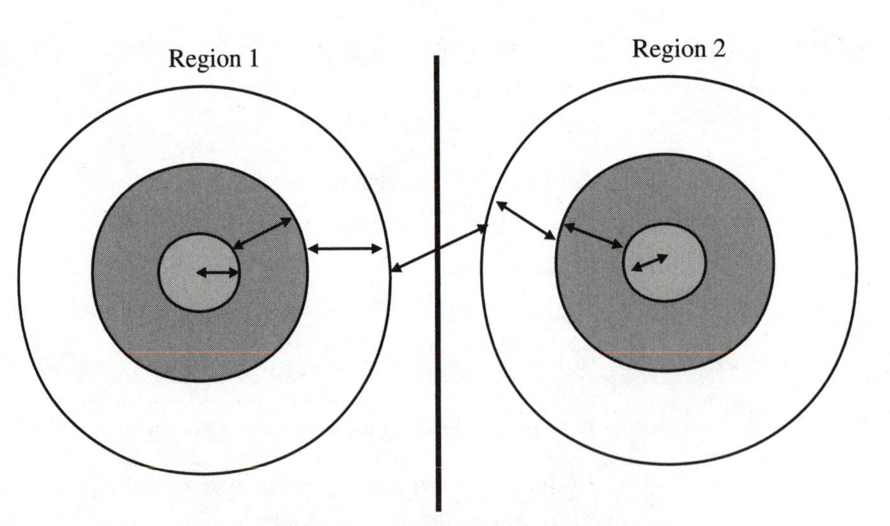

Network Signaling and Its Applications

One of the binding forces of network interconnection is signaling. The network signaling I am referring to is the language in which service providers communicate with other service providers and with network elements within their own network. A more technically precise definition would be that signaling is the exchange of information in a telecommunications network (public and private) that establishes and controls the connection of a call or communication between subscribers or computing systems and also the transfer of subscriber/end-to-end and management (subscriber and network) information. In Chapter 1, "What Is a Telecomunications Network?", I focused on describing *Multi-Frequency* (MF) signaling and *Signaling System 7* (SS7) because they are the dominant network signaling protocols that are used by public telecommunications networks today. The reader should not be swayed into believing that there are no other signaling protocols, however. In fact, there are a number of industry-standardized signaling protocols. At this time, the most popular signaling protocols are as follows:

- *Multi-Frequency* (MF)
- *Signaling System 7* (SS7)
- *Asynchronous Transfer Mode* (ATM)
- *Transfer Control Protocol / Internet Protocol* (TCP/IP)
- Frame Relay
- *Integrated Services Digital Network* (ISDN)

Other types of signaling are just as important. In general, signaling can be defined in the following ways:

- Circuit Condition
  - *Address* These signals convey call destination information, or the digits that are dialed by the calling party. There are several types of address signaling, including Dial Pulse and *Dual-Tone Multi-Frequency* (DTMF).
  - *Supervisory* This type of signaling indicates the status of the line or trunk. In other words, supervisory signaling indicates whether or not there is a busy (off-the-hook) condition or an idle (on-hook) condition. Several types of supervisory signaling exist in the landline telephony world, including loop, reverse battery, and E&M supervision.
  - *Alerting* These signals alert the user to an incoming call. In other words, this ringing is what the user hears and is also known as power ringing in the landline world.

  - *Call Progress* These are the tones that the calling party hears when making a call. These tones inform the calling party of the various events that are occurring during the course of the call.
  - *Control* Control signals are used for special auxiliary functions that are beyond a service provider's network. These signals communicate information that enables or disables certain types of calls. One example would be call barring.
  - *Test* Test signals are used by carrier personnel to check for network health and to check the network's status.
- Subscriber/User Initiated
- Intersystem
  - Intranetwork interswitch and internetwork interswitch

These subcategories are largely associated with the traditional telephone network. Every network, however, supports some form of circuit signaling to maintain system heartbeat, intercomputing system communication, or interface with a user.

The following sections will demonstrate some of the commonalities that I have just described. The reader should view the next sections with a broad view toward developing the understanding that from a logical/conceptual view, all networks behave similarly and have common attributes.

We shall first address the most common signaling category that is of interest to all types of carriers: intersystem signaling. Not all systems support the same type of intersystem signaling, but they do need to communicate with each other if the user groups of different networks hope to exchange information.

Intersystem Signaling

Multifrequency (MF) Signaling

MF signaling uses tone pulsing in the voice frequency range. MF signaling's telecommunications history begins in the wireline telephone world. MF signaling (tone pulsing) transmits digits by combining two of six frequencies. Each combination of tones represents a single digit. As indicated, MF tones are pulsed in the voice frequency range, which means that the tones are sent over the talking path. MF signaling is known as in-band signaling.

Both the voice and the telephone call's control signaling and the called telephone number are transmitted over the same physical transmission facility.

MF signaling utilizes tones in the 700 through 1,700 Hz range. When a MF tone is transmitted (representing a digit), two frequencies are sent simultaneously for each digit. The tones I refer to are not audible to the subscriber who is making the telephone call.

The following table illustrates the MF signaling code. See Table 4-1. Note that supplementary tones called KP and ST are also supported. KP stands for request for Key Pulsing Sender, which refers to the beginning of a pulsing sequence. ST (Start) stands for the completion of keying and the start of circuit operations. There are additional frequency combinations used to support certain types of telephone calls. See Table 4-2.

As you can see from Table 4-1, 10 of the frequency combinations are used to support dialed digits zero through nine. Additional types of KP and ST tones also exist, as Table 4-2 describes.

The KP2 tone is used in a non-North American telephone system. The same frequency combination for KP2 is also used by ST2prime tones for an

Table 4-1

MF signaling code

Digits	Frequencies (Hz)
1	700+900
2	700+1100
3	900+1100
4	700+1300
5	900+1300
6	1100+1300
7	700+1500
8	900+1500
9	1100+1500
0	1300+1500
KP	1100+1700
ST	1500+1700

Table 4-2

Frequency combinations—additional tones

Digits	Frequencies (Hz)
1	700 + 900
2	700 +1100
3	900 + 1100
4	700 + 1300
5	900 + 1300
6	1100 + 1300
7	700 + 1500
8	900 + 1500
9	1100 + 1500
0	1300 + 1500
KP	1100 + 1700
ST	1500 + 1700
KP2 (ETSI)	1300 + 1700
ST2prime (ST'')	1300 + 1700
ST3prime (ST''')	700 + 1700
STprime (ST')	900 + 1700

operator services system. The STprime and ST3prime tones are used to support operator services activities and coin telephone operations.

For wireless carriers and Internet service providers, this information is not necessarily useful. The wireline CLECs that need to support interconnection to the ILECs and maybe even manage their own coin telephones, however, need to support this type of signaling. Not every carrier is capable of supporting SS7 or some other type of common channel-signaling scheme.

Multi-Frequency (MF) Applications MF signaling is used to connect wireline switches in an in-band signaling environment. MF signaling is not DTMF signaling, which will be described in later chapters. MF is an interconnectivity signaling protocol. The reader might consider the use of the word protocol in describing MF signaling as going too far. The reader, however, should remember that a protocol is a communication procedure, a language, and a set of directions. MF signaling communicates information in the form of tones, rather than in the form of bits, bytes, or computer code. MF signaling only transmits address information. This address information is a cornerstone for applications that are supported by MF signaling.

Figure 4-1 describes how MF is used.

Most service providers are converting their respective networks to a network that is based on common-channel signaling. Advantages exist to separating the voice and control signaling; however, there are advantages to and applications for MF signaling.

The following applications are supported by MF signaling. The applications do not exist because of MF; rather, the applications are all tone based —and interswitch MF signaling is tone based.

- Plain Old Telephone Service *(POTS)* As unusual as this fact might sound, the reader must remember that there are millions of people who just want to purchase basic voice services. The challenge for all carriers is to increase revenue from this type of subscriber by selling services over and above basic voice.
- *Computer modems* The modems in home computers and laptops convert digital information into tones (i.e., analog sine waves) that can be sent over the standard telephone transmission facility. This statement does not cover computers using ISDN or other types of protocols.
- *Operator Services* Inband interaction with the automated menu systems are used by many carriers and are called keypad menus.
- *Directory Assistance* Inband interaction with automated menu systems are used by many carriers and are also called keypad menus.

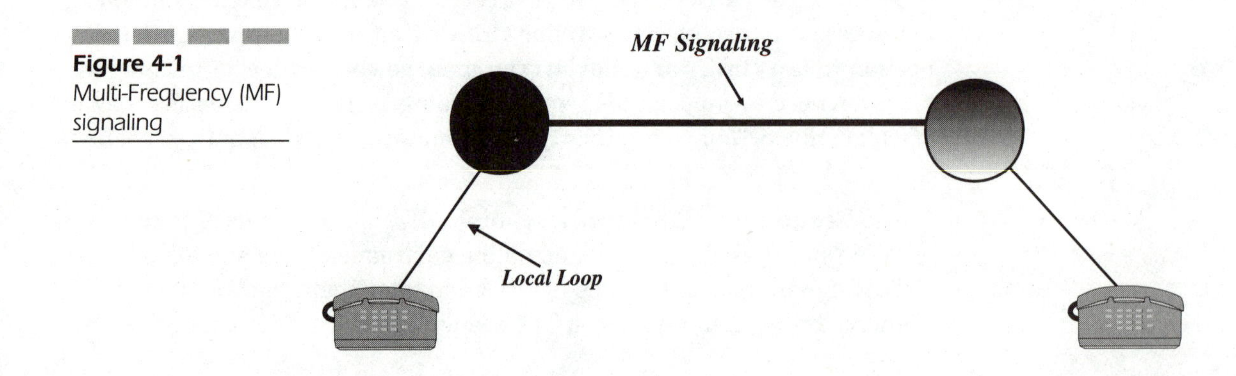

Figure 4-1 Multi-Frequency (MF) signaling

- *Non-caller identification numbering services* Such services include Call Forwarding, switch-based voice mail, Call Waiting, and Call Hold (terminal support is needed).
- *Terminal-based services that support subscriber interaction via tones* Customer-based voice mail systems are accessed and navigated by the user via the tones that are generated by touching the keypad of the telephone device. These tones are called *Dual Tone Multi-Frequency* (DTMF) tones, but they are carried over MF trunks (transmission facilities). DTMF tones are the signaling tones between a customer's telephone set and the switch. DTMF will be explained later in this chapter.

Although MF signaling networks are being replaced with common channel-signaling networks, there are future applications for MF beyond what I have described. These future applications are as follows:

- *Interconnection* Interconnection is a current application; however, the reader should keep in mind that there are a number of small wireline carriers that will not have common channel signaling-supported switches for years. These small carriers will still need to interconnect to larger carriers; therefore, the larger carriers will need to maintain MF support.

We are not being facetious. The fact is that there are many people (definitely on the order of thousands of people) who do not even own a telephone. In many cities, there are businesses that cater to people who make long-distance and local voice calls on a pay-by-the-minute basis. In fact, in large cities a single switch or multiple switches will serve an entire neighborhood that cannot even afford more than basic voice. There is nothing bad or wrong about providing only voice service. You should remember that the subscriber base's requirements will drive change in the network. In this case, the incentive to change the switch will have to be based on either future expectations or on operating costs. Operating costs of a MF system might rise as time passes and as vendor support diminishes in lieu of supporting newer and higher profit-margin technologies.

The disadvantages of MF are that MF signaling cannot support many of the SS7 services in the marketplace today. These services are all of the caller identification types of services and all of the SS7 database capabilities.

MF is a signaling method that can continue to support a variety of services, as long as the economic drivers to change the network do not exist.

Signaling System 7 (SS7)

Signaling System 7 (SS7) is a *Common Channel Signaling* (CCS) protocol and an out-of-band signaling scheme. SS7 is the dominant form of CCS in North America. The signaling used in previous examples is circuit-associated/in-band signaling. MF pulsing is tone based, while SS7 is actually a computer message-based signaling protocol. In MF pulsing, the signals that are used to set up facilities between the called and calling party occur on the same physical path as that of the conversation. Under SS7 signaling, the information associated with setting up the facility path for the conversation is different than the conversation path. Not only do new subscriber services exist, but network efficiencies are also enabled by SS7—which we will address later in this section.

Prior to SS7, there was *Signaling System 6* (SS6) or CCIS 6-supported toll traffic (i.e., long distance) in the pre-divested AT&T days. SS6 is a CCS protocol that supports 4.8 kbps transmission and also represents the first packet data application in the public telecommunications network. The development of SS7 enabled wireline carriers to increase the data transmission speed to 56 kbps. SS7 also extended the strengths of CCS to the user in the form of real-time, subscriber-controlled services and caller identification services.

The network architecture is associated with SS7 systems (refer to Figure 4-2).

SS7 Benefits As I had indicated, CCS—specifically SS7—is the dominant form of CCS that is used in North America. SS7 is currently being installed in wireless and wireline CLEC networks in order to interconnect to other networks and to provide services to their respective subscribers. There are specific network benefits to deploying an SS7 network, however. These benefits are not obvious to a subscriber, and they include the following:

- *Faster call setup* Subscribers who are calling within a local calling area (e.g., the other side of a town or a city, or the other side of a state) will not perceive a decrease in the amount of time needed to establish a connection between subscribers. In the case of a coast-to-coast call, the users can expect between a three to four-second decrease in call setup from the time that the calling user dials/enters the last digit to the time that the calling user hears ringing on the telephone. Users might not believe this statement, but when one compares how long it took to hear audible ringing (the ringing that the calling user hears) prior to the early 1980s, call setup is much faster today.

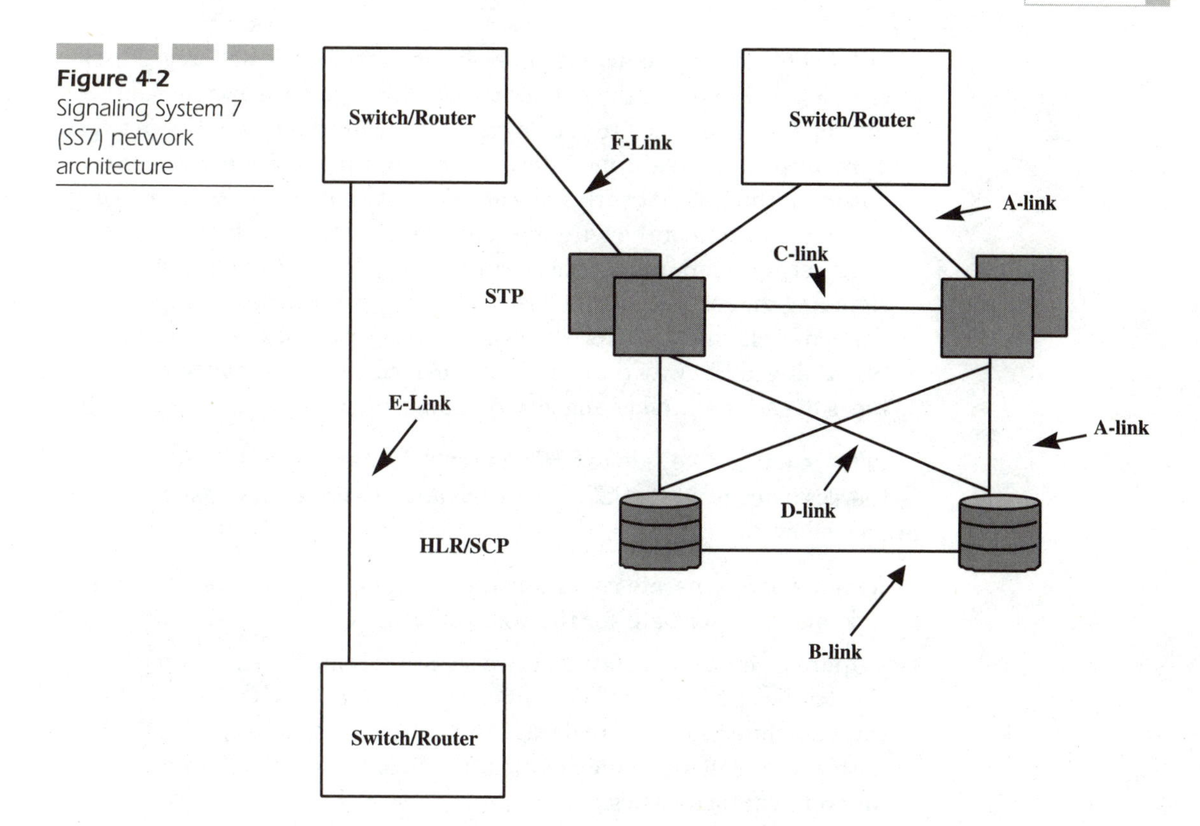

Figure 4-2 Signaling System 7 (SS7) network architecture

- *Optimized use of transmission facilities* Prior to SS7, end-to-end CCS did not exist in the public telecommunications network. The lack of end-to-end CCS meant that a transmission facility had to be physically set up between the originating point and the terminating point. In order to determine whether the called subscriber was on the line, the transmission facilities between the calling parties had to be completely set up. If the called party was busy, the originating party's switch would be informed via the line status signaling tone, and the calling party would hear a busy tone. SS7 avoids this call setup scenario by enabling the carrier to look ahead via the data link—thereby avoiding the unnecessary and wasteful expenditure of time and resources in setting up transmission facilities (to support the voice path) across the nation.
- *Database querying and management* Using databases in a wireline telephone environment existed prior to SS7. Databases had stored customer service profile information, billing information, and internal carrier operation information. However, none of this information could

be used to execute an activity in real-time fashion (at will, regardless of time). SS7 enabled the carrier to use all of the aforementioned information to be used to set up calls and to even manage the network in real time. The real-time information accessing and management nature of the SS7 network was considered to be a revolutionary event in some quarters of the wireline (public telecommunications) network engineering community. Others simply consider SS7 or out-of-band signaling an event that was destined to occur. Regardless of one's opinion of the nature of SS7's revolutionary or evolutionary nature, SS7 changed the way that carriers provided services, changed the types of services, and changed the management of the network.

Is there a negative side to SS7? As revolutionary or evolutionary as SS7 is, there is a down side to SS7. The disadvantages or negative aspects of SS7 are as follows:

- Requires two separate communications paths: one path for the data link, and another path for the voice/information content
- Requires traffic engineering of two separate communications paths. The traffic engineer needs to understand that each of the two paths is carrying different types of data. One path might be voice. The other path will be call setup information. Voice and call setup data have different characteristics.
- Maintenance of two separate physical transmission networks, which requires additional time and money on the part of the carrier
- Additional network elements to engineer, install, and maintain. These network elements are the *Signal Transfer Point* (STP) and the *Service Control Point* (SCP).
- SS7 does not enable high-speed data. The maximum transmission speed of an SS7 data link is 56 kbps. As a high-speed data link, a SS7 link is a poor choice.
- SS7 does not support the new large-bandwidth multimedia services that are envisioned.

SS7 was really an evolutionary event in the growth and development of the public telecommunications network. Out-of-band signaling existed for some time before the first commercial deployment in the wireline local loop. It was only a matter of time before the ILECs could economically justify the enormous capital expenditures in the local loop to support out-of-band signaling in the loop. During the early and mid−1980s, the various ILECs were struggling to deploy SS7; business cases could not support

SS7 on services alone; network efficiencies did not provide the economic incentives; and visions of the future were not enough. The fact is that a number of carriers made strategic choices to deploy based on a combination of all of the aforementioned factors—plus an understanding that the industry was undergoing some type of change that no one could entirely understand.

The local market deployment of out-of-band signaling was influenced by the following elements:

- Market need for new services
- Availability of switching equipment that could support out-of-band signaling. The equipment existed for long-distance switching; however, the economic drivers that were commercially available in the local switching business had not existed for a long period of time.
- Need for greater network efficiencies
- Computing technology had grown and was advanced enough to enable the creation of switches that had tremendous computing power, causing the switches to become software-based as opposed to hard-wired. The switch's software architecture facilitated the development of new types of services.
- A changing regulatory environment, which created uncertainty and opportunity. The opportunity facilitated the growth of competition, and the competition encouraged innovation and more change.

I have just described a cyclical process and vicious cycle that fed on itself and grew. The telecommunications industry was (and still is) in this environment of change. SS7 does not represent the end of the network's development. There are faster and more robust signaling protocols, but one should not overlook the following facts and contributions:

- SS7 deployment is now widespread.
- SS7 introduced CCS into the public network.
- SS7 meets the needs of the mass marketplace for voice communications.
- SS7 is the foundation of the Intelligent Network.

SS7 is the basis for all intelligent network platforms. Wireless systems (cellular and PCS) are installing enhanced-service platforms, many of which require SS7 as the basic signaling protocol for their networks. I use the word "enhanced" to describe a network or equipment platform that is designed to add value or capabilities to a carrier's existing primary-switching matrix. Enhanced systems have some type of intelligence; therefore, the terms "enhanced platforms" and "intelligent networks" tend to be used interchangeably although they are not precisely the same. The Intelligent Network will be described in greater detail in Chapter 5, "Applications."

The following section will describe the SS7 network architecture. These concepts had been briefly discussed in Chapter 1. We will continue the discussion with an eye towards gaining an understanding of how the SS7 network served as a basis for the current evolutionary state of the network of networks, which supports intelligent networking. Such a discussion will assist the reader with gaining a greater understanding of how and why networks evolve.

SS7 Network Architecture The SS7 network architecture consists of a set of network elements that are generally referred to as nodes. These nodes are also called *Signaling Points* (SPs). A switch is a SP, and these SPs are interconnected via point-to-point signaling data links that support transmission rates of 56 kbps.

Figure 4-3 illustrates how two separate SS7 networks can interconnect The signaling links shown are as follows:

- *Access link (A-link)* Used for *Switch-Signal Transfer Point* (STP) signaling connections and STP-HLR (Home Location Register)
- *Bridge link (B-link)* Used to connect STP-STP in other networks
- *Cross link (C-link)* Used to connect diverse STPs of the same hierarchical level
- *Diagonal link (D-link)* Used to connect STPs of different hierarchical levels or diverse STP-HLRs/SCPs (Service Control Points)/databases combinations
- *Extended link (E-link)* Links between switches and non-home STPs
- *Fully Associated (F-link)* Optional links between switches

The architectural configuration in Figure 4-3 illustrates how two separate networks can be interconnected via SS7. The SS7 network is composed of the following three key network elements:

- The STP relays messages at the network layer between switches. STPs are paired and are a crucial part of SS7 network design. STPs contain the routing templates used by carriers to route calls, and these templates are called decision matrices.
- The SCP is the database that contains the subscriber profiles. SCPs can also provide assistance with routing a call.

Figure 4-3
Interconnected SS7 networks

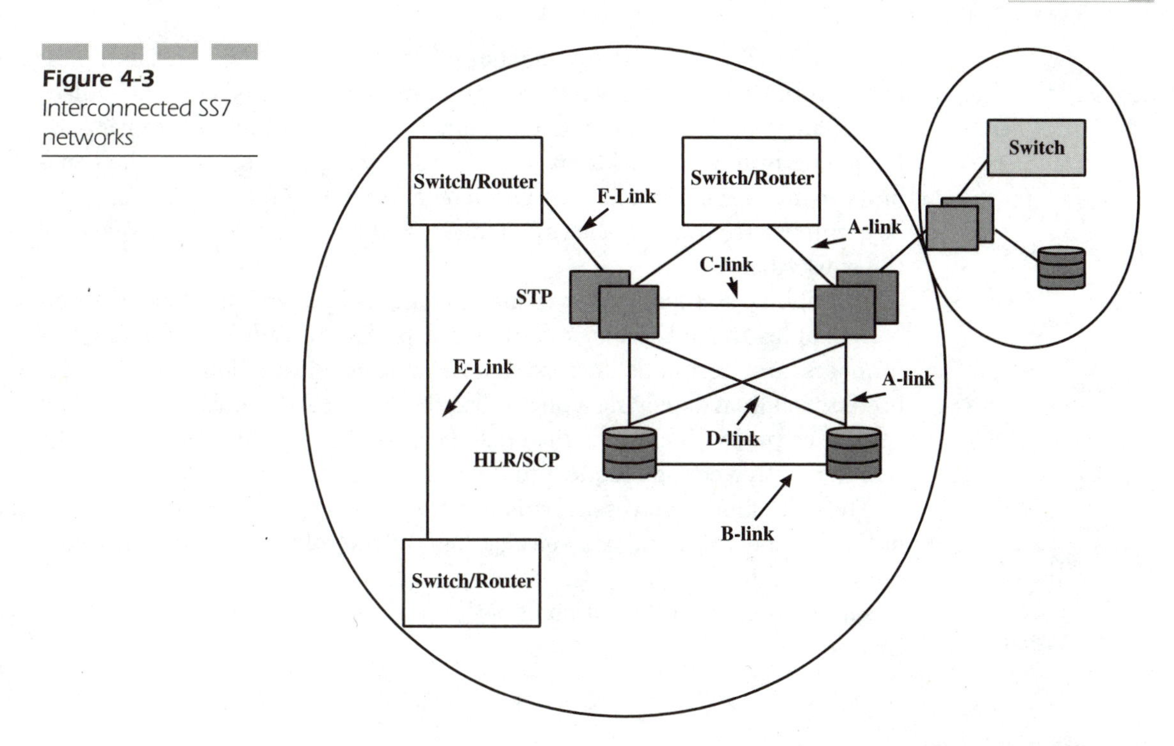

- The SSP is the switch (or routing element) in the SS7 network. The switch would be a Class 5 central office in the wireline network (LEC). In the cellular or PCS world, the SSP would be the wireless switch/mobile switching center.

The STP and SCP are shared network elements. In other words, a carrier usually has a number of switches that are subtended from a single SCP. One pair of STPs will be interconnected to multiple switches. The STP has a similar function in the SS7 network, as does a wireline carrier's tandem switch. The STP serves as a traffic cop for network data messages.

The SCP is an outboard database that serves multiple switches. The SCP is more than just a repository of customer data; it is also a repository of switching instructions. The SCP plays a key role in routing calls. Unlike its predecessors, internal switching databases, SCPs are the intelligence of the SS7 network. The intelligence I refer to has nothing to do with the capability to switch or process a call. The intelligence I refer to is the customer profiles, routing tables, time of day and day of week routing instructions, and billing information stored in the SCP. One can argue that the predecessors of the SCP were also the intelligence of the network; however, the intelligence in those databases was a combination of hard-wired circuits, nailed-up circuits, and dedicated circuits and could only support routing by telephone number translation. The SCP's routing tables enable the capability to make routing decisions based on a number of factors, such as time, day, number from which the call is made, and real-time decisions made by the subscriber.

The ability of a subscriber to make changes in a screening list of numbers enables the subscriber to either accept, bar, or redirect calls to other numbers and terminal devices. This real-time interaction was a major breakthrough in deploying a new subscriber service. Up until that point, all service changes—including a service such as Call Forwarding—had to be handled with a change request to the telephone company.

The SSP simply processes both signaling messages for the data portion of the SS7 network and the voice content of the telephone call or wireless telephone call.

Figure 4-4 is an illustration of both the signaling and voice links in the SS7 network.

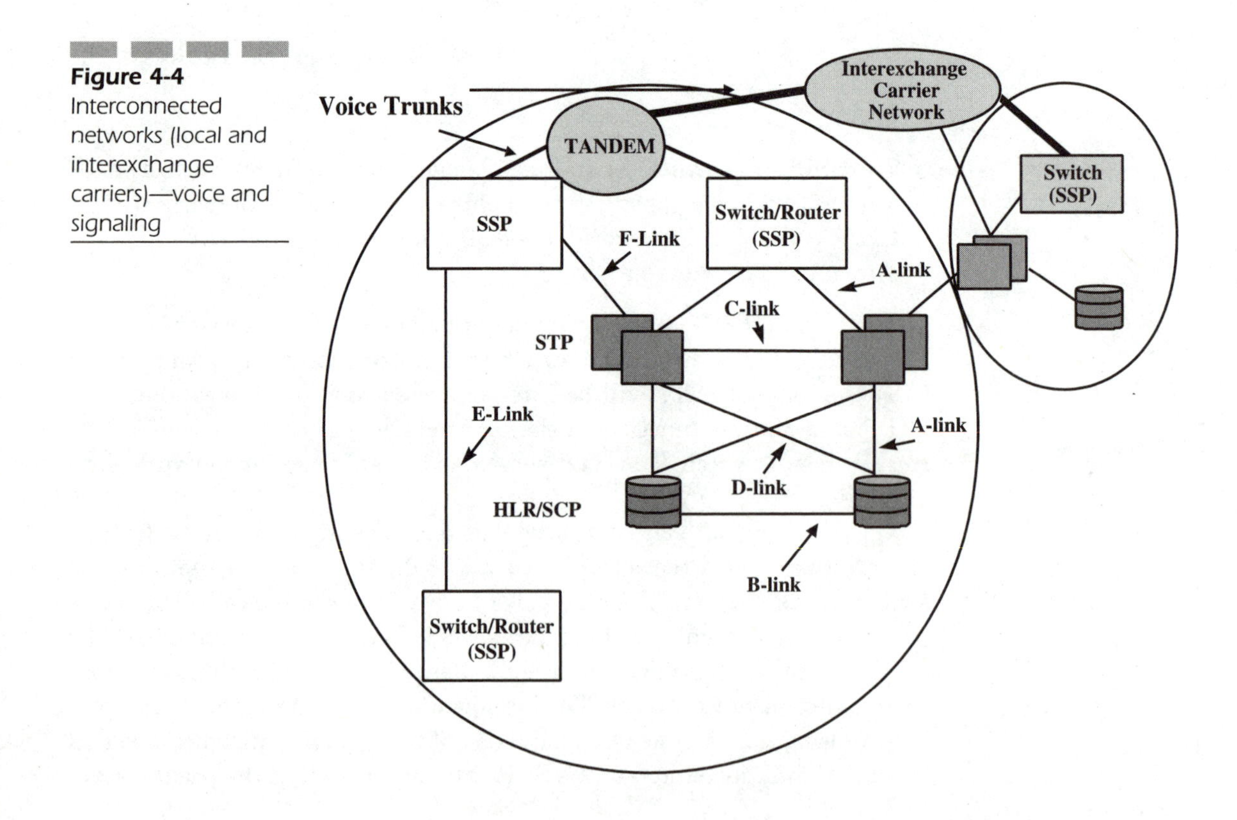

Figure 4-4 Interconnected networks (local and interexchange carriers)—voice and signaling

SS7 networking is hierarchical in nature, similar to the traditional in-band signaling network. There is an orderly way in which calls are processed and transmitted and received, and more detail will be provided in Chapter 6, "Wireline Telephone Networks."

The next section addresses *Asynchronous Transfer Mode* (ATM), which is a high-speed out-of-band signaling protocol that to some represents the next evolutionary step in the growth of the public telecommunications network.

Asynchronous Transfer Mode (ATM)

Asynchronous Transfer Mode (ATM) is a broad-band network protocol. ATM is a part of the larger set of broad-band ISDN procedures that are standards in the industry. ATM can exist without ISDN, however. ATM receives a great deal of attention, because it is considered to be the future of broad-band and integrated data, voice, and video in the public telecommunications network. ATM is currently being deployed, but like so many of the network signaling protocols that are in use today, its development began in the days before the divestiture of the AT&T Bell System.

ATM had been envisioned as a way of interconnecting ISDN customers. Given the complexity of developing ATM, therefore, SS7 became the signaling network envisioned to interconnect ISDN customer groups. A brief explanation is needed to explain my view of the complexity behind ATM development.

ATM is a backbone-signaling protocol designed to interconnect networks; however, ATM involves the asynchronous transmission of signaling information. The public telecommunications network was synchronous and connection oriented in the late 1970s and largely still is today. To deploy a network that did not even support the most basic aspect of signaling transmission would mean a fundamental change in the way in which calls were transmitted, received, and processed. For many, asynchronous transmission was simply too radical a change in the public telecommunications network. Any change in the public telecommunications network would result in financially costly change in every switching system in the public telecommunications network. The market and financial needs to drive more bandwidth in the network created an overwhelming wave of support to accelerate ATM development and deployment. ISDN no longer served as a driver for ATM; instead, the general need to create value both from a consumer perspective and from a carrier financial perspective are what drive ATM today.

Asynchronous Transfer Mode (ATM) Signaling Description As I had noted, ATM is a network signaling protocol that was developed in support of *Broadband-ISDN* (B-ISDN) and combines packet-switching concepts with traditional switching concepts. ATM is one class of packet technologies that relays traffic via an address contained within the packet's header. ATM transports information by using short, fixed-length packets or cells. These cells are asynchronous time-divisioned multiplexed. ATM is both connection oriented and connectionless oriented.

Packet switching is a method of information delivery in the network. A packet switch sends the information in packets of data. The information is dissembled into discrete packets of information/data. Each packet has a unique identification, and each packet carries its own destination address; therefore, each packet can travel through the network independently of each other. Each packet of data will travel to the final destination/termination point via different routes. Because each packet will travel different routes, the likelihood is high that the packets will arrive at different times and will arrive out of sequence at the destination.

Packet switching essentially establishes a virtual connection between the originating and terminating points. Packet switching is connection oriented. Because large files are broken down into packets of data and the data is transported in this virtual manner, packet switching is an efficient way of transporting large data files between networks. Early packet networks use the X.25 transport protocol. Because the X.25 was and still is a standardized industry protocol, both public and private packet-switch networks were able to communicate with each other. ATM is replacing X.25 as the protocol of choice (for packet networks).

Connection-oriented calls are information transactions such as the typical voice call, ATM calls, Frame Relay calls, ISDN calls, packet-switched calls, and all circuit-switched calls. Connection-oriented calls are characterized by the following steps/processes:

- *Call Setup* Includes a request for the circuit and acknowledgment of call receipt
- Information/Call/Data transmission
- Call Teardown/Release

Connectionless-oriented calls are information transactions that do not involve a formal acknowledgment of call completion/answer supervision/circuit establishment. Delivery of the information is not guaranteed or reliable. Internet transmissions are connectionless transactions.

ATM combines the best of both packet switching and connectionless switching. ATM is similar to SS7 because it is a backbone network signaling protocol designed to transmit information at high speeds. ATM has the potential of binding the network of networks into one virtual and integrated network. The question is, then, "Why would anyone even think of applying ATM in such a manner?"

ATM Advantages The following are perfect reasons for supporting ATM. In a world where there is a multiplicity of networks, ATM appears to be a good candidate for integrating these networks and their services, including wireless networks and services.

- *Granular control and network flexibility* The processing of cells provides greater network control over routing, error control, flow control, copying, and assigning priorities. Note that in an ATM network, each cell can be assigned a delay or loss priority; therefore, the service provider can exercise control over one traffic class relative to another class.
- *Dynamic bandwidth allocation* ATM has the capability to dynamically allocate bandwidth to all types of cell traffic.
- *Service Transparency* ATM is service application-transparent. The cell size is a compromise between the short, repetitive frames of voice and the long frames for data. Therefore, ATM enables mixing data, voice, and video within an application. ATM's granular control of cells enables the network to be customized to a specific application.
- *Scaleability* An ATM cell can be carried over a 45 Mbps link to a switch and can be changed into a 2.4 Gbps SONET link. ATM is blind to elements such as rates or framing.

ATM Disadvantages There are reasons why ATM has not been aggressively deployed as of this writing. The reasons are in part due to inherent aspects of ATM. The disadvantages of ATM are as follows:

- ATM lacks congestion control. Congestion control is the capability of the network to manage large volumes of traffic in a manner that avoids network information congestion.
- ATM can also suffer from variable cell delays caused by random queuing delays at each switch. ATM cells are stored in queues in the ATM switch, awaiting cell assignment.

There are other issues involving ATM that are not problems, but they are matters/concerns that should be addressed from a non-design perspective. This perspective could be business and implementation-related. The following list describes these other matters:

- Remember that ATM stands for Asynchronous Transfer Mode, and asynchronous refers to the protocol's non-periodic behavior. Voice transmission in the public telecommunications network is synchronized with the carrier's switching system. The use of ATM switching in the existing network infrastructure requires integration via interoperable network interface standards. This synchronization matter has also been addressed within the framework of ATM.
- ATM has the capability to transmit information and interconnect networks on a point-to-point basis. The size of the bandwidth/ transmission speed, however, is about 155 Mbps. The question many corporations and carriers are facing is, "Are there less-powerful or complex solutions to connecting networks?". The answer is, "There are less-robust and less-financially expensive ways to achieve high-speed transmission." This scenario involves balancing capability/capacity with need.

Note that all decisions to select a particular technology over another require a careful and methodical investigation into the cost-benefit of one technology over another. Still, the overriding question is, "Why ATM?"

One way to answer this question is by understanding the benefits or uses of ATM.

ATM Cost-Benefit Discussion ATM can be used to connect networks of all kinds, just like SS7 or MF. ATM's advantage over SS7 and MF, however, is the transmission rate and service types that can be supported. These services types include the following:

- Voice
- Video with audio
- Audio (music recording industry-quality)
- Continuous data/information transmissions
- Graphics (high-resolution, three-dimensional renderings, etc.)
- Interactive multimedia (combining voice, video, audio, and user interaction)

SS7 and MF systems carry data at transmission rates of 0 bps to 56 kbps. These transmission speeds are largely enough for most mass-market applications (residential and small to medium-size businesses), but they are not sufficient for the newer video, audio, and multimedia applications envisioned—which will require transmission bandwidth at the OC−12 rates. ATM can be used to support interconnection between different networks, just like MF and SS7.

Most service providers do not need to support data rates at the OC−12 level at this time. If the intention is to get into video and data services provisioning, however, consideration should be given to larger bandwidth on the transmission facility side. ATM will have the capability to integrate a multitude of envisioned services for transport throughout the network.

Essentially, by using ATM, the service provider can more efficiently utilize his or her network resources for multiple-service applications. In a world in which economies of scale are a critical factor, ATM is an attractive choice. Current ATM switching might be considered costly by some service providers, but when you consider the skyrocketing operating costs of running a network, the short cost might be worth the long-term savings. Another way of looking at the cost of ATM is that as demand increases, the cost of deployment will decrease.

From a cost perspective, one argument against installing an ATM network is a cost-benefit analysis. Many service providers have only recently completed the installation of their SS7 networks. Interconnection with other networks might mean a speed reduction for service providers that support only ATM switching. In this case, the full value of ATM cannot be realized. Assuming that existing carriers and CLECs (of any kind) will suddenly change their networks to gamble on projected services that do not yet exist is unreasonable. When you consider the entire picture, however, the argument against installing and operating an ATM network falls short. Large service providers who intend to grow with the industry and move into new areas of the information business should consider the tools that are needed to be in the business of information provisioning. The CLECs and incumbent carriers (wireless and wireline) must balance the cost of the system with the projected value. Network planning is a gamble on the future, and it is part science and part art.

SS7 networks are simply not capable of delivering the broad-band video and data services that are being envisioned. At this time, a brief explanation of how ATM works is in order. ATM is so different from current signaling methodology. ATM's asynchronous nature is radically different from current primary-signaling methods that are used by the public telecommunications networks.

ATM Mechanics STM stands for *Synchronous Transfer Mode*. You can look at ATM as the complement of STM or as the other side of the coin. STM is used by the traditional voice telecommunications networks (landline carriers) to carry voice and packet data. STM is a circuit-switched networking capability, where a connection is established between two end points before data transfer commences and is torn down when the two end points are finished. Thus, the end points allocate and reserve the connection bandwidth for the entire duration—even when they might not actually be transmitting the data. Another way of looking at STM is in terms of in-band signaling.

In an STM network, the transmission facility bandwidth is divided into transmission time slots or buckets. The following transmission facilities and rates illustrate this statement:

- T1
- T3
- DS1
- DS3

These time-slot buckets are organized into a train that contains a fixed number of buckets and are labeled from 1 to *N*. Each bucket is a train car. The train is the entire set of train cars. The train itself repeats periodically, during every time period *T*, with the buckets in the train always in the same position with the same facade. In other words, each train car is in exactly the same position as before. There can be up to *X* different trains labeled from 1 to *X*, all repeating with the time period *T*, and all arriving within the time period *T*. The parameters *N*, *T*, and *X* have been determined by ANSI and ITU standards bodies.

Another example is as follows: You have a set of railroad cars, each carrying some type of widget. Each widget is different but is associated with the construction of Big Widget A. The train might have to make several trips, but each trip supports the construction of Big Widget A.

As the train pulls into the station, it drops its load and moves on. If one car is empty, it stays empty. As a matter of fact, the train does not stop at this station to get filled. To get filled, the train goes to another station.

Now, replace widgets with data. Now you have a data connection. Figure 4-5 illustrates the ATM data stream structural format.

On an STM link, once the connection is established, the link stays up for the duration of the call. In an environment of multimedia services, an STM protocol has the difficulty of trying to provide services in which each service has different bandwidth and transmission characteristics. Even traffic characteristics are different. If you compare voice, video, and data services to people, you will find that these services are different not only in appearance but also in personal characteristics. Each service is obviously different, and each service has a different requirement for transmitting and receiving the information. The services have their own subscriber

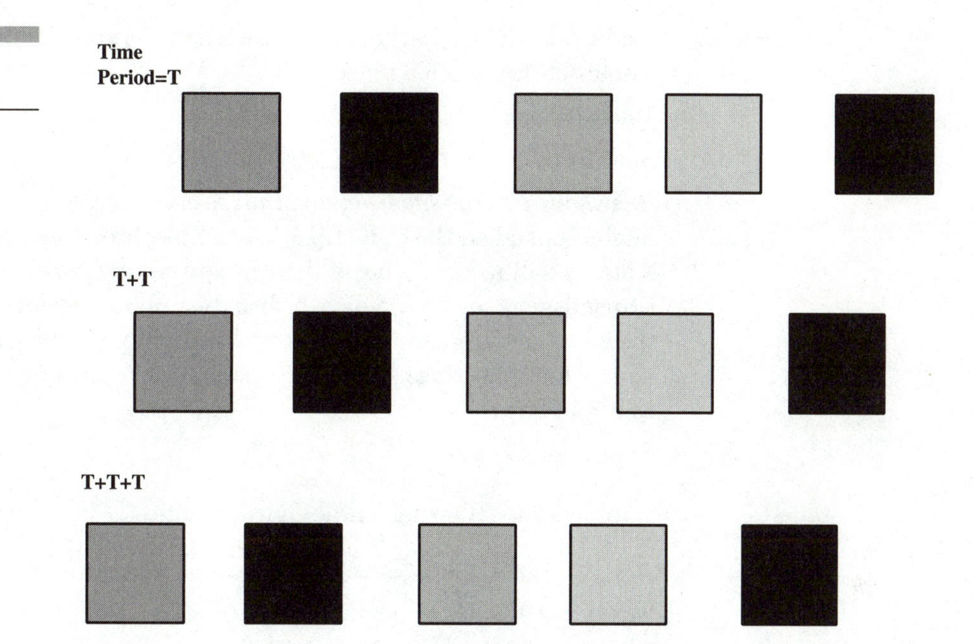

Figure 4-5
ATM data format

groups, and each subscriber group has its own traits and needs. One example is voice telecommunications, which has the following characteristics:

- Continuous (i.e., non-bursty transmissions)
- Real-time conversation
- Specific bandwidth needs (normal voice is within the 8 Khz range)
- *Busy hour for calls* Busy hour refers to a traffic characteristic—the time of day in which the volume of calls is at its largest. I am not sure about the history behind the term, but I believe the term "busy hour" was used because of the time period for which traffic engineers engineered switch capacity. One hour was used because of the need to engineer around a specific unit of time. There might be more than one busy hour for a switch, however.
- *Tolerates loss but not delay* The loss I am referring to is information loss, whether as a result of noise or poor transmission facilities.

While voice has these characteristics, video has the following characteristics:

- *Real time* This term assumes that the subscriber is not seeking slow-motion video. I do not consider slow-motion video real time. If you have ever seen a slow-motion video transmission (over an ISDN line), you might find the appearance so disturbing that you will demand high-resolution full-motion video.

- Tolerates some loss but not delay
- Specific bandwidth needs
- *Busy hour for transmission and maybe even reception* This factor might depend on the industry. For example, broadcast television stations will receive some of their programming via satellite transmission one or two days before the public broadcast. If you are in the video conferencing business, you will probably have a day-long busy hour, where meetings start at the beginning of the day and can continue forever.
- Continuous

Data has the following traffic characteristics:

- *Bursty* Data can be sent in large bursts of data, and transmission might be random.
- Tolerates delay but not loss of data
- Specific bandwidth needs
- Real time is not a requirement.

Combining all three types of traffic or information on single set of transmission facilities, using STM, would be difficult. If a service provider were to use STM to transmit voice, video, and data over the same network, the STM would be choking its own network. In such a scenario, it would be similar to running sets of 20 mile-long trains, one after the other, without a break. Some of the railroad cars would have information, and some would not. Some of the train cars would be different sizes, and some would be the same length. The railroad yardmaster would be working hard to make sure that each set of trains ended up where it was supposed to arrive. STM's switching requirements would be pushed past its limits. ATM is designed to handle multi-gigabit information streams.

The telecommunications industry trend is multimedia networks that combine voice, full-motion video, and high-speed data on a single network by using a single switching platform.

Combining these three types of information requires some type of protocol that enables the service provider to maximize his or her transmission network facilities without losing any data or without delaying the transmission and reception of the data. In order to ensure voice quality, the service provider has to have fixed-length information streams. Combine all of these types, and it looks worse than alphabet soup. In the case of ATM, the soup consists of information packets.

Given the nature of the information mix, ATM identifies the train cars of information via an identifier called a *Virtual Circuit Identifier* (VCI). The VCI is carried in the header of each packet of information. At the end of the information stream, the packets are assembled by using the VCI as the means of associating each train car.

ATM sounds and looks like a fast packet protocol and is designed for switching fixed-length packets. Given the current economics of ATM, it makes sense to use ATM in situations involving the transport of large volumes of data over long distances. This situation leads to the Internet. The TCP/IP protocol that is used by Internet networks can be supported by an ATM provider. The ATM network would serve as a national backbone, with the Internet data packet being broken into ATM 53-byte cells (including the headers). The overall network protocol stack would look similar to the following:

- Data
- TCP
- IP
- ATM adaption
- ATM data
- Physical layer

Figure 4-6 is a pictoral representation of the ATM Protocol Stack. Envisioning signaling as a stack reminds us that signaling protocols are composed of functional layers that build upon the strengths of the other functional components.

At the time of this writing, ATM is fairly new to the commercial market. Manufacturers and service providers are working together to develop applications for an ATM network. At this stage, many of the services are in the visionary category. A service provider can view ATM deployment as a major investment in the future. If the service provider wishes to compete in the broad-band services market and to simply be in a position to provide future services, ATM is a step in the right direction.

ATM and SS7 are considered to be network investments that have no obvious and immediate return on investment. Like SS7, ATM is not only an access technology but is also an enabling technology for new services. ATM is the next step beyond SS7.

The next section discusses the specific protocol used to support the currently hottest telecommunications market commonly called the Internet.

Figure 4-6
ATM Protocol Stack

Just as ATM can serve as a backbone signaling protocol, *Transmission Control Protocol / Internet Protocol* (TCP/IP) is the network signaling protocol of the Internet.

Transmission Control Protocol/Internet Protocol (TCP/IP)

The TCP/IP protocol suite was originally (and still is) used for the internetworking of *Local Area Networks* (LANs). The Internet is a collection of different networks and providers that have different software and hardware technologies. The Internet is quite literally a hodgepodge of LANs, servers, small Internet service provisioning companies, large Internet service provisioning companies, and search engines. The Internet is, in reality, a major step towards bringing communications and information to all people, however. The Internet is possibly a super information highway but is most definitely a place where information about the most obscure topic can be found by merely a stroke of a computer keyboard (and you can even file your income taxes online and save postage). Of course, accessing the Internet is not free, but the Internet is the first application of telecommunications that has opened up the world of information so rapidly. The Internet will be discussed in greater detail later in this book.

All of the signaling protocols used by the Internet are part of the TCP/IP protocol suite. The TCP/IP protocol suite was developed as a result of work that was first begun by the United States Department of Defense's *Advanced Research Project Agency* (ARPA) in 1957. ARPA's objective was to develop science and technology in response to the military threat posed by the Soviet Union (at that time). You can find more information about the history of the Internet later in this book.

The TCP/IP protocol suite is composed of multiple protocols. The protocol suite is layered—more so than SS7 or even ATM. TCP and IP are just two of the protocols in the suite of Internet protocols. The term TCP/IP refers to this family of protocols. The TCP/IP protocol suite's multiple layers facilitate future development of new Internet protocols. Whether or not this layering occurred by design is not relevant for this discussion; however, it is fortunate that the suite was architected in this manner, because it has enabled software engineers across the globe to find new applications for the Internet.

TCP/IP Protocol Architecture There are four layers to the TCP/IP protocol suite:

- Physical/Link
- Network
- Transport
- Application

The Physical/Link layer (as it corresponds to the OSI model) is also known as the Network Interface layer and manages and routes the exchange of data between the network device and the network. This data or information includes header information/overhead information.

The Network layer (as it corresponds to the OSI model) is also known as the Internet layer. This layer is responsible for managing the *Internet Protocol* (IP). The IP provides the Internet addressing for routing and is a connectionless protocol that provides datagram service. A datagram is a method of transmitting information. The datagram is broken into sections and is transmitted in packets across the network.

The Transport layer, which corresponds to the same Transport layer in the OSI model, transports the data. The *Transmission Control Protocol* (TCP) is run at the Transport layer.

The Application layer is responsible for managing all services. This layer corresponds to the Session, presentation, and Application layers of the OSI model.

The way in which the TCP/IP protocol suite is designed will make it capable of supporting a variety of different protocols. Figure 4-7 through 4-9 describe in greater detail, the concept of protocol/functional layering.

Figure 4-7 illustrates the layering concept, while Figure 4-8 shows an illustration of the kinds of protocols that can be supported. The abbreviations stand for the following protocols:

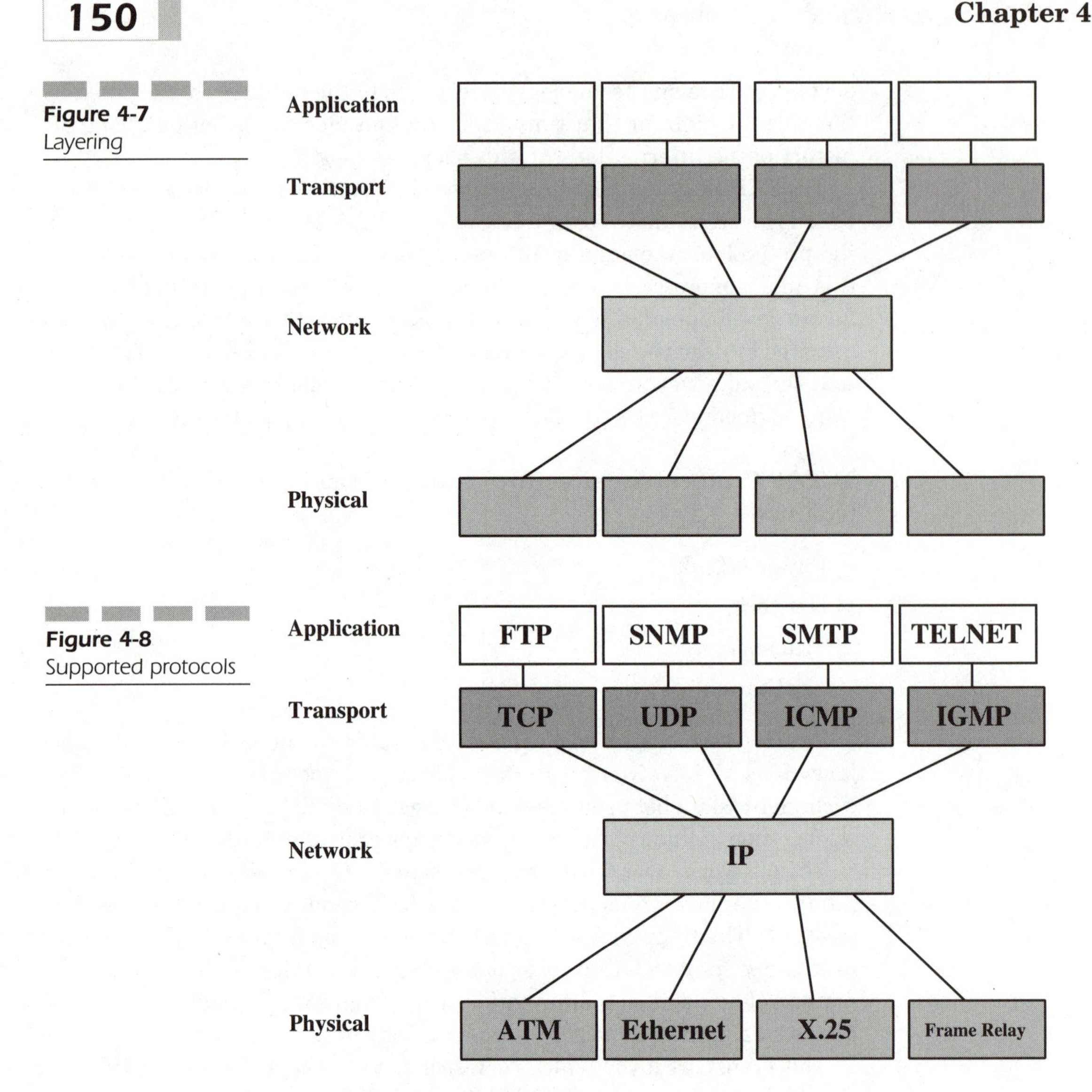

Figure 4-7 Layering

Figure 4-8 Supported protocols

- *TELNET* Telnet is program that enables logging on to a computer remotely.
- *FTP* FTP stands for File Transfer Protocol and supports the transfer of files between computers that are remote.
- *SMTP* SMPT stands for Simple Message Transfer Protocol and supports e-mail transmission and reception.
- *SNMP* SNMP (Simple Network Management Protocol) supports network management.

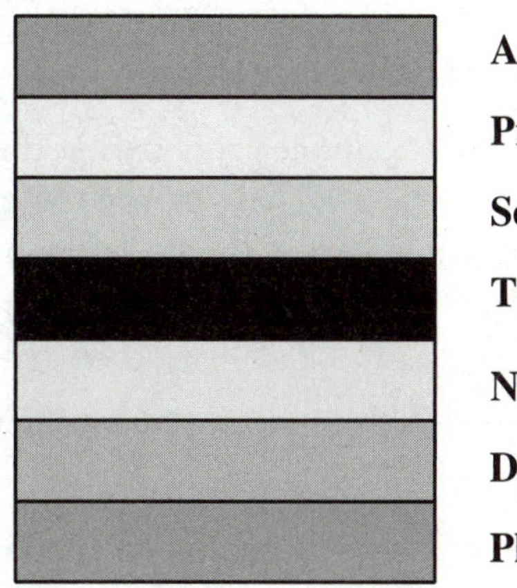

Figure 4-9
OSI model

- *TCP* TCP stands for Transmission Control Protocol.
- *UDP* The *User Datagram Protocol* (UDP) is normally bundled with the IP. UDP supports connectionless transmission.
- *ICMP* The *Internet Control Message Protocol* (ICMP) supports diagnostic functions.
- *IGMP* The *Internet Group Management Protocol* (IGMP) supports group management on a router.
- *ATM* ATM stands for Asynchronous Transfer Mode.

The Internet will be described in greater detail later in this book.

Frame Relay

Frame Relay is a packet-switching signaling protocol that is similar to X.25. X.25 is a packet protocol that enables connectivity between computing devices and a packet network. The best way to explain Frame Relay is by starting with X.25.

X.25 was first deployed in the mid–1960s by companies wishing to interconnect employees with their large mainframe computers at various corporate offices. X.25 is a packet data protocol. The X.25 protocol is a Data Link layer protocol; specifically, X.25 uses the *Link Access Protocol Bearer* (LAPB) portion of this layer. X.25 is also supported at the Network layer for packet switching on the D and B channels see Figure 4-9.

One concern with X.25 is the enormous amount of overhead. This overhead needs to perform the following actions:

- Each network node must examine every packet that passes through to determine whether the packet of data can be routed to the next node.
- Every node checks the data packet for errors.
- If a node detects an error, retransmission occurs from the last node that passed that data.
- The process of error checking introduces delays in the transmission and routing of data.

The overhead drives up the cost of equipment, making X.25 an expensive necessity. Frame Relay is a Data Link layer protocol that provides the user with the same capabilities as X.25 with a fraction of the overhead messaging. Frame Relay does not provide any error detection, error correction, or real-time flow control between terminals—thereby making Frame Relay a less cumbersome and less expensive way of providing packet service. Frame Relay is a low delay and high-speed packet solution and is slower than ATM but much faster than SS7 (supporting transmissions of up to DS−3 rates).

Frame Relay takes all of the transmitted data and protocol headers and packs them into a frame. This frame of data is sent over a Frame Relay-conditioned network. The protocol headers can be from other network signaling protocols. Frame Relay is essentially protocol agnostic. Despite the fact that Frame Relay itself has little overhead messaging, it leaves the overhead messaging embedded within the transported data undisturbed. Frame Relay can serve as an interconnect protocol between different types of networks, and more importantly, Frame Relay can be brought directly to the customer premises by using ISDN terminals and circuits. Frame Relay can be run over any type of physical, metallic, or optic transmission facility.

The frames can be variable in size (up to 4,096 octets). Frame Relay supports packetized voice and compressed video. Given the lower overhead of Frame Relay, building a Frame Relay switching network is an attractive prospect. Virtual networks such as Frame Relay networks are alternatives to leased-line networks, which are used by many companies and carriers. Figure 4-10 illustrates the leased-line scenario. In this scenario, multiple transmission facilities are leased from the incumbent LEC. In this network configuration, the carriers are using standard facility multiplexers in which the T1 is multiplexed into 24 channels.

In Figure 4-11,the Frame Relay network enables the carriers to lease fewer transmission facilities. Because the Frame Relay network is a packet network and is therefore a virtual circuit network, the carriers that are

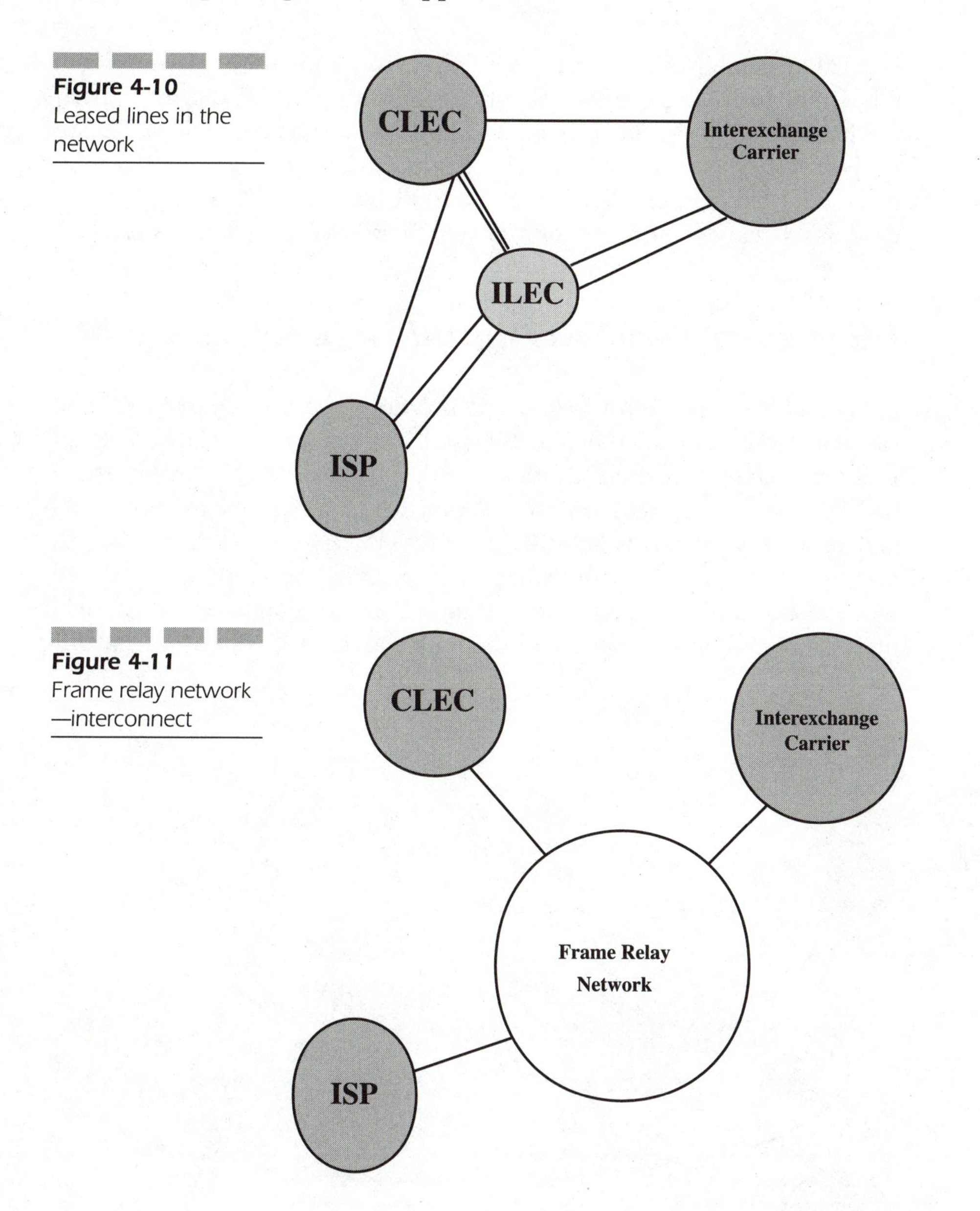

Figure 4-10 Leased lines in the network

Figure 4-11 Frame relay network —interconnect

using standard multiplexing and switching methods can reduce the number of transmission facilities. As Figure 4-11 illustrates, a Frame Relay network can be used to interconnect multiple service provider networks. Frame Relay is not the fastest network signaling protocol to use for network interconnect, however, the point I am making is that it can be done.

ATM, in combination with Frame Relay, is a formula for a powerful and robust network. This type of network configuration is capable of supporting high-speed data, video, voice, and multimedia applications. ATM can support the network backbone, and Frame Relay can serve as the local access and exit ramps. Figure 4-12 is an illustration of how ATM can be used as a backbone network for a series of local Frame relay networks.

Integrated Services Digital Network (ISDN)

Integrated Services Digital Network (ISDN) development began in 1976 and was not completed until the mid-1990s with the deployment of the Regional Bell Operating Companies' National ISDN−1. Until that moment, there had only been small and scattered deployments. The problem with ISDN had absolutely nothing to do with the concept but had everything to do with timing. Although ISDN is a Switch-to-User signaling protocol, its overall place within the context of CCS and its role in bringing common-channel network signaling capabilities to the user warrants its discussion in this section.

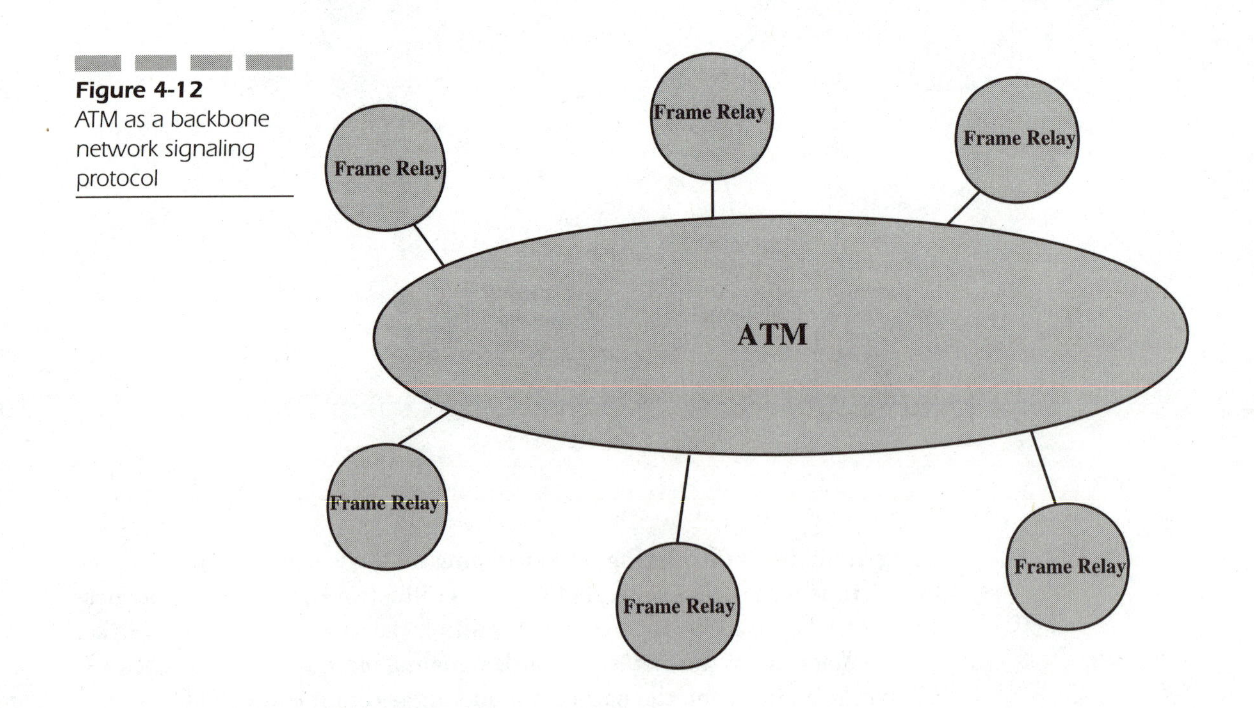

Figure 4-12 ATM as a backbone network signaling protocol

ISDN was envisioned to provide the subscriber with voice, video, and data on a single pair of wires. The goal of ISDN was to bring voice, video, and data to the user's desk. In its early days, ISDN was radical and was the first protocol to introduce out-of-band signaling in the local loop—and was the first protocol to bring this set of services to the home and business user. Unfortunately, in some telecommunications technology efforts, the vision caused the reality to become unrealistic. In the mid-1980s, carriers were selling ISDN services before terminal equipment had been developed and sold. ISDN terminals had been developed, but no terminal equipment existed that had standard interfaces for the user or network interconnect. This fact didn't stop ISDN services and terminals from being sold; however, it did create a problem with the customer in the areas of customer care, repair, installation, and training.

Although ISDN was meant to function as an end-to-end service (people to people), it could only do so within an LEC's central office. The LEC network at that time could not support an out-of-band signaling interswitch intranetwork. Therefore, without an SS7 network, the full capabilities of ISDN could not be realized. The vision of ISDN overwhelmed the technical, business, and market realities and created mayhem in the development of the market and technology. ISDN and SS7 were meant to function together, but unfortunately they were treated as separate LEC planning efforts and deployment efforts. The disconnect in planning and deploying hurt the rollouts of both technologies. During the 1980s, ISDN was considered the subscriber interface with power for the SS7 network. Ultimately, because SS7 could only be provided by a few large equipment manufacturers, the industry standardization discussions were far more orderly to manage. In the case of ISDN, however, there were dozens of small, mid-size, and large manufacturers that were providing ISDN terminal equipment. So much time and money had been invested in terminal devices in the manufacturing community and the subscriber base that bringing the industry together was a daunting task.

A nationwide *Regional Bell Operating Company* (RBOC) effort was launched in the early and mid-1990s to unify ISDN development (in the standards arena) and to promote ISDN in the marketplace. So important was this effort that even the interexchange carriers worked with the LEC community in the standards arena to bring ISDN back to life. By the time this effort came to life, however, it was already too late. ISDN costs in the network and terminal sides of the industry were already skyrocketing. There was a massive amount of ISDN terminal equipment in the business place and in the warehouses of manufacturers that complied to different standards (or that simply were vendor interpretations of a standard). Subscribers still had to be educated in the value of services such as Caller Identification and Caller Identification Restriction, and they had to deal with concerns about privacy. ISDN was the first time that the LECs launched a coordinated technical and marketing (promotion and subscriber education) effort to promote a digital service. ISDN costs were reduced to the end user in order to increase market penetration. ISDN services were marketed not as ISDN services but as individual services.

Part of the problem with the early marketing efforts was the technology focus. Subscribers do not care how the service is provided; rather, they simply care about whether the service has value and whether it works consistently. Unfortunately, by the time that all of the marketing efforts came about, the industry and the subscriber base had new technology alternatives to provide them with voice, video, and data. These new alternatives, such as cable TV and the Internet, have caused the economics of ISDN to change. This change in the economics of ISDN has caused small vendors to drop out of the marketplace; remaining vendors to reduce the price of the equipment to move inventory; and carriers to reduce the price of ISDN-based services to penetrate the market and generate profit through volume. This situation does not mean that there is no value to ISDN; rather, it simply means that ISDN had a bad start and is still plagued with non-compatible terminal devices.

ISDN Mechanics ISDN is offered in two forms: ISDN BRI, which is the ISDN Basic Rate Interface, and ISDN PRI, which is the ISDN Primary Rate Interface. ISDN BRI supports two Bearer channels and one Data Channel (2B+D). In North America, ISDN PRI supports 23 Bearer channels and one Data channel (23B+D). Outside North America, the *European Telecommunications Standards Institute* (ETSI) version of ISDN PRI (23B+D) is ISDN PRI (30B+D).

The throughput of ISDN looks similar to the following: the Bearer channel (B channel) operates at 64 kbps, and the Data channel (D channel) operates at 16 kbps. The term *throughput* is typically used rather than the term *speed* when referring to data transmissions. The reader might wish to interchange the term with simply "speed" or "rate." The B channel supports transmission of voice/data/slow-motion video, and the D channel supports network signaling. ISDN is provisioned to the customer premises via a pair of wires.

ISDN is supported by several different protocols that were originally based on a series of *International Telecommunications Union* (ITU) recommendations. These recommendations support ISDN at the Physical, Data Link, and Network layers:

- *Network* Q.931 for Call Control on the D channel
- *Data Link* Q.921 LAPD (Link Access Protocol D channel), I.465/V.120 Circuit switched information on the B channel, and LAPB (Link Access Protocol B)—with X.25 for packet switching on the B channel
- *Physical Layer* I.430 *Basic Rate Interface* (BRI) and I.431 (*Primary Rate Interface*, or PRI). Also supported is X.25 packet switching on both the B and D channels.

ISDN signaling operates across multiple layers of the OSI protocol stack yet is only supported by specific standards within each layer. The reader should remember that the OSI model provides a conceptual design framework and is not a protocol standard or a set of protocol standards. Therefore, when looking at ISDN, the reader should note that the protocol—like other protocols —has specific functions that are defined explicitly against the OSI framework.

The B channel (Bearer channel) is used to transmit information, voice, fax, packet data, circuit-switched data, and slow-motion video. Note that full-motion video cannot be supported because there is not enough bandwidth. The D channel (Data channel) supports control signaling for the B channel. The D channel supports a rate of 16 kbps. Despite the fact that the D channel has specific responsibilities to ensure B channel transmissions, the D channel can be used to support low-speed data. Low-speed data applications might include system health monitoring or basic telemetry applications such as home security monitoring.

ISDN Services At its inception, ISDN subscriber services included the following:

- Call Waiting
- Calling Line Identification Presentation
- Calling Line Identification Restriction
- Closed User Group communication
- Conference Calling
- Video Conferencing
- Simultaneous voice and data transmission

ISDN also has other applications. This list represents the carrier communities' retail perspective of ISDN. Corporate users of ISDN might have their own applications of ISDN as it relates to data.

Despite the confusion that ISDN development created for itself, ISDN standards work caused the telecommunications community to categorize services in such a way that the old wireline companies never had. The categories are as follows:

- *Bearer Services* Bearer Services include voice, slow-motion video, data, and applications of data. The Bearer Services can be connection oriented, connectionless, or packet switched.
- *Teleservices* Teleservices include computing device-to-computing device communication, file transfer, and terminal access to remote computing systems.
- *Supplementary Services* Supplementary Services are defined as services over and above basic voice, including Call Waiting, Call Forwarding, Caller ID, Remote Call Forwarding, Conference Calling, Call Hold, Video Conferencing, etc. Supplementary Services are activated and are operated within the switching system. ISDN supports all of these services. Call Forwarding and Call Waiting existed before ISDN Supplementary Services; however, the original allures of ISDN were Calling Line Identification Presentation, Conference Calling (the subscriber's ability to add parties to a conference call in real time), and video conferencing.

Network Applications

Network signaling applications can be viewed in a technical and business model. The business model is broken into two basic categories: retail and wholesale.

The model shown in Figure 4-13 illustrates the retail perspective. In this model, the carrier (regardless of what kind of carrier) supplies service to the individual subscriber or corporation. The carrier can provide basic voice or services that are embedded, activated, and operated from within the carrier's switching system.

The model shown in Figure 4-14 describes the wholesale business relationship. In this business model, the carrier is typically providing another carrier with access to and use of call-processing capabilities with the switch, access to and use of business operations support (billing or customer care), access to and use of operational support systems (network management, call detail recording, etc.), leasing of transmission facilities, leasing of towers, leasing of real estate, and leasing of electric and water utilities.

The technical perspective can be broken into the following categories: Interswitch/Internetwork network signaling and Switch-to-User signaling.

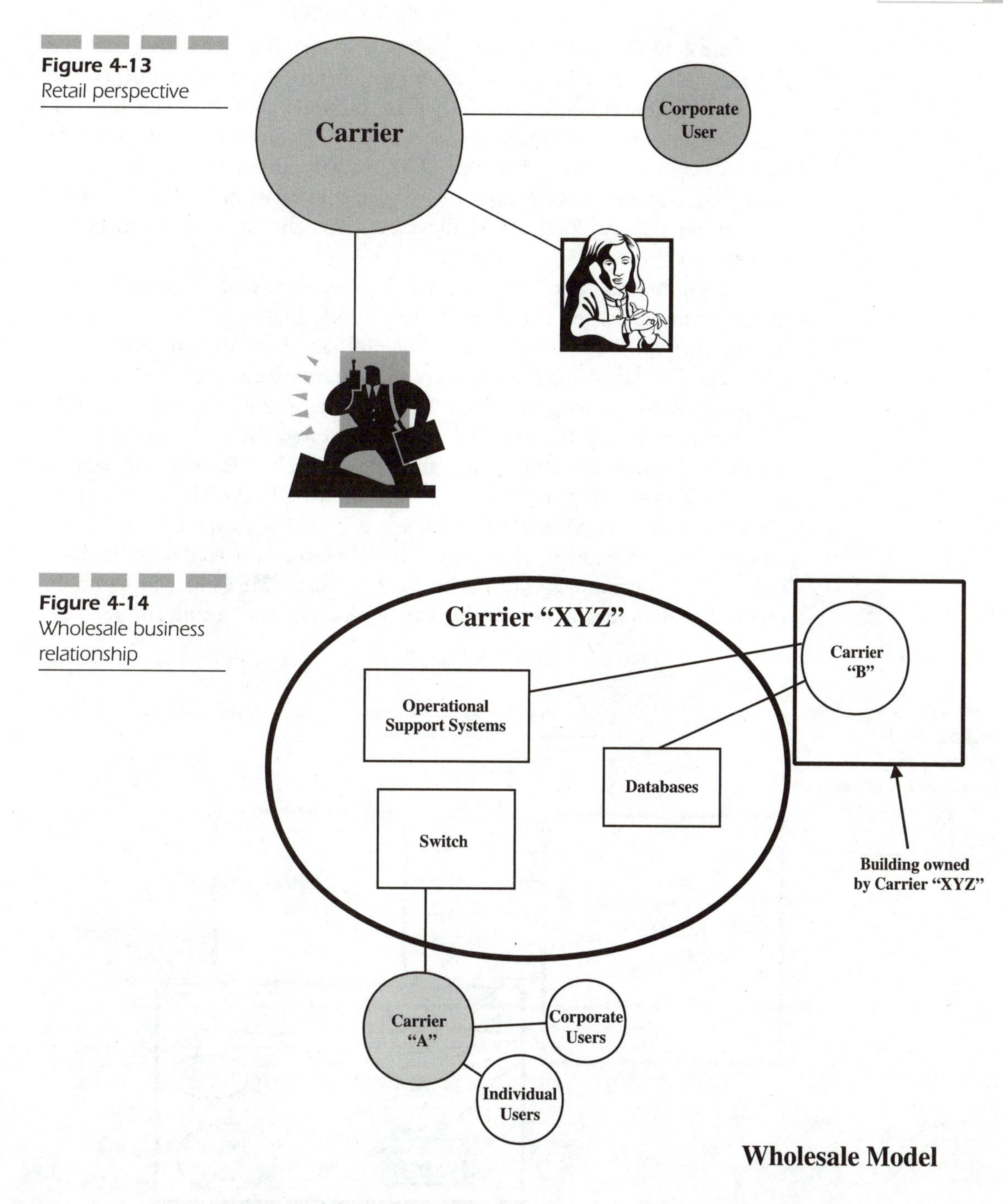

Figure 4-13 Retail perspective

Figure 4-14 Wholesale business relationship

Each category of signaling enables a carrier to provide capabilities to support one or both of these business models.

Figure 4-15 illustrates the relationship between different carriers' switches. The signaling supported by the carriers will dictate the relationship between the carriers. One example is as follows: In the world of *Multi-Frequency* (MF) signaling, interconnect between carriers took on the flavor of dominant carrier and subordinate carrier. This type of MF-based (one-sided from a business view) arrangement was typical of the relationship between the ILECs and the cellular carriers. As Figure 4-15 illustrates, switches can interconnect in a variety of ways.

The original signaling between wireless carriers and the wireline telephone companies was Dial-Line interconnect. In the Dial-Line arrangements, the wireless switch was translated in the wireline switch as a customer *Private Branch Exchange* (PBX). This PBX was not even MF; rather, it was local loop signaling. This original arrangement between the cellular carriers and the wireline carriers dictated the personal views and business arrangements for years, and many wireline folks simply considered wireless switches to be nothing more than big PBXs. The advent of MF did not change that view. MF is both one way and two way in nature. MF's two way attribute means that the old ILEC's interconnected with the independent (old terminology) LEC's as peer networks. MF's one way signaling attribute refers to the way in which the wireless switch could simply origi-

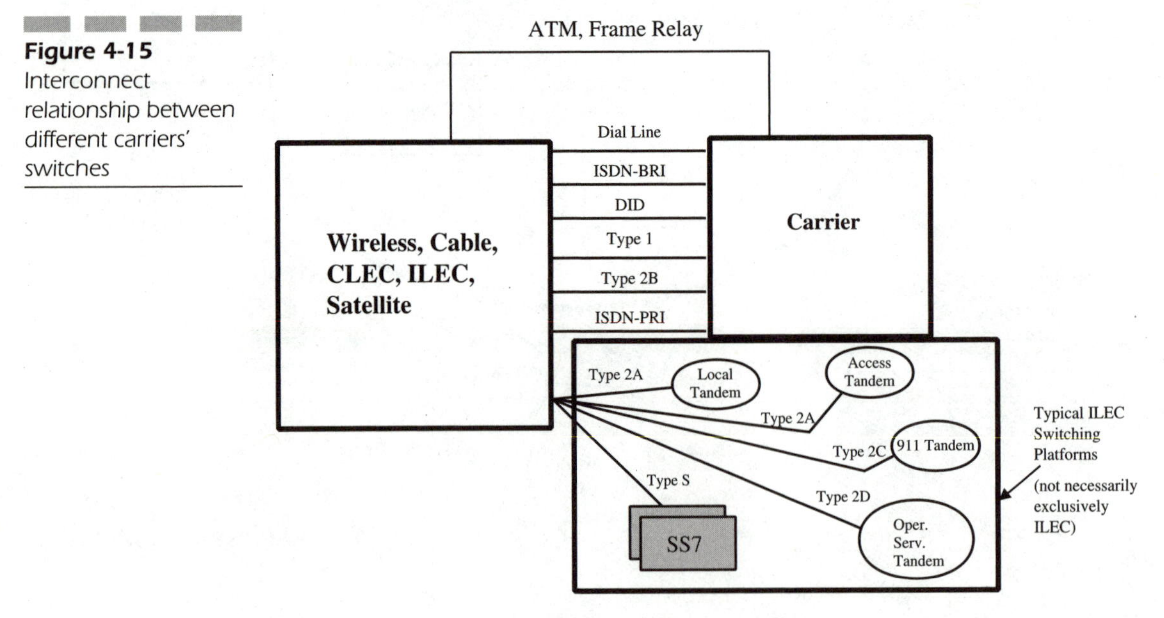

Figure 4-15 Interconnect relationship between different carriers' switches

nate and terminate in a wireline network via the wireline carrier's central office. This configuration meant that the wireless switch was interconnected to the PSTN as a PBX. The reasons for the continued views were part wireline carrier lack of acknowledgment, part limited wireless carrier understanding of what was possible, and part limited wireless switch capabilities. Although the wireless switches could support MF signaling and the same dialing sequences as the wireline carriers, it was not enough to change the perception among wireline carriers. CCS enabled the wireless community to force a change in the perception of many people in the wireline carrier community and to recognize the wireless carriers as being peer carriers in both a technical and business sense. SS7 is a two-way signaling protocol only and does not recognize one-way signaling.

Figure 4-16 illustrates Switch-User signaling. ISDN (which was illustrated in Figure 4-15, although it was displayed as user signaling) can be provided to the wireless carrier today via one of the ISDN network interconnects. I included ISDN in Figure 4-15 and in Figure 4-16 to acknowledge the peer relationship of the wireless carriers and to highlight the fact that even a Switch-user signaling can be wholesaled as a service to another carrier.

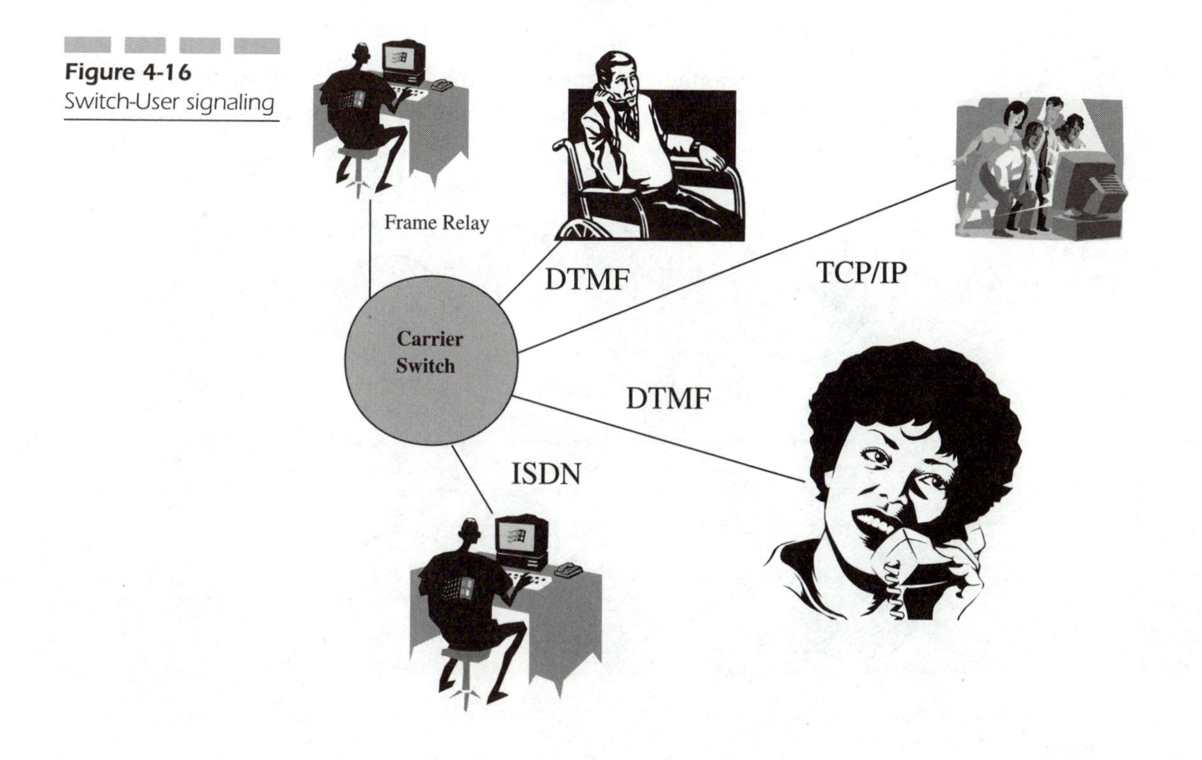

Figure 4-16 Switch-User signaling

SUMMARY

The word *network* once referred to the big network owned and operated by the pre-divestiture AT&T Bell system. Today, the word *network* is not an appropriate description; rather, we should describe the network as a *network of networks*. The plethora of new carriers (of all kinds) requires careful planning and execution of a network signaling plan. The fierce competition among small and large carriers demands not only technically functional signaling, but also network signaling that brings value to the customer in the form of services and capabilities. The applications can be switch-controlled services, bandwidth, or signaling protocols capable and flexible enough to support new types of services.

The wisest actions to take in terms of network signaling planning are as follows:

- First, decide what business the carrier is in, which includes knowing the market segment to be targeted.
- Second, decide what services the carrier will provide to meet the needs of the market segment.
- Third, decide what technology and network signaling will enable the business plan.

Figure 4-17 is a pictoral representation of the current network environment; multiple providers and multiple networks. The current telecommu-

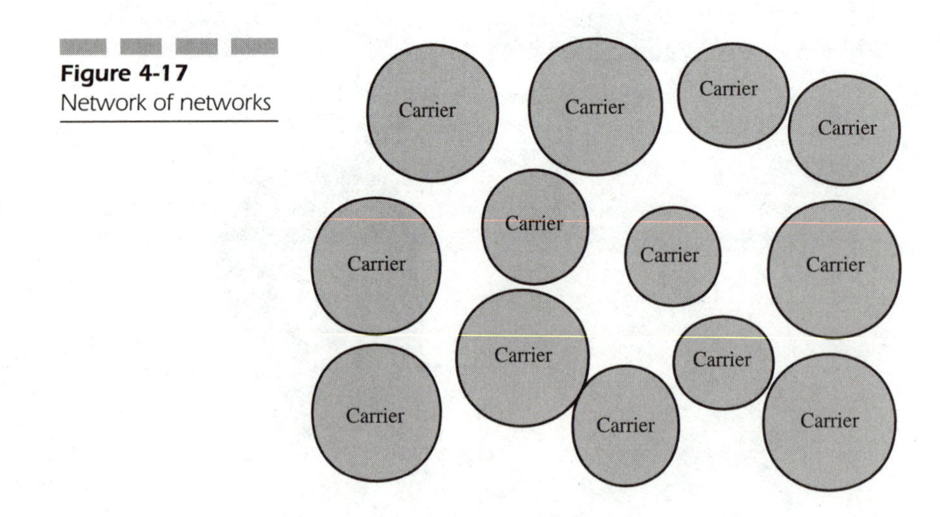

Figure 4-17
Network of networks

nications marketplace is no longer one network but rather multiple networks; a network of networks.

Focusing on the business first is necessary. In the days when there was one dominant player and no hope for change or opportunity, technology could drive the business. Technology as a driver means that change takes place for the sake of new technology (trying out the "new toys"), rather than implementing technology to meet the needs of customers or the projected needs of customers. Today's market environment might appear to be filled with big mergers; however, the environment really involves constant change. Companies market new products, but customers pay for the products and services. Change is being driven by the customer.

The next chapter will focus on specific types of applications and the business opportunities that are available.

CHAPTER 5

Applications

To a great extent, the words *application* and *product* are synonymous. An application does something useful or carries out a useful task, while a product is sold to a customer. An application is not necessarily a product, but a product is always an application. Conversely, a service is always a product, and a service can be an activity or a system feature. An application can be a service, and a service is always an application. In addition, an application can be an internal activity that only impacts the user and no one else; therefore, the application is not a product. The author's view of an application not only acknowledges all of these relationships but also takes a broader view. Applications that are internal to the carrier's operation can, in fact, be sold as a product to subscribers of all kinds, including individual and corporate subscribers.

Applications can be viewed from a variety of perspectives. However, the principal ways to view telecommunications applications are

- Technical
- Market/business

Technical perspectives are internal and external. To a carrier, one example of an internal application is a new network interface that enables the switch to communicate with an operational support system. To a carrier, an external application is a new network interface that enables the switch to communicate with another carrier's switch or with a customer's network. The application is external, because it has a direct and obvious impact on those who are outside the carrier. The application can be hardware or software based. For the purposes of this book, we will focus on the external applications (i.e., products and services).

Marketing/business perspectives impact the customer. The applications are provided to the carrier's customer, which can be an individual, a corporation, a family, or another telecommunications carrier.

Figure 5-1 illustrates this relationship.

The technical and marketing/business applications can be further divided into a series of other categories. These categories will be necessary in order for the reader to gain an understanding of how a carrier can use its full capabilities to generate revenue. These other categories are as follows:

- *Network centric* Applications that are network based, network activated, or network controlled
- *Subscriber terminal device* Applications that are embedded or that reside only within the terminal device. These devices can communicate information across a network interface, but the actions of initiating, terminating, and controlling reside in the terminal device. A terminal device can be a laptop computer, desktop computer, telephone device, mobile handset, PDA, or even a television set.

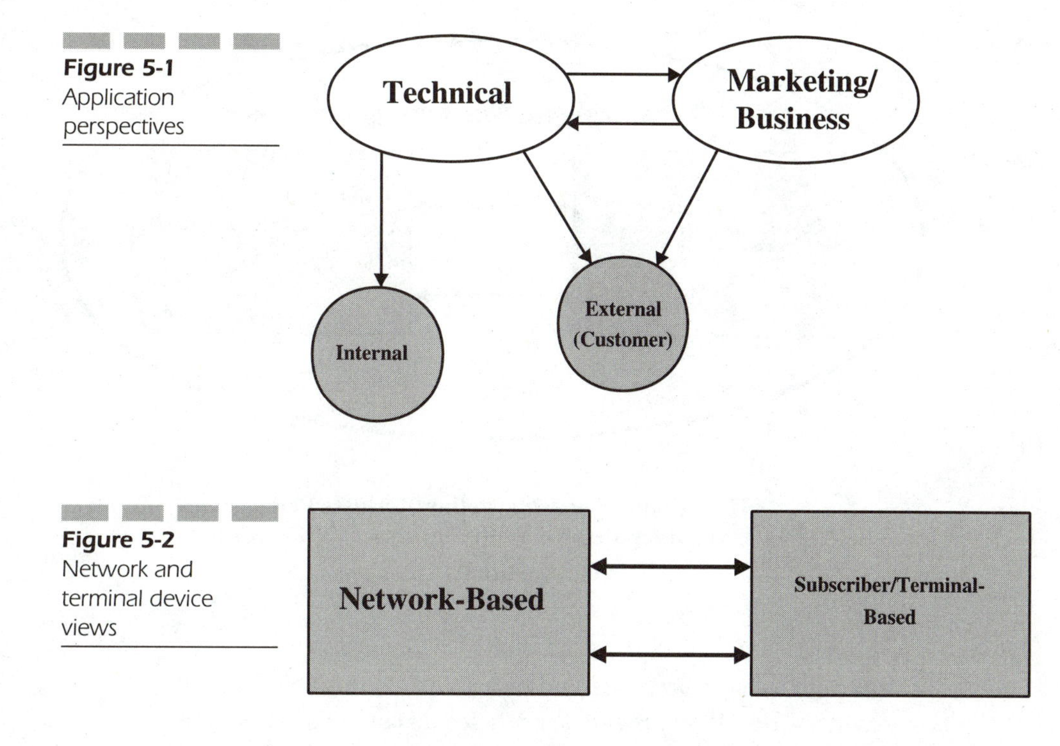

Figure 5-1 Application perspectives

Figure 5-2 Network and terminal device views

Applications in both environments still requires communication between the environments.

Figure 5-2 illustrates these network and terminal device views.

Figure 5-3 describes the relationship between technical and marketing/ business decisions and the type of technology and solutions that are selected.

Business relationships impact the way in which services are provided to the subscribers. The following business relationships describe how the services are provided by the carriers to the subscribers. These relationships dictate how the business is managed:

- *Retail* Retail suppliers of telecommunications services provide services directly to the subscribers. In other words, the carrier uses its own brand name and renders the bill. Third parties do not provide services to the carrier.
- *Wholesale* Wholesale suppliers of telecommunications services provide services to other carriers. Wholesalers typically do not sell

Figure 5-3 The Inter-relationship of technical and marketing/business perspectives

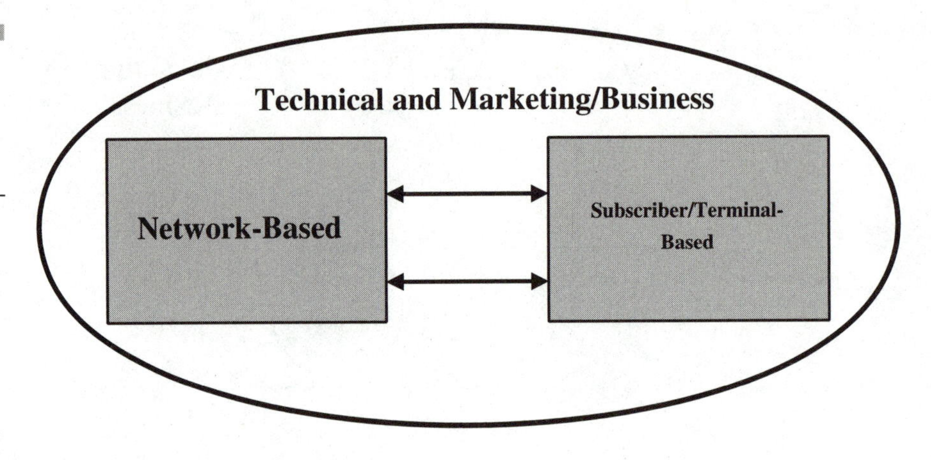

directly to the subscribers/end users; rather, they sell through resellers. Resellers provide the wholesale carrier's services to the users, and branding of the wholesaler might still be permitted under the terms and conditions of the reseller. The wholesaler does not render bills to the customer. Wholesalers seek every opportunity to leverage the network interconnect.

Theoretically, subscribers should not see any difference in quality—assuming that every carrier strives to provide the highest quality of service (the services that the subscribers desire). The fact is that every carrier claims to be different in terms of types of services provided and the level of service quality.

CLECs that resell from the ILECs do care whether the subscribers know and see a difference in the types of services and service quality that they provide. The challenge for the CLEC when reselling network-based services and when using the ILECs switching systems is to provide network-based services that are of a higher service quality than the ILEC. CLECs in such a situation secure customers by advertising lower rates. The challenge for these CLECs is maintaining the business on pricing alone. At some point, a CLEC will want to control its own destiny by creating its own services. In order for a CLEC to create its own services, it needs its own switching systems. If the CLEC cannot own a switch, then at least the CLEC should consider owning and operating a network element such as the database that contains the routing tables for the call and the service profiles of the subscribers. Figure 5-4 illustrates the CLEC/ILEC relationship. In the case

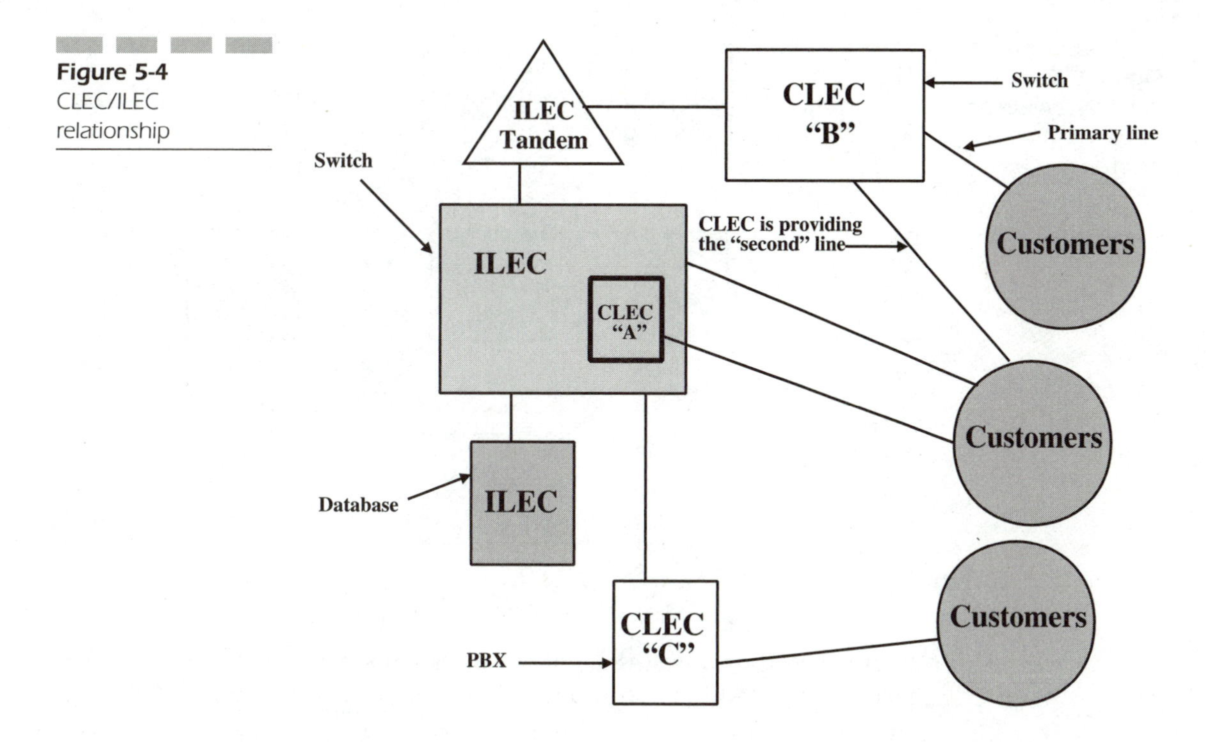

Figure 5-4 CLEC/ILEC relationship

where the CLEC owns the switch and is directly connected to the subscriber, the CLEC comes closest to owning (controlling) the customer than in any other situation. This same case assumes that the CLEC is providing all of the functions necessary to support the customer—everything from customer care to directory assistance.

The Process of Creation

The carrier that wishes to provide services to a subscriber needs to perform an analysis of the marketplace. This analysis should be followed by examining what the carrier can physically and technically do in terms of meeting those needs. Figure 5-5 highlights the overall process of creating services and meeting customer needs. The carrier needs to approach the development of services and provisioning of services from both a macro and micro level.

First, the marketing department needs to identify an opportunity, and then it needs to examine the marketplace to determine whether the proposed service matches projected customer needs. This analysis should be

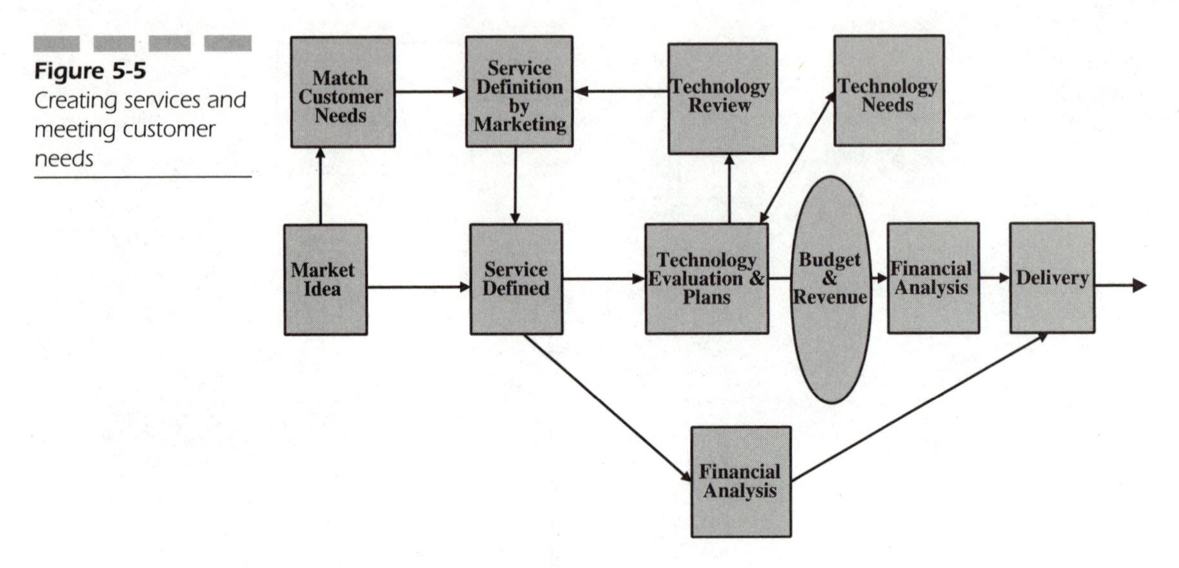

Figure 5-5 Creating services and meeting customer needs

then followed by a well-thought-out proposed service definition. The technology people will simultaneously have input into the service definition. Once the service has been defined and has been agreed upon by everyone, the technology folks determine what equipment needs to be purchased and installed. The technology process is run parallel to the whole service definition process. Once budgets have been established and revenue analyses have been conducted, the service is delivered to the customer.

Marketing services was not a major concern of the ILECs (pre-divestiture of the AT&T Bell System). Services needed to be introduced, but competition was not as fierce—and the level of sophistication and power that was needed to market new services was not as intense as it is today. There are basic concepts, however, which have and always will apply: market segmentation, marketing mix, and product life cycle.

Market segmentation is the process of analyzing the markets in terms of discrete parts or segments. Market segmentation assumes that different parts of the marketplace have different needs and requirements. The challenge is to identify the market segments properly. The market segments can be broken into any number of categories, such as the following:

- Residential
- Residential, with two parents working outside the home
- Residential, with one parent working outside the home
- Residential, with one parent working outside the home and one parent going to night school

- People who are 30 to 36 years old
- Two-income families that together make no more than $50,000 per year

These variables are called demographics. Demographics include gender, age, income, occupation, family size, religion, gender of any children, etc. The carrier can gather an enormous amount of information about the overall marketplace. How the information is used, however, determines how the segments are identified. The factors that are used to identify market segments are as follows: demographics, geographic location, lifestyle, service types used, spending habits, level of service usage, etc. If the customer is another carrier, then the factors are as follows: demographics of the carrier customer's subscribers, geographic location of the carrier, regional or national nature of the carrier customer, level of service usage, etc. The variables are similar between the individual users and the carrier customers.

The marketing mix is comprised of those variables and factors that are controlled by the carrier to plan, introduce, and sell the product. These factors are called the four Ps:

- *Product* Refers to the physical nature of the product and the perceived value of the product. The product also includes the conduct of the vendor, the guarantees that vendor offers, and the vendor's reputation.
- *Price* Refers to the product's purchase price, terms and conditions of the payments, and discounts
- *Place* Refers to how the carrier will sell and distribute the product. The carrier can sell directly to the subscriber base, and the carrier can also act as a wholesaler of its own network services and capabilities to another carrier.
- *Promotion* Refers to advertising and selling

These factors do not describe a serial process; rather, they are interrelated and can be illustrated as follows (refer to Figure 5-6). These factors are applicable to all customer groups.

Product life cycle refers to the life of the product, and telecommunications services have a long product life cycle. The life of a telecommunication service is on the order of years and decades. Call waiting, call forwarding, and three or four-digit extension dialing are examples of services that have been around for years and will continue to be around for many decades to come. A product's life cycle is composed of four stages, which are all affected by the four Ps. The stages in the life of a product are as follows:

- *Introduction* Refers to the product's initial entry into the marketplace. At some point, the product enters a growth phase. In the

Figure 5-6
The Four Ps

telecommunications business, the introductory phase includes an intense period of selling, marketing, and educating.

- *Growth* Refers to the period of time in which the product has caught on with the public, when education is no longer a priority, and when sales begins to grow
- *Maturity* Product maturity refers to the period of time in which the product's sales growth levels off. In such an environment, telecommunications services sales might level off because of zero population growth, no new interest in the service, change in customer lifestyles, a new type of service emerging in the marketplace, similar products being introduced, changing needs of the subscribers, etc.
- *Decline* Product decline occurs when sales begin to decline. The decline is the result of product obsolescence or product replacement.

These stages typically look similar to a bell curve. Refer to Figure 5-7.

The Unique Nature of Telecommunications Services Pricing

Pricing in the telecommunications business is regulated by the government and is subject to what the market will bear. Wireline CLECs are required to

Figure 5-7
Stages in the life of a product

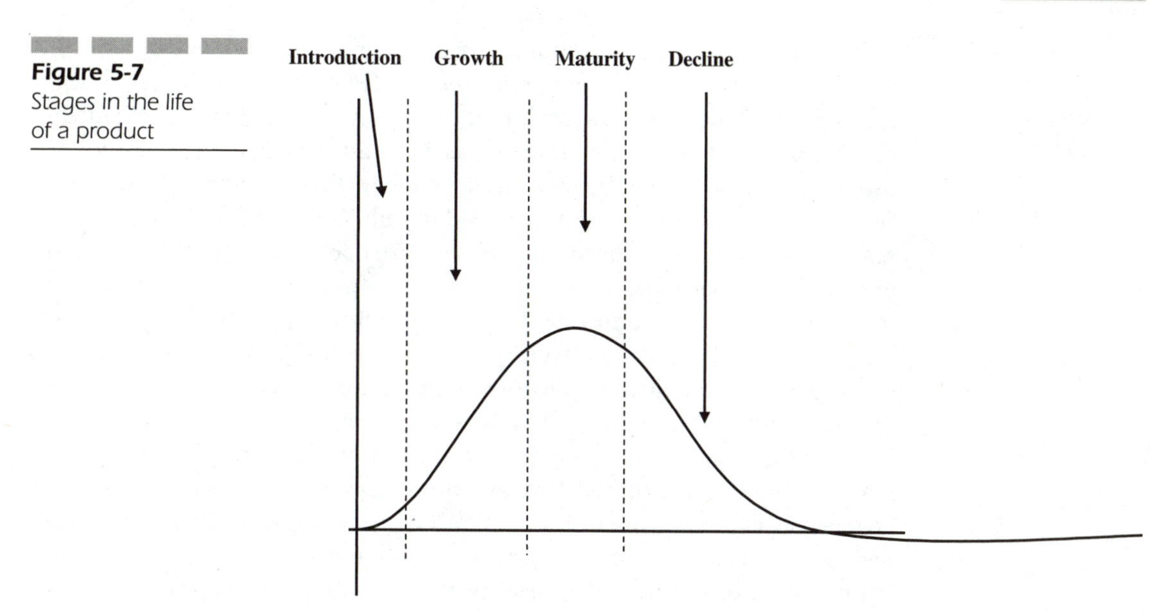

file tariffs and submit to the same state and federal regulatory authorities as the ILECs. The CLECs are given some latitude in order to encourage competition, but they are required to pay into the Universal Service Fund and provide *Enhanced 911* (E9-1-1) access to subscribers.

Pricing in the traditional wireline world is based on the peg count. In other words, every call generates a dime in revenue. Pricing based on usage is an old model that has proven to be useful and profitable. Given the competitive nature of the market, however, volume discount models have been given a chance to flourish. Another unique aspect of CLEC pricing is the required discounts that are mandatory under federal and state regulatory law. These discounts are applied by the ILEC as a way for the CLEC to launch operations in a given state. These discounts also enable a CLEC to resell ILEC services to subscribers. These ILEC discounts are from 20 percent to 30 percent for subscriber types of services. The discounts can go even higher for certain types of services that are provided by the network, and some of the discounts can go as high as 60 percent.

Pricing in this unregulated environment has taken on an unusual role. Pricing to CLECs from the ILECs has been regulated downward. The same pricing that the ILEC charges for the same types of services is regulated higher. This level of competition is both open; yet, it is still regulated. This situation is known as regulated competition.

Pricing in other telecommunications sectors such as paging, satellite, cellular/*Personal Communications Service* (PCS), and cable are largely unregulated. These sectors are largely price unregulated to the extent that specific pricing to the individual subscriber and corporate user are not dictated by a state or federal regulatory agency. Public telecommunications services provisioning mandates a level of public responsibility. Surcharges and fees are applied to those carriers that are operating as CLECs (wireless or wireline). These fees and surcharges are applied to efforts such as the maintenance of the wired local loop and the funding of the Universal Service Fund. The Universal Service Fund is a federally mandated fund used to support the beneficial application of telecommunications in underserved markets and in the nation's educational system.

Providing services in the public sector requires a level of public responsibility, which is mandated by the federal government. Regardless of the government's involvement, there is still a pervasive attitude among many people that voice communications is a public right. Such attitudes have translated into specific activities throughout the nation to protect the perceived rights of the wireline telephone customer. Some of these legislative acts are Lifeline type services for the elderly and indigent, three-month notification of potential shutdown of residential telephone services for nonpayment of the bill, multiple notifications of non-payment before termination of service, and acceptance of partial payments in lieu of full payments. The objective of these acts is to ensure that the user has the ability to maintain basic communication with the world. What has driven the creation of these acts and attitudes are incidences that have occurred in the subscriber base. One example occurred more than a dozen years ago in a large, metropolitan, United States East-coast city. Telephone service was lost in a lower middle-income neighborhood as a result of a fire in the wireline telephone company's central office. Service to 40,000 people was lost. The telephone company's response was instantaneous and massive. More than 350 people were sent into the field to assess the damage and to perform repairs. More than 300 pay phone setups were established for the sole purpose of providing the customers with phone service while the switches were out of service. Subscribers in the neighborhood could call anywhere in the world without time limitations at any time of day, and this service was provided free of charge. Security for the people and for the setups was paid for by the telephone company. Why did the telephone company respond in this manner? The answer is simple. The residential population in the United States expects nothing less than a caring telephone company. The term *The Telephone Company* is more than just a few generic words. Despite complaints, the term has connotations of expected quality, care, reliability, and 100 percent availability to the majority of the population.

Subscribers have expectations of their telecommunications services providers. Depending on the telecommunications segment, the subscribers might expect a dial tone 100 percent of the time, dialing into their ISP late at night because that is the only time that they can log on, noisy/static-filled conversation over their wireless handset, pagers that occasionally receive pages, or maybe even frequently dropped calls over a wireless network. In some instances, the subscribers expect poor quality in order to obtain low-cost voice service. Getting service at low rates is an important way of gaining a foothold in a community, but at some point, utility of the service and the quality of service eventually becomes the most important matter. At some point, every deep-pocketed carrier will be charging the same low rates.

The next section will address the specific products that carriers can provide.

Getting Down to Business: Applications

There are a number of different types of applications that a service provider can provide to another service provider. These applications can be broken into the following categories:

- Network-based subscriber services
- Network-based carrier services

Network-Based Subscriber Services

Network-based subscriber services are those services that are provided to the carriers' individual or corporate users. Network-based carrier services are those services that are provided to the carriers' carrier customers. The subscriber service enhances the way that the subscriber makes and receives calls and is billed for calls. In some cases, the subscriber service even limits who you can call or where you can call. A network-based subscriber service is one that is provided by the communications service provider's infrastructure equipment, as opposed to customer-premise equipment. Centrex is an example of a network-based service. Centrex is an AT&T-trademarked, central office-based, abbreviated dialing system that gives the subscriber the feeling and appearance of a PBX.

The following sections describe a representative example of the more popular network-based subscriber services.

Call Delivery (CD) *Call Delivery* (CD) permits a mobile subscriber to receive calls to his or her directory number while roaming in a wireless network.

Call Forwarding-Busy (CFB) *Call Forwarding-Busy* (CFB) permits a called subscriber to have the network send incoming calls addressed to the called subscriber's directory number to another directory number (forward-to number)—or to the called subscriber's designated voice mail mailbox—when the subscriber is engaged in a call or service.

Call Forwarding-Default (CFD) *Call Forwarding-Default* (CFD) permits a called subscriber to send incoming calls that are addressed to the called subscriber's directory number to the subscriber's designated voice mail mailbox or to another directory number (forward-to number) when the subscriber (wireline or wireless) is engaged in a call, does not respond to paging, does not answer the call within a specified period after being alerted, or is otherwise inaccessible (including no paging response, the subscriber's location is not known, the subscriber is reported as inactive, call delivery is not active for a roaming subscriber, etc.).

Call Forwarding-No Answer (CFNA) *Call Forwarding-No Answer* (CFNA) permits a called subscriber to have the system send incoming calls that are addressed to the called subscriber's directory number to another directory number (forward-to number) or to the called subscriber's designated voice mail mailbox when the subscriber fails to answer or is otherwise inaccessible (including no paging response, the subscriber's location is not known, the subscriber is reported as inactive, call delivery is not active for a roaming subscriber, Do Not Disturb is active, etc.). CFNA does not apply when the subscriber is considered busy.

Call Forwarding-Unconditional (CFU) *Call Forwarding-Unconditional* (CFU) permits a called subscriber to send incoming calls that are addressed to the called subscriber's directory number to another directory number (forward-to number) or to the called subscriber's designated voice mail mailbox. If this feature is active, calls are forwarded regardless of the condition of the terminating end.

Call Transfer (CT) *Call Transfer* (CT) enables the subscriber to transfer an established call that is in progress to a third party. A third party is defined as an individual who is not involved in the initial call. The call to be transferred can be an incoming or outgoing call.

Call Waiting (CW) *Call Waiting* (CW) provides notification to a controlling subscriber of an incoming call while the subscriber's call is engaged in a two-way call. Subsequently, the controlling subscriber can either answer or ignore the incoming call. If the controlling subscriber answers the second call, call waiting can alternate between the two calls. Theoretically, call hold is an inherent capability of call waiting.

Calling Number Identification Presentation (CNIP) *Calling Name Identification Presentation* (CNIP), commonly known as Caller ID, provides the number identification of the calling party to the called subscriber. One or two numbers can be presented to identify the calling party.

Calling Number Identification Restriction (CNIR) *Calling Number Identification Restriction* (CNIR) restricts presentation of that subscriber's *Calling Number Identification* (CNI) to the called party. When CNIP was first deployed on a commercial basis, CNIR was just as popular among wireline customers. This situation has eventually shifted to where CNIP is far more popular. One can only assume that people who have nothing to hide do not mind having their numbers displayed.

Calling Name Identification Presentation (CNaIP) *Calling Name Identification Presentation* (CNaIP) provides the name identification of the calling party to the called subscriber. CNaIP is not a popularly deployed service at this time.

Calling Name Identification Restriction (CNaIR) *Calling Name Identification Restriction* (CNaIR) restricts the presentation of the calling party's name to the called subscriber.

Conference Calling (CC) *Conference Calling* (CC) provides a subscriber with the capability to establish a multi-connection call (i.e., a simultaneous communication between three or more parties or conferees).

Three-Way Calling (3WC) *Three-Way Calling* (3WC) provides the subscriber with the capability to add a third party to an established two-party call so that all three parties can communicate via a three-way call.

Do Not Disturb (DND) *Do Not Disturb* (DND) prevents a called subscriber from receiving calls. When this feature is active, no incoming calls will be offered to the subscriber. DND also blocks other alerting, such as the Call

Forwarding-Unconditional abbreviated (or reminder) alerting and Message Waiting Notification alerting. DND is also a feature on some telephone devices, rather than a network-based feature.

Message Waiting Notification (MWN) *Message Waiting Notification* (MWN) informs subscribers when a voice message is available for retrieval. MWN can use the pip tone, an MS indication, or alert pip tone to inform a subscriber of an unretrieved voice message. MWN does not impact a subscriber's capability to originate calls or to receive calls.

Selective Call Acceptance (SCA) *Selective Call Acceptance* (SCA) is a call screening service that enables a subscriber to receive incoming calls only from parties whose *Calling Party Numbers* (CPNs) are in an SCA screening list of specified CPNs. Calls from CPNs that are not on the SCA screening list—and calls without a CPN—shall be given call refusal treatment while SCA is active.

Short Message Service-Point-to-Point (SMS-PP) *Short Message Service-Point-to-Point* (SMS-PP) provides bearer service mechanisms for delivering a short message as a packet of data between two service users, known as *Short Message Entities* (SMEs). SMEs are SMS endpoints that are capable of composing or disposing of a short message. One or both of the service users can be a mobile station. The data packets are transferred transparently between two service users. The network or destination application generates negative acknowledgments when it is unable to deliver the message as desired. The destination application might respond with an automatic acknowledgment and might include application-generated or user-provided information.

Messaging Delivery Service (MDS) *Messaging Delivery Service* (MDS) permits pending voice messages to be attempted for delivery to subscribers on a periodic basis until the subscriber acknowledges receipt of the messages.

Paging Message Service (PMS) *Paging Message Service* (PMS) permits paging messages to be attempted for delivery to subscribers (via the SMS) on a periodic basis until the subscriber acknowledges receipts of the message.

Voice Mail (VM) *Voice Mail* (VM) provides the subscriber with voice mail services, including not only the basic voice recording functions but also time-of-day recording, time-of-day announcements, menu-driven voice recording functions, and time-of-day routing. VM is a service that was first deployed in an ILEC switch during the early 1980s. The popularity of the answering machine limited VM's growth. VM has seen an enormous growth as a value-added service in the wireless (cellular, PCS, and paging markets) industry, however.

Fax Mail (FxM) *Fax Mail* (FxM) provides the subscriber with fax mail services, including the capability to store faxes that are received and to forward faxes that are recorded based on a number of subscriber-set parameters.

This list of services is a generic and typical list of network-based services. CCS enables the service provider to separate the signaling information from the content information (e.g., voice). Some of the services require a CCS schema such as SS7 to provide the functionality expected of them. SS7 provides the service provider with the opportunity to control the signaling information in such a way that the service provider will have the capability to perform the following tasks:

- Manipulate the routing of the call
- Add information to the signaling stream
- Enhance the signaling stream with additional functionality by using parameters embedded within the call itself, and take advantage of the information that is embedded within the SS7 protocol
- Use the embedded signaling information for other types of applications

Figure 5-8 is a high level representation of the general nature of the SS7 protocol. The overall structure of the protocol lends itself to building and creating new applications.

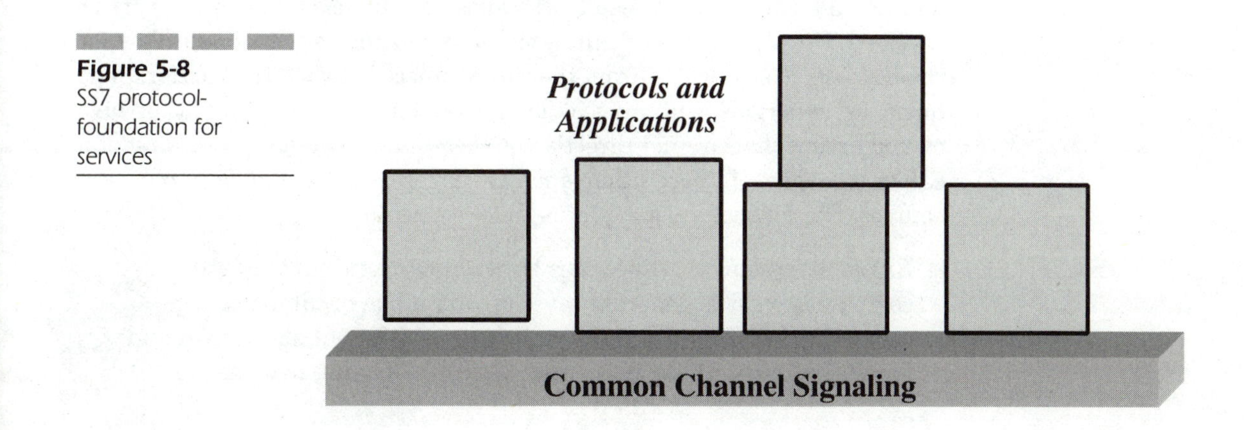

Figure 5-8 SS7 protocol-foundation for services

There are other types of signaling, but the foundation for future networking within the existing public communications lies with SS7. As the existing public communications networks evolve, there will be a migration from SS7 to some other form of CCS. The point is that all future applications will be based upon some form of CCS.

Network-based carrier services are another aspect of service provisioning. The following section addresses services that are provisioned to another carrier.

Network-Based Carrier Services

Everything inside your network should be viewed as a business opportunity. This statement is the basis of philosophy for a service that is provisioned by a carrier to another carrier and is called a carrier's carrier view of service provisioning, or simply a wholesale view of providing services. Interconnect today is being used by most carriers simply to gain visibility (access) between subscriber groups. Interconnect can also be used as a way of easing old tensions and establishing a business relationship. The interconnect can also be used to physically support a variety of services, however. Interconnect support for services comes in two ways:

- Business relationship
- Technical relationship

The business relationship engendered by physically interconnecting the networks of two or more carriers causes the carriers to seek ways of gaining more value from the basic business relationship of network interconnect (such as purchasing a Type 1 or Type 2A or by leasing T1 facilities). The additional business value results in additional financial gain for the parties involved. The interconnect facilitates discussion and serves as a catalyst for the carriers to seek value from the interconnect between the carriers. The business synergies will involve business objectives and might involve owned assets that are not directly connected to the actual processing of a call or transport of information (such as real estate) or common purchasing sources. The following list explains these concepts:

- *Business objectives* Objectives between carriers can vary. All companies wish to generate revenue and make profit; however, companies (whether a carrier or not) have different short-term and long-term goals. These goals might involve opening new market territories, new product lines, selling off unprofitable divisions, or buying into new markets. The list can be extensive.

- *Owned assets* Assets can include equipment, tools, vehicles, office equipment, office space, buildings, billing systems, or network management systems. If the carriers are working together already, then one can take the relationship farther and look at leasing, purchasing, or bartering these assets.
- *Common purchasing sources* Purchasing sources is not an obvious base upon which to further a business relationship. Consider the situation in which a carrier is providing wholesale services to another carrier via the wholesale carrier's network, however. The carriers can conceivably have exclusive relationships with one another in which one carrier will only purchase from the other carrier. In this case, the carriers can approach vendors and demand volume discount purchases of infrastructure equipment. This process has occurred successfully in the industry.

A number of factors play into creating a business relationship between carriers. Figure 5-9 illustrates the key factors in establishing carrier business relationships.

Figure 5-9 depicts the business relationship between the carriers as overlapping interests. The commonality in interests should be examined by any carrier that is entering into an interconnection agreement with another

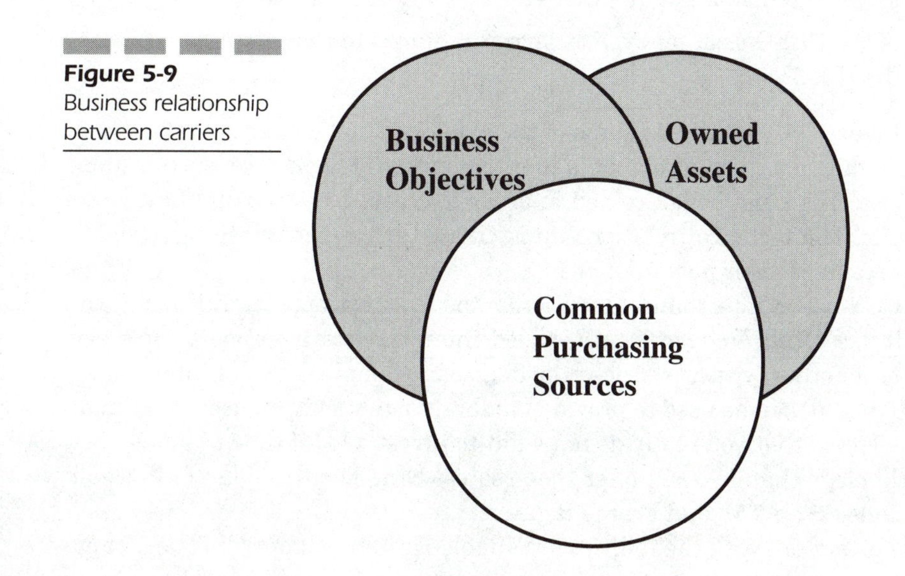

Figure 5-9 Business relationship between carriers

carrier. The ILECs recognize this situation and enable this type of synergistic growth to take place in an ILEC interconnection agreement. The CLECs that are entering the marketplace today need to take advantage of whatever the ILEC has to offer. The CLEC or other type of carrier needs to examine the ILEC interconnect agreement from the following perspectives:

- What does the entire agreement enable? Many ILEC interconnection agreements include provisions for wholesale services. The CLEC should read the entire agreement. CLECs can be wireless or wireline.
- What rates are being offered?
- What rates can the carrier negotiate?
- What value does the interconnect have beyond simple connectivity?
- What are the time limitations of the agreement?
- Can the CLEC or other type of carrier transport information from its own customers without having to log and report every transaction to the interconnected ILEC?
- How much visibility does the ILEC have into the CLEC or other type of carrier network? The network interconnect enables the transmission of information, which means that at a minimum, the ILEC should have the capacity to see the size of the traffic volume, traffic peak times, and directionality of the traffic. This feature does not sound important, but the fact is that the information provides clues that another carrier can use as intelligence on the carrier.
- Is the CLEC or other type of carrier required to report traffic types to the ILEC?

Figure 5-10 highlights some of the questions and concerns that carriers face when creating agreements to interconnect networks. These questions or concerns must be addressed in order to create balanced and fair agreements. The technical relationship is created when the carriers physically interconnect their networks.

CCS-based interconnects will offer the greatest opportunities. You can still use *Multi-Frequency* (MF)-based interconnects to support some services for other types of service providers. For example, a Type 1 interconnect (MF based) can be used to provide landline carrier switch-based voice mail. The fact is that you really do not want to invest a lot of time or money in a technology that does not offer the greatest bang for the dollar. CCS would include SS7, ATM, and Frame Relay.

Certain network capabilities are enabled as a result of CCS. These capabilities are the basis of network services that can be provisioned by carriers

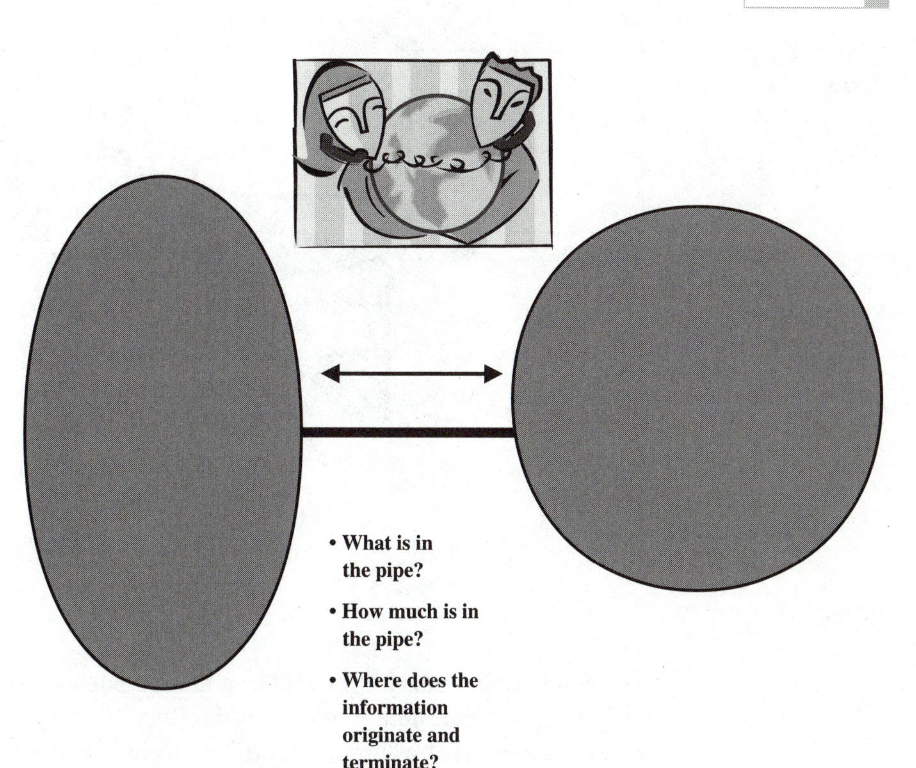

Figure 5-10 The Puzzle of network interconnection agreements

to other carriers and are opportunities that affect the commercial side of the telecommunications business. There will be a more detailed explanation of the commercial versus non-commercial aspects of network-based carrier services later in this chapter. The following list of basic capabilities are provisioned as services to other carriers. These capabilities represent the basic network functions needed to provide service:

- Database querying
- Call processing/call routing
- Service triggers
- Subscriber services

Figure 5-11 is a pictoral representation of the aforementioned functions. The functions represent opportunities to generate business for a carrier in both the retail and wholesale marketplace.

Database Querying Database querying does not sound glamorous, and the term does not even sound like a network service. Database querying is

Figure 5-11 Basic network capabilities

not a subscriber service, but it does enable subscriber services. In a CCS switching environment, there is a database that contains all of the subscriber service requirements (profiles). Typically, cellular carriers, PCS carriers, and wireline telephone companies use these databases to store all subscriber information, which will be used for routing purposes, for storing information on the specific subscriber's services, and even for call routing restrictions (if any).

The databases employed by service providers do vary functionally and even by name. These databases still share one common trait, however: they store information. Service providers—for a variety of reasons—use the information stored.

For example, call routing instructions, authorized lists of subscriber services, special subscriber billing needs, information to authenticate a call, calling card information, passwords or personal identification numbers, call forward-to numbers, intelligent network triggers (more on this subject later), restrictions on calling areas and services (i.e., no international calling, restricted area codes, restrictions on telephone calling services such as dial-a-joke or dial-up-a-companion, etc.), time-of-day restrictions, mission-critical information for network operations, etc. In summary, databases are store houses of information. The information can take on so many different forms such as; subscriber information, call routing instructions, and billing information.

The key point to note is that the database is a repository of critical information. A service provider will query the database for information. How the database responds and what information it responds with determines the response of the network. The network's response could be forwarding the call to another number or to voice mail. The response might be an announcement that the calling party is unavailable, or it could be a link to the Internet or to a menu of data services to peruse. The response could be a request for identification information, or it could be rerouting the call to another service provider for special services. The response could also be rerouting the call to a party who has a special time-of-day or location-based routing option recorded in their home service provider's database. How you use the information is up to you.

The databases can go by various names, such as the following:

- *Home Location Register* (HLR)
- *Visitor Location Register* (VLR)
- *Service Control Point* (SCP)
- *Authentication Center* (AC)
- Subscriber Service Profile Storage
- Some vendors create brand names of their own.

Functionally, all of these databases have different roles. Nevertheless, the databases store information. The information residing in the databases is essentially your products. Find out what types of databases are in your network and identify what type of information resides in them.

The database can be used between carriers. A carrier providing carriers' database services would be performing the following activities:

- Storing service profiles
- Storing routing information
- Storing authentication information
- Storing billing information
- Storing time-of-day restrictions
- Storing passwords

As I have indicated, the database would do more than simply store information. The databases respond to inquiries that are directed from specific authorized parties. The databases are interactive network elements that can be partitioned so that carriers can have secure access to their own information.

Call Processing/Call Routing A call can be either a voice call or a non-voice call (a data call). The databases that are employed in telecommunications networks contain routing information that is used to support call routing. This information can be summarily described as routing instructions for a variety of call types, routing instructions for specific subscribers, and decision-tree matrices used to support the call routing process. Call processing refers to the process of call setup, transmission facility establishment, and routing of a call or information transaction.

Network elements such as *Signal Transfer Points* (STPs) participate in the routing of the call. Tandem switches also participate in the routing of a call. Tandem switches are the sector and regional switching elements for large geographic areas. Cell site towers can be used by multiple service providers. Service nodes are used to process and route intelligent network-based transactions/calls. Today, switches tend to be dedicated to specific industry segments. For instance, Class 5 central offices are used by wireline carriers. Cellular carriers use mobile switching centers. ISPs use routers and servers. The telecommunications industry is moving towards multi-functional switches that are capable of serving all of the industry segments. These multi-functional switches are a step towards deploying technology that serves a converged industry.

Services Triggers The term *trigger* is normally used when speaking about intelligent networks. A trigger is "something" that can cause "something" to occur in the network. That "something" might be an event, a type of call, a time of day, a subscriber location, a day of the week, a calling location, a piece of terminal equipment, etc. A trigger causes the network to take some type of action, and triggers are used to support services.

For example, a calling subscriber decides to call another subscriber on a Monday. The called subscriber has the service provider set up the network to route all calls made to him to his home in Montana. The same called subscriber has his profile set up so that any call received on a Tuesday is sent to his home in California.

These types of triggers are called *terminating triggers*. Most triggers are terminating triggers, where calls are given special routing treatment at the terminating location. Originating triggers would include the location of the originating party or the type of terminal equipment that originates the call. An example of an originating trigger would be a subscriber making a computer-based telephony call. The network identifies that the terminal equipment is a computer. The subscriber's service profile has been set up to so that all computer users who are recorded on a specific list are alerted to the subscriber's location. Furthermore, the subscriber's service profile has been set up to send all calls to that subscriber to voice mail with a specific announcement. The trigger was a terminal device type.

Subscriber Services Subscriber services are those services that are provided directly to the individual end user/subscriber. Services such as call forwarding are provided directly to the end user, and services such as database querying are not end user services. These same services, however, can be provided by carriers that use another carrier's switching system. This scenario is the technical basis of CLEC resale. In such a resale environment, the CLEC's value is the reduced rates to the subscriber.

The following list is representative of subscriber services. You will find that there are only so many subscriber services that one can provide. What differentiates one service provider from another are price and the way in which providers enable their subscribers to invoke, activate, and use the services.

- *Call Delivery* (CD)
- *Call Forwarding-Busy* (CFB)
- *Call Forwarding-Default* (CFD)
- *Call Forwarding-No Answer* (CFNA)
- *Call Forwarding-Unconditional* (CFU)
- *Call Transfer* (CT)
- *Call Waiting* (CW)
- *Cancel Call Waiting* (CCW)
- *Calling Number Identification Presentation* (CNIP)
- *Calling Number Identification Restriction* (CNIR)
- Calling Name Identification Presentation
- Calling Name Identification Restriction
- *Conference Calling* (CC)
- *Three-Way Calling* (3WC)
- *Do Not Disturb* (DND)
- *Message Waiting Notification* (MWN)
- *Remote Call Forwarding* (RCF)
- *Remote Feature Control* (RFC)
- *Screen List Editing* (SLE)

- *Selective Call Acceptance* (SCA)
- *Selective Call Forwarding* (SCF)
- *Selective Call Rejection* (SCR)
- *Short Message Service-Point-to-Point* (SMS-PP)
- *Messaging Delivery Service* (MDS)
- Voice Mail
- Fax Mail
- Computer-based telephony
- Data services (broadband, narrowband, bursty, etc.)

These network capabilities represent one aspect of the network-based carrier service. Figure 5-12 illustrates how the author has segmented the network-based carrier services. The segmentation is important for understanding the product components of the carrier.

As Figure 5-12 illustrates, I have further segmented network-based carrier services into two distinct areas:

- *Commercial services network support* This category of network support involves the following network elements and support systems and includes the previously mentioned capabilities (database querying, call processing/call routing, service triggers, and subscriber services). The following list represents those parts of a carrier's network that directly impact the way in which a carrier provisions services to the subscriber:
  - Switch
  - Router
  - Servers

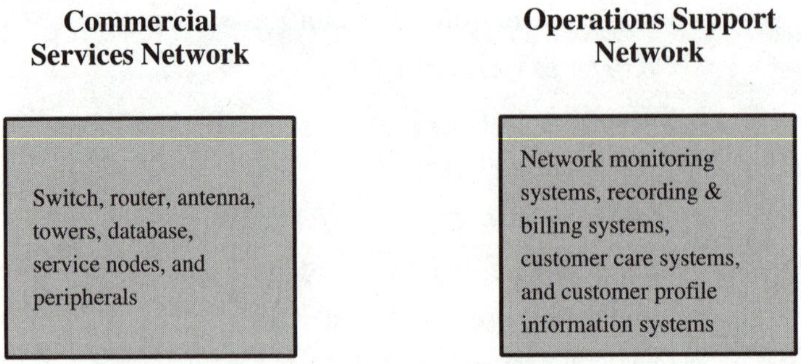

Figure 5-12 Network-based wholesale: commercial services and operational support services

Commercial Services Network	Operations Support Network
Switch, router, antenna, towers, database, service nodes, and peripherals	Network monitoring systems, recording & billing systems, customer care systems, and customer profile information systems

- Antennas
- Towers and roof tops
- Databases
- Service nodes
- Peripherals

The commercial services network support aspect of service provisioning has already been described.

- *Operations support network services* Operations support is normally defined as methods, procedures, and systems that directly support the operations of a telecommunications service provider. Another way of looking at operations support is that it involves all of the back-room systems, methods, and procedures that the subscriber does not see but that enable the carrier to run as a business. An analogy would be a department store in which the product on the floor represents the commercial side of the retail business, while the stockroom and warehouse represent the operations side of the business. This category includes the following systems:
  - Network management systems
  - Recording and billing (also known as call detail recording systems)
  - Customer care systems
  - Electric power
  - Water
  - Heating ventilation and air conditioning
  - Building space/floor space

The following section addresses the subject of operations support network services.

Operations Support Network Services

The following list consists of operations support network services business opportunities:

- Data warehousing
- Recording and billing support

- Network management
- Customer care

Operations Support can be broken down into functional areas. These areas can be translated from back-office functions into external business opportunities. As we have noted, by analyzing the carrier's network so that discrete layers of functions are identified, we can discover business opportunities. Figure 5-13 illustrates this point and these opportunities.

Data Warehousing There has been a great deal of industry buzz about data warehousing. Simply put, data warehousing is information processing. Running a company today—whether or not the company is a carrier—requires the company to process massive amounts of statistical and logistical data. Data warehousing is process of processing this type of information for the purposes of running the business more efficiently, more cost effectively, and in a manner that optimizes its competitiveness. The explosive growth of computer-based systems has created a slew of new ancillary businesses. As businesses computerize and electronicize their businesses, the number of different computer systems grows within a company. In a typical business of 5,000 employees, for example, you will find several different computer/information systems that all support different functions

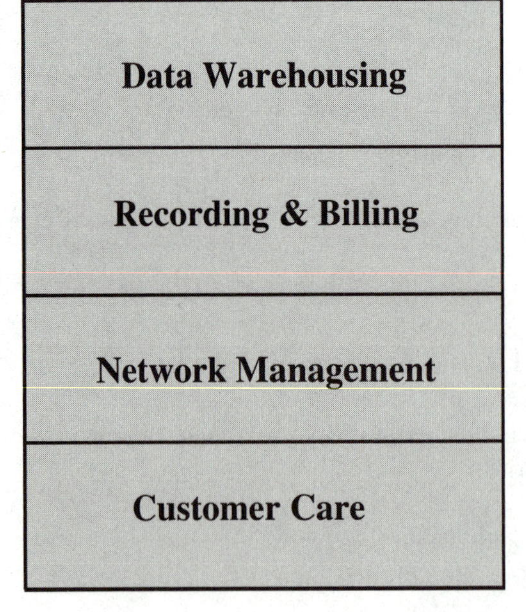

Figure 5-13 Operations business opportunities

within a company. For example, you will find information systems that support the following functions:

- Payroll
- Benefits
- Internal accounting (employee expenses)
- Asset management (furniture, office supplies, equipment used to support the business, etc.)
- Internal information systems (corporate e-mail/intranet)
- Customer information
- Marketing information
- Sales information
- Technology design
- Product design

The fact is that almost every department or organization within a company has a computer and probably has its own unique information-processing needs. To add complexity to the mix, companies will change or upgrade systems only when absolutely required. The result is that some organizations will have information-processing systems (a.k.a., computers) that might be several generations out of date. Furthermore, the sheer number of systems and the dynamics and complexity of the marketplace require some way for companies to process this information. The need for processing this information might involve improving operating efficiency, better understanding market trends, improving product performance, improving product safety, improving customer service, etc.

The amount of information compiled by a carrier can be used in a way to optimize the financial performance of the carrier. The amount of information recorded and maintained by a carrier about the market and the subscriber is enormous and can be used to for a variety of internal and external purposes. Figure 5-14 is an illustration of the storehouse of data in the carrier.

A carrier needs a way to view and process all of the information that is embedded within the multiple computer systems in the company. If a company did not attempt to get even the barest top-down view of the information that resides in its computers, then that company would be missing the chance to maximize its revenue. A data warehouse has this function. The data warehouse is a repository of information from transaction systems. The warehouse can audit, analyze, generate reports, examine the overall

Figure 5-15 Information: maximizing value

- Payroll
- Benefits
- Internal accounting (employee expenses)
- Asset management (furniture, office supplies, equipment used to support the business, etc.)
- Internal information systems (corporate email/intranet)
- Customer information
- Marketing information
- Sales information
- Technology design
- Product design

Processing & Compiling Information: Maximizing Value

health of the company's computer systems, and grind market information. The company conducting the data warehousing activity is also said to be performing data mining.

Large companies and carriers have the capability to perform this activity for themselves. Whether or not they wish to perform such an activity for others might be more of an issue of potentially helping a competitor.

Recording and Billing This term is somewhat self-explanatory. Companies that generate bills for their own customers might have so much excess computer processing capacity that this capacity could be used to support other providers' billing needs. This function might also include bill stuffing (sending the bill to the customer). You might not want to help your competitor, but then again, who says you cannot go to someone else's backyard to market your services?

Billing can be provided on a wholesale basis. A company that maintains its own billing systems and ancillary support functions can perform the following activities that are needed to support the billing function:

- Bill stuffing
- Call detail record processing

- Maintaining reports on customer account status
- Data delivery
- *Reconciliation* Clearinghouse function for multiple carriers. Interconnected carriers terminate calls in one another's network. Under such circumstances, the carriers are entitled to some form of compensation. Subscribers of both networks might even be entitled to some form of repayment. This compensation can be in the form of money or in some other form of compensation. The mechanism needed to manage this process is called reconciliation. In some cases, reconciliation between carriers is on the order of tens of millions of dollars annually.

Network Management Network management is a broad term. Some people use the term to describe the procedures that are associated with operating the commercial side of the service provider's network. Other folks use the term to describe the physical systems that are employed in the operation of the network. Network operation would include network monitoring of all commercial systems and maintenance of these systems. I view both positions as valid, because both are critical activities that I do not see as indistinguishable. There are service providers today that have created subsidiary companies to perform consulting in the area of establishing a process for managing the network. The same service providers have created companies to perform the network management activity. Carriers do possess the capability to segment their network operations center's monitoring capacity. At a minimum, even the real estate (building space) can be shared.

Network management and operations functions/capabilities can be extended to other carriers. In order to perform such a feat, the service provider (in this case the wholesaler) must establish a direct connect to the customer carrier. The customer carrier is quite literally treated like a network element within the wholesaler's network. Figure 5-15 illustrates this point.

Customer Care: Taking Care of the Subscriber and the Carrier Typical customer care includes handling customer complaints, signing up new customers, customer services trouble handling, managing customer service requirements, and removing customers from the carrier's service lists. Today, you will find companies that have established themselves as service bureaus for such activities. An expanded view of customer care would also include handling the concerns of the carrier customer as well as the concerns of the individual subscriber. In either case, the need for timely and courteous response to questions and troubles is paramount.

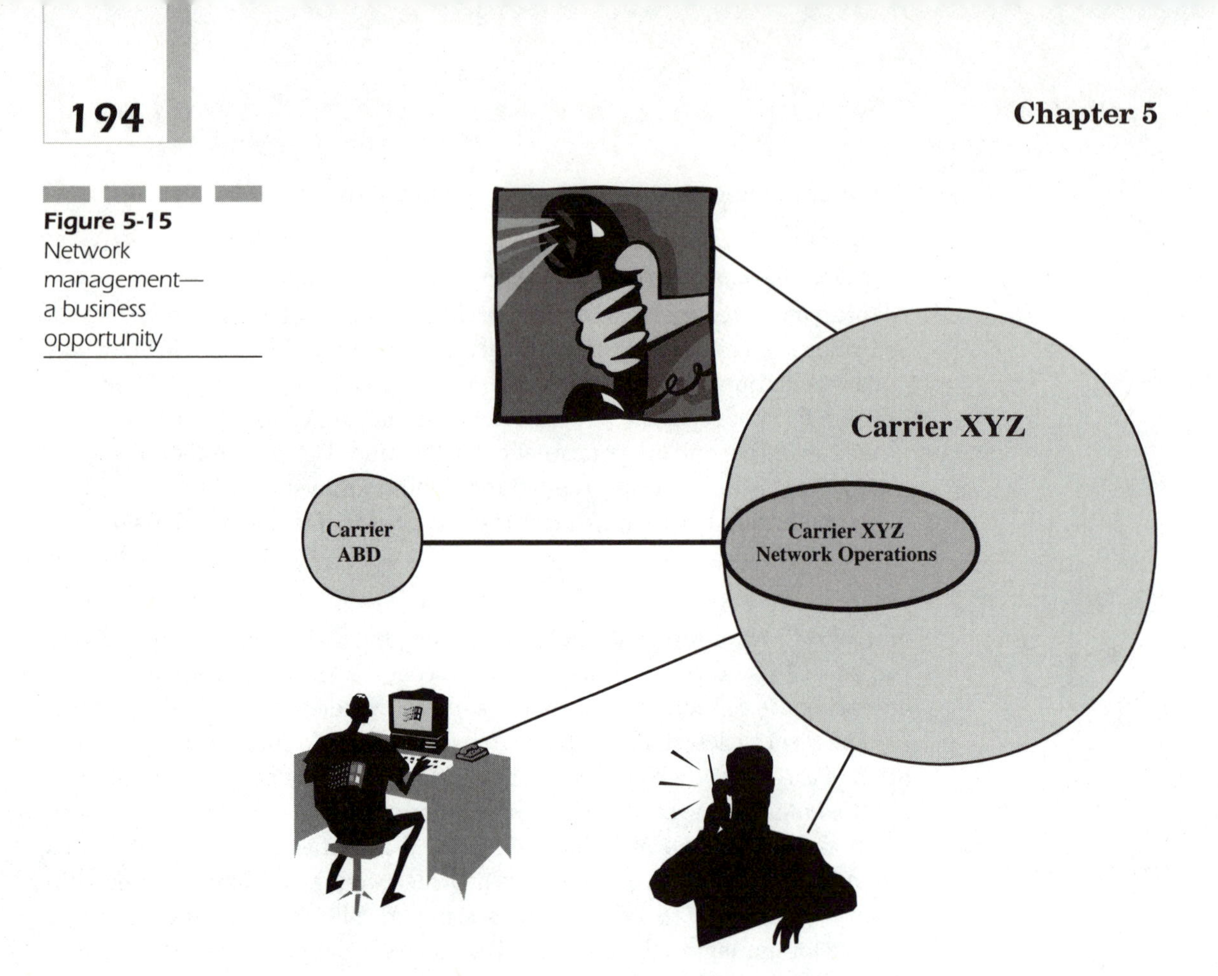

Figure 5-15 Network management—a business opportunity

If a CLEC chooses an outside service bureau, the challenge faced by the CLEC is finding a competent firm. The generic questions that a CLEC should ask of any service bureau, whether the service bureau is supporting directory assistance, operator services, network operations, or field repair are as follows:

- How long has the bureau been in business?
- Who are the principals of the service bureau company, and what are their resumes?
- How many employees are on staff?
- What are the staffing levels of the company, from management through technicians?
- What training is provided to the staff?
- Who does the training (if any)?
- What are the salary ranges of the field forces? This question might be considered too intrusive, but the fact is that carriers who are

experienced in dealing with outsource centers know that the question is critical for understanding the potential capabilities of the employee. This salary range has absolutely nothing to do with the type of person who is filling the position; rather, the question has everything to do with the level of investment and commitment that the company has placed in the employee and in the business. "Fly by night" operations will do just enough to make things look good and no more. This question cuts past the smoke and mirrors of the presentation of the company. The author's personal experience has been that no service bureau has ever refused to answer the question. These service bureaus understood the subtleties of the question. Nothing spells commitment better than the money, time, and energy invested by the service bureau.

Figure 5-16 illustrates the point that the customer is the focus of the service provider. This care can be outsourced and provided by specialized firms or by other carriers.

SUMMARY

Network interconnect between carriers involves more than just buying the privilege to terminate calls in the other carrier's network. To gain the

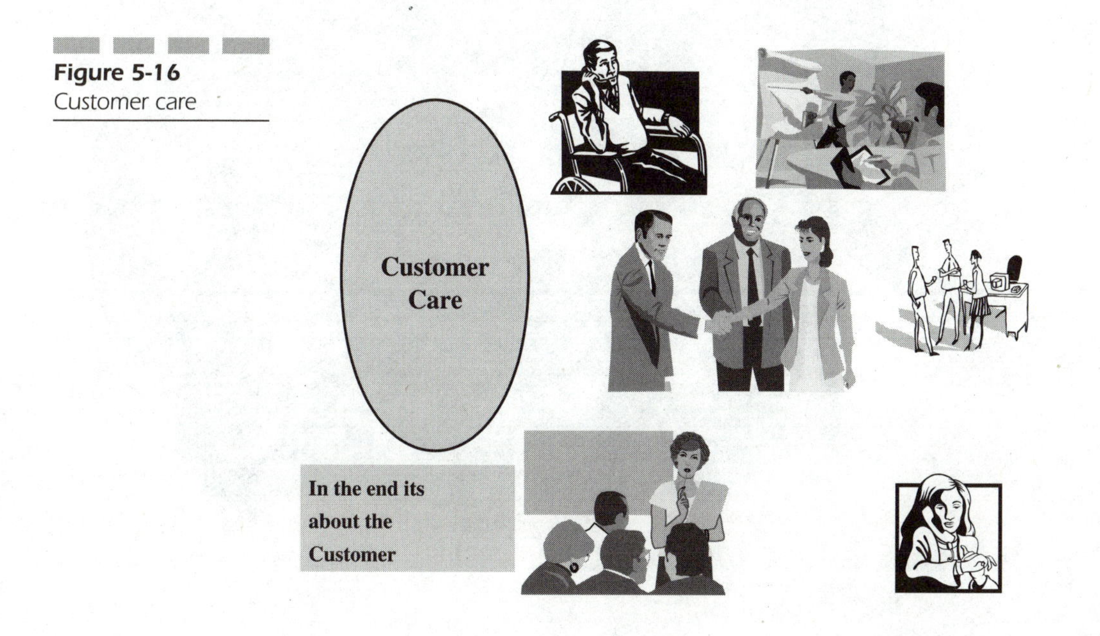

Figure 5-16
Customer care

total value of the network interconnect, the carrier needs to explore the less-obvious opportunities that are made available by the network interconnect. Leveraging the network interconnect is about technology opportunities, connecting subscribers, finding synergistic opportunities, and exploring opportunities outside the currently accepted business paradigm.

Figure 5-17 illustrates the nature of the industry today. The industry is filled with many competitors and opportunities. The picture appears confusing, and to many inside and outside the industry, the environment is also confusing.

Leveraging the network interconnect in the fashion that I have described was unthinkable as recently as the mid-1990s. Growing competition, mergers and acquisitions, a nearly unpredictable market landscape, and the need to be flexible to meet the needs of the marketplace, however, demands a carrier to be open to change. Treating network interconnect as a business tool is a necessity for survival. Network interconnect is an opportunity. Figure 5-18 illustrates this point.

The next section of this book will examine the types of telecommunications networks that are currently in operation today.

Figure 5-17 Telecommunications industry overview

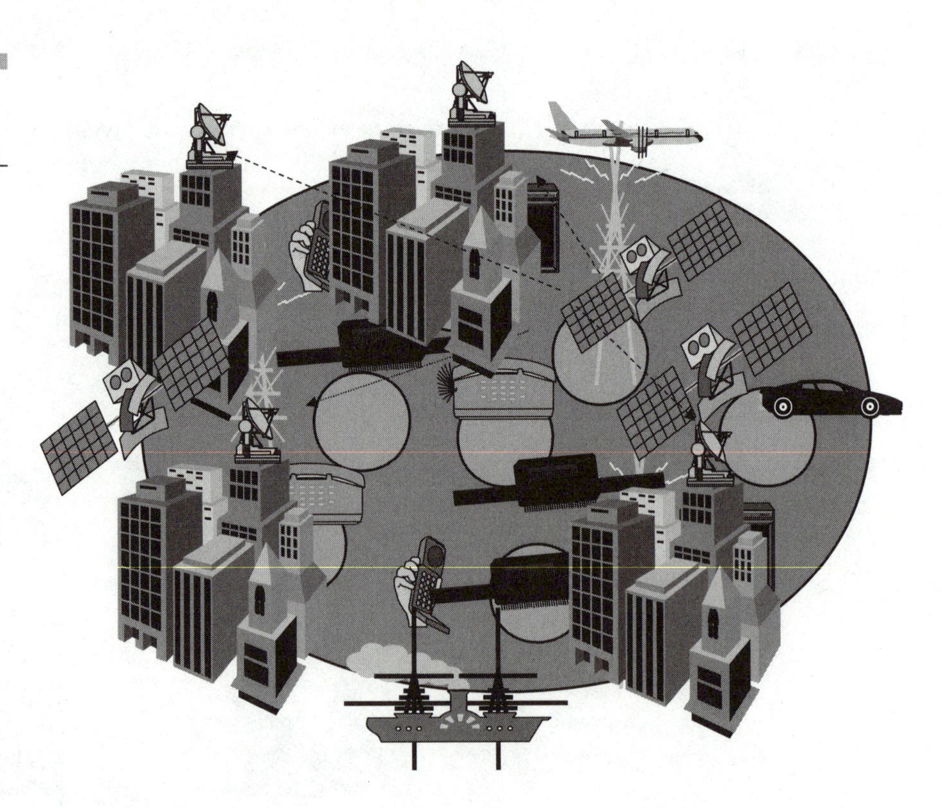

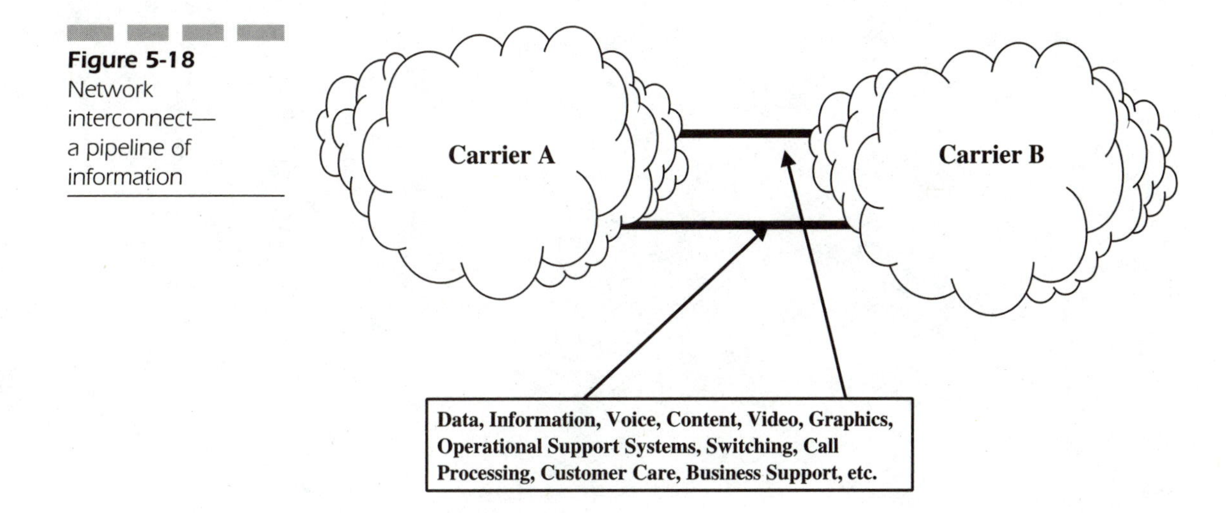

Figure 5-18 Network interconnect—a pipeline of information

Wireline Telephone Networks

To set the stage, let's start with a brief history of telephony in North America. It all began in 1876, when Alexander Graham Bell received a basic patent on his "talking machine." With investment in hand, he formed the American Bell Telephone Company. In 1885, the *American Telephone and Telegraph* (AT&T) company was established as a subsidiary of the American Bell Telephone Company to operate the long distance connections among the rapidly growing local Bell telephone companies. These local Bell telephone companies operated throughout most of the nation.

In 1900, AT&T was reorganized into a holding company, becoming the parent of the Bell companies and making Western Electric the exclusive manufacturing arm of the Bell system. After a series of federal government investigations and court challenges throughout many years, the Bell system was broken up. On January 1, 1984, the divestiture went into effect, with AT&T providing long-distance services and equipment manufacture and sales, and seven regional holding companies comprising local telephone companies with separate, "unregulated" subsidiaries for competitive activities providing local telephone service.

During the more than 15 years since divestiture, the "baby Bell" companies have tried to protect their respective markets. Ultimately, the baby Bells sought permission to expand into new markets (including long distance), whereas the long distance service providers sought to leverage their way back into the local exchange market. The result was the Telecommunications Act of 1996. The divestiture of AT&T's local telephone properties and the Telecommunications Act of 1996 have served as opportunities for change in the entire telecommunications industry. The breakup of the wireline telephone companies has served as the single most significant event in the last century of the telecommunications industry. Whole new opportunities have arisen as a result of the breakup of the telephone companies. These opportunities include video, data, Internet, and satellite. If you think about what the industry was like before the AT&T divestiture, you will remember that change occurred only when AT&T allowed the change. You may remember the joke that the single most significant event was when subscribers could now get a telephone in a color other than black—change occurred that slowly. When the AT&T "dike" had a hole punched into it, a flood of new opportunities came. To those without a personal history in the old Bell system, the changes appear slow and difficult to achieve. Those with a history in the Bell System, no matter how brief, will see the changes as rapid.

The industry today is a mixture of those with Bell system history and those who have no conscious memory of the Bell system. The skill sets and perspectives in the business today are far-ranging. However, you should not forget how much of basic telecommunications traffic engineering, philosophy, and business practices are derived from the wireline telephony business. Remembering and understanding how things were derived will give the reader an understanding of how interconnected the underlying telecommunications principles and practices are.

I have classified *local exchange carriers* (LECs), *competitive local exchange carriers* (CLECs), *interexchange carriers* (IXCs or IC), and *competitive access providers* (CAPs) as wireline carriers. It's a fairly loose view of the wireline carrier. In reality, a wireless carrier or an *Internet Service Provider* (ISP) can also be a CLEC or even an interexchange carrier. The Telecommunications Act of 1996 has opened the telecommunications market to all comers. Classifications that had their roots in the wireline carrier community are now applicable in all industry segments. The reader should approach this chapter with an open mind toward the historical aspect of telecommunications in the world; it all began with a wired communications device called the telephone. Much of the telecommunications traffic theory in use today was the result of studies in queuing theory, mathematical probability studies, and the old Bell system traffic studies.

The original pre-divestiture AT&T Bell System had one goal: to provide basic telephone service (voice) at an affordable price to everyone in the nation. The original organizational structure of the Bell system was based on functional tasks; each organization and department was organized around tasks that had to be performed. Local telephone companies were established to address the needs of the communities that they served. A long distance company was formed to coordinate activities and connectivity on a national scale. The Bell system meant one set of standards, one technology, and one goal.

The landline telephone network is comprised of the following basic elements:

- Class 5 end offices (local switching office)
- Class 4 and 4X switches
- Class 3 switch
- Class 2 switch
- Class 1 switch

The "class" designations had far more meaning and relevance when the old Bell system dominated the wireline telephone network. Generic terms such as local switching office, local end office, and tandem are used today. The current generic terminology is

- Local switch
- Local tandem
- Access tandem

Network/Traffic Structure

The pre-divestiture AT&T Bell system network had come to be known as the *Public Switched Telephone Network* (PSTN). The PSTN had been organized around two different but very interdependent network structures: the local network and the toll network.

Local Network Structure

The local network consists of customer/subscriber terminal equipment, which used to simply consist of a rotary dial telephone set or even the hand-cranked units that required manual operator intervention. The subscribers were and still are connected to the local switching equipment by loops of metallic wire. A local switch would serve a specific set of customers. This set of customers was geographically close to the local switch to reduce loop costs and to ensure quality voice transmission. There could be several switches in a building serving the customers in a geographic area. For example, 50,000 customers would be served by 2 switches capable of handling 25,000 customers each. The building in which the switches were located was called a wire center. The Figure 6-1 illustrates the wire center concept.

The area in which the wire center is located is known as an exchange area. Normally there are multiple wire centers in an exchange area. In medium and large metropolitan areas (and sometimes in small metro areas), a wire center contains multiple switches (see Figure 6-1). Before the AT&T divestiture, a local call was any call between parties in the exchange area and in which there was a single set of charges for calling within the exchange area. Therefore, an exchange area referred to a local calling area. Before divestiture, single area codes covered an entire state and one could call anywhere in a state without any long distance charges. After divestiture, a local call was determined based on legal/divestiture court-defined geographic boundaries called *Local Access Transport Areas* (LATAs). Years before divestiture, an exchange area was enormous in size. Now the

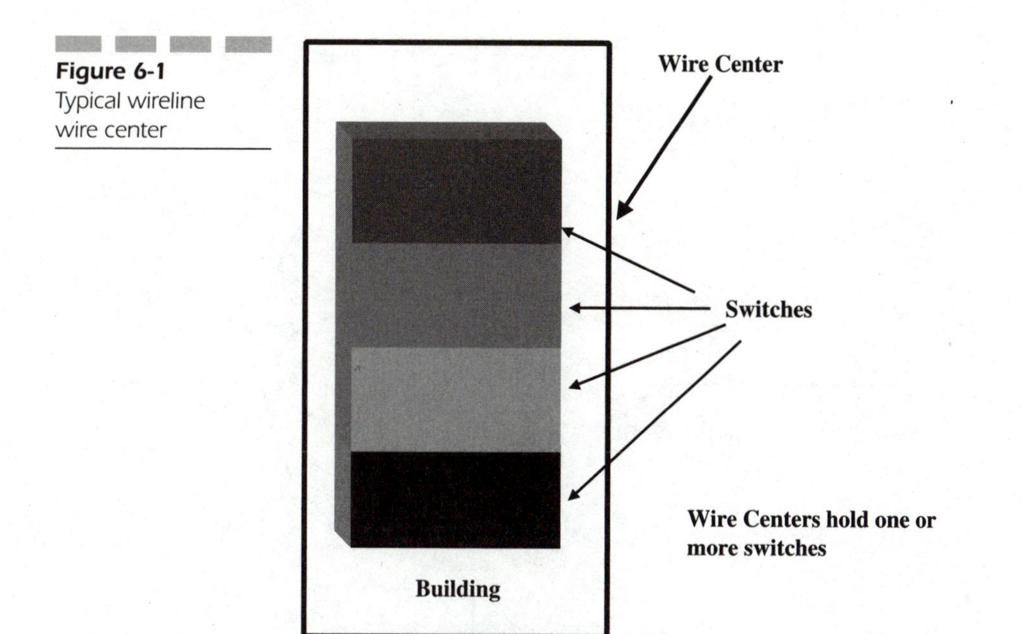

Figure 6-1 Typical wireline wire center

exchange area refers to just the three-digit central office code (also known as the exchange) that operates within the local telephone company LATA. This is why one will hear pre-divestiture telephony engineers appear to interchange the terms *exchange* and *central office* code.

Figure 6-2 illustrates the exchange area as it was before divestiture. Figure 6-3 illustrates the redefinition of an exchange area.

As the reader can see from Figure 6-2, calls between wire centers in the same exchange area were treated as single rate calls/local calls. The calls between exchange areas were often treated as toll calls. When the Bell system was a single and intact telephone system, the terms made sense. Today, the lexicon of pre-divestiture times can be confusing.

In today's post-divestiture environment, we have LATAs (local access transport areas). Setting up and completing calls within LATAs is the responsibility of the local exchange carrier. Setting up calls between LATAs is the responsibility of an interexchange carrier. In an inter-LATA call scenario, the LEC sets up the call to an interexchange carrier tandem, which transports the call across LATA boundaries to another tandem (owned by the same interexchange carrier) in the called LATA (let's call this LATA 2). The interexchange carrier tandem in LATA 2 that receives the terminating portion of the call sends the call to a LEC tandem or local switch. The LEC

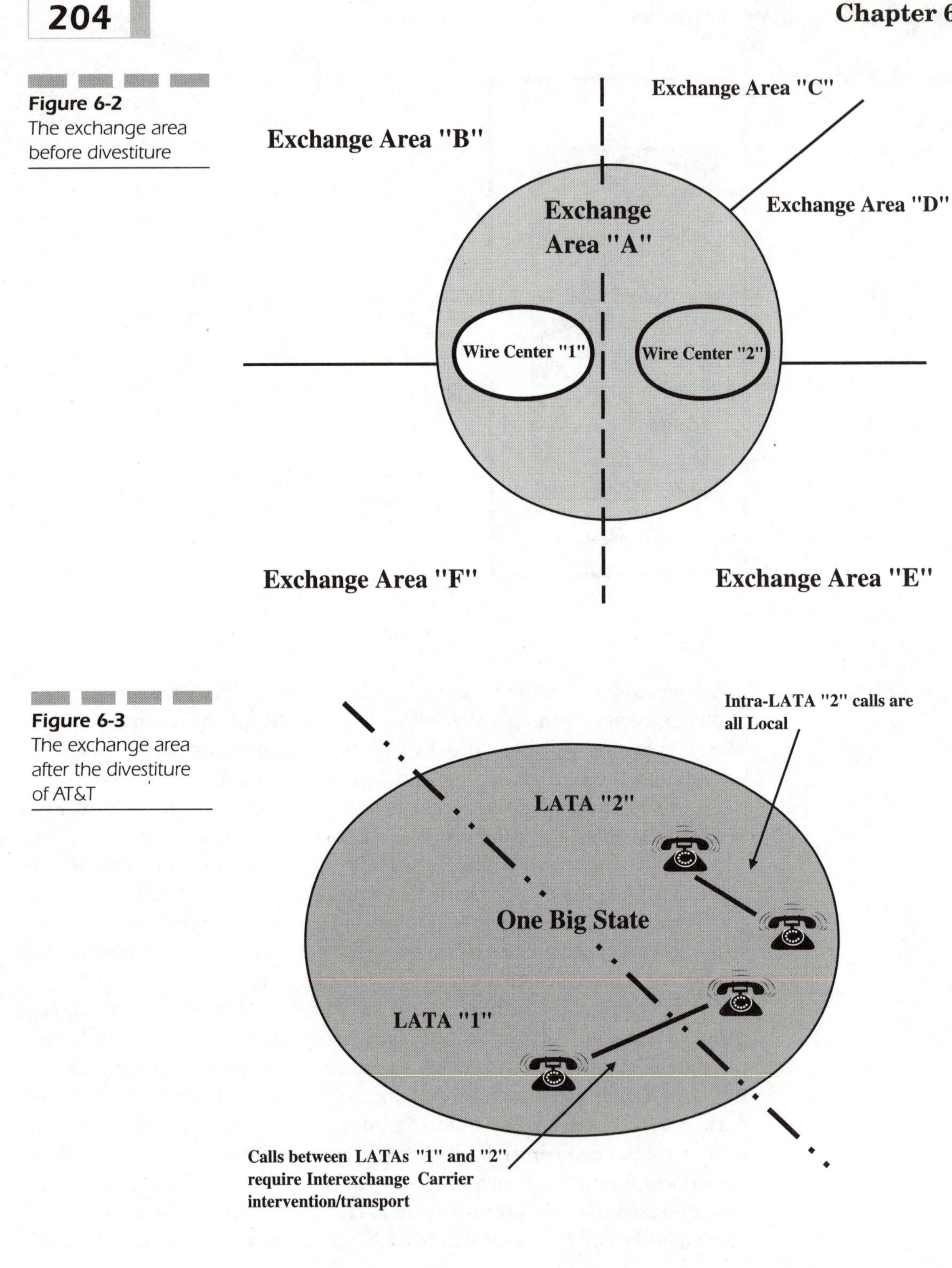

Figure 6-2 The exchange area before divestiture

Figure 6-3 The exchange area after the divestiture of AT&T

then carries the call to the subscriber. Figure 6-3 illustrates this new defined concept of the exchange area.

As I had noted in Chapter 1, the numbering plan used within the United States of America is administered by the North American Numbering Plan Administration. Each nation has its own numbering administrator. Overall global administration is handled by the *International Telecommunications Union* (ITU) Telecommunications Standardization Sector, also known as ITU-T. The current 10-digit numbering used in the United States was created in the mid-1940s. The original 10-digit plan used a 3-digit area code. However, the remaining seven digits were a alphanumeric combination—two letters and five numerical digits. You might have seen 1950s TV shows where the actors would say things like "Give me Murray Hill–12345." Due to the explosive growth of telephony, a shortage of alphanumerical designations occurred. This shortage led to a conversion to all number codes and designation, which was completed nationwide in 1980. Numbering was and is still the primary means of making routing decisions. The current numbering scheme is shown in Figure 6-4. At some point in the near future, there will be a need to expand the telephone numbering scheme beyond 10-digits. This occurrence will have a profound affect on all current wireline

Figure 6-4 Current 10-digit numbering scheme

Country Code (CC)+National Significant Number (NSN)

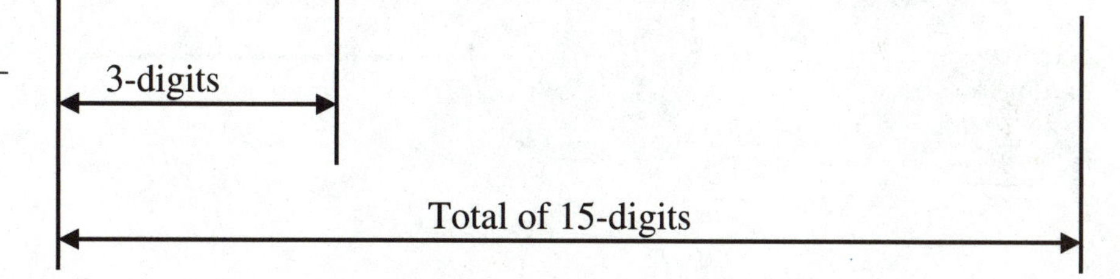

North America and the United States

- NPA-NXX-XXXX
  - Area Code - Central Office Code + Station Number
- NPA= 3 digits; 2-9
  - NXX= 3 digits; N=2-9 and X=0-9

switches, for most will be incapable of processing a number string greater than 10 digits without the help of some form of outboard processor

As I had noted in Chapters 1 and 2, switches are connected to each other via direct high-usage trunks or tandem trunks. The size of the traffic volume determines whether switches are connected through direct trunk groups or tandem trunk groups (see Figure 6-5).

Direct trunk groups are employed between wire centers when there is a strong community of interest (in other words, high traffic volumes). When the community of interest is low, a tandem trunk group is used (see Figure 6-6).

Figure 6-6 also highlights the two-level hierarchy used in the local network. Level 2 is comprised of local tandems and Level 1 is comprised of end offices and remote switches. This is also illustrated in Figure 6-7.

Intermediate-sized traffic volumes may warrant a combination of direct and tandem trunk groups. Basically, this solution would be the most economical. The reader will recall that this traffic configuration was described in Chapter 2, "The Telecommunications Hub—Creating Value." This configuration is supported by alternate routing techniques. Under an alternate routing schema, a switch will first attempt to pass the call over a high usage trunk group. The high usage group is often a direct trunk group, but it does

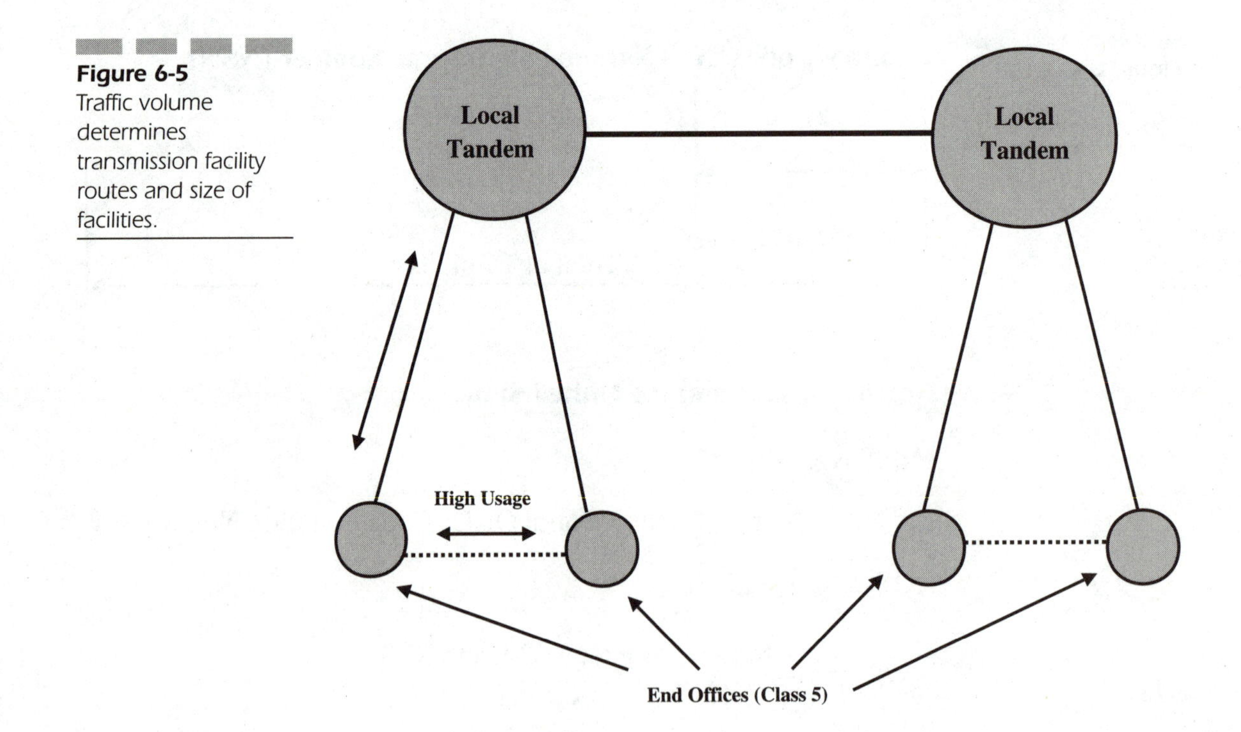

Figure 6-5 Traffic volume determines transmission facility routes and size of facilities.

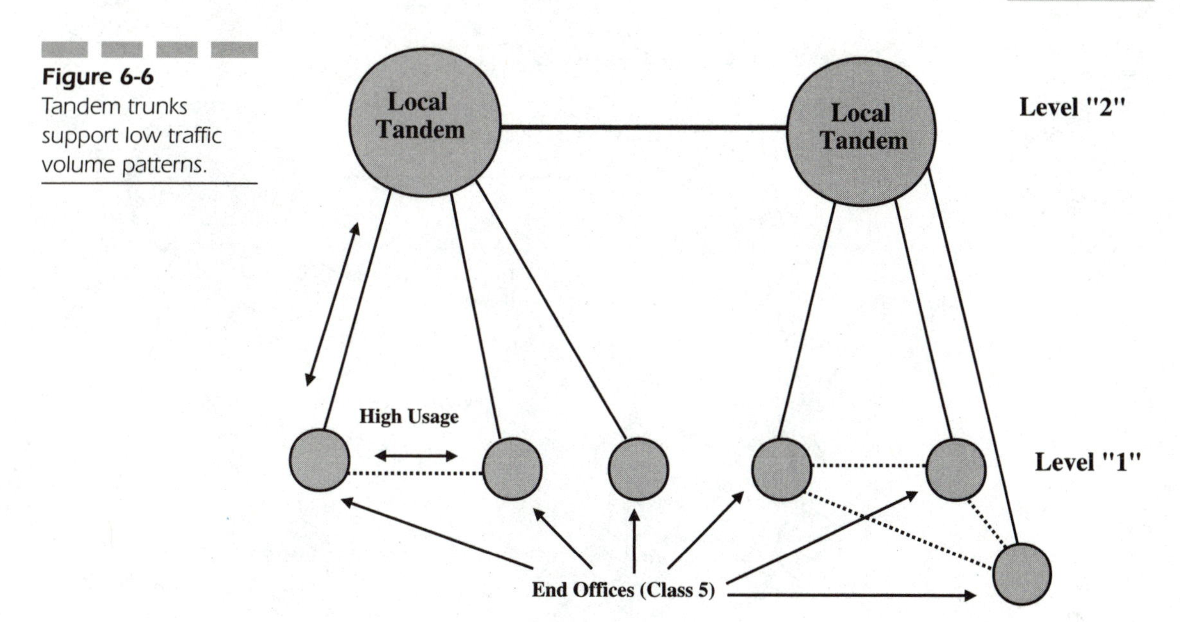

Figure 6-6 Tandem trunks support low traffic volume patterns.

not have to be. If the high usage group is busy, the call is routed over the final trunk group. A final trunk group is the final route. If all trunks in the final trunk group are busy, the subscriber is sent a reorder tone and the call is not completed (see Figure 6-7).

As I noted, a direct trunk group is not necessarily a high-usage trunk group. Economics drives a network configuration and the manner in which the traffic is managed. Network efficiency is also sometimes cited as the driver, which is also correct. Network efficiency is nothing more than efficient use of network resources, which boils down to economics. Efficiency/economics is a major factor in the design of networks.

The local network structure is comprised of a two-level switching hierarchy: local end office and local tandem office. The local tandem serves as a way of aggregating traffic for multiple end offices in a given local calling area, serving as the interconnect point for high usage and final trunk groups.

Toll Network Structure

Pre-divestiture, the long distance network was referred to as the *toll* network. The term is still used by millions of people in the United States when referring to long distance calls. The toll network structure defines a methodology for connecting/transmitting calls in an economical fashion

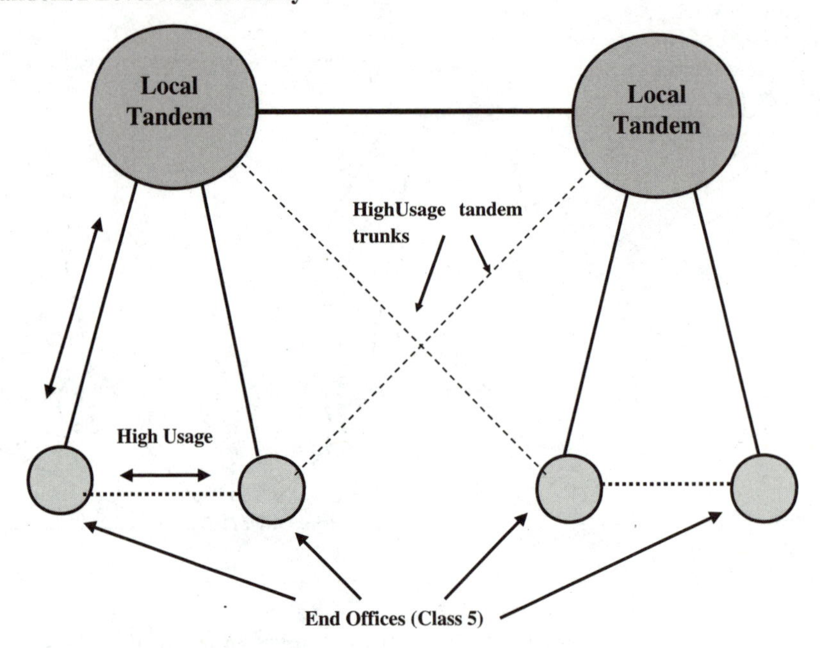

Figure 6-7
The call is routed over the final trunk group if the high-usage group is busy.

between widely spaced communities. Traffic studies of voice communications have determined that the volume of traffic between two widely spaced communities does not warrant the use of direct trunk groups between the two locations. One example is the number of calls between New York City and Los Angeles. As large as the volume of calls is sometimes, it is not nearly large enough to spend the money and resources to build a single trunk group between the two cities. The most cost-effective manner to transmit calls is through trunk groups that will be shared by multiple communities on each coast.

The need to cost-effectively connect the entire country led to the development of the five-level switching hierarchy used for decades. Figure 6-8 illustrates the original five-level switching hierarchy. In a very high traffic area usually associated with very dense population areas, local tandems can be interconnected to each other via a local sector tandem.

Figure 6-9 illustrates a current network hierarchy. Interexchange carrier and LEC areas of responsibilities are highlighted.

Today, the long distance companies have compressed their portion of the hierarchy. Some use mega-tandems to handle the traffic in large areas of

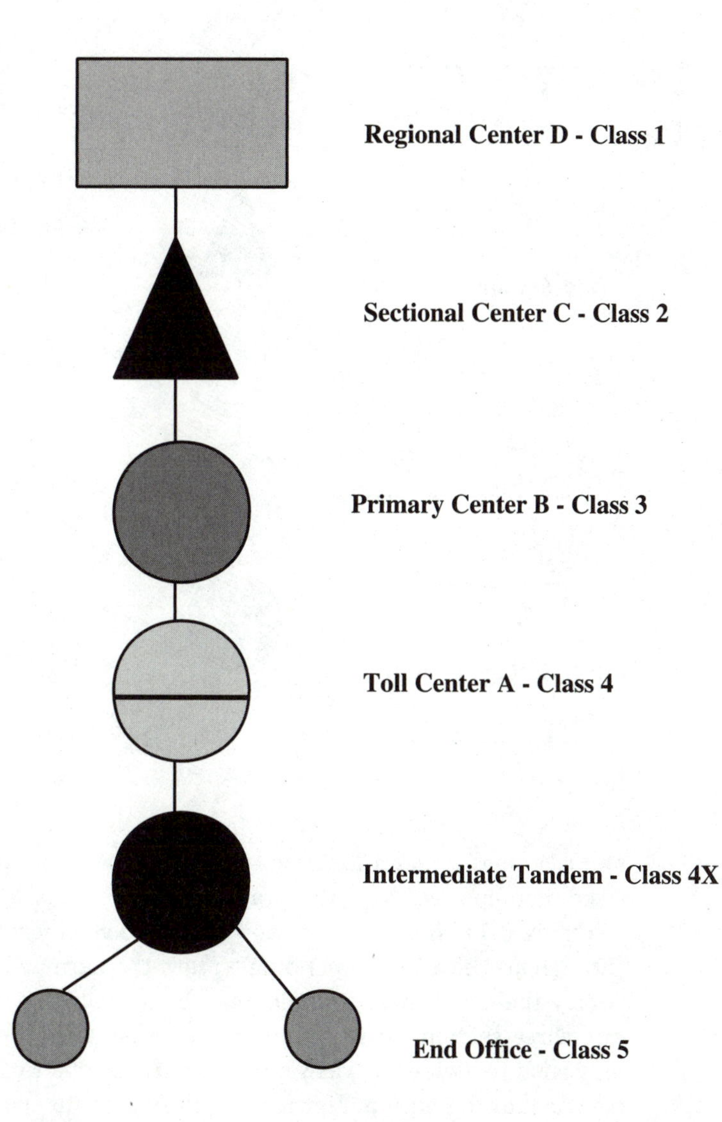

Figure 6-8
The original five-level switching hierarchy

the nation, reducing the need for sectional or primary centers. The current ILECs still employ two-level switching hierarchy. The following diagrams describe both the local and long distance switching hierarchy. I have noted local sector tandems. In some very high traffic and high density areas, tandems are used to aggregate traffic from local tandems that are supporting dozens of end offices each. The switching hierarchy today is different than the one supported by the consolidated Bell system. The long distance carriers do not need to support the multilevel hierarchy. In fact, the regional

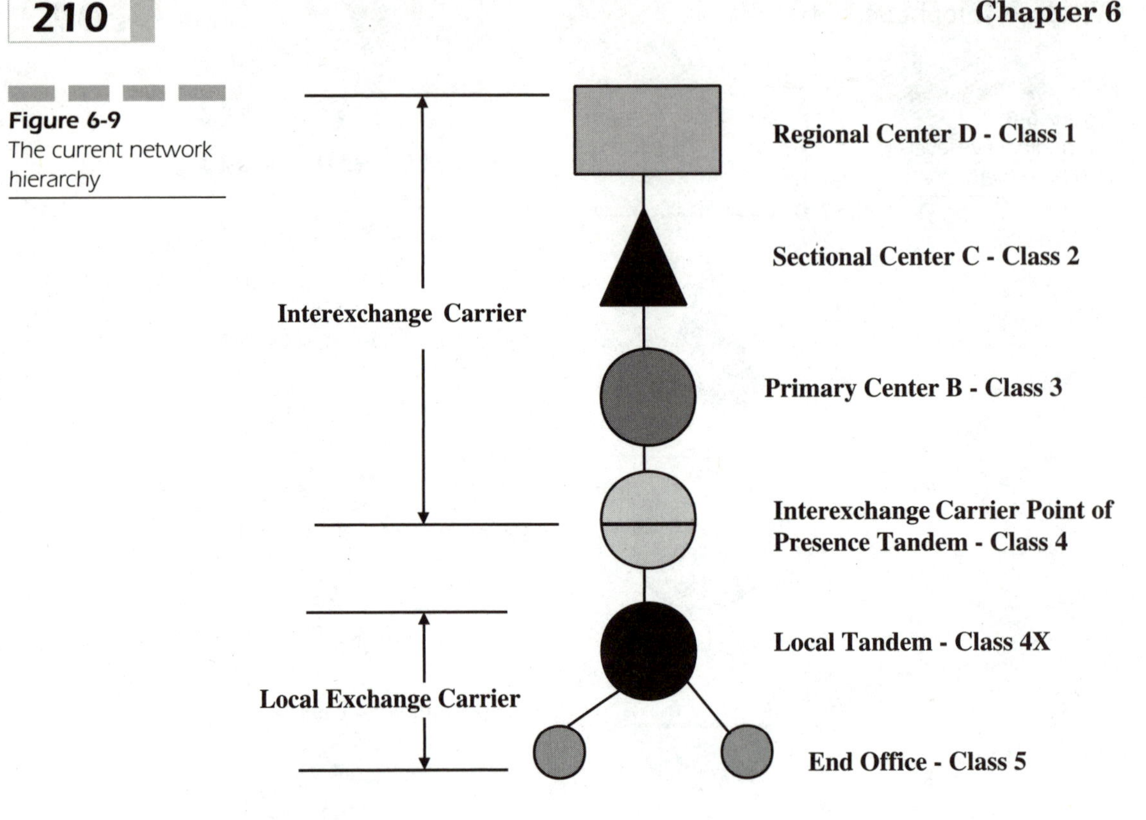

Figure 6-9 The current network hierarchy

and sectional tandem functions have been consolidated into single super-sized tandems, as I have illustrated in Figure 6-10.

Figure 6-11 offers a macro view of how such a hierarchy could be applied throughout the USA. Depending on how the various long distance carriers deploy their switches, a single long distance carrier may divide the USA into three sections: west, midwest, and east. Other long distance carriers may wish to divide the USA into two parts so one regional center can handle one half the nation. Figure 6-11 illustrates this point.

As Figures 6-8 and 6-9 show, the local end office (Class 5 switch) and the local tandems are part of the local switching network. The old toll network and current long distance network still interconnect into the local network at the local access tandem. The local access tandem is equivalent to the Class 4X switch, whereas the Class 4 is equivalent to a long distance carrier's *point-of-presence* (POP). The long distance carriers still employ the use of (or equivalent) Class 1, 2, and 3 switching centers to manage traffic throughout the nation. Discussion on Alternate Routing and Dynamic Routing techniques aside, the basic rule of call routing is to route and complete the call at the lowest level of the switching hierarchy. Completing a call

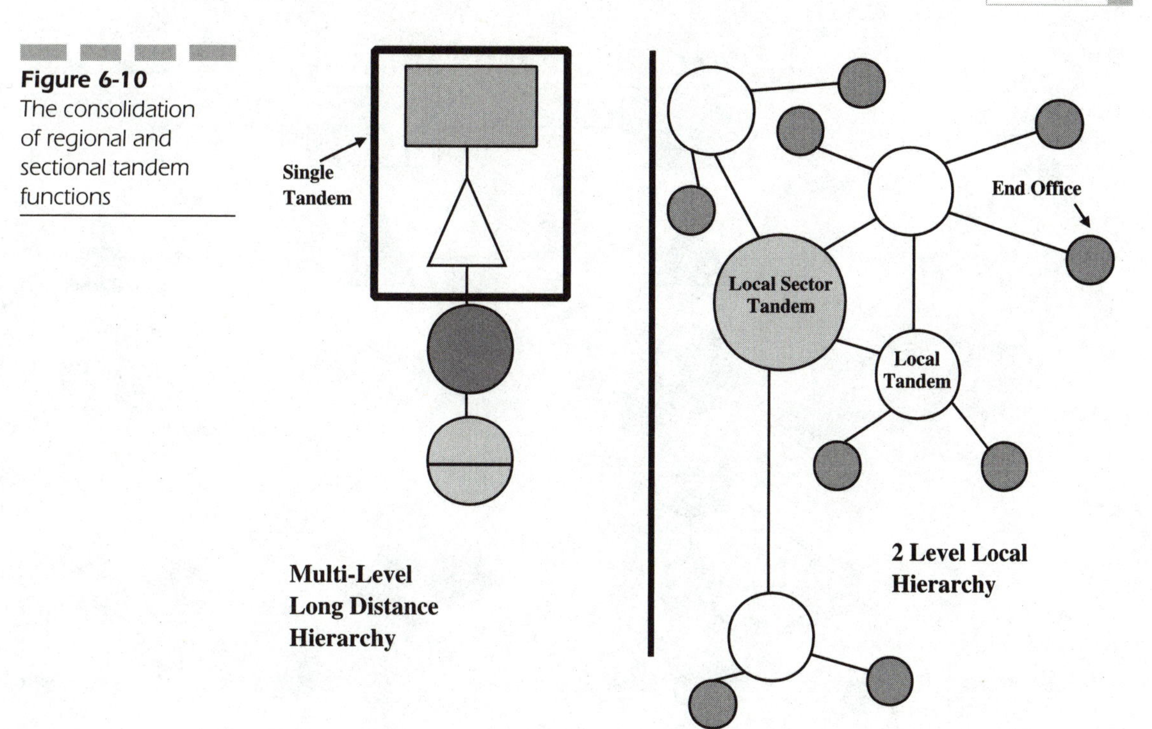

Figure 6-10 The consolidation of regional and sectional tandem functions

becomes expensive when interswitching system transmission facilities (also known as trunks) and their associated Class 4 and higher switching systems get involved in transmitting the call. Therefore, taking a route that uses the fewest trunks is the most desired.

Switch Types

The typical switching entities used in a voice wireline network are described in greater detail below.

End Office The end office/local switch (popularly known by the predivestiture telecom professional as the Class 5 switch) provides the subscriber access to the telecommunications network. The end office is the user's point of interconnection into the larger communications network. The end office is capable of routing a calling subscriber on a selective basis. End offices typically have more than one route to send out a call. End offices can direct connect each other on high usage trunk groups when economically appropriate. The end office is a component of the local loop.

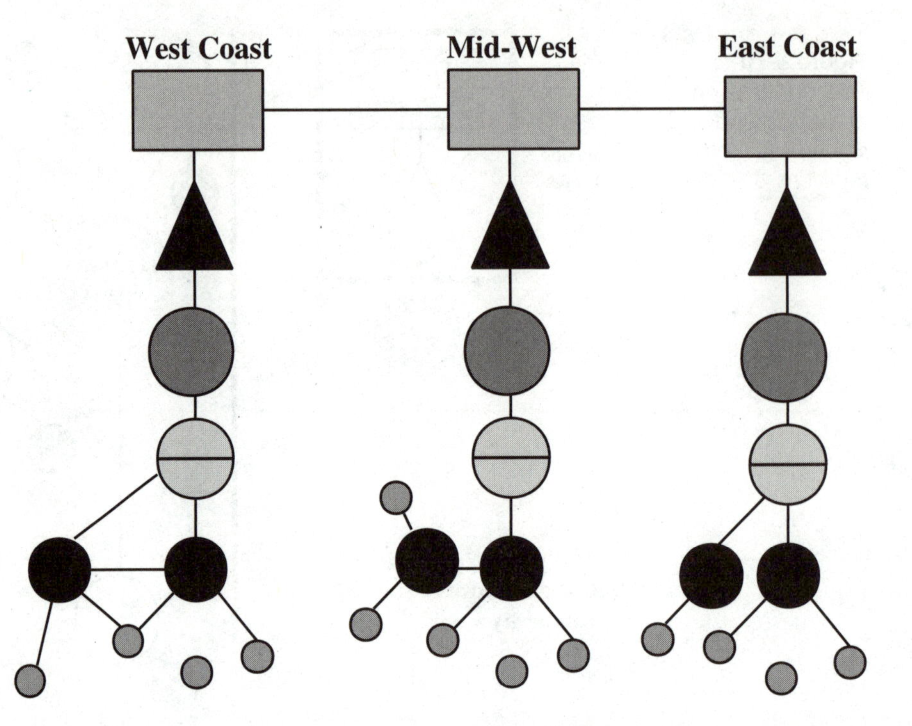

Figure 6-11 A long distance carrier might divide the United States into three zones.

Tandems The tandem switch (Class 4 and higher) is a network component used to interconnect end offices when direct trunk groups cannot be economically justified and used to connect tandems to each other. The tandem is like a "gateway" to other networks. There are two basic tandems types: local tandem and access tandem.

In a traditional wireline configuration, the tandem performs the following functions:

- Connect end offices to other end offices
- Connect other tandems
- Connect other networks
  - Access interexchange carrier networks
  - Access operator positions
  - Access wireless networks
- Connect centralized function

The local tandem routes traffic from several end offices to different areas served by other local tandems. The access tandem is used to provide access

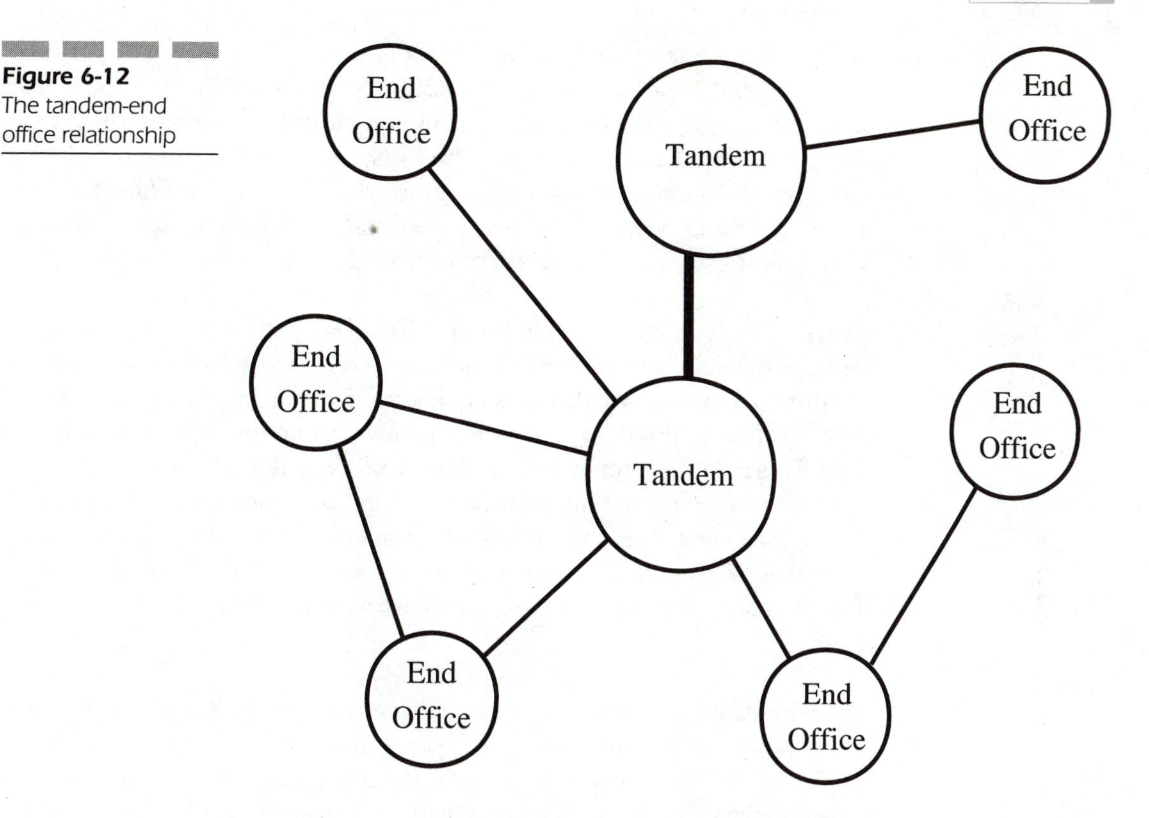

Figure 6-12
The tandem-end office relationship

to other service providers, normally the interexchange carriers. Figure 6-12 is a simple illustration of the tandem-end office relationship.

Applying Wireline Network Structures

The pre-divestiture network structures I have described represent a hierarchical switching methodology. This hierarchical switching methodology is still in use today regardless of the carriers involved. As I noted in Chapter 2, hierarchical network switching design and alternate/dynamic routing capabilities enable a service provider to collect and distribute traffic to all switching points. Therefore, it does not matter whether a AT&T Bell system is in place today because hierarchical switching is a necessity. When you look at the wireline telephone network today, you can still see a network of networks that supports three basic transmission facilities networks. As old as the concepts and structures that I have described are, they all still apply.

It does not matter that there are CLECs, wireless carriers (PCS, cellular, dispatch, and so on), cable television providers, or satellite carriers; hierarchical switching will always apply. As I just indicated, there are three basic transmission facilities networks: the local transmission, interswitch, and intersystem/network transmission networks. The wireline facilities networks are described next. The reader will find that the concepts and structures described apply to other service provider types.

Loop The loop, whether it is wireless local loop or wireline loop, is the connection between the subscriber and the serving switching system. In the landline telephony world, the loop is a physical facility that connects the central office building in a large feeder cable via an underground cable system. This cable system branches off into small feeder cables and then even smaller feeder cables that eventually extend the loop to a point near each subscriber's home. In the case of wireless local loop, the last mile or hundred feet is wireless. Figure 6-13 is an illustration of the local loop. As the figure shows, the loop extends itself all the way down into the customer premise.

Interswitch Networks Interswitch networks refer to the facility connection between two or more switches within a given service provider's network. Generally, you could classify them into two groups: trunks and special-service circuits. The users of the service provider's network share trunks. Special-services circuits are dedicated to those subscribers who have a long-term requirement to reach specific distant locations in order to support specific service groups like *Integrated Services Digital Network* (ISDN). Figure 6-14 illustrates how a single provider's network can interconnect its own switches.

Intersystem/Network Connecting the networks of multiple service providers involves physically interconnecting the networks, identifying the demarcation point between these networks, and establishing common network signaling protocols. Given the current competitive environment in which the telecommunications network functions in, we essentially have a network of networks. This network of networks is capable of functioning because of interconnection.

Figure 6-15 depicts all three of the aforementioned basic transmission structures. When you take a broader view of the following diagram, you will see that the overall structure is applicable to all network types. Figure 6-15 illustrates the intersystem/network concept. As the figure notes, the networks of different providers will interconnect at some agreed point. Usu-

Figure 6-13
The local loop

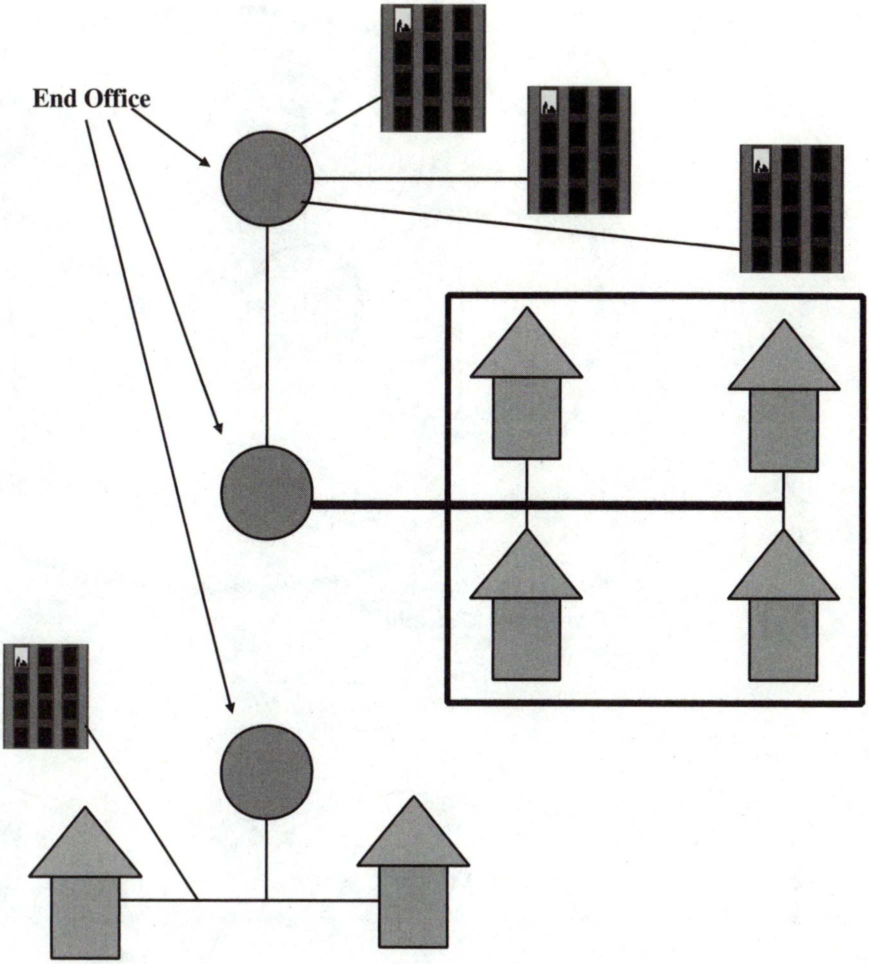

ally this point of interconnect is a tandem or some type of major switching element.

The Hub

The hub of a wheel is the center of a wheel. A bicycle tire, automobile tire, or even a truck tire has a center. Sometimes these tires have spokes that radiate out from the center to connect the center (the hub) to the outer rim of the wheel. The wireline network is a series of hubs, all interconnected through a series of loop, interswitch, and intersystem connections. The

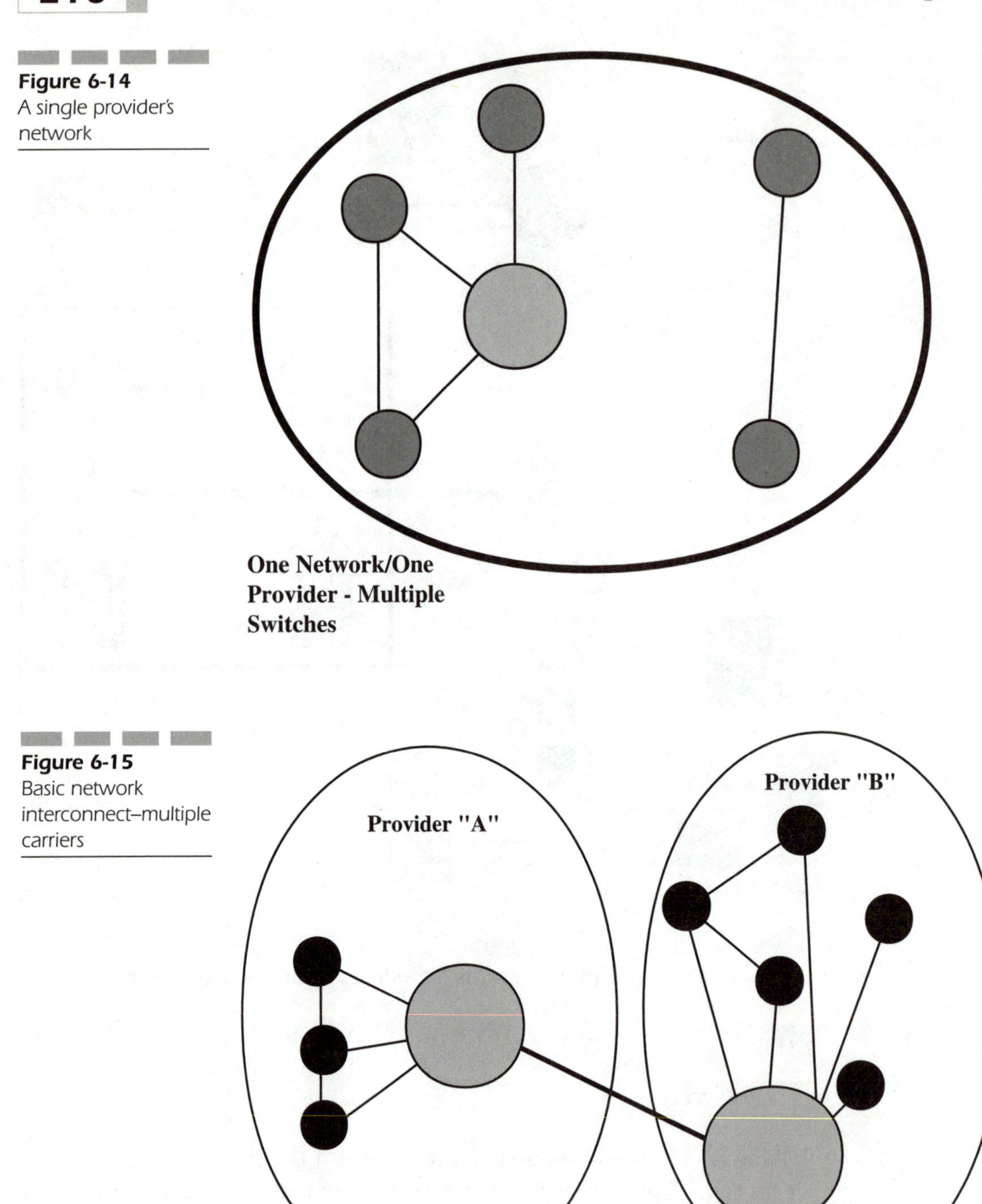

Figure 6-14
A single provider's network

Figure 6-15
Basic network interconnect–multiple carriers

wireline ILEC networks were and still are well positioned to reach the entire population of the nation. The ILECs' two greatest advantages over other wireline carriers are

- The local loop
- The ability to switch communications traffic throughout a state or region. Everyone, including the interexchange carriers, must come to an ILEC point-of-presence in order to reach a market or even a local subscriber.

The hub configuration facilitates the ILEC's ability to interconnect people and communities with other people and communities. The hub configuration was created because it enabled the ILEC to cost-effectively manage the flow of traffic. The hub is a network traffic configuration that is essentially a network advantage.

The ILECs' network advantage is also an advantage as it relates to provisioning of other types of services. The ILEC's local presence means that the ILEC is literally in every home and business in a community. Note that there are homes that do not have any telephone service and some businesses that use CLEC service. The ILEC's "home court" advantage is overwhelming. The ILEC's ability to switch traffic easily throughout a market and state means that the ILEC controls the flow of all traffic in that state. This is an obvious tactical advantage. Figure 6-16 is an illustration of how entrenched the ILEC's are in the local marketplace.

The CLEC today is working to obtain a local presence. The CLEC still needs the ILEC for access to the local loop and still needs the ILEC for switching outside of a local area.

The ILEC is in the middle of the entire local and regional switching hub. It is ideally positioned to enter a variety of different types of telecommunications businesses. Figure 6-17 is a high level pictoral representation of the ILEC's level of subscriber base penetration.

Applications/Products/Services

Applications of the wireline network were once restricted to internal activities that impacted the carrier. Products and services were restricted to "things" sold to a subscriber. The new wireline carriers, which would include CLECs, cable television providers, new attitude ILECs, and other types of wireline carriers, recognize the need to expand the definition of applications, products, and services. As I indicated in Chapter 5 "Applications," an

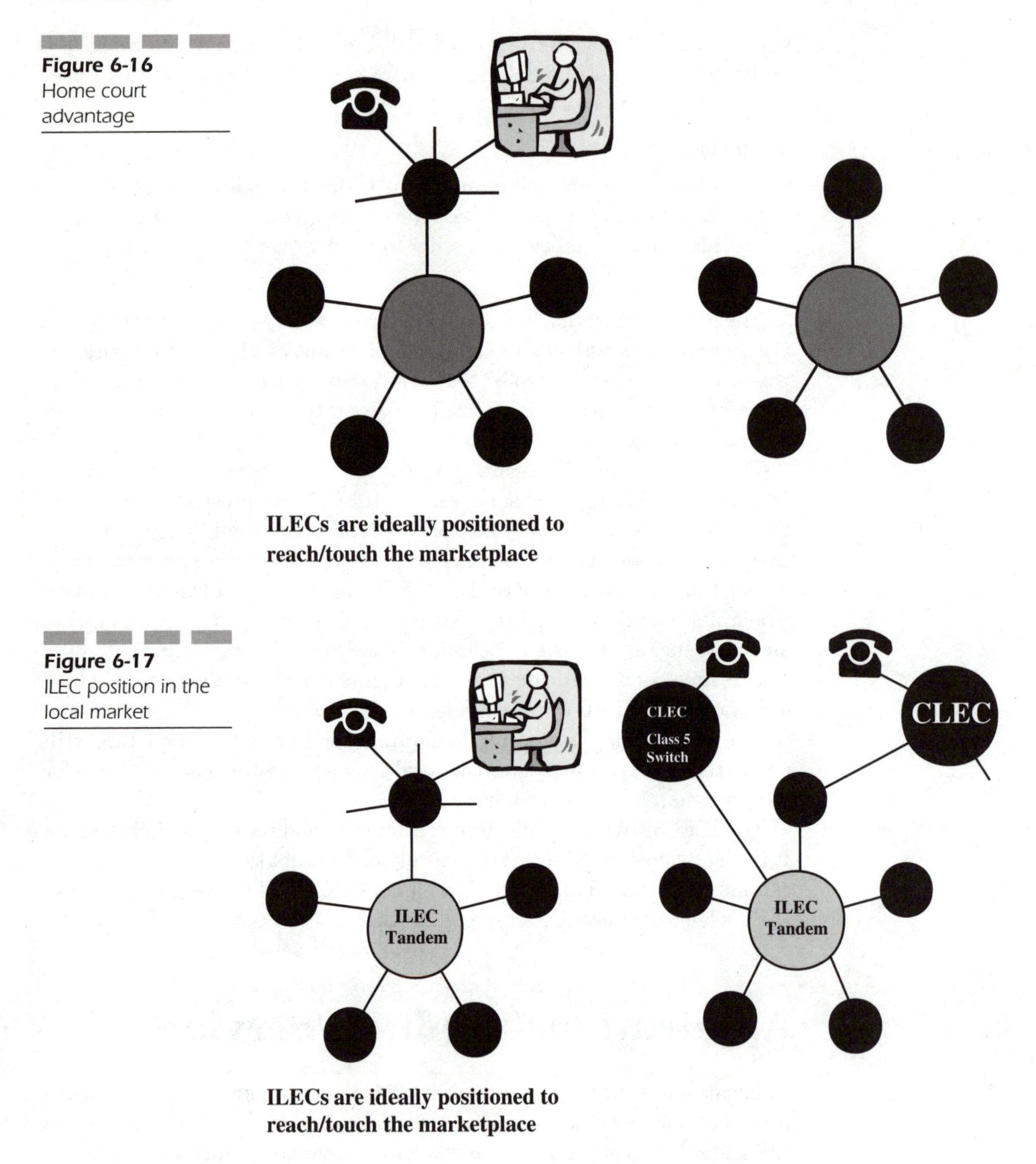

Figure 6-16 Home court advantage

Figure 6-17 ILEC position in the local market

application is something that does something useful or carries out a useful task. A product is the item that is sold to a customer; an application is not necessarily a product, but a product is always an application. A service is always a product. A service can be an activity or a system feature. An application can be a service, but a service is always an application. An application can be an internal activity that only impacts the user (the carrier) and no one else; therefore, the application is not a product. My view of an "application" not only acknowledges the aforementioned, but also everything else that can possibly be envisioned. Applications that are internal to the carrier's operation can in fact be sold as a product to subscribers (of all kinds —individuals and corporate). To a great extent, the words "application" and "product" are synonymous.

ILECs that have successfully transitioned into the new competitive marketplace look at their networks as revenue generators. However, this view goes beyond network platforms providing new services to subscribers. The new ILEC view of the network is one of total revenue generation, where each part of the network is a profit center. This model I am describing sounds like a carrier's carrier business model. On the one hand, this is accurate; on the other hand, it is only partially true. Such a carrier should seek to provide services to both the subscriber and the service provider. The reader should note that the service provider customer does not need to service the same subscriber base or even be in the same telecommunications segment. This type of carrier should be called a "full service carrier."

A wireline carrier is capable of supporting the following services. The reader will note that the following services are in fact functions an ILEC performs for itself today.

- Database querying
- Dataehousing (this is new to most landline carriers)
- Call routing
- Service triggers (known to landline carriers as intelligent network triggers)
- Subscriber services
- Billing support
- Network management
- Customer care

Figure 6-18 illustrates a network configuration that can support the aforementioned services.

Figure 6-19 is an additional illustration of the aforementioned.

Many of the existing wireline carriers have their own directory assistance and operator services bureaus. Directory assistance provides help

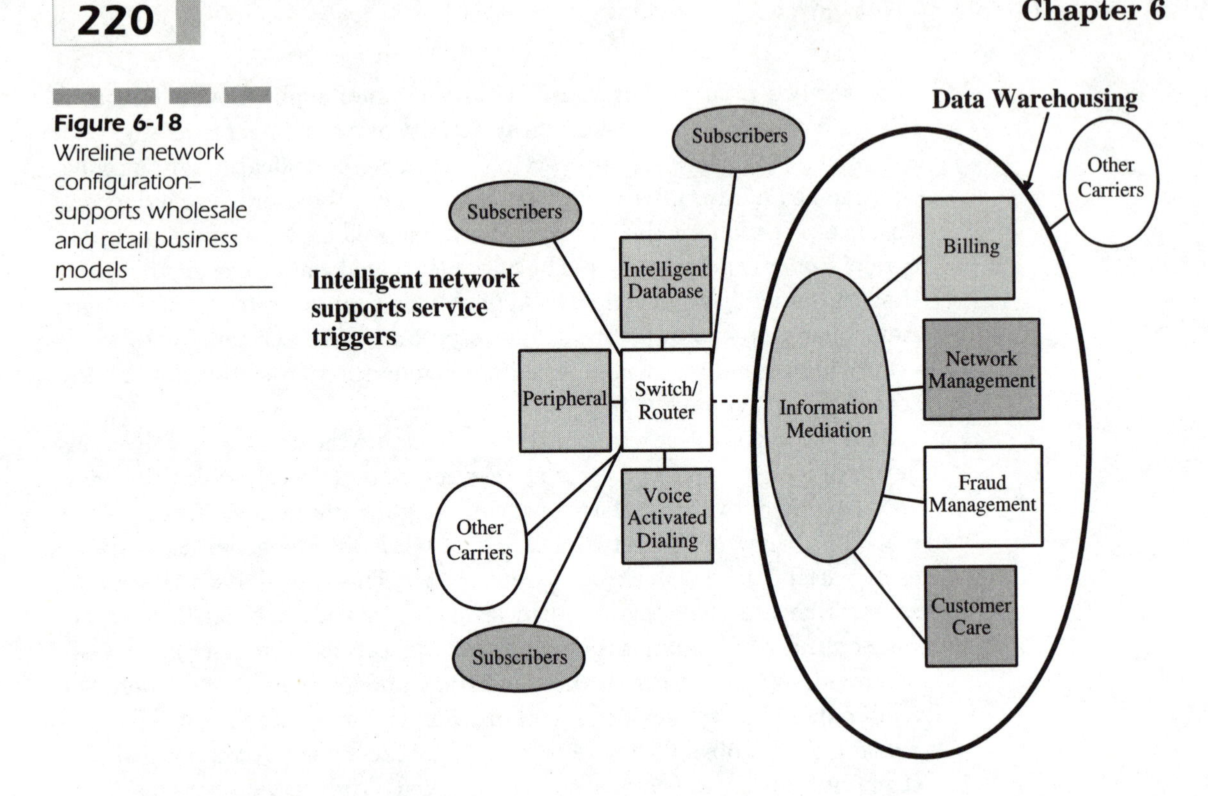

Figure 6-18 Wireline network configuration–supports wholesale and retail business models

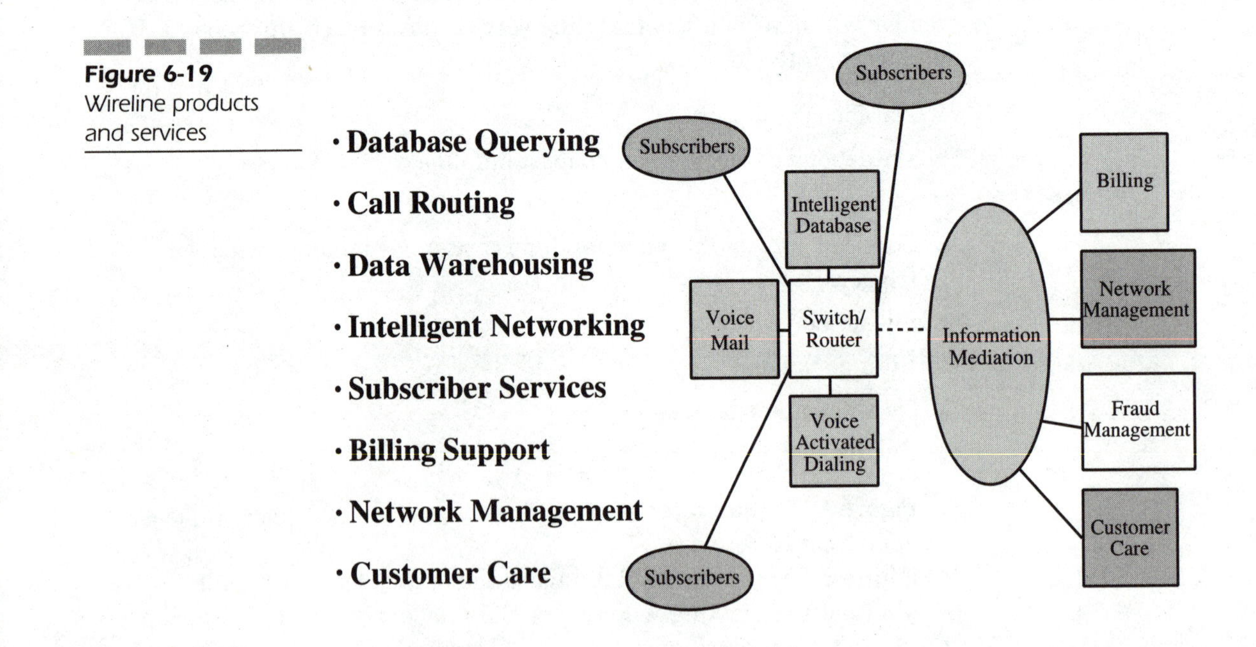

Figure 6-19 Wireline products and services

searching telephone number listings. Operator services is closely related but is treated functionally and in many cases operationally separate. Operator services is responsible for providing customer services assistance and call completion services. Directory assistance can also provide call completion services. However, the legacy of operator services included directory assistance, billing calling cards, collect calls, third-party billing, and repair. The operator encompassed all aspects of direct one-on-one customer interaction. The operator today is still viewed as the "face" of the service provider. Many of these functions have been broken out into special N–1–1 dialed number services. Not so long ago, the subscriber simply dialed zero and was connected to the operator. Figure 6-20 is a high level view of the directory assistance and operator services functions.

The following sections will give the reader a closer look at the wireline networks and how even the ILECs and other wireline carriers have becomes customers of smaller niche market players. The following sections will provide the reader a "flipside" view of the marketplace.

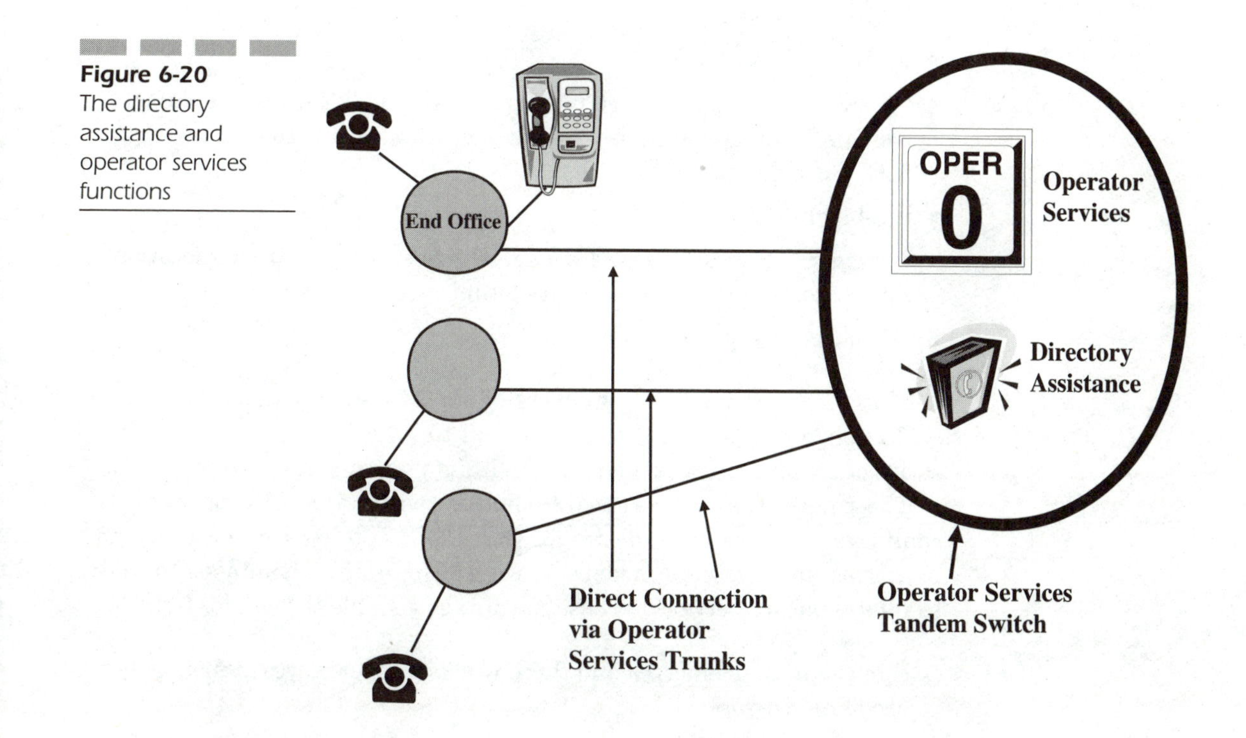

Figure 6-20 The directory assistance and operator services functions

Networks within Networks: Layering Opportunities

Wireline networks (traditional voice and low-speed data) cannot support all types of telecommunications without modifications. The point to note is that the wireline ILEC is well positioned to support a variety of different telecommunications business opportunities. The wireline network today, currently dominated by the ILECs, is a series of different networks. To date, most ILECs have focused on creating new opportunities through the deployment of new and additional networks. Some of the networks have no visibility beyond a select customer group, whereas others support ILEC operations. Interconnecting the networks was not an objective. The ILEC networks can be described as follows:

- *Public voice* The traditional public switched telephone network supported voice only at first.
- Low speed data
- Narrowband data
- Wideband data
- Video
- *Closed user groups* Closed user groups in a PSTN sense are in fact "tie lines;" direct voice connects between two specific customers. The ILECs provide such a service.
- Intelligent network
- *Transport* Provisioning of leased facilities between customer locations. The customer may use the facility for internal data transactions.
- Operational support system networks

These networks are in fact layers of opportunities. When you view the ILEC network as layers, you can see that the wireline network is in fact a series of opportunities or a business within a business (see Figure 6-21).

These networks exist as separate entities within the ILEC network. The challenge or maybe even desire for the ILECs is to converge the network platforms and business models. However, there are advantages to maintaining separate networks. These advantages enable the carrier to:

- Run separate businesses and thereby avoid appearances of monopolistic use of its networks.

Figure 6-21 The multiple networks that make up the ILEC network

- Avoid the encumbrances of legacy network equipment. This would include the enhancing of network elements. Better to deal with the most current systems instead of older systems.
- Less complex network management.
- Easier to administer new opportunities. If the network opportunities are discrete, it is easier for the ILEC to manage the marketing and sales of the product than if they were mixed with existing products and businesses.

The disadvantages of maintaining separate networks are also the advantages of convergence. The benefits of network convergence are:

- One company means one network and reduced administrative expenses and reduced management staff.
- Common network platforms facilitate lower operating costs, which includes network management expenses, fewer technicians, and probably less building space. One of the assumptions I am making is that a multifunctional switching platform will require less space than five separate network switches.

- Common network platforms facilitate the creation of value-added services.
- Fewer network interconnects for a large holding company to manage. This assumes that the common network platform is allowed to utilize the full bandwidth of a transmission facility.
- Common billing system
- Common network management system
- Common operational support systems
- Common service creation platforms

The layers of opportunities are in fact a partitioning of the ILEC business. The CLECs have the Telecommunications Act of 1996, AT&T's divestiture, and the Carterfone Decision to thank for this new environment. The Carterfone Decision of 1968 permitted the direct connection of customer-owned terminal equipment to the Bell system network. Carterfone is considered by many to be one of the pivotal events that led to the breakup of the Bell system.

Today, you can find a number of providers specializing in all of the above activities. Rather than the ILEC providing its own network infrastructure as a support mechanism for other carriers, the ILEC's functions have been assumed by non-carriers as a new business opportunity. This is a twist in the business that had not been foreseen by most industry pundits in the early 1980s. This "twist" is called "outsourcing." Outsourcing will be discussed more later in this chapter and in this book. Non-carriers have essentially assumed a role in the operation and provisioning of one of the many network opportunities. Figure 6-22 illustrates how directory assistance and operator services has been outsourced to non-carriers. These services bureaus represent a new way of doing business.

ILEC Challenges and the Future

As strong as the ILEC's position appears to be as it relates to the provisioning of new types of telecommunications services, the ILEC's face their own challenges. Some of these challenges are

- *Network element modifications* ILEC switching software may need to be provided to provide video, mobility, data, and operational support

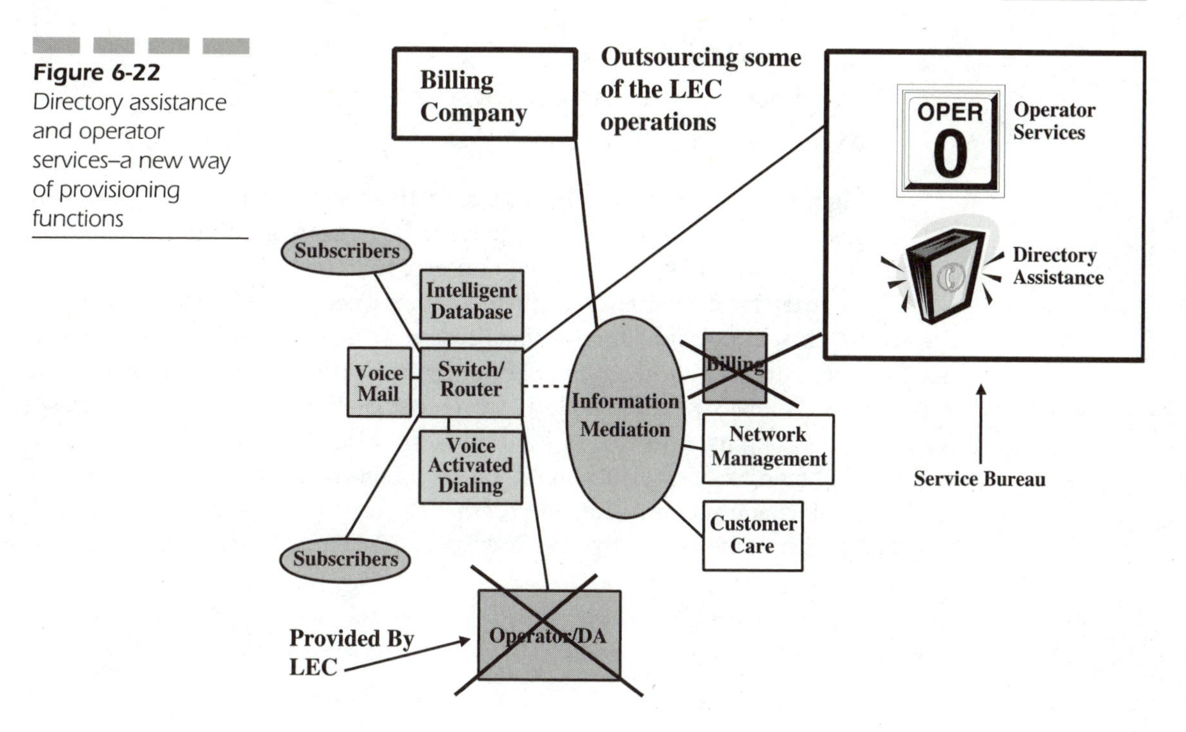

Figure 6-22 Directory assistance and operator services—a new way of provisioning functions

for the aforementioned. This takes time and money to execute. Some examples are

- Video will require high-speed data processing and transmission facility bandwidth. A wireless environment requires mobility support.
- Mobility support means the ILEC would need to support the following:
  - Call delivery to a mobile subscriber (home and visiting in a network)
  - Roaming (supporting calls to and from visiting subscribers)
  - Authentication
  - Validation
  - Handoff
  - Path minimization (efficient routing of a mobile call)
- Create new billing algorithms
- Modified traffic engineering practices
- Network reliability and availability
- Interoperability with legacy systems

- Common billing system
- Common network management system
- Common operational support systems

Figure 6-23 illustrates the fact that the new network platforms will appear the same but are in fact different. These network platforms must be flexible, capable of supporting old and new functions.

In order for convergence to have business value, a telecommunications network will not only need to facilitate new businesses, but also need to support service hybridization. Service hybridization is a concept in which a service spans/covers multiple network environments (wireline, wireless, Internet, paging, and so on). Examples would include cable TV telephony dialing using a television remote control, displaying a paging text message on a television set, or having an *interactive voice response* (IVR) unit repeat the text message verbally over a wired telephone. Figure 6-24 illustrates the concept of hybridization.

Unfortunately, most of the major telecommunications projects currently underway in 2000 inside an ILEC are not focused on creating value from

Figure 6-23 Creating common platforms

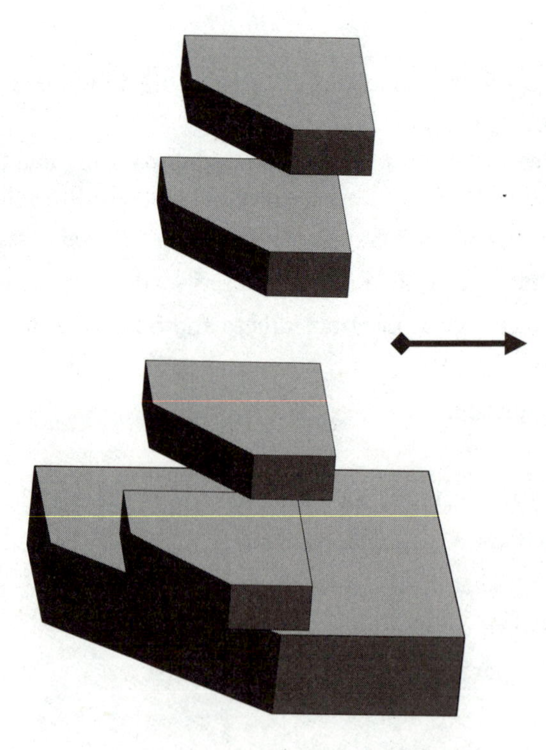
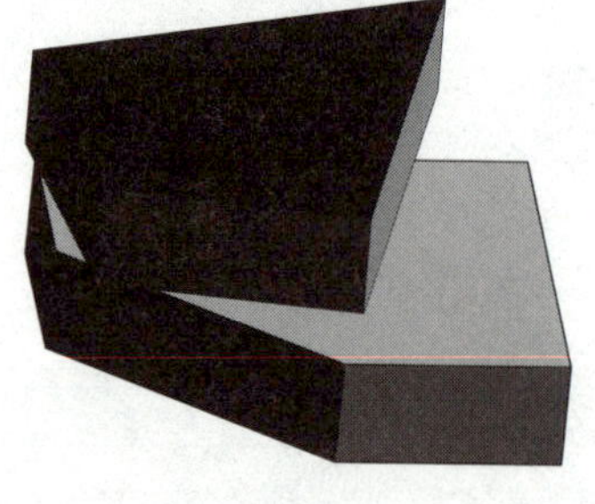

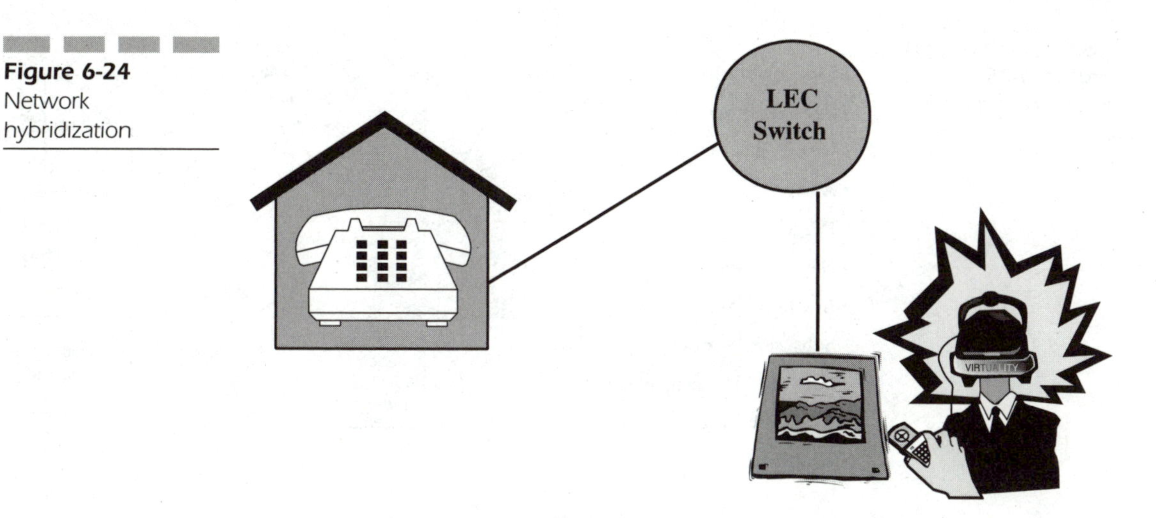

Figure 6-24 Network hybridization

the existing wireline network; rather, they are focused on building additional networks with new switching equipment. Some carriers do rent floor space to other carriers. Some do it for profit and others because the Telecommunications Act of 1996 supports colocation; therefore, space provisioning is more about being a good telecom "citizen." Some carriers rent rooftop space out to wireless tower companies. Some carriers do monitor and manage the data networks of large corporate customers; the sheer size of the customer warrants managing the facility network as a value-added service. Many of the ILECs will handle billing for the CLEC. Although the efforts appear minimal, in 1990, all of this was unheard of. Wireline carriers will need to compete with each other on all fronts. The wireline carriers will need to seek ways of providing services to the marketplace. The same wireline carrier will need to broaden their view of the marketplace. Figure 6-25 illustrates this point.

Up until this point, I have described the carrier's carrier business model. The carrier's carrier model can apply to all types of telecommunications carriers and users of all types. As the marketplace grows, the wireline carriers will need to find ways of adding shine to their businesses to maintain an appearance of freshness and to maintain competitive leads. There are only so many ways one can package services such as call forwarding, call waiting, voice mail, and caller identification. Rates can only get so low unless the carrier believes a business can be maintained by providing the aforementioned services free of charge.

The customers of the wireline carrier can be both customer and partner. By finding synergistic relationships, the wireline carrier will be able to

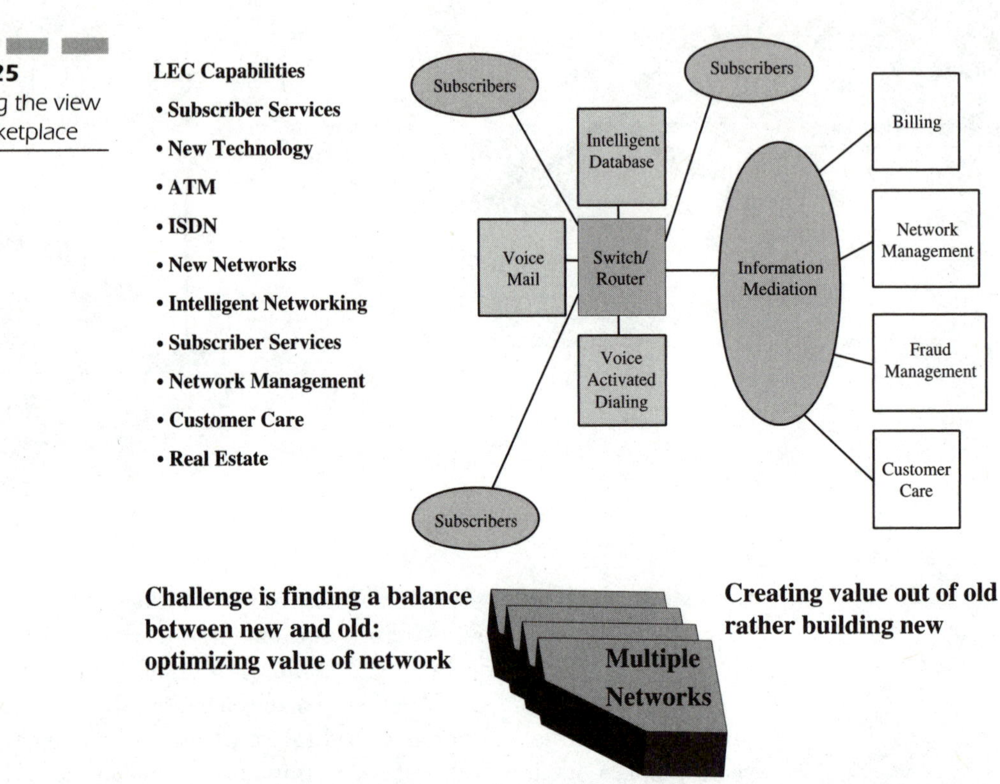

Figure 6-25 Broadening the view of the marketplace

grow the carrier by leveraging relationships and creating new opportunities. The most obvious example of the wireline carrier finding new opportunities is the merger. However, after the merger, what can a company do? One answer might be to find a way to add value to the existing product set by combining services or adding functionality to existing services. The future of the network will be discussed later in this book.

The next section discusses outsourcing as a new opportunity for the noncarrier. From a carrier's perspective, outsourcing serves as a way of controlling costs and creating new revenue.

Outsourcing

As a way of staying competitive, carriers have been cutting costs. To many, cutting costs involves reducing the size of the workforce or improving the efficiency of work processes. Sometimes changing work processes results in a reduction of the workforce. Telecommunications carriers have been changing their work processes in order to reduce their workforces. Some telecommunications carriers have also simply reduced their workforces and then worried about changing work processes to manage the outcome of the workforce reduction. In the later case, spot/targeted workforce reductions have been implemented (meaning only certain departments are offered early retirement or incentives to leave). Targeted reductions allow a carrier to manage the outflow of its workforce. Some carriers have experienced such drastic and unanticipated reductions that they have been left without resources and experience. No matter how the workforce has been reduced, outsourcing has been a result of these workforce processes. Figure 6-26 is a pictoral of the carrier functions that can be outsourced.

Outsourcing/Cutting Costs/Managing Costs Many people believe outsourcing is nothing more than a matter of hiring a consultant. However, outsourcing is far more than simple consulting. Outsourcing is taking a work process of a company and hiring another company to perform the activities of that process. Outsourcing can be likened to a landscaper mowing a homeowner's lawn. The homeowner, for whatever reason, has decided

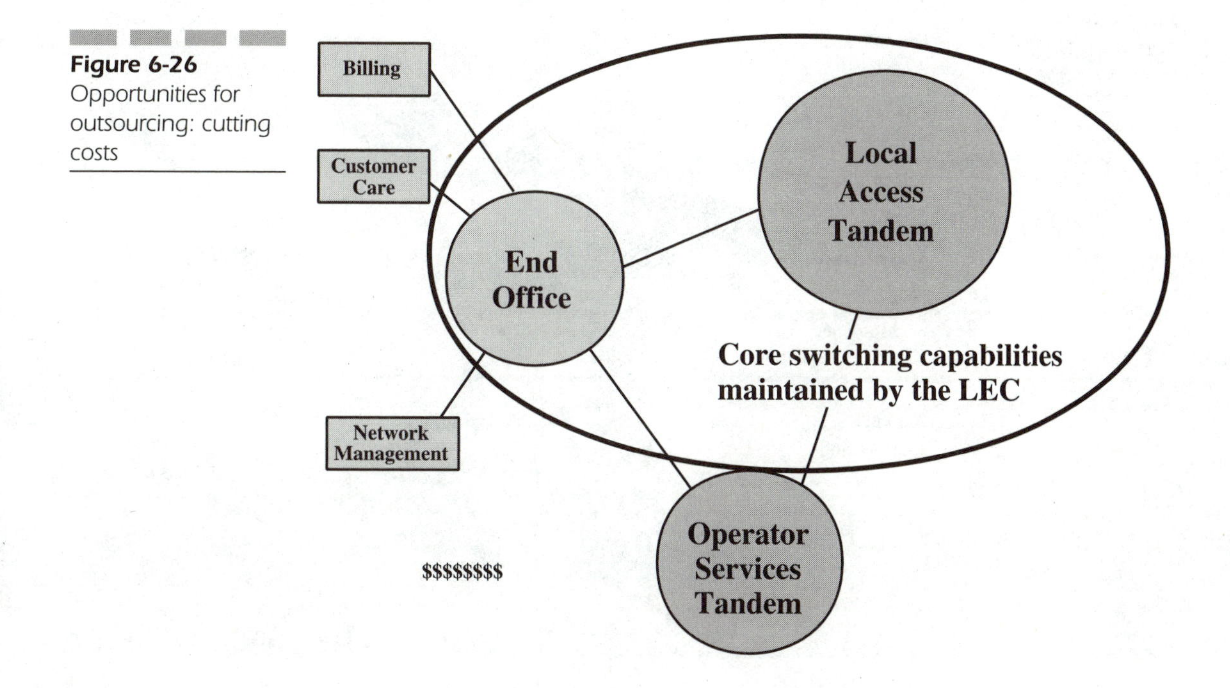

Figure 6-26 Opportunities for outsourcing: cutting costs

not to mow their own lawn and has decided that someone else can do the job either better or because the homeowner wants to save time and energy.

Telecommunications carriers can outsource many of its activities. Figure 6-27 is a representation of the pieces of the business. To date, the most popular functions to outsource are

- Directory assistance
- Operator services
- Bill processing
- Network maintenance
- Office maintenance
- Human resources
- Network engineering

Directory Assistance/Operator Services Directory assistance is a telephone number search activity, which includes white pages (residential) and yellow pages (business) telephone number searches. Directory assistance may also include call completion. The directory assistance call completion

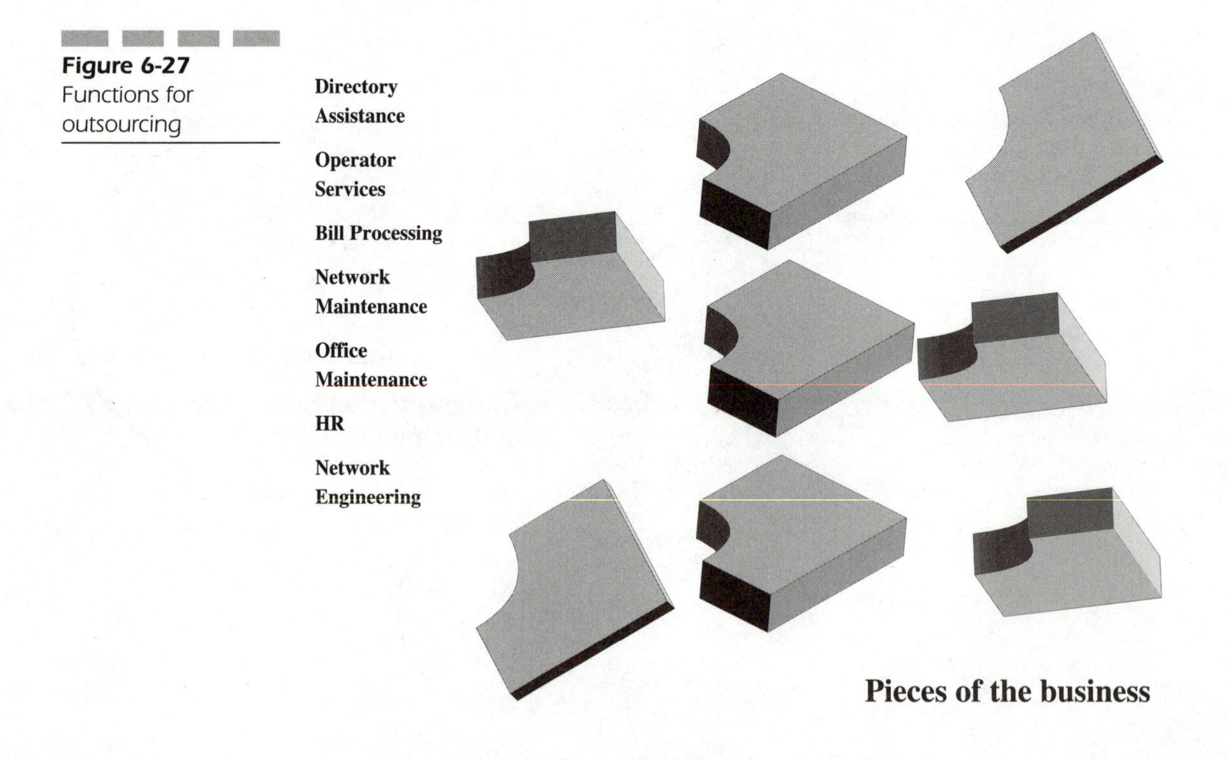

Figure 6-27 Functions for outsourcing

function searches for a telephone number and then connects the subscriber to the telephone number requested. Operator services is a function that includes the directory assistance function. Operator services includes a number of sub-functions, including:

- *Toll / long distance* Involves the completion of long distance calls. In today's environment, this means that a long distance/interexchange carrier will assist in the completion of long distance calls. This would include collect, third party, and person-to-person calls on the long distance. Collect and person-to-person may require the participation of the local telephone company's operator.
- *Billing* Non-automated (personal) assistance is rendered when a calling party is having difficulty making a call using a credit card, calling card, public coin phone, or public non-coin phone.
- *Number searching* Number searching is directory assistance. Directory assistance can include 4–1–1 calls, 555–1212 calls, or in some cases a carrier-specific telephone number (free).
- *Intercept* Intercept is concerned with the handling of calls that have been made to unassigned telephone numbers, disconnected telephone numbers, or changed telephone numbers.
- *Rate / Route / Assistance* Rate, route, and assistance are three inter-related activities. The operator can provide the calling party with rate information, connect the callers, and special services such as call backs, emergency connects (for example, forcing a call through a busy station), and working with long distance operators.

Many of the previous functions have been divided between the various types of wireline carriers and are even duplicated among the wireline carriers. To a subscriber, calling an operator for help has become an ordeal. Instead of going to one person for the majority of their questions, a subscriber now has to navigate through automated voice-driven and push button menus and multiple people to get questions answered. The operator was and still is the "human face" of the carrier. This is one point a CLEC must remember. Subscribers are lost when the help they expected through the phone is not forthcoming. Angry subscribers take their anger and frustration out on the operators. The lack of satisfaction the subscriber feels (through no fault of the operator) is usually blamed on the operator's inability to assist. No matter how low the price, the subscriber will terminate their service if customer service becomes bad enough. Operators can only perform the activities they are empowered to perform; unfortunately for the carriers, the dissatisfied subscriber could care less.

Service bureaus perform directory assistance and some aspects of operator services. The service bureaus for directory assistance are often located in areas not even serviced by the carrier. Note that the location of the bureau affects the wage the bureau pays its directory assistance personnel and results in lower costs to the carrier. These bureaus identify themselves as the subscriber's carrier. The subscriber's telephone number appears on a console and the telephone number is automatically identified as carrier "XYZ's" subscriber. The bureau maintains an electronic copy of the carrier's directory listings. Many subscribers have discovered that the directory assistance operator is not from the subscriber's calling area via a variety of amusing methods:

- Regional voice accents
- Unfamiliarity with popular local stores, streets, and restaurants
- Unfamiliarity with local news
- Unfamiliarity with today's local weather. This may sound silly, but many subscribers carry on casual conversations with the operators: a testimony to operator training.

As of today, the operator service functions typically outsourced are intercept and routing. These functions have been replaced with automated systems. If a subscriber mis-dials a number or dials a disconnected number, she will hear an automated announcement that may give them either directions to follow or a simple announcement that the number is no longer in service or is unlisted. The automated system is normally run by the carrier but can be handled by the outsource directory assistance bureau. Sometimes the outsource bureau rents building space in the territory of the carrier in order to integrate itself into the carrier's network for call-handling purposes. Bureaus integrated in this fashion can perform the whole range of operator functions. See Figure 6-28.

Bill Processing In the wireline world, the activity known as *billing* is in fact comprised of two distinct activities: recording and billing. Recording is the act of collecting the billing details used to create a bill. Billing is comprised of multiple acts: applying the appropriate rates to a call, taxes, surcharges; formatting the bill that the subscriber receives; and mailing the bill to the subscriber.

Although billing is a mission-critical function, carriers have found that the cost of billing has escalated so much they will trust an outside company to handle the revenue collection portion of the carrier business. The recording function is still performed by the carrier because it is an integral part

Figure 6-28
Operator services

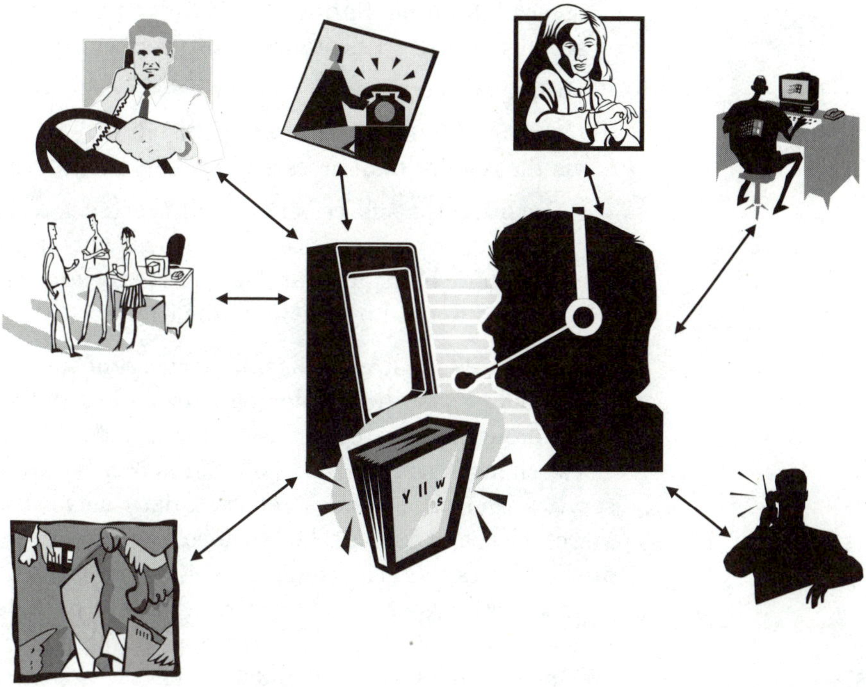

of the switch's capabilities. In an outsource scenario, the carrier still needs to perform some basic system integrity checks of the recording mechanism. The primary objective of outsourcing the billing function is to obtain and maintain a low cost-to-revenue ratio.

The outsource billing company receives the recorded data via a number of ways: magnetic tape, CD-ROM, system-to-system electronic download/transfer, or paper. The most popular method today is magnetic tape, followed by CD-ROM, paper, and then finally electronic transfer feed. The outsource billing company is responsible for the following billing functions:

- *Reading and processing all records* This activity includes identifying calling telephone numbers, called telephone numbers, date of call origination, time of call origination, and time of call termination (not duration of the call). Call duration is calculated as part of the processing of the records.
- *Rating the call* Rating a call includes calculating the distance between the calling and called switches, applying local/state taxes, applying local/state surcharges, applying federal taxes and surcharges, and applying the appropriate rate based on time of day, date, distance,

and call duration. Rating of calls that employ the use of special services such as directory assistance is also performed.

- Posting of charges on the subscriber's account
- *Creating and rendering a bill* The bill may be sent to the subscriber via the regular mail process or even by the Internet.
  - Crediting the subscriber's account includes receiving payment and any carrier-provided credits. Payment may be made via check, automatic bank account debit, or credit card. The Internet may also be used to send payments to the carrier.

Figure 6-29 illustrates the billing aspect of outsourcing.

The billing process is undergoing revolutionary changes. The changes are being brought about by the convergence of telecommunications services. These changes address how to bill a subscriber for voice, video, and data services over the carrier's Internet, data networks, and voice networks. Voice, video, and data can be provided over all of the various networks operated by the carrier. The challenge the carrier faces is determining how and what the subscriber shall be billed for: time or packets of data. The question is not a simple one. To illustrate my point, here is one example: The carriers are required by federal and state laws to apply certain types of charges for voice calls over the traditional voice network. When the laws were cre-

Figure 6-29 Outsource billing

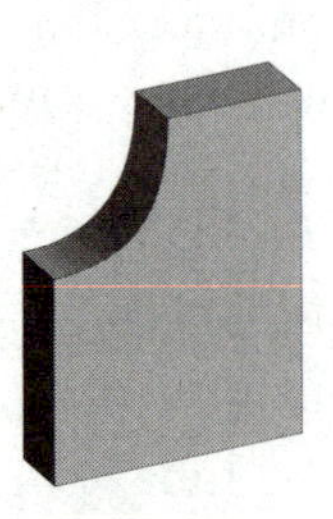

ated, the idea of the Internet handling voice services was not envisioned. Today, the Internet can carry not only voice but also other types of data simultaneously. The carriers wish to be paid for the services their network is supporting even if the support was not the result of a telephone company switch but the result of an Internet Service Provider interconnected to the wireline network. In the past, much was made of how an Internet user could access the Internet and be charged only a small flat fee by the telephone company for regular flat-rate voice call. This was not lost on any wireline carrier. The carriers are in business to make money, not give it away. The wireline carriers are seeking ways to be compensated for the interconnection between ISPs and long distance carriers. At this time, the ISPs have lost and are now required to compensate the wireline carriers for such access charges. However, as regulatory actions go, the ISPs will continue to fight this decision.

In addition to developing billing methodologies for the new telecommunications services, the carrier community is seeking to find a way of using common billing platforms. Common platforms would entail common hardware and software. Common billing platforms supports the convergence of the network technologies by taking all telecommunications services ranging from voice to entertainment and creating a common bill that can be understood by the customer. Converging the billing process also means converging all service order processing, repair and installation tickets, and all aspects of bill processing onto a common system to be administered by a common staff. The carriers are seeking to outsource billing, but in order to stay competitive, the carriers wish to retain billing companies that can address the carriers' concerns about convergence. The carriers do not wish to retain billing firms that require multiple processes and groups of people to manage the billing of all of the carriers' telecommunications services. Billing firms that do not change will not stay in business for long; their costs will escalate as they continue to use outdated systems to process billing information for their carrier customers. The need for convergent billing platforms is illustrated in Figure 6-30.

Billing will be discussed in greater detail in the later chapters.

Network Maintenance Network maintenance addresses the overall management of the network's health. Maintenance is not a glamorous attention activity in the carrier world. However, maintenance is crucial to the business. Aspects of network maintenance have been outsourced to external companies. The driver behind these moves is cost reduction. Network maintenance can be divided into two categories:

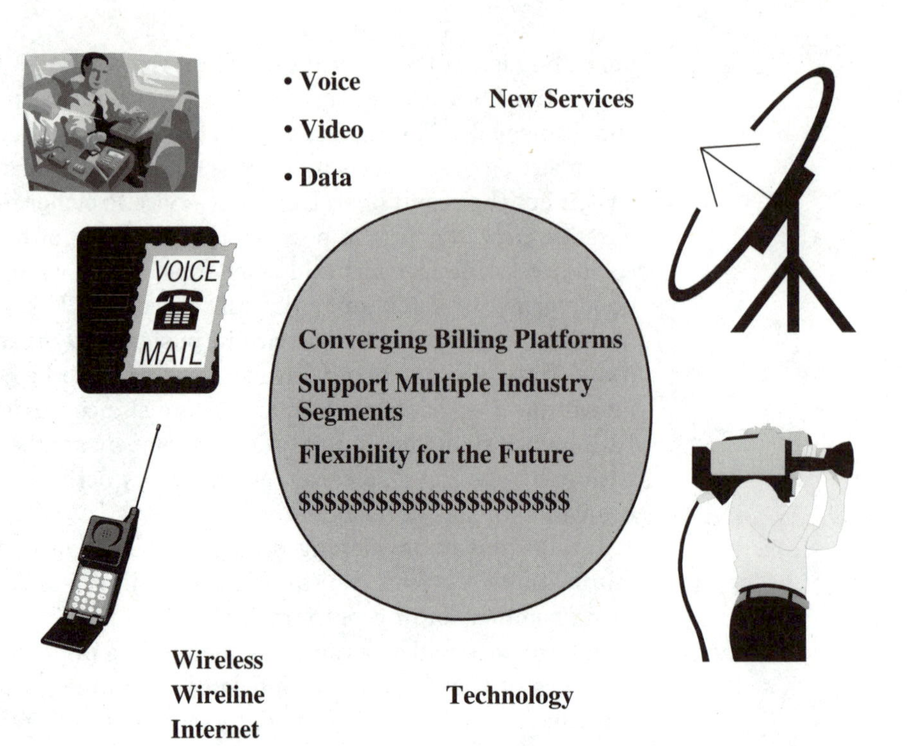

Figure 6-30 Converging billing platforms

- *Preventive* Preventive maintenance involves procedures to detect potential trouble conditions. This form of maintenance activity is part of a periodic routine carried out by a technician(s).
- *Corrective* Corrective maintenance includes a variety of procedures and functions as described below:
  - *Trouble detection* Trouble detection recognizes when a trouble condition exists, as defined by the carrier. The troubles normally reported would involve customer/subscriber station equipment and troubles in the local loop (as described by the customer). These troubles are reported to the wireline carrier by the subscriber/customer.
  - *Trouble notification* This is an internal carrier response whereby the appropriate field technician force is notified of a trouble condition.
  - *Trouble verification* Trouble verification is the determination of a continuing trouble condition. Most trouble conditions are transient in nature. Therefore when a trouble verification exists, that trouble receives the highest priority in resolution.
  - *Trouble location* Trouble location is the process of determining the location of the "trouble." The trouble can be in the end office, the local

loop, the interoffice transmission facility, or the customer premise equipment. After the trouble has been located, the appropriate technical force can be mobilized to correct the trouble.

- *Trouble repair* Trouble repair is the process of correcting the trouble. Trouble repair can take anywhere from a few minutes to several days.
- *Service verification* Service verification is the process of verifying the restoration of service. Service verification is both an internal and external process. The field technician verifies the trouble has been fixed and service restored. More importantly, the customer verifies that the trouble has been fixed and service has been restored.

The customer attendants who receive the customer complaints are in-house employees. However, the carriers have outsourced the field technical workforce function. Many times, CLECs and others will pay the ILEC to run "wire" underground and on top of telephone poles for them. However, the CLECs and cable TV companies will send its own technical force out to the customer's premise. Carriers that outsource these functions to another carrier or to a non-carrier firm must be prepared to execute a process that facilitates communication between the carrier's processes and an outsource firm's processes.

Office Maintenance Office maintenance is a common consideration throughout the business world. Unlike most businesses, the old Bell system maintained its own in-house office maintenance personnel. These people perform janitorial functions and the like. Office maintenance is not a small expense item. The cost of cleaning, routine repairs, and disposal of waste is a high cost. There are hundreds of firms today that perform these functions. It is a highly competitive and lucrative business. All carriers outsource this activity.

Human Resources Aspects of *human resources* (HR) have been outsourced. This is new to the carrier community and especially the ILECs (the old Bell system began executing this in the mid-1980s). Administration of human resources is not completely outsourced. The following can be and in some cases have been outsourced to external firms:

- Creation and administration of medical plans
- Administration of life insurance plans
- Administration of employee conflict resolution plans

- Administration of employee performance evaluation programs
- Administration of all employee training as it relates to employee conduct under federal affirmative action and anti-discrimination laws.

Outsourcing the previous items does not absolve a carrier from discharging its duties to the employees and to the company. There will always be a HR staff that will oversee implementation of all HR policies. However, the HR department no longer needs an army of personnel to manage the various programs it makes available to the employees or any in-house HR trainers.

Network Engineering This has been a fairly new development (late 1990s). The outsourcing of network engineering has come about for a variety of reasons. Established carriers do not need armies of engineers because the level of growth has pretty much flattened out. Rather than maintain a hundred engineers on staff, the ILECs maintain a staff one-third of that size and retain the design services of their infrastructure vendors for design support. New carriers, especially the wireless and the wireline CLECs, would hire large groups of engineers for their initial "build-outs," but would lay off these people upon completion of the build-out. A small staff would then be retained for "engineering design maintenance" purposes.

The new trend for new carriers is to hire a engineering firm to design the entire network for the carrier. This sounds like the highly lucrative engineering consulting business; however, it is not. This takes technical consulting to another level. Typically, technical consultants are retained for very narrowly defined design efforts. The trend is to hire these consulting firms to design every component of the network and in some cases make critical policy decisions. These policy decisions are related to what blocking levels will be established, channel allocation plans to follow, vendor short-listing, small equipment vendors, equipment procurement policies, capacity plans, network topology layouts, and even real estate to be leased. Figure 6-31 is a depiction of the network functions that can be outsourced.

Common Technical and Business Precepts

There are fundamental activities that must be addressed to ensure the carrier's ability to provide service. These generic activities cut across the technical and business areas of a carrier:

- Forecasting market needs
- Creating new services based on forecasted needs or trends

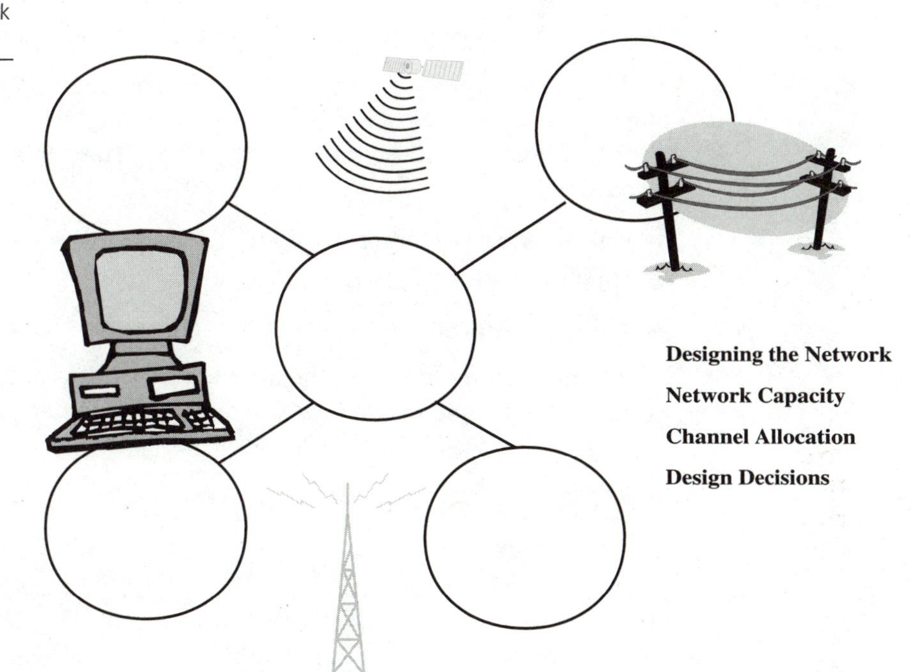

Figure 6-31 Outsourcing network design

- Creating a service that the marketplace believes has utility
- Creating a service that works
- Listening to the customer/subscriber

Note that the following specific areas of concern match the above generic areas.

As I noted in Chapter 5, the marketing/business sides of the carrier are concerned with four factors. These factors are called the four P's:

- *Product* Product refers to the physical nature of the product and the perceived value of the product. The product also includes the conduct of the vendor, the guarantees offered by the vendor, and the reputation of the vendor.
- *Price* Price refers to the purchase price of the product, the terms and conditions of the payments, and discounts.
- *Place* Place refers to how the carrier will sell and distribute the product. The carrier can sell directly to the subscriber base. The carrier

can also act as a wholesaler of its own network services and capabilities to another carrier.

- *Promotion* Promotion refers to advertising and selling.

There are a series of common wireline engineering principles that are applicable to all types of carriers. The specific engineering formulas maybe different but the philosophies are the same. The goals of an engineering department are

- Customer service needs are met.
- Quality of service objectives are met.
- Sufficient capacity in the network to meet growth and usage needs.

As I had noted previously, the marketing/business and technical areas of a carrier have common goals. In the end, the customer comes first.

The State of Switching/Multifunctional Switching

This will be discussed later in the book. Briefly, the wireline CLECs are seeking ways of cost-effectively competing in the telecommunications business. Aside from outsourcing and more efficient technologies, the CLECs are also deploying multifunctional switching platforms capable of supporting the call processing algorithms of multiple telecommunications industry segments. These segments include wireline, wireless, and Internet.

The competitive nature of the wireline industry has forced the CLEC community to seek competitive edges in all areas. Multi-functional switching is a one–size fits all approach to designing the network. See Figure 6-32 for an illustration of this point.

SUMMARY

The wireline carrier community is changing constantly. New competitors, new opportunities, and new technologies. The wireline carrier is no longer a *Plain Old Telephone Service* (POTS) business providing voice only. The wireline telecommunications segment has become blurred with the data, Internet, and entertainment. Convergence made its first public appearance in this industry segment. The AT&T divestiture and, more importantly, the Telecommunications Act of 1996 had been the primary drivers behind these

Figure 6-32
Multi-functional switching

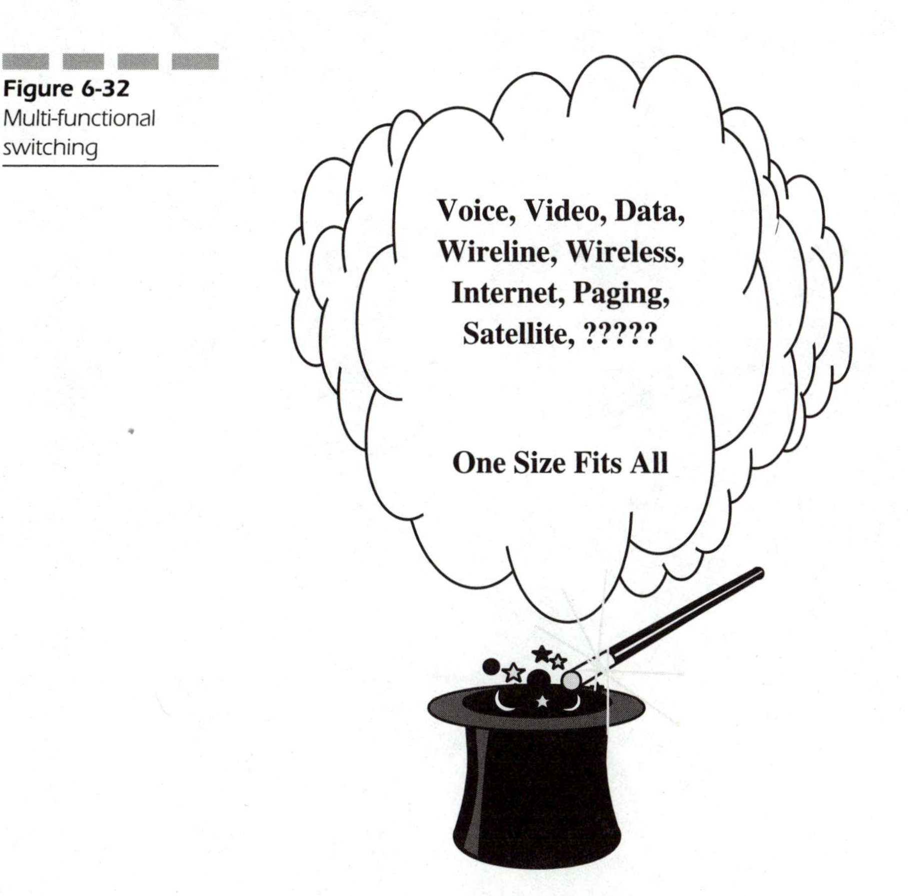

new opportunities. The telecommunications industry is a mass of opportunities that involves not only wireline but also wireless communication of all types and the Internet.

The following chapters continue our examination into the wireless and Internet segments, reviewing the basics, the commonalties, and the opportunities.

CHAPTER 7

Wireless—Cellular and Personal Communications Services Networks

Forms of wireless communication includes cellular, paging, *personal communications services* (PCS), land mobile radio, air-to-ground, *local multipoint distribution service* (LMDS), wireless local loop, and so on. The current dominant (in terms of usage and market penetration) wireless or mobile services are cellular, PCS, and paging. Cellular and PCS are closely related; therefore, they will both be discussed in one chapter. Paging will be discussed in Chapter 8. Figure 7-1 is a high level view of the cellular and PCS network.

History

In 1946, two-way mobile radio service was introduced. Soon after its introduction, the disadvantages of two-way mobile service became apparent. From a customer's or engineer's standpoint, the disadvantage was competition for RF channels and interference. The technical challenges were giving the subscriber a larger pool of RF channels to make their calls and reducing interference between subscribers. The simple solution could have been not to worry about giving subscribers more RF channels and to physically separate the radio coverage areas to ensure that there would be no overlap.

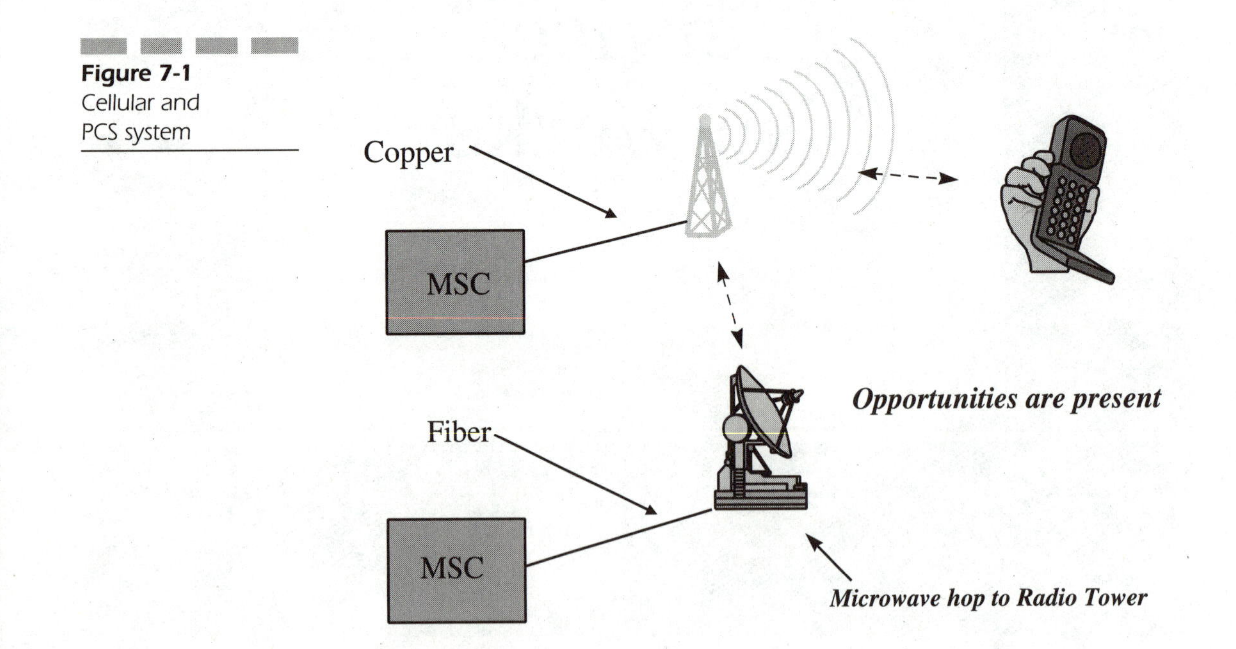

Figure 7-1 Cellular and PCS system

Instead, engineers from Bell Telephone Laboratories began to explore a concept that would reuse frequencies in small radio coverage areas. These coverage areas (called cells) would be linked together using a switch that would enable calls to be made while the telephone was moving. Computer and switching technology had to improve before mobile radio service became commercially viable. Availability was another issue, however; that was due to regulatory delays. Cellular telephones became commercially available in the United States in 1983. Today there are two basic radio technologies that are commercially available in the cellular world: analog cellular and digital cellular. Figure 7-2 is an illustration of the cellular reuse concept.

The analog cellular systems, which were the original cellular systems, employ *frequency modulation* (FM). Digital cellular became commercially available in 1991. Digital cellular employs two different radio technologies: narrow band and wide band. *Time Division Multiple Access* (TDMA) and *Code Division Multiple Access* (CDMA) have been standardized by the *Telecommunications Industry Association* (TIA). TDMA and CDMA represent the narrow band and wide band technologies respectively. There has been an attempt to standardize various flavors of these technologies. However, from the perspective of a fixed network switching engineer, the differences are not an issue. The fixed network includes transmission facilities (for example, backhaul) and all network elements that support the processing of the calls, backbone network signaling, and signaling between the switch and other network elements.

Figure 7-2
Cellular reuse

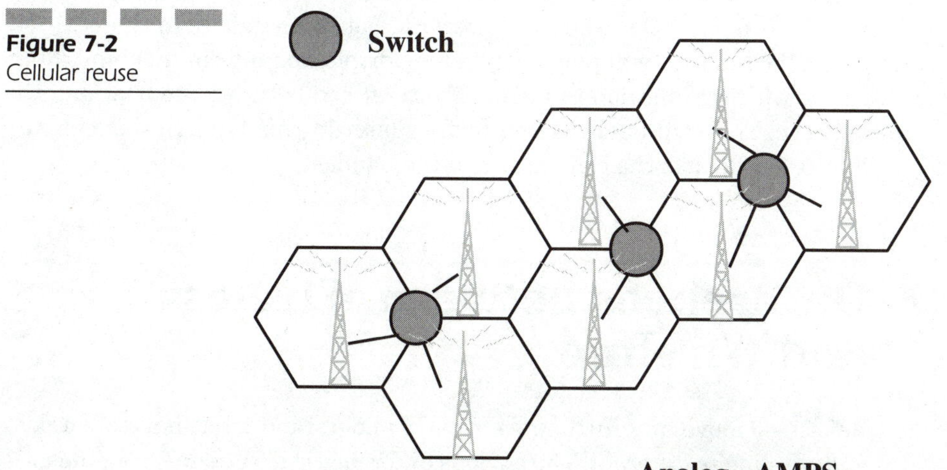

Figure 7-3
The Handset and the network

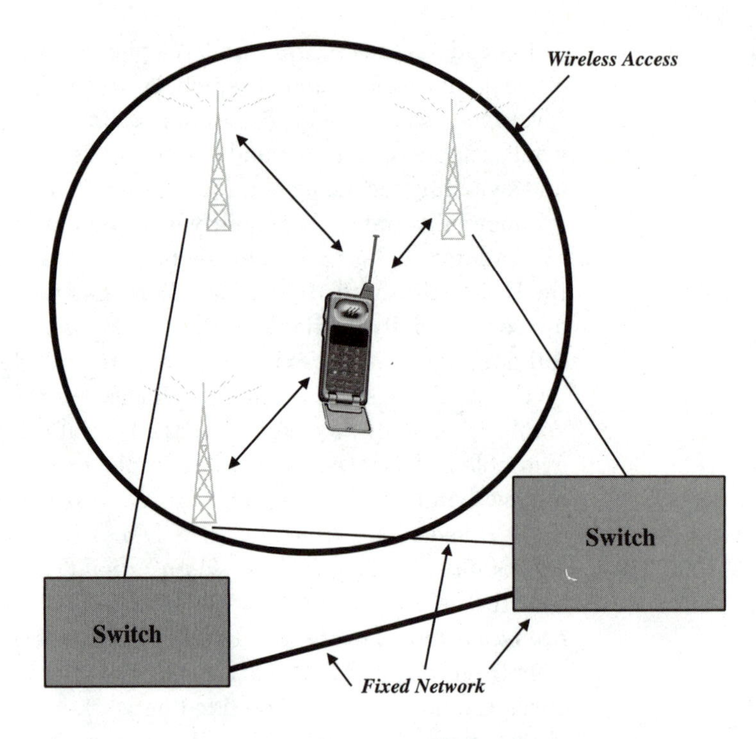

Fixed network engineers perceive radio technology as an access issue. In other words, to a fixed network engineer, the mobile handset is nothing more than a device used to access the network. Radio engineers may disagree with this view. However, at this time, all core call processing and data storage resides within the fixed network, not the handset. In the next few years, the handset will play a key role in displaying information and interacting with information to facilitate interaction between the user and the network. This will be discussed further later in this later in this chapter. Figure 7-3 depicts the central role of the handset.

Differences between Wireless and Wireline

Before we move any further, we should understand what makes wireless and wireline so different. The obvious difference is that in one case, the user uses a wireless handset and in the other case, the user typically does not. In the wireline side of the business, the user can use a cordless handset (cordless—another way of saying wireless). However, the cordless handset has limitations that a so-called "wireless" handset does not. Cordless handsets can be used only near the home cordless base station. They do not allow roaming. Cellular and PCS allow roaming, or more specifically, they support mobility. Cordless uses unlicensed spectrum. Cellular and PCS systems use spectrum licensed by the FCC. Figure 7-4 illustrates the unlicensed and licensed aspects of providing wireless telecommunications service.

The greatest attribute of cellular and PCS is mobility. Mobility can be described in a number of ways. You could call cordless a mobile device because it is wireless. On the other hand, a cordless user cannot leave the vicinity of their home (that is, the cordless base station), which limits range and therefore possibly limits the device's utility. Mobility has the following attributes:

- Call delivery to a mobile subscriber (home and visiting in a network)
- Roaming (supporting calls to and from visiting subscribers)
- *Authentication* Authentication is the process of user identity confirmation. Identity confirmation can involve checking handset/ terminal device identity by interpreting "secret" keys/data messages. If

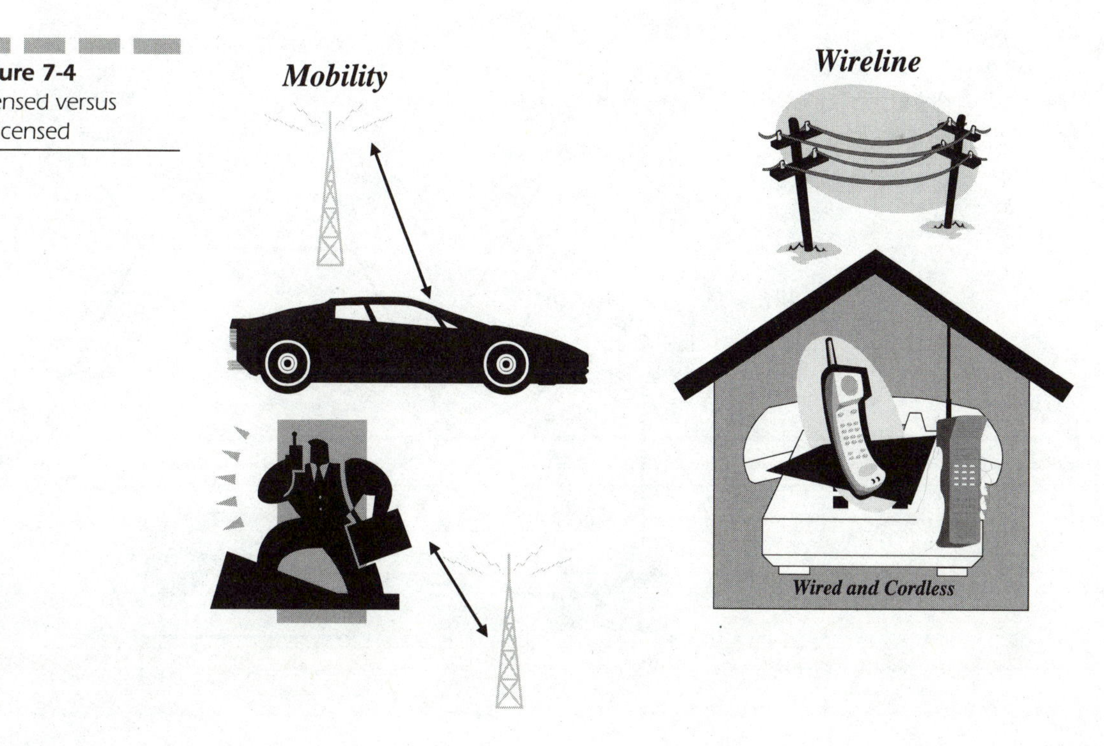

Figure 7-4 Licensed versus unlicensed

the data keys/data messages have been altered or do not show a specific format, the call will not be completed.

- *Validation* Validation is often confused with authentication. Whereas authentication essentially certifies the user as a either "real" or "fake" and either "good" or "bad." Validation certifies the "permission" to complete the call. An example: A user calls another party using a mobile handset. The carrier certifies that the handset is the real one and is not a clone being used by a criminal. After the handset has been given the "thumbs up," the carrier checks if the call is allowed under the user's billing plan. This next step is validation (the final "greenlight" to complete the call).
- *Handoff* Handoff is the process of reassigning subscriber handsets to specific radio channels as the handsets move from cell site to cell site.
- *Path minimization (efficient routing of a mobile call)* Path minimization is the process of efficient fixed network (the non-wireless portion of the call) routing of wireless call tables in the wireless carrier switches.

Figure 7-5 illustrates the aforementioned concepts.

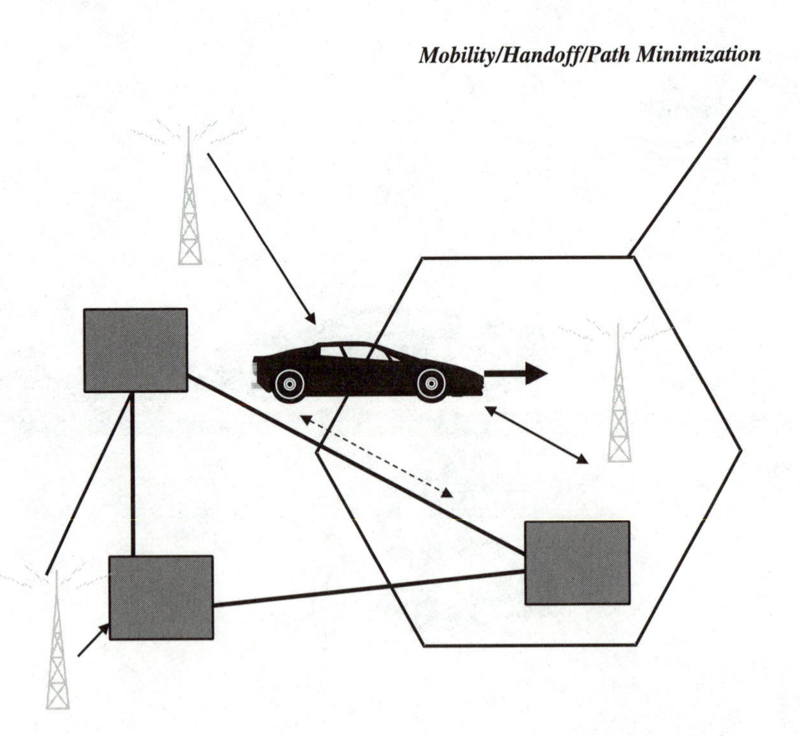

Figure 7-5 Mobility/handoff/path minimization

Mobility: Personal and Terminal

Mobility comes in two flavors (that is, types): personal mobility and terminal mobility. Personal mobility is defined as the ability of a subscriber (user) to make and receive calls regardless of location and the type of terminal equipment the call is being made from or to. Personal mobility supports independence from specific terminal devices and locations. At this time, most users of telecommunications services are tied to a specific handset or fixed termination point (for example, termination at a home or business). Personal mobility requires the use of personal numbers or some other form of personal identification (for example, smart cards) to identify the user. Terminal mobility is defined as the ability of a subscriber (user) to make and receive call regardless of location. However, in this case, the user is associated with a specific handset or terminal device. This is how cellular and even PCS work today, and how wireless will work the next several years. See Figure 7-6.

If I may digress for a moment, one of the original views of PCS was that cellular could only support terminal mobility and that PCS would support

Figure 7-6 Personal mobility versus terminal mobility

both personal mobility and terminal mobility. Personal mobility was considered the ultimate objective for mobility users. No technically substantive reason why cellular could not support personal mobility existed; at the time, the reasons were more politics and business ballyhoo than anything else. To the subscriber, this a non-issue.

As with most products' initial or early product lifecycles, the subscriber only cared about the following:

- Price
- Performance
- Coverage

Recently, wireless carriers have been working to differentiate themselves in order to keep existing customers and attract new subscribers (who would not have purchased wireless even with outrageously low prices). Product differentiation typically occurs when a product has matured enough that customer education is no longer the primary objective and price is no longer the primary driver for customer churn or growth. The challenge for wireless is to find a way of creating differentiation that adds value in a telecommunications environment that appears to be dominated by Internet growth and wireline carrier de-regulation. I will discuss differentiation further later in this chapter.

Mobility is the key to the growth and evolution of wireless. Mobility, in fact, has a role in the convergence of the telecommunications industry segments. Personal mobility can be implemented in both the wireline and wireless environments. The implementation of personal mobility in the two operating environments will bring these two sides of the business closer together. Personal mobility enables seamless service. Seamless service can be defined as the ability of the user to make and receive calls regardless of technology, service provider, and location. In order for personal mobility to work, the carriers must be capable of supporting nationwide roaming for all services; independent of terminal device type. Terminal mobility can also be implemented in both a wireline and wireless environment. One example of terminal mobility is the continuing use of terminal devices by individuals for all aspects of their telecommunications services. Wireless carriers today provide limited access to Internet and other data services. Some users will simply use a single terminal for all of their communications needs. Figure 7-7 is a pictoral of mobility's role in the growth and evolution of wireless.

The next section of this chapter will take a look at the basic components of the cellular and PCS networks.

Figure 7-7 Mobility—Key to growth and evolution

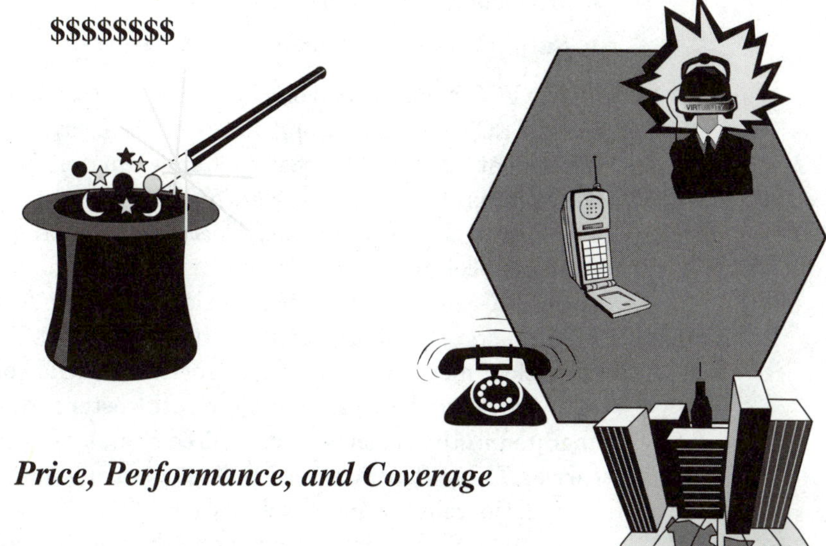

Cellular and PCS Carriers—Regulatory Operating Environment

Cellular and *Personal Communications Service* (PCS) systems are radio systems operating at different radio frequency bands. The existence of these systems, which are operated by cellular and PCS service providers, is a result of two *Federal Communications Commission* (FCC) decisions made in 1982 (service was first launched in Chicago in 1983) and 1994 respectively. These decisions created the regulatory framework in which these carriers would operate.

Cellular Carrier—Regulation

The FCC decision that created the cellular carriers created a structure in which there are two carriers per designated area. The FCC issued licenses to two carriers per geographic service area. The geographic service areas are broken down into two categories:

- *Metropolitan Statistical Area* (MSA)
- *Rural Service Area* (RSA)

The FCC had awarded one license to the *Incumbent Local Exchange Carrier* (ILEC) and the second license to a carrier not associated with the incumbent (dominant) local exchange carrier. Note that the ILECs are the dominant wireline carriers in an area; in many areas of the USA, this means the *Regional Bell Operating Companies* (RBOCs). The ILEC-supported cellular carrier is known as the B-side carrier. The non-wireline/non-ILEC cellular carrier became known as the A-side carrier. Prior to the recent mergers and acquisitions, the distinction of B-side versus A-side had some clear meaning. However, today the artificial boundaries created by the FCC are blurred. As an example, Southwestern Bell Corporation originally operated in the southwestern United States as both wireless and wireline carrier. Today, Southwestern Bell is operating in the southwest and northeast as the cellular and wireline carrier. Southwestern is also operating as both the PCS and wireline carrier on the Pacific coast. The changes are part of the convergence we have discussed up to this point. Figure 7-8 illustrates the operating boundaries of the various cellular carriers.

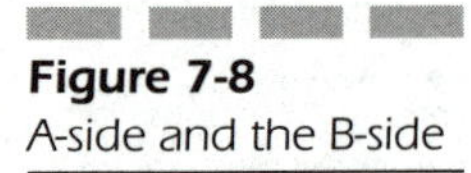

Figure 7-8
A-side and the B-side

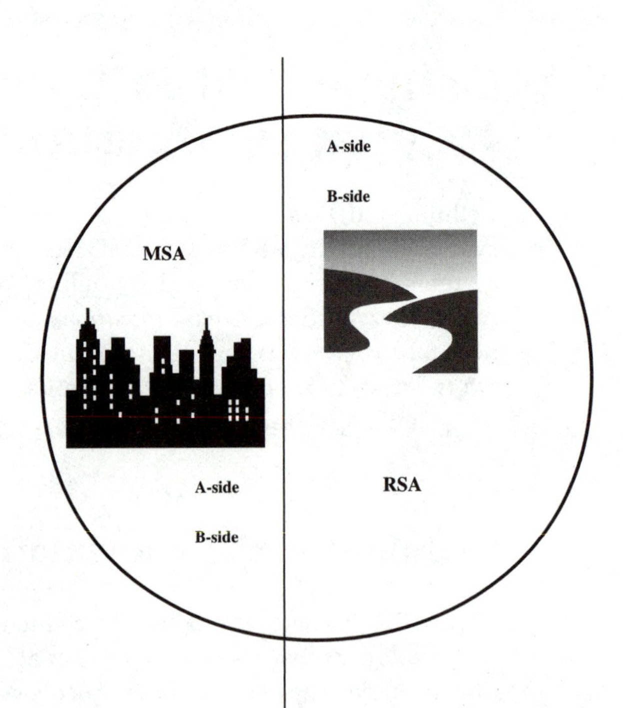

The confusion I have alluded to is only apparent to those with the telecommunications community. Those folks working within telecommunications see a flurry of activity and are not sure who is working for whom. However, the subscriber/customer doesn't see a technical difference, at least not one that would cause the subscriber annoyance. Regardless of the business structure in operation in a service area, the oligopoly in operation remains in effect: two licenses per MSA and two licenses per RSA. Note that MSAs and RSAs do not overlap. MSAs and RSAs are fairly large in size and can cover huge areas of a single USA state. RSAs are physically larger and cover areas of less dense population than the MSAs. MSAs and RSAs do not match the federal, state, and local governmental maps that we know. In other words, a RSA could cover three counties and half of another county, or even half of one town and an entire adjoining town. MSAs and RSAs are based on statistical population information used by the United States Census Bureau. There are more than 300 MSAs in the USA and more than 200 RSAs in the USA. Note that the MSAs and RSAs also do not match the LATA boundaries established by the FCC for the ILECs as they relate to interexchange carrier relationships. This is confusing to those working in the telecommunications community and to some extent the subscribers also.

The following figure illustrates the relationship of MSA and RSA to the LATA. The relationship does not exist. See Figure 7-9.

A cellular licensee may provide service in the entire MSA or RSA or only a portion. Figure 7-10 illustrates how cellular system construction can occur. This may occur for a variety of reasons:

- Phased build-out of the network
- Insufficient number of subscribers to warrant capital expenditures in the MSA or RSA
- No money to build out
- No permits to construct radio towers. This results in a carrier attempting to provide cellular coverage to an area by installing focused directional antennas in hopes of providing coverage.

The area in which the cellular carrier is actually providing service is called the *Cellular Geographic Service Area* (CGSA). The service area can cover multiple towns and communities. MSAs and RSAs even overlap states. Given the fact that divestiture did not play a driving role in the creation of the coverage areas, cellular carriers are not governed by the divestiture agreement and the *Modification of the Final Judgment* (MFJ). Therefore, the cellular carrier can transmit across wireline LATA boundaries and even state lines. One key point to remember is that radio waves

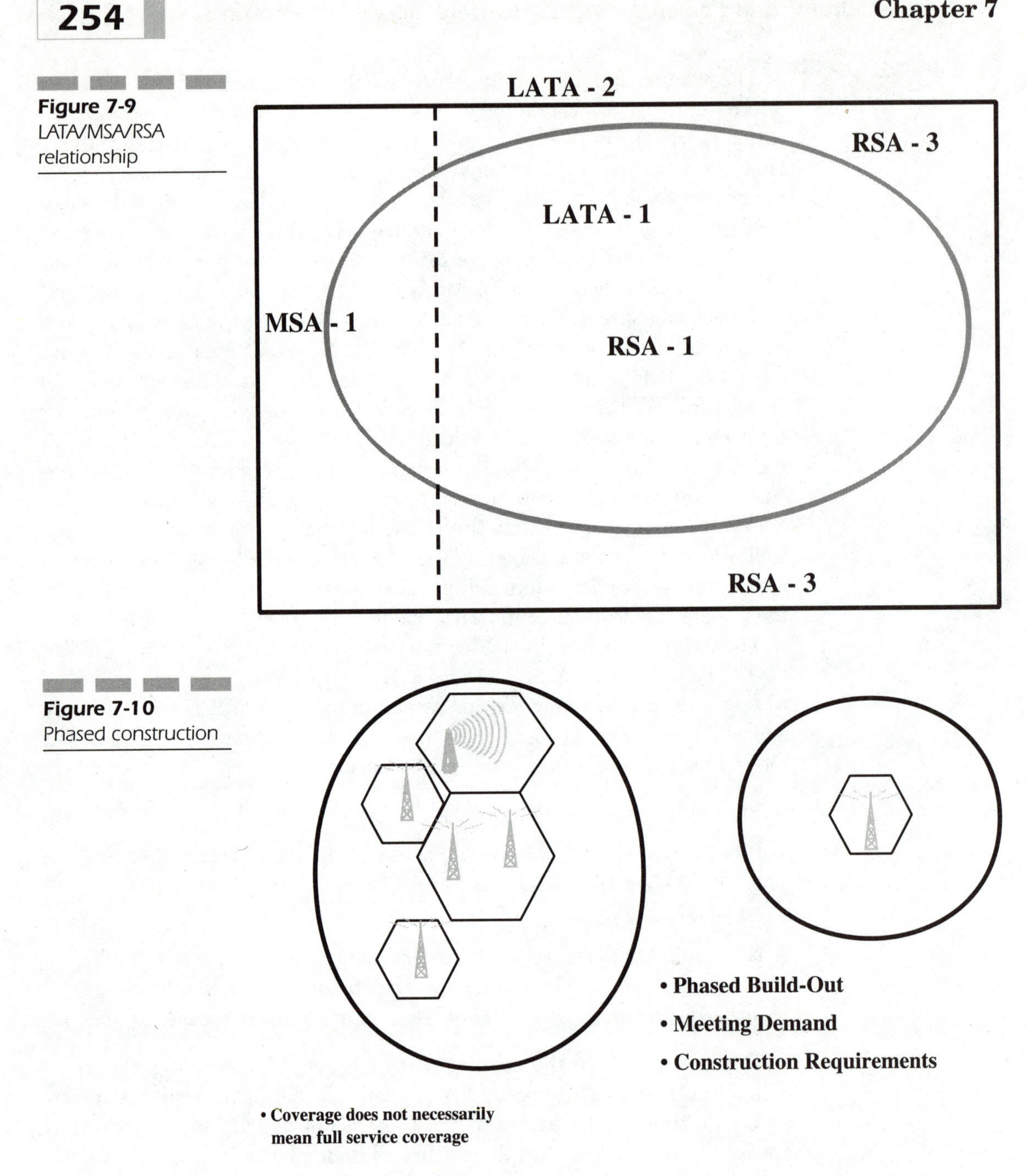

Figure 7-9 LATA/MSA/RSA relationship

Figure 7-10 Phased construction

do not know geographic boundaries, hence the logic in not creating additional artificial regulatory boundaries like LATAs to dictate how cellular carriers transport and interconnect.

Cellular Frequency Channels

The FCC allocated a total of 50 MHz of spectrum within the 825-890 MHz frequency band for cellular service. Originally, the FCC allocated 20 MHz for each carrier licensee. An additional 5 MHz per cellular carrier was allocated in 1985 to meet growing capacity demands. Cellular carriers manage their spectrum by using different frequencies for transmission and reception between base stations and the mobile handsets.

The original cellular systems were all analog. Today, these analog systems are being slowly phased out by digital systems. The original analog systems were based on *Advanced Mobile Phone Service* (AMPS) technology work performed by AT&T Bell Laboratories conducted prior to the AT&T divestiture. The analog system has also become generically known as AMPS. The original AMPS system employed *frequency modulation* (FM). FM techniques enabled the cellular carriers to divide the frequency bandwidth into 30 KHz channels, which yielded a total of 832 analog channels, or 416 channels pairs (transmit and receive) per cellular carrier. The cellular carriers tend to manage the 416 channel pairs differently; however, the general rule of thumb is to assign 21 channel pairs for call control functions. Current cellular systems use *Frequency Division Multiple Access* (FDMA). FDMA is a multiplexing technique that divides the frequency bandwidth into 416 channels pairs as well. However, FDMA can be used to support analog and digital transmission.

Figure 7-11 illustrates the concept of channel allocation and management. The 416 channel pairs are divided and managed through the carrier's network. You will not see all of the 416 channel pairs in one cell site. Instead, the channel pairs are spread throughout the network and even repeated.

Personal Communications Service (PCS)—Regulation

Personal Communications Service (PCS) was and is supposed to be the next generation of cellular, operating in what is known as the 1800 MHz band. However, what is apparent today is that PCS is cellular engineering applied to a frequency band different than that used by the cellular carriers. In the beginning, PCS was supposed to be everything that cellular was not and could not be. The reality is that technology advancements make cellular

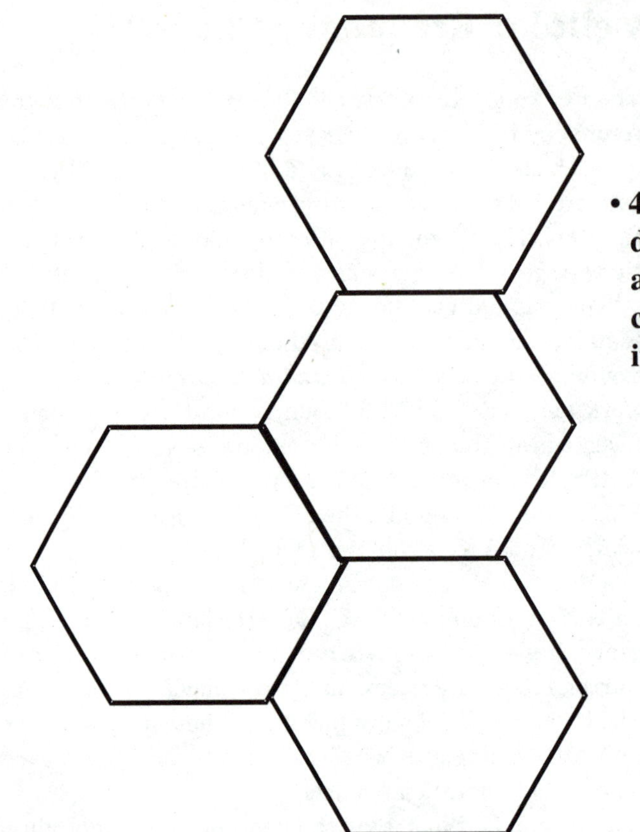

Figure 7-11
Channel allocation

competitive with PCS. One of the advantages PCS supposedly possessed that cellular did not was legacy equipment; that is, PCS operators did not have to worry about maintaining the "old" analog radio systems that the cellular carriers had to worry about. This means the cellular carriers had to spend money to maintain and upgrade equipement that the PCS carriers did not have to spend money on. The thinking of the time was that if the PCS carriers did not have to spend money on maintaining older/legacy systems, it was one less capital expenditure and expense item than the cellular carriers had to worry about.

To the subscriber, cellular and PCS are just simply words. The subscriber only cares that he or she is getting the service for which he or she paid.

In 1994, the FCC issued a decision that allocated frequencies for the express purpose of creating new wireless carriers that would bring new types of wireless services to the consumer. The underlying goal was to find a way to bring better and lower-cost wireless service to the consumer; that is, competition to the cellular industry. The FCC's decision set aside 120 MHz of spectrum; however, unlike the original cellular allocations, the ILECs would not be given "carte blanche" frequencies. The FCC held a series of auctions. Depending on your point of view, these auctions were either successful or a complete disaster.

The term *Personal Communications Services* (PCS) was originally envisioned to encompass a broad range of services designed to allow people to access the *public switched telephone network* (PSTN), regardless of their physical location. Today, most people believe it be digital cellular, something better than cellular (but they don't know why), or perhaps cellular without the old baggage.

In the mid-1990's, many envisioned PCS as having features that supported terminal personal and service mobility. PCS was supposed to combine many emerging "intelligent network" capabilities of the public networks (CCS, ISDN, and IN) with sophisticated wireless access technologies and related radio network mobility control capabilities. The biggest differences between PCS and cellular are

- PCS must contend with high capital costs for equipment and deployment.
- PCS faces competition from the beginning. Cellular carriers in their infancy did not have much competition to worry about.
- The competitors are both new and established.
- PCS carriers are installing the latest technology. Cellular carriers are doing the same thing, but still have to maintain the established equipment base.
- PCS carriers have a fraction of the time needed to make their networks commercially operational. In other words, PCS carriers must work hard and fast to capture market share. The PCS carriers have the disadvantage of building a new network infrastructure as rapidly as possible in order to make money as quickly as possible.
- PCS carriers spent a lot of money to obtain their licenses.

From the Perspective of the Cellular Carriers

The cellular carriers have an existing customer base. However, the problem with being the "big dog" on the block is that there are always other dogs nipping at your tail.

- Cellular carriers must deal with the burden of existing infrastructure, whereas PCS carriers are deploying new infrastructure (brand new "stuff").
- The cellular carriers have a need to upgrade existing equipment to compete with new carriers coming into the marketplace.
- The PCS carriers have the burden of generating enormous expenditures in order to create a profit-making network.
- The PCS carriers are dealing with a two-edged sword. The cellular carriers can make the case that they are doing the same: upgrading aging equipment while deploying new infrastructure to meet the needs of an expanding marketplace and new coverage areas. In the case of the cellular carrier, the two-edged sword is not as big or nasty.

The major differences between cellular and PCS have more to do with business/regulation and less to with technology.

As I had noted, the PCS licenses were awarded to six carriers per service area. Rand-McNally has defined 51 MTAs and 487 BTAs in the United States of America. The service areas were broken down into two categories:

- *Major trading area* (MTA)
- *Basic trading area* (BTA)

As the reader will note, MTAs and BTAs do not match the contours of MSAs and RSAs. Furthermore, they do not match the contours of LATAs. Figure 7-12 highlights the fact that the licensed coverage areas of PCS do not match either the cellular or wireline licensed/FCC mandated coverage areas.

Figure 7-13 is an illustration that further highlights the differences in coverage area contours. Note that the service areas can and in many cases overlap.

As of today, a large number of PCS licensee "winners" have shut down their operation. There are a number of reasons for this:

- There can be as many as six PCS licensees per MTA and BTA. A local economy can only support so many carriers.
- Some carriers paid enormous auction fees per POP (population—one POP equals one person).
- Some carriers underestimated the cost of building out their networks and went bankrupt.
- Some carriers did not have the budgets to market appropriately.
- Some carriers could not obtain the long-term financing to build and operate.

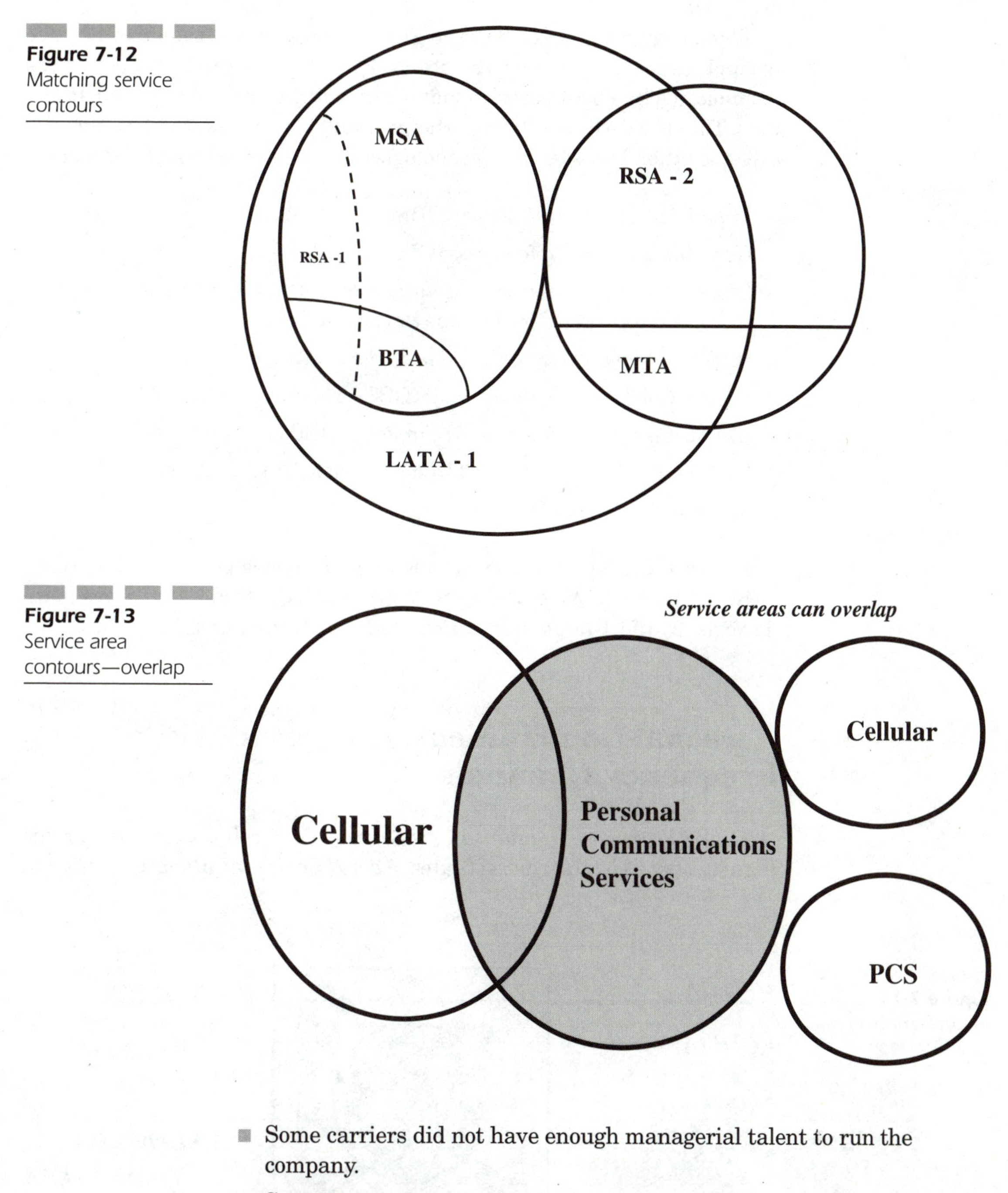

Figure 7-12 Matching service contours

Figure 7-13 Service area contours—overlap

- Some carriers did not have enough managerial talent to run the company.
- Some companies ran out of money before their first installation.
- Some carriers did not have a business plan.

Technology was the least of the PCS carrier's worry. A number of radio technologies were vying for the attention of the new carriers. Each of the technologies had both positive and negative technical and financial attributes. This book will not discuss the pros and cons of one radio technology over the other. However, the technologies that had been considered were

- *Time Division Multiple Acess* (TDMA)
- *Code Division Multiple Access* (CDMA)
- *Global System for Mobile Communications* (GSM). Some may recall that GSM once stood for Groupe Special Mobile.
- *Personal Access Communications System* (PACS)
- *Digital Cordless Telephone-1800* (DCT−1800)
- *Digital Equipment Cordless Technology* (DECT)
- *Wideband CDMA* (Wideband Code Division Multiple Access)
- Composite TDMA/CDMA

Figure 7-14 highlights the various radio technologies reviewed.

Note that each one of the aforementioned radio technologies was standardized by the *Telecommunications Industry Association* (TIA).

Personal Communications Service (PCS) Frequency Channels

Officially, *Personal Communications Service* (PCS) refers to the set of radio licenses offered by the United States *Federal Communications Commission*

Figure 7-14
Radio technologies in wireless

(FCC) in several auctions from 1994 to 1996. Licenses were auctioned in six different radio bands generally known as bands A, B, C, D, E, and F.

Band	Spectrum Block Size	Frequency Range (MHz)	Coverage Area
A	30 MHz	1850–1865/1930–1945	Major Trading Area
B	30 MHz	1870–1885/1950–1965	Major Trading Area
C	30 MHz	1895–1910/1975–1990	Basic Trading Area
D	10 MHz	1865–1870/1945–1950	Basic Trading Area
E	10 MHz	1885–1895/1965–1970	Basic Trading Area
F	10 MHz	1890–1910/1975 –1990	Basic Trading Area

Figure 7-15 illustrates how densely packed the PCS field could become. As of today, the numerous license winners have either sold out to larger carriers (thereby fueling consolidation of frequencies) or have simply gone out of business.

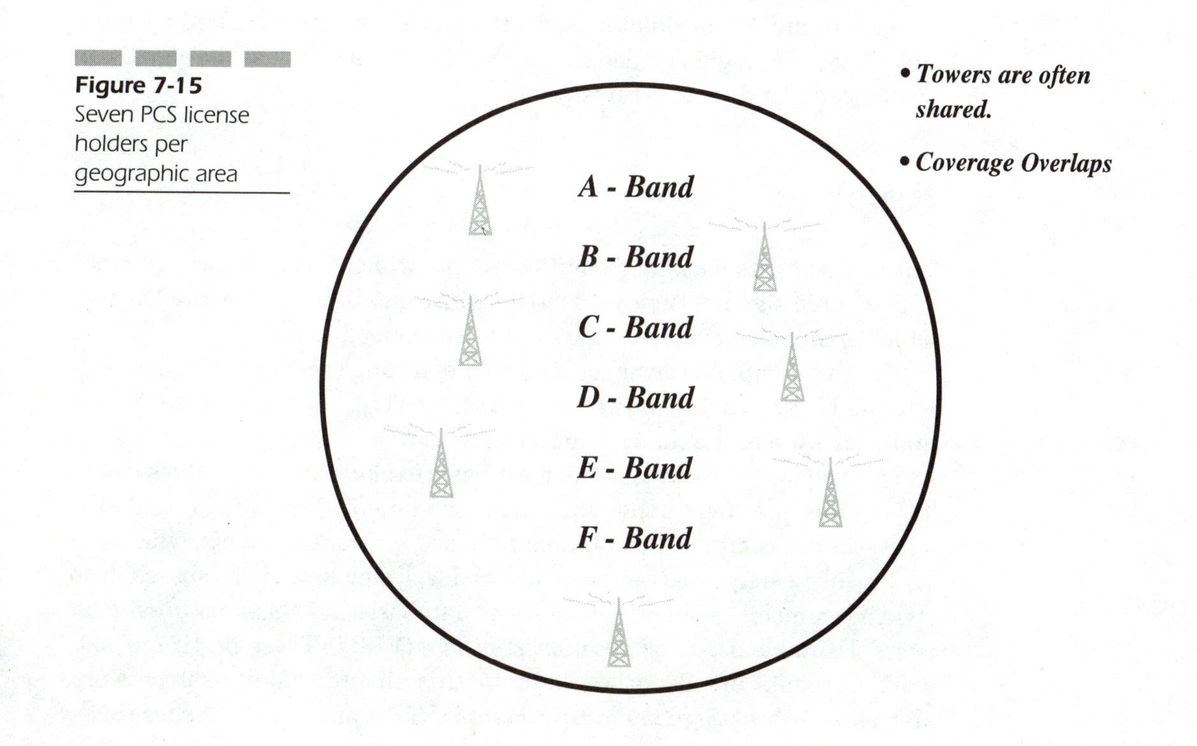

Figure 7-15 Seven PCS license holders per geographic area

Cellular and PCS Network Concepts

Cellular and PCS systems are radio systems. Radio was first postulated in 1873 by Maxwell, demonstrated in 1888 by Hertz, and used for practical communications in 1895 by Marconi. Radio is an electromagnetic phenomenon and radiates as photons. It belongs to the same family of radiation that includes X-rays, light, and infrared. The different categories of radio differ in frequency.

All practical radio systems can be reduced to the following basic scheme: transmitter, receiver, and modulator.

Transmitter

The transmitter consists of two basic parts: a modulator and carrier. A radio frequency generator generates the radio energy that will carry the signal. This generally consists of an oscillator (which produces the initial signal) and a number of amplifier stages (which amplify the level to that required at the antenna). A modulator mixes the signal to be transmitted with the radio frequency signal (called the carrier) in such a way that the signal can be decoded at a distant receiver.

Receiver

The receiver gets a signal from its antenna, which also receives a number of unwanted signals. The tuned circuit tunes out all but the wanted signal, which is then demodulated (decoded) by the demodulator.

The very simple receiver consists only of a tuned section and a demodulator and is known as a *tuned radio frequency* (TRF) receiver. Figure 7-16 is an illustration of the receiver concept.

The previous definition is the most basic explanation of a radio system. Taking this one step further, the reader should note that cellular and PCS networks are comprised of a number of fixed network elements, which can be broadly categorized as switches, radio systems, and databases. The switch is a mobile switching center. The databases are *Home Location Registers* (HLRs) and *Visitor Location Registers* (VLRs). These particular network elements are described later in this chapter. Additional network elements such as *Signal Transfer Points* (STPs) and *Service Nodes* (SNs)

Figure 7-16
Receiver concept

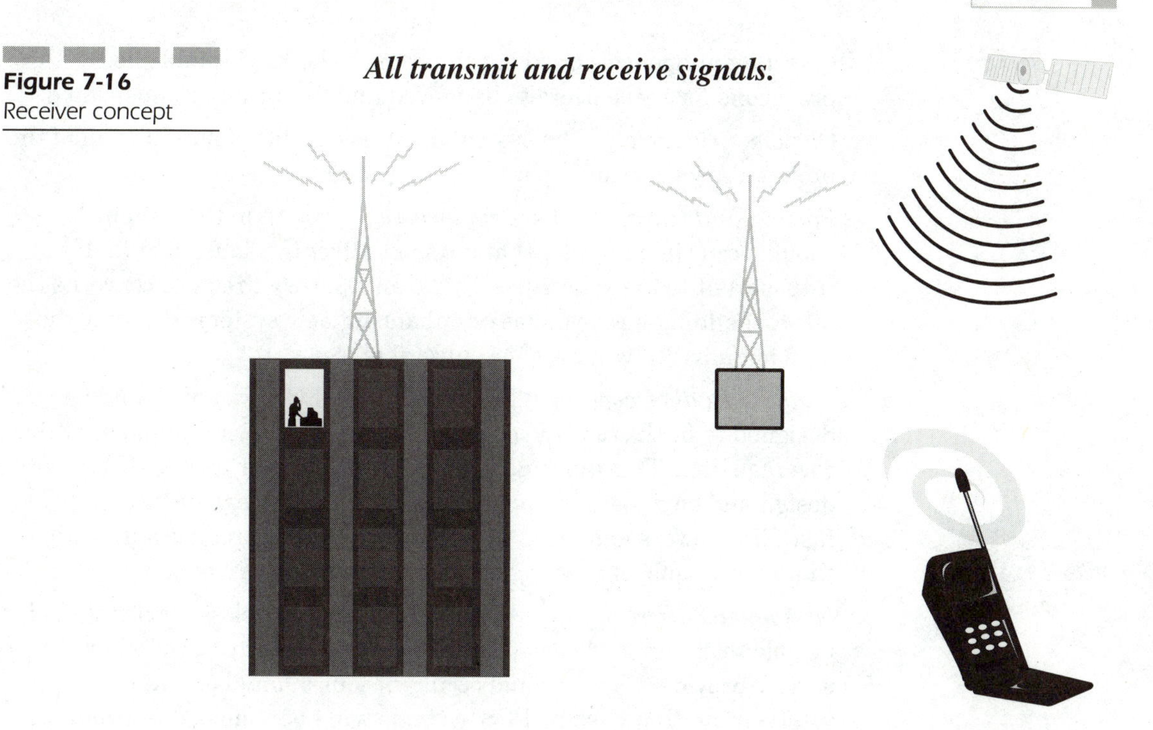

can also be used. However, the most basic components needed to make or process a call are the three basic network elements. Network concepts are also very similar. The next sections will focus on network engineering concepts employed in both industry segments (cellular and PCS) and a detailed explanation of the network elements employed in both cellular and PCS.

Cellular and PCS System Concept Review

Both cellular and PCS have the same basic objectives:

- *Efficient use of the spectrum* Useable radio spectrum is not as plentiful as one would believe. In order to create the new PCS carriers, some of the 1800 MHz bandwidth had to be cleared of state and local users. Many underestimated the number of users operating at those frequencies. During the original comment period of the FCC PCS notice of proposed rulemaking, hundreds of companies and local government agencies cried out in protest. The FCC established provisions for these users to be moved and financially compensated.

- *Customer Capacity* The system must be capable of handling all of the projected users including anticipated visitors.
- *Smooth and transparent system growth* Growth in the system should occur in a manner that does not affect the subscriber and is transparent to the subscriber. This transparency I refer to concerns the affect of adding new systems or enhancing new systems in a way that does not adversely impact the subscribers.
- *Tailored Radio Coverage* The cellular and PCS system should be designed to fit the *radio frequency* (RF) topological environment of the coverage area. This tailoring will enable the carrier to cost effectively design and engineer a system that meets the carrier's needs. Simply installing towers and radio equipment will likely produce a network that is incapable of meeting the needs of the subscribers.
- *Nationwide Accessibility* A subscriber of one wireless carrier should be able to obtain service within the operating territories of other service providers. This should occur without manual operator intervention. Today, some PCS systems will not enable a user to roam onto a cellular system. Some carriers are using tri-mode handsets that enable subscribers to roam from one carrier to another and between frequency bands (that is, 800 MHz and 1800 MHz).
- *Affordability* Cellular and PCS service should be affordable to the subscriber. This is a simple concept. Carriers want people to buy their service; therefore, it should be low enough in cost to entice people to buy wireless service.
- *Acceptable Quality of Service (QoS)* During the original AT&T Bell Laboratories study, researchers had felt that the quality of wireless service ought to be similar to the quality of service that a user expects from wireline service. However, as many wireless users have found, wireless carriers do not have an "industry standard" for quality of service. In fact, many subscribers may feel the QoS for wireless is substantially lower than that of wireline.

Design Concept

Both cellular and PCS networks design their networks around the same basic design premise: channel reuse. All wireless systems must be designed with the constraint of limited frequency allocation. No matter how much of the spectrum one obtains, frequency allocation will be a limiting factor.

Wireless (cellular and PCS) systems are designed around the cellular configuration. This means that the radio part of the network takes the geographic serving area and divides the area into cells (hexagonal shaped coverage zones). A cell is a geographic area within which wireless handsets are served from centrally located transmitter/receiver stations (antennas, towers, and base station equipment). In this type of network configuration, growth is accomplished by splitting the cells smaller cells equals more cells. The splitting of cells would be meaningless and have no effect if it were not for "channel reuse."

Channel reuse was revolutionary in the early 1980s. Prior to present day cellular and PCS, mobile phone service (the car phone) was possible only if the carrier installed a single antenna at a high elevation. This single antenna had an effective radiated power of 500 watts. The single antenna allowed for only a small number of handsets to operate simultaneously in the antenna's coverage area. Each handset required between 25 KHz to 30 KHz of bandwidth. Only a limited spectrum was available. The old two-way mobile/*Public Land Mobile Service* (PLMS) operated at both the *Very High Frequency* (VHF) and *Ultra High Frequency* (UHF) bands. VHF service occurred in the 152 MHz frequency and at the 157–158 MHz band. UHF service was provided at both the 454 MHz frequency and at the 459 MHz frequency. Only 11 to 12 channels existed.

Modern day cellular had to increase systems capacity if it were to become popular. The concept of channel reuse was created in order to enable high system capacity in a geographic area with limited spectrum availability. Channel reuse refers to the practice of independently using radio channels that have the same radio frequency to cover different coverage areas. This assumes that the mobile units are operating at power levels lower than their PLMS predecessors. Reduced handset power output is a must in order to address the interference issues envisioned by having hundreds and thousands of handsets operating in a small area.

In the cellular network configuration, multiple transceiver sites are established throughout the designated service area. Each site serves a small area of no more than one to two miles in diameter. The systems transmit at low power levels (less than 100 watts). This also means that the mobile handsets have to be relatively close in order for the site equipment to process calls from and to the mobile handset. Because the serving sites are so small and power transmission levels are so low, there is an opportunity to reuse the channels of one serving site in another serving site. Reusing channels would only occur in sites that are not adjacent to one

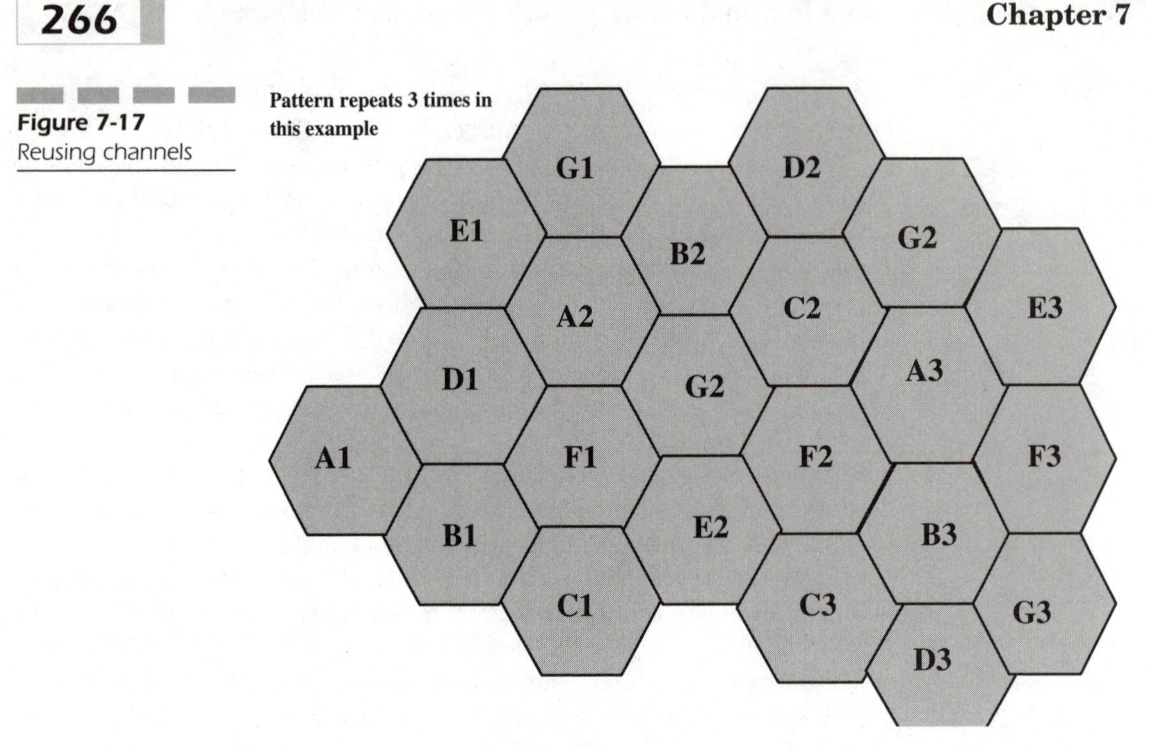

Figure 7-17 Reusing channels

another; this would reduce any chances of co-channel interference. Figure 7-17 illustrates the pattern repeats in non-adjacent sites.

Some common engineering procedures and practices are used by both PCS and cellular network engineers to determine the optimum ratio of serving site separation distance and serving site radius. The serving site is assumed to be almost circular in nature—radio waves radiate away from omni-directional antennas in circular patterns. In fact, the cell site RF coverage patterns overlap. The reader will note that the hexagonal shaped cells used by carriers to illustrate system coverage can be used to design and engineer coverage. In fact, the hexagonal shapes are better than the overlapping circular shapes for calculating ideal power coverage areas. The hexagonal shape is an engineering convention and not a representation of the real world; the hexagon simplifies planning and design. Figure 7-18 depicts the hexagonal planning concept.

The optimum ratio of separation *distance* (D) to serving site (cell) *radius* (R) is based on system design parameters such as signal to noise ratio, propagation characteristics, environmental characteristics, and so on. The ratio also known as the channel reuse ratio is D/R and varies from four to seven in analog and digital systems. The ratio is an overall representation of the

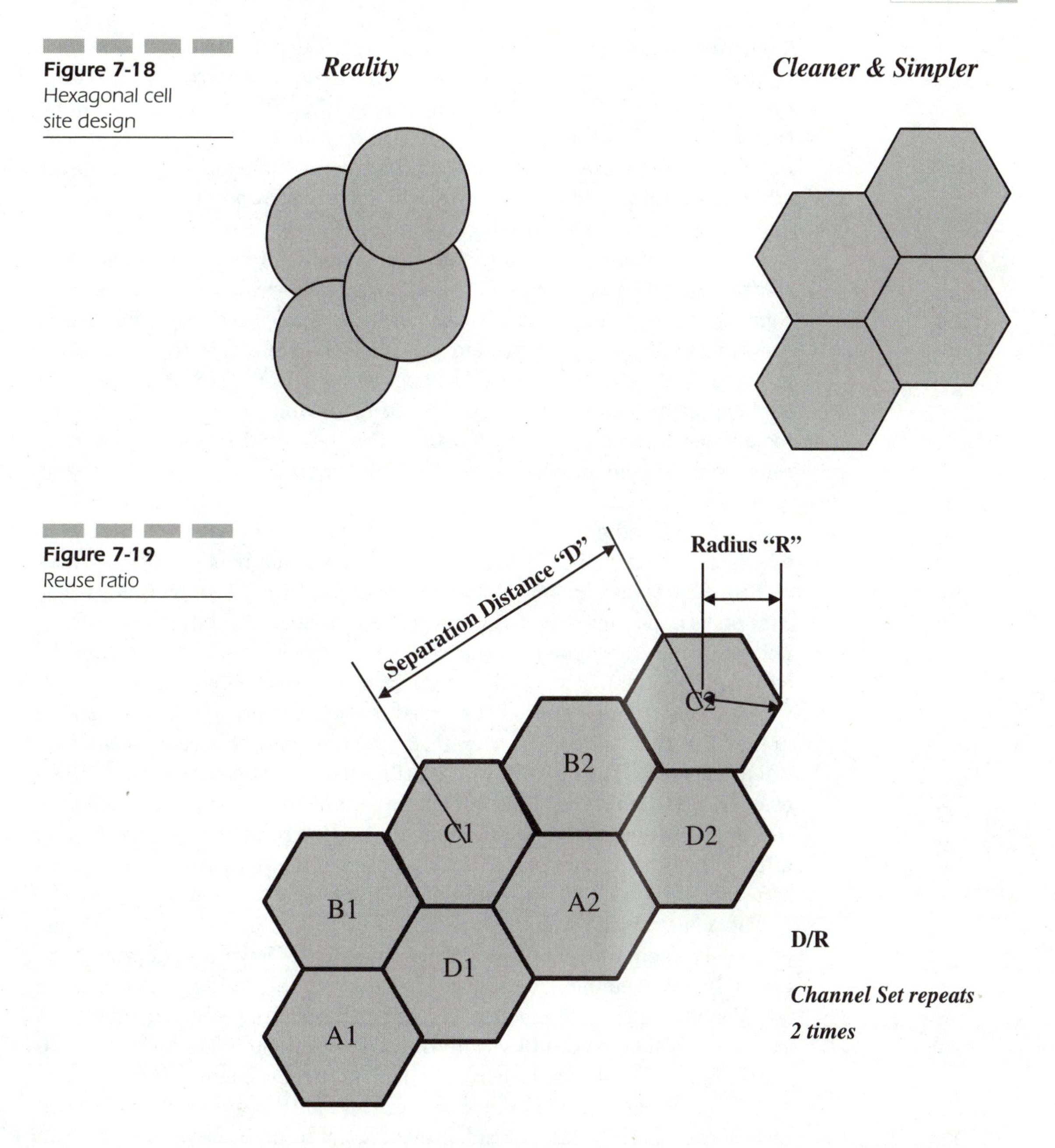

Figure 7-18
Hexagonal cell site design

Figure 7-19
Reuse ratio

repeat pattern. The carrier's design configuration will also play a role in the reuse ratio. Figure 7-19 illustrates the how the reuse pattern is calculated.

Channel Attachment and Cell Splitting The previous scheme enables a carrier to reuse channels; however, the issue of growth has not been totally addressed. The carrier's radio system capacity can be increased not only by increasing the D/R ratio but also by cell splitting and cell attachment. I will not address the pros and cons of one radio technology over the other. The intent of this section is to examine generic approaches to increasing a carrier's network capacity.

Cell attachment is simply the process of adding a new cell site to a carrier's network. Cell attachment is an appropriate way of increasing system capacity if the main goal is to cover new geographic areas. Cell splitting is the process of increasing system capacity when the objective is to meet growing demand within an existing set of sites. Cell splitting is the most challenging and interesting way of meeting growing demand. Cell splitting techniques include adding new cells within cells and sectorizing existing cells using unique antenna array. Cell splitting offers unique engineering challenges.

To understand cell splitting, one must examine a realistic scenario in which the carrier has a fixed allocation of radio channels for a given MSA or RSA. The total number of channels is divided into channel groups. The channel groups themselves contain a fixed number of channels. The channel groups are assigned to the cell sites. The channel groups may be assigned in a manner in which two are assigned to each high traffic cell site, whereas one channel group is assigned to each cell site in the lower traffic areas. Growth outside of the overall serving area may be accommodated via cell attachment. The new cell would utilize a different channel group. However, growth within the high traffic cells cannot be handled in this manner. Cell splitting enables the carrier to maintain the reuse pattern except that after cell splitting, the carrier is able to reuse channels more often in the same area. Figure 7-20 illustrates the channel splitting concept.

The reader should note that the high traffic area may be within the heart of the city—cell attachment would be impossible. New channel groups are probably not available. The solution is to take a cell site and split it into a set of smaller sites. This requires revising the existing cell site boundaries so smaller cell areas can fit within the existing single cell area. The resulting configuration allows the carrier to run more radio channels for the same overall macro area. Some may consider cell splitting just cell attachment plus re-engineering techniques: This would not be an incorrect assessment. Figure 7-21 illustrates cell attachment.

The radius of the new cells is less than the original cell area. The overall D/R ratio remains the same, but the separation distance is reduced.

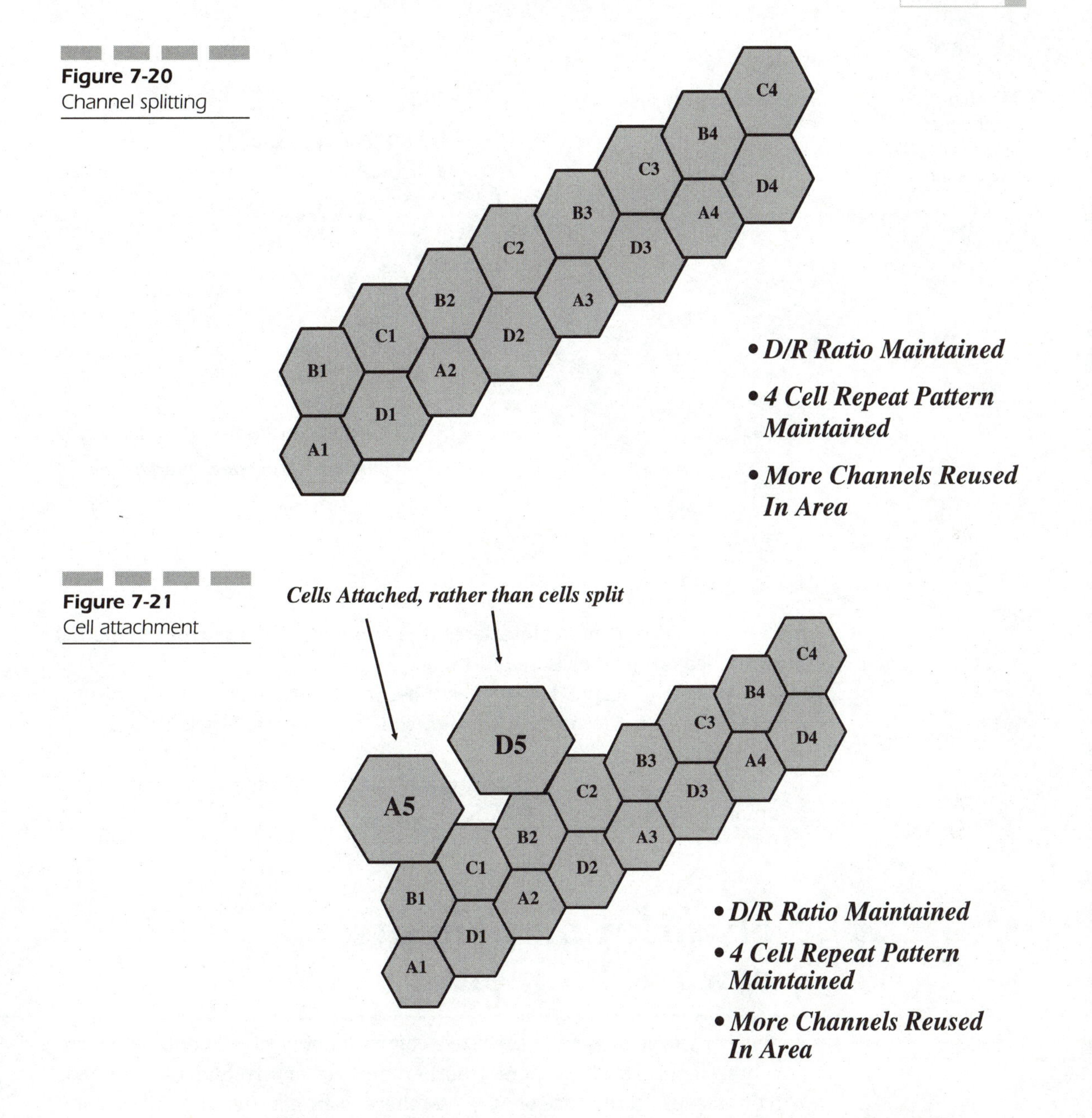

Figure 7-20 Channel splitting

Figure 7-21 Cell attachment

As Figures 7-19 through 7-21 illustrate, the total number of channels is divided into channel groups A through D. Each channel group is available twice throughout the MSA or RSA. Then the carrier splits the cells so the channel groups are repeated four times. However, the groups in each cell

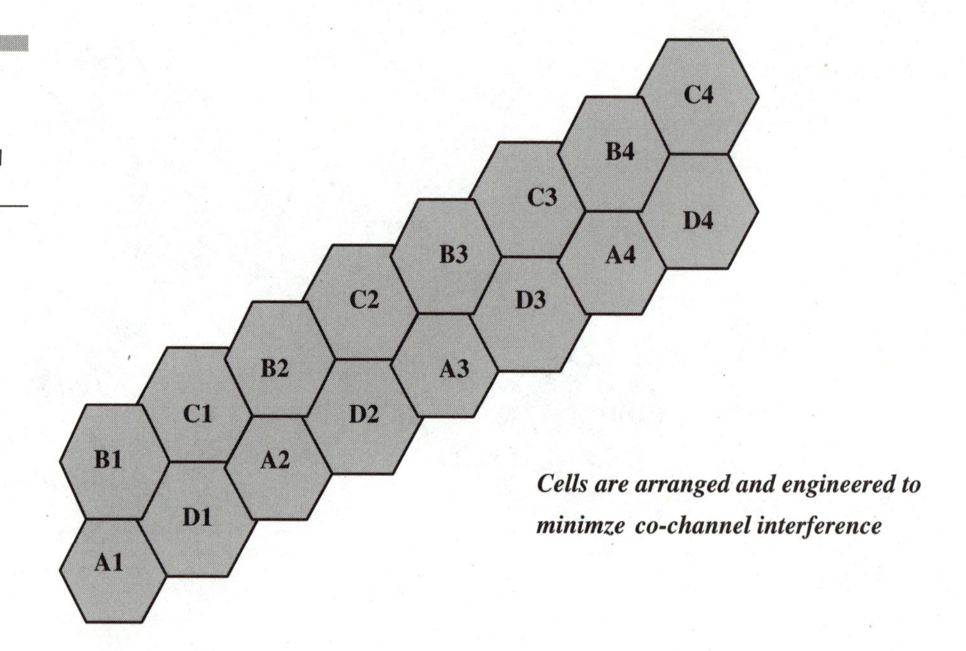

Figure 7-22 Minimizing co-channel interference via cell site engineering

are arranged so co-channel interference is minimized. Minimizing co-channel interference is a goal as noted in Figure 7-22.

The design concepts I have described are applicable in both cellular and PCS. In fact, the previous design concepts are used by both industry segments.

The following section will address network elements and components common to both cellular and PCS.

Network Components/ Network Elements

As with any type of carrier, there are common network elements and components. Generally all carriers, whether wireless or wireline, use switches and databases. In the case of wireless, there is an additional network component: the radio equipment. Radio equipment includes the transmitter, receiver, and modulator (these components have already been described). In addition to these pieces of radio equipment is the antenna. Figure 7-23 illustrates the major network components of the cellular and PCS networks. The following represents the other common cellular and PCS network components, which include

- *Mobile Switching Center* (MSC)
- *Home Location Register* (HLR)
- *Visitor Location Register* (VLR)
- Radio equipment

Figure 7-23 depicts the network elements as physically separate elements. However, in reality, the networks elements can physically reside within the same hardware package. Normally, the logical functions are illustrated as separate boxes.

PCS and cellular are more regulatory constructs than technology constructs. Both cellular and PCS utilize the same design and planning concepts. The only obvious and significant difference is the spectrum band in which the two industry segments operate. During the period prior to FCC docket 90-314, which established the rules of new personal communication services, PCS "wannabes" of the wireless industry claimed that PCS was

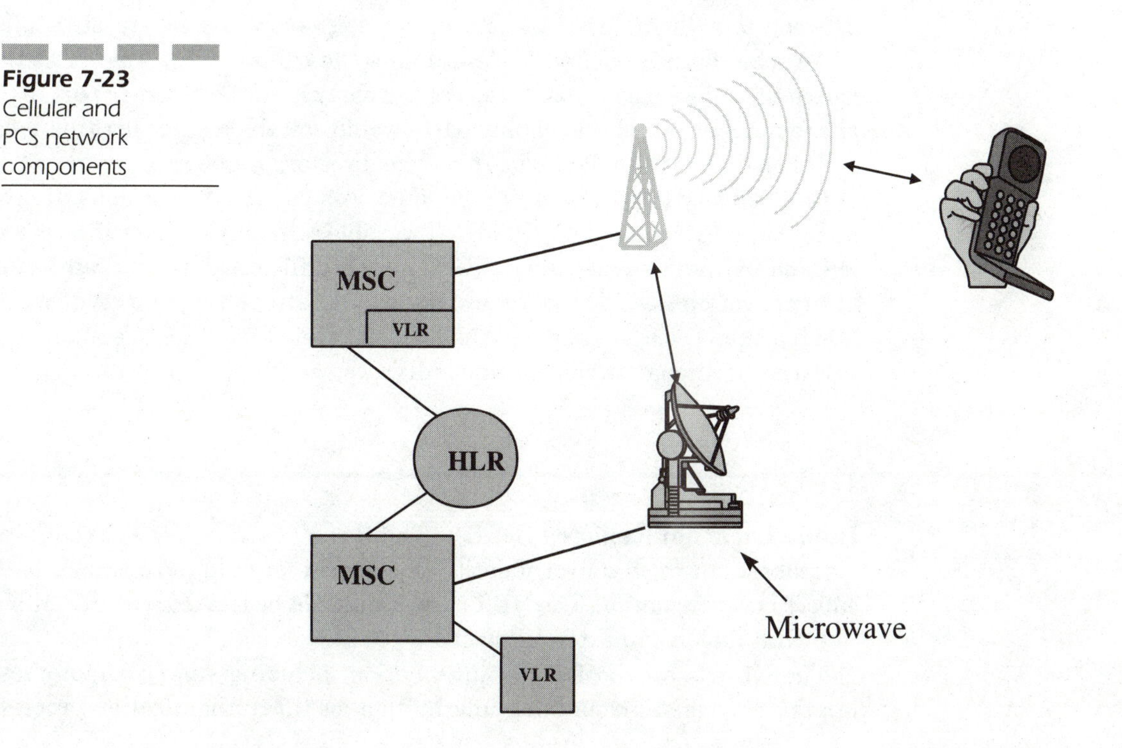

Figure 7-23 Cellular and PCS network components

different because it used different equipment, PCS and cellular had nothing in common, PCS used better technology, PCS engineering was better, and so on. The list made no technical sense, but was only politics in action. Today, you will find that the only obvious differences are that PCS and cellular operate in different frequency bands and that cellular carriers may use older technology. To the subscriber most of the industry talk amounted to a lot of industry noise over insignificant issues.

Mobile Switching Center (MSC)

The *Mobile Switching Center* (MSC) is a highly sophisticated call processing system that has only trunk connections. The MSC does not support lines, as in lines in a local loop. Instead, the MSC supports trunk connections between itself and cell sites, between itself and other MSCs, and between itself and other switches from other network providers.

The MSC is a call-processing device. During the cellular industry's early days (circa 1982–1985), MSCs from different manufacturers could not communicate with each other directly. Subscribers of cellular systems roaming between their home systems and visited systems could not receive calls while roaming in a visited system unless the calling party knew exactly where the called party was. If the calling party knew the roamer port telephone number to call, the calling party would call the roamer port number and then dial in the called party's mobile directory number.

Call delivery to parties in this manner was possible but tedious. It created a ripe environment for fraud and variable call completion success rates affected by radio environments and handoff difficulties (partly radio and fixed network based). The cellular industry was attempting to provide wireless telephone service that had the look and feel of wireline telephony but with the additional attribute of mobility.

Databases

Home Location Register (HLR) The *Home Location Register* (HLR) is a database to which a user identity is assigned for data purposes such as subscriber information. The HLR may or may not be located within a MSC. The HLR may be an external device.

The HLR is a depository of information including subscriber profiles, location information, enabling information for the authentication process, and features. The HLR logical function can be implemented so a single HLR device can support multiple MSCs.

Visitor Location Register (VLR) The *Visitor Location Register* (VLR) is the location register (database) other than the HLR used by a MSC to retrieve information for handling of calls to or from a visiting subscriber. The VLR is essentially a temporary memory storage area. VLRs are typically internal within the MSC.

Radio Equipment

The basic components of the radio portion (transmitter, receiver, and modulator) have already been described. The only component of the radio system not described yet is the antenna. Antennas radiate electromagnetic energy—radio energy. This energy is radiated in a pattern at a certain power level.

The pattern, gain, height, and tilt of the antenna all affect cellular and PCS design. The antenna pattern can be omnidirectional, directional, or even engineered to assume any shape in the horizontal and vertical planes. The pattern plays a role in optimizing coverage in and around buildings, structures, trees, roadways, hills, mountains, and various geographic and topological features. Antenna gain is a function that compensates for actual transmitted power. Optimizing the gain of the signal helps the carrier maximize the signal to noise ratio. Antenna tilting is done in order to reduce the interference from other nearby cell sites. Antenna height plays a role in achieving coverage and optimizing the shape of the coverage area.

In general, you can say that all the previous factors play a role in the design and engineering of a cellular and PCS system. Figure 7-24 illustrates how antenna height affects the propagation of radio waves.

Network Interconnect

Cellular and PCS carriers interconnect their respective networks to the PSTN in the same manner. The interconnect is governedprimarily by the ILEC specifications defined in Bellcore specification GR−145-CORE, formerly known as TR-NPL−000145. As for carriers that function as carriers for other carriers, although not intended for this purpose, carriers' carriers

Figure 7-24
Antenna height and radio propagation

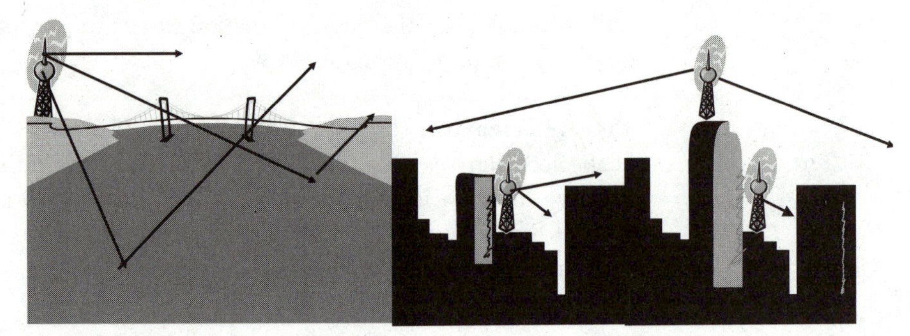

Antenna height affects the propagation of the radio waves. Buildings and geography can affect coverage.

have gravitated toward using the the *Telecommunications Industry Association's* (TIA's) standard, which is known as IS−93. As I have indicated in past chapters, the two documents are similar to one another. What is important for the reader to note is that the interconnect used by cellular is also used by PCS; see Figure 7-25.

Network Signaling

Wireless carriers, like the wireline carriers, use a variety of network signaling applications to support different services and business segments. This section will address the primary network signaling applications currently used by both cellular and PCS to support wireless services.

The North American cellular industry employs a network signaling application that is supported by *American National Standards Institute Signaling System 7* (ANSI SS7). This signaling application is called *Interim Standard 41* (IS−41). The IS−41 standard was created by the ANSI-accredited *Telecommunications Industry Association* (TIA) TR45 committee. The IS−41 standard provides a standard protocol for the operations that enable subscriber mobility between MSC serving areas. IS−41 specifies the signaling communications that occur between MSCs, location databases, specialized network databases, and specialized service nodes/bureaus (authentication centers, fraud management centers, short message services centers, and so on).

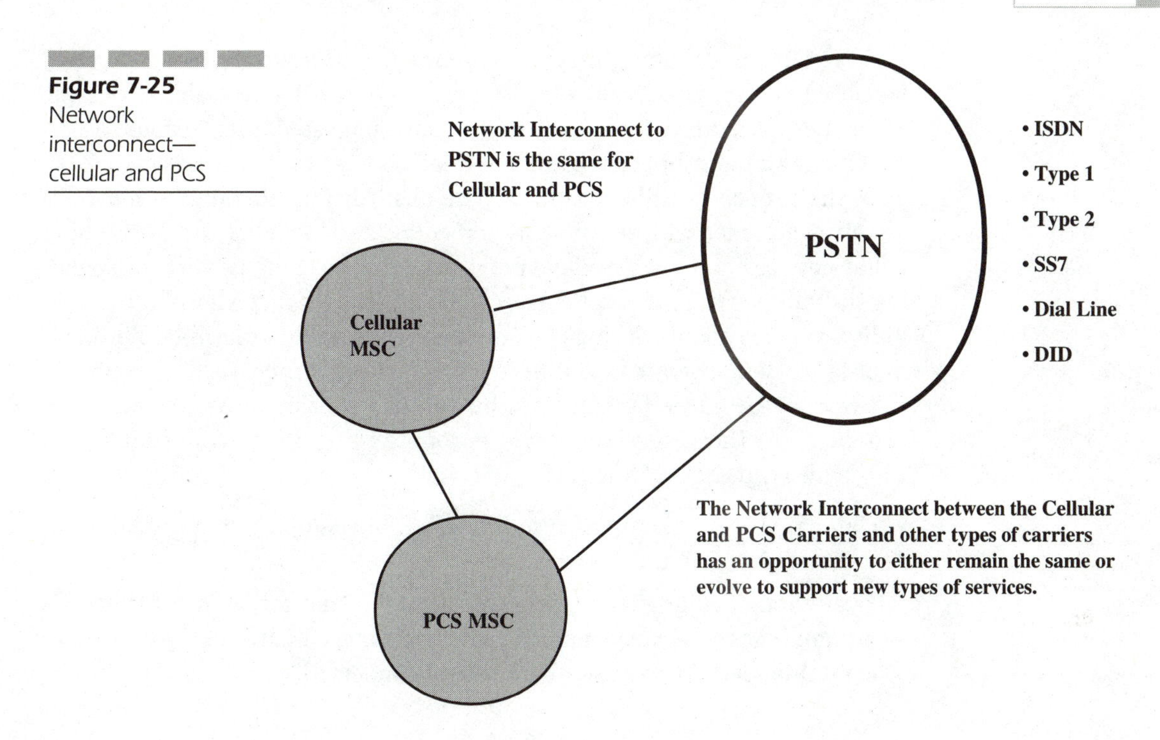

Figure 7-25 Network interconnect—cellular and PCS

The non-ANSI flip side of IS–41 is GSM (formerly *Groupe Special Mobile,* now known as *Global System Mobile* communications). GSM is currently functioning within the North American market supporting PCS systems. Cellular systems are using IS–41 for now. GSM is actually a suite of standards that includes OA&M and radio standards. GSM is used in many non-North American cellular markets. IS–41 was developed separately from the radio standards used in North America. You can argue the merits for each. The only thing that matters is who can get to market—I believe they both have merit.

Both IS–41 and GSM *Mobile Application Part* (MAP) are used to support internal wireless carrier operations and wireless carrier-to-wireless carrier transactions. IS–41 and GSM MAP do not replace the internetwork signaling protocols that were described in Chapter 4. The internetwork signaling protocols described in Chapter 4 are needed to ensure that the wireless carriers can communicate with the larger wireline networks. As a side note, during the early days of PCS, many envisioned using network signaling protocols that were incompatible with the established signaling protocols used in North America. Fortunately for the PCS industry, these individuals did not prevail. If these pundits had succeeded in establishing such a policy position within the community, PCS subscribers would have been unable to complete calls to the wireline network or cellular networks. Some telecommunications executives had suggested the various North American signaling networks be modified so that they can be compatible with the proposed GSM networks. The absurdity of the suggestion completely overlooked the fact that the embedded and established multi-billion dollar network of networks could not practicably be changed without spending the value of the national debt on such an effort. The policy of incompatibility was not about serving the subscriber but about creating differences in order to demonstrate how different PCS was: in other words, politics.

Prior to IS−41 or GSM MAP, the cellular service providers in North America and the rest of the world encountered the following challenges; this is also illustrated in Figure 7-26.

- Call delivery to roaming subscribers could be supported but validation was not supported.
- Even though call delivery was possible, it was not automatic. Prior to automatic call delivery, roamer ports had to be called first. However, even this method was not ubiquitously supported.

Figure 7-26 Network incompatibility and the impact on mobility

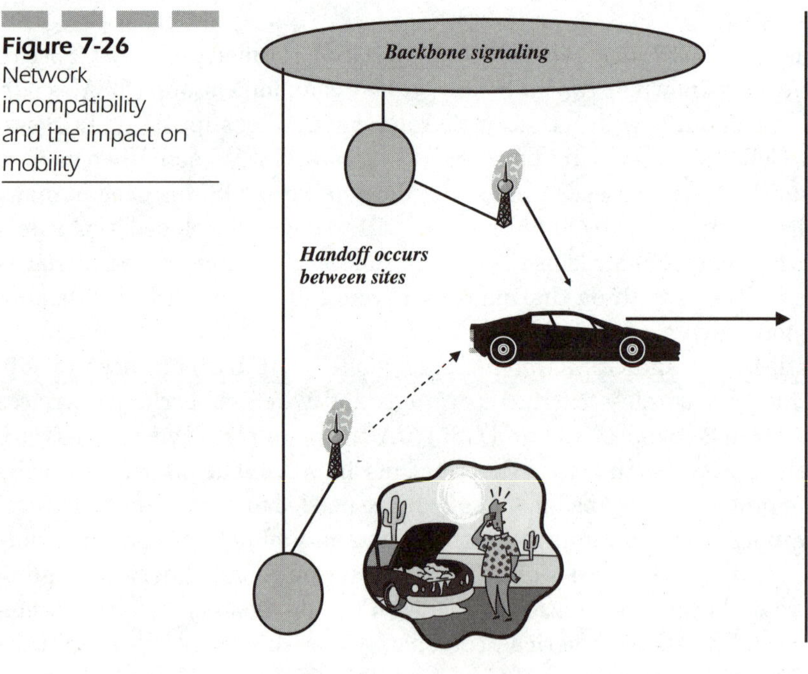

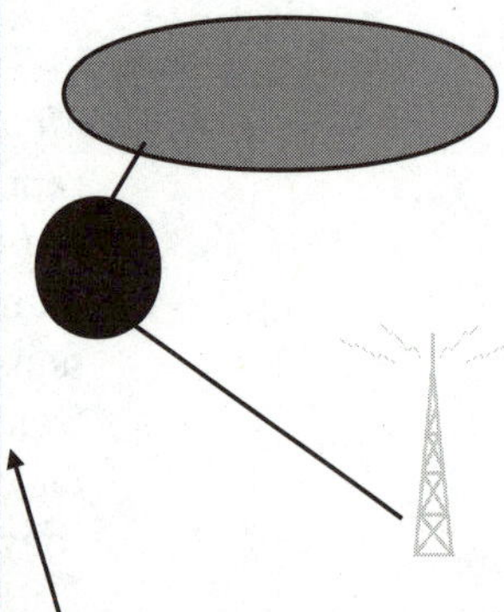

- Handoff between adjacent serving systems was not necessarily possible if the service provider used different vendors' MSCs.
- Non-standard network signaling protocols were being used with standard network signaling protocols.
- Subscribers could not be authenticated in a near-real-time fashion.
- Call origination by roaming subscriber was problematic. Service providers did not necessarily allow calls to be originated by roamers.

In the very early days of cellular, none of the preceding considerations were a problem. However, as the number of subscribers grew, the previous issues became major problems. Call delivery in a mobile environment brought challenges that have been largely overcome by the creation of an intersystem operations standard called *Interim Standard* (IS)-41.

GSM was created for the same reasons. The technical and business environment was different in Europe when the GSM suite of standards was developed, but the standards accomplish the same goals as IS−41. Both GSM MAP and IS−41 are standards that support the following capabilities:

- *Roaming* Roaming is an activity. When a mobile subscriber is originating or receiving calls in a cellular service provider system (or even PCS network) other than the one from which service his was subscribed, he is roaming.
- *Authentication* Authentication is the process of user identity confirmation. Identity confirmation can involve checking handset or terminal device identity by interpreting "secret" keys or data messages. If the data keys or data messages have been altered or do not show a specific format, the call will not be completed.
- *Call Delivery* Call delivery permits a subscriber to receive calls to her directory number while roaming. Cellular service providers once treated call delivery as a subscriber feature. Today, call delivery is a required feature in the sense that this capability is expected.
- *Handoff* Handoff comes in two flavors: hard and soft. *Handoff* is the seamless transfer of an in-progress wireless call from one base station to another base station as the wireless party travels from cell site boundary to another. In the GSM suite of standards, the term is *handover*. One can argue the merits of one term over the other, but such a discussion is not meaningful for this book.
    - *Hard Handoff* Hard handoff is a "break-before-make" form of call handoff between radio channels. In this scenario, the mobile handset

temporarily (time measured in milliseconds) disconnects from the network as it changes channels. The radio protocols, AMPS, TDMA, and GSM support only hard handoff.

  - *Soft Handoff* Soft handoff is the reverse of the hard handoff scenario. Soft handoff is a "make-before-break" form of call handoff between radio channels, whereby the mobile handset temporarily communicates with both the serving cell site and the targeted cell site (one or more cell sites can be targeted) before being directed to release all but the final target cell site radio channel. Currently only CDMA supports soft handoff.

- *Messaging for subscriber services* Messaging for a variety of subscriber features as shown in the list of services in Chapter 4.
- *Path Minimization* Path minimization refers to the efficient network routing of a call. To those of the wireline telephony industry, this would appear to be a "no-brainer." However, until six or seven years ago, the concept of efficient routing was non-existent in the cellular world. The only reason for efficient routing is if the service provider had a large network with lots of traffic. Back in the late 1980s and early 1990s, the cellular networks were in their adolescent stages and did not see a need to save time, energy, or money in wasteful network routing practices. Path minimization requires the following:
  - Routing plan
  - Efficient routing translation tables

The following figures illustrate hand off and path minimization. Figure 7-27 illustrates basic handoff. Figure 7-28 looks at multiple handoffs.

Network Applications and Services

At this time, most wireless carriers are working to meet subscriber growth by undercutting their competitors' pricing plans. The deployment of new subscriber services is beginning to move toward the forefront of carriers' market plans. Wireless carriers (cellular and PCS) are competing with one another furiously. The following subscriber services are the typical fare of a wireless carrier:

Figure 7-27
Basic handoff

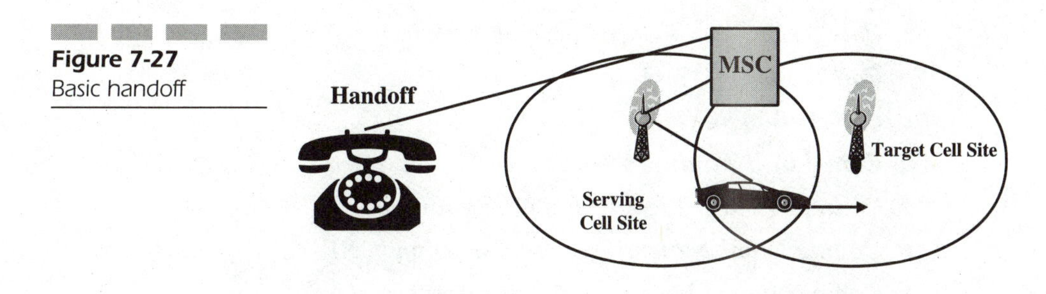

The Target Cell Site represents the cell site the automobile is driving towards. To maintain the call connection the Serving Cell Site must handoff the management and connection of the call to the Target Cell Site. If the handoff does not occur, the Serving Cell Site would eventually drop the connection due to loss of coverage.

Figure 7-28
Multiple handoffs

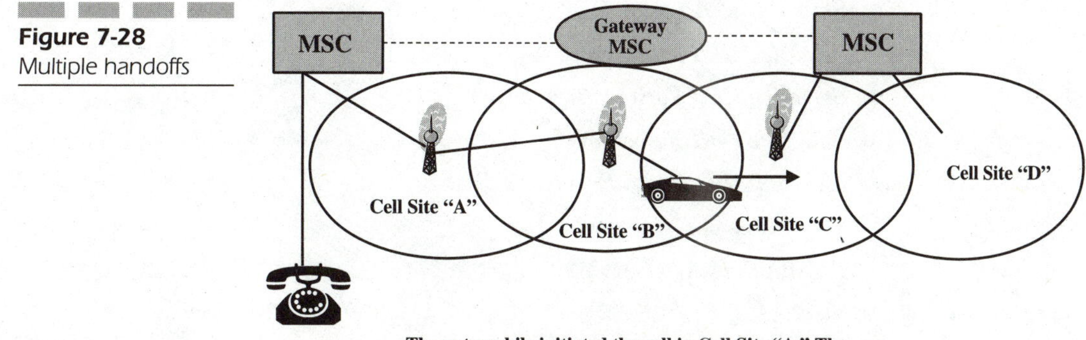

The automobile initiated the call in Cell Site "A." The automobile was handed off to "B." As the automobile travels to its ultimate destination of Cell Site "D," the call is physically routed through the fixed network to another MSC via a tandem called the Gateway MSC. The call remains up and the second MSC is handling the call.

- *Call Delivery* (CD)
- *Call Forwarding—Busy* (CFB)
- *Call Forwarding—Default* (CFD)
- *Call Forwarding—No Answer* (CFNA)

- *Call Forwarding—Unconditional* (CFU)
- *Call Transfer* (CT)
- *Call Waiting* (CW)
- *Cancel Call Waiting* (CCW)
- *Calling Number Identification Presentation* (CNIP)
- *Calling Number Identification Restriction* (CNIR)
- Calling Name Identification Presentation
- Calling Name Identification Restriction
- *Conference Calling* (CC)
- *Three-Way Calling* (3WC)
- *Message Waiting Notification* (MWN)
- *Mobile Access Hunting* (MAH)
- *Password Call Acceptance* (PCA)
- *Remote Call Forwarding* (RCF)
- *Remote Feature Control* (RFC)
- *Screen List Editing* (SLE)
- *Selective Call Acceptance* (SCA)
- *Selective Call Forwarding* (SCF)
- *Selective Call Rejection* (SCR)
- *Messaging Delivery Service* (MDS)
- Short Message Service
- Paging Message Service
- Voice Mail
- Fax Mail
- *Single Number Service* (SNS)

One challenge wireless carriers are facing is the appearance of sameness. Subscriber services attract new subscribers and in many cases keeps existing subscribers. However, in a mature wireless market environment, subscriber churn and subscriber growth are affected primarily by

- Pricing
- *Coverage* This includes both physical coverage and the quality of the coverage.

As I had noted, the above assumes a mature (stable and flat) market environment. In order to continue growing the wireless carrier's base or maintaining competitiveness in the market, the wireless carriers are seeking ways of enticing new subscribers and keeping their existing subscribers. These new marketing plans involve differentiation via new technology and existing technology. New technology applications will be explored in this chapter. The use of existing technology involves the bundling of existing services possibly in conjunction with new technology deployments or simply re-bundling existing services differently. The carriers are also seeking ways of maintaining cost competitiveness. Cost competitiveness is not a product, but it is a business necessity. Cost competitiveness involves reducing and controlling costs. For most carriers, cost control means first reducing labor expenses and then finding more cost-effective ways of performing tasks within the carrier.

The reader will recall that I had defined an application as something that does something useful or carries out a useful task. The applications that the wireless carriers are seeking to market are similar to those being deployed by the wireline carriers. The application efforts include the following and all involve convergence of technology and business:

- The wireless local loop
- The Internet
- Intelligent networking
- Service bundling
- Cost control

Figure 7-29 illustrates how telecommunications convergence can occur via wireless carriers.

Note that subscriber interface with these services (in other words, subscriber access) will present a challenge as well. Subscribers like to use as few terminals as possible. Marketing research in this area will be a necessity.

Cost Control

Cost control is simply the process of controlling and managing operating costs. The service provider is like any other corporation; it gauges its financial effectiveness via a series of typical financial mechanisms. These mechanisms involve activities such as tracking profitability, rate of return, and return on investment. The aforementioned are corporate-wide high-level

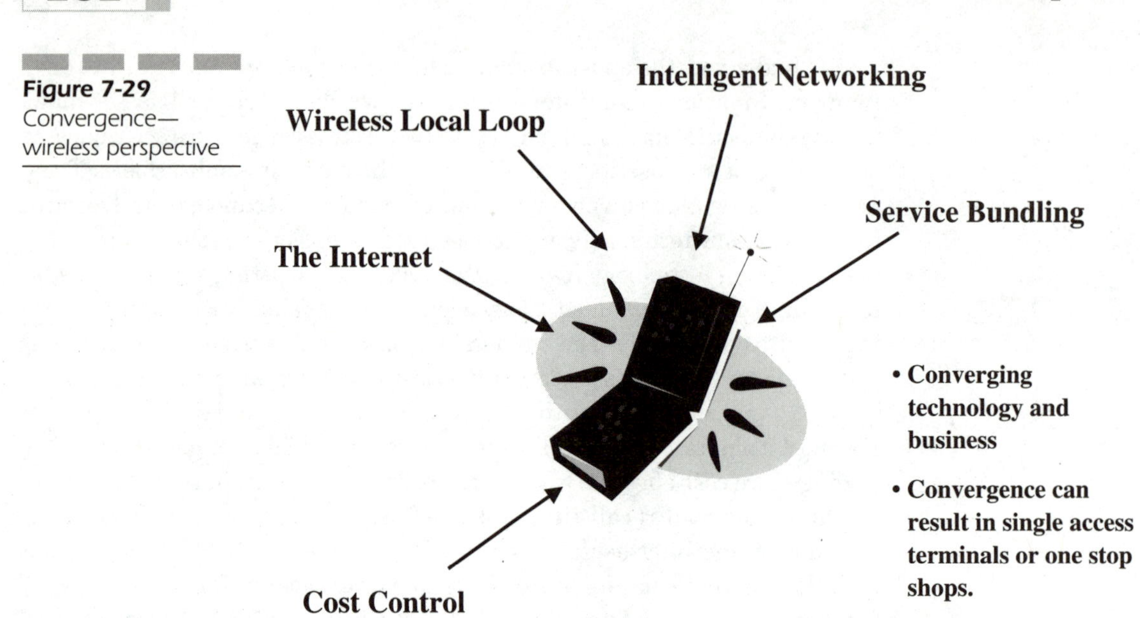

Figure 7-29 Convergence—wireless perspective

financial measurement mechanisms. As in any company, there are also lower-level budgetary management processes that enable an organization within a company to manage its own costs. A provider of service of any kind utilizes financial measurement systems that measure performance as perceived by the customer and internally within the company. These mechanisms are respectively called service measurements and performance measurements. Service providers use these customer-oriented measurement mechanisms to ensure that they are meeting customer expectations. The specific mechanism details will vary among different types of service providers. What is important to understand is that a service provider must have such mechanisms.

All of the aforementioned financial measurement systems are used as part of an overall analysis of company performance. The importance of financial mechanisms cannot be stressed too much, especially as the wireless carriers work to include services such as the Internet. The ability to measure financial performance is as important as the network.

Service Bundling

Telecommunications service bundling is the act of combining services in a variety of ways that entices subscribers to buy service from the carrier. It has taken wireless carrier several years to embrace service bundling. This is not unusual—it also took wireline carriers several years to market services as parts of subscriber service packages. The thinking among carriers had been that subscribers would pay for basic voice and then pay incremental fees for each individual service purchased. However, subscribers could not see the value in paying more and more money for services that had only perceived minimal value. The wireline carriers initially tried to sell the following services as individual services:

- *Voice mail* In the early days of network-based voice mail, carriers had an enormously difficult time marketing this service in a market that was and still is filled with customer premise answering machines or systems.
- Call forwarding
- Conference calling
- Three-way calling
- *Call waiting* (CW)
- *Cancel call waiting* (CCW) In the early/mid-1980s, this was conceived to be a separate service. In other words, the subscriber would have to pay for the right to shut off the call waiting tone. Needless to say, this service was among the first to be bundled. In fact, when you purchase call waiting, you immediately receive, for free, cancel call waiting.
- *Remote call forwarding* (RCF)

The subscriber penetration levels for these services by themselves was less than 5%. The wireline carriers realized that these services ought to be treated as value-added services. A value-added service is a service that is sold as part of a package of services. Subscribers were open to purchasing a package of services that had overall value to them. The subscriber might use voice mail once or twice a month, not often enough to warrant paying a monthly fee. However, if the carrier bundled the service as part of a set of services, the subscriber would be paying for the value of the whole package. In other words, the package has value, whereas the individual services do not.

The wireless carriers have also successfully marketed services in this manner. Bundling services has greater perceived value than the individual service. The subscriber may in fact be paying as much or even more for the package of services rather than the total cost of paying for each individual service. In fact, the subscriber may even have services in the package they do not want. What is important is that the subscribers feel they are getting more for their money. Rate plans figure prominently in the marketing of

Figure 7-30 Service bundling—wireless perspective

Service bundling is about value perception: Getting more for one's money.

service packages. Service bundling is about value perception; this is depicted in Figure 7-30.

Intelligent Networking

As I had noted in Chapter 1, "What Is a Telcommunications Network?," *intelligent networking* (IN) enables users and carriers to more easily configure new services. However, there are other reasons for intelligent networking. The following are applicable to both wireless and wireline:

- *Real-time rating of a call* Real-time rating is the application of rates to a call for a subscriber or the carrier. The rate of a call may be used as a decision point for a subscriber or even a carrier attempting to determine whether a calling card has sufficient credit to cover the cost of the call.
- *Real-time checking of a customer's billing record* A "deadbeat check" of the calling subscriber before the call is allowed to go through.
- Per call payment authorization
- *Number portability* Especially important to those wireless carriers working to merge with other types of carriers. Number portability can support convergence. Subscribers do not like having to remember multiple ID numbers. One carrier goal is to make it easier for a

subscriber to use their services. Forcing a subscriber to remember multiple telephone numbers, mobile numbers, or passwords makes a service unattractive. The concept of number portability can be used to support identification portability (alpha and numeric in nature).

- *Traffic Management* Especially important as it relates to managing the wireless accessed Internet traffic. Internet traffic has different characteristics than a typical wireless call. Internet usage involves alphanumeric data, long holding times, audio, and video. The switching matrix needed to support this must be capable of ensuring that all traffic within the network is flowing. The avoidance of traffic congestion is always a carrier objective. This objective becomes more difficult to attain as different types of telecommunications traffic traverse the network. The next generation switches, multifunctional switches, are a first step towards building a network that is capable of managing a multiplicity of telecommunications traffic. Managing the traffic across the network and across different networks is a challenge for intelligent traffic management systems.

Figure 7-31 is an illustration of the wireless intelligent network.

Figure 7-31 Wireless intelligent network—real time value

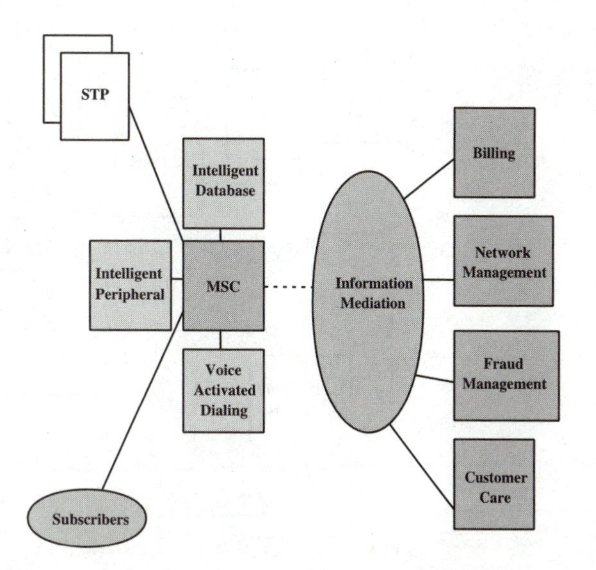

Real time value to the subscriber and to the wireless carrier

Within the wireless carrier community, work on something known as the wireless intelligent network has been proceeding. The *Wireless Intelligent Network* (WIN) is a network concept that supports the use of intelligent network capabilities to provide seamless terminal services, personal mobility services, and advanced network services in a mobile or wireless environment. WIN was envisioned to support seamless roaming between wireless and wireline environments. The wireline carrier community had wanted to extend the capabilities of the wireline IN and the Class 5 central office, which served as a primary driver for WIN. The support for WIN has wained a great deal; being supplanted by the various data, Internet, and IN based efforts for telecommunications convergence. Like the wireline intelligent network, WIN also includes functional capabilities that support creation and execution of service logic programs. These programs are resident outside of the central switching system, but work collaboratively with the switching equipment based on a common definition of call models and protocols. The service logic programs may utilize data resources and physical resources, which also reside outside of the switching equipment. Figure 7-32 illustrates the basic functions and components of the wireless intelligent network.

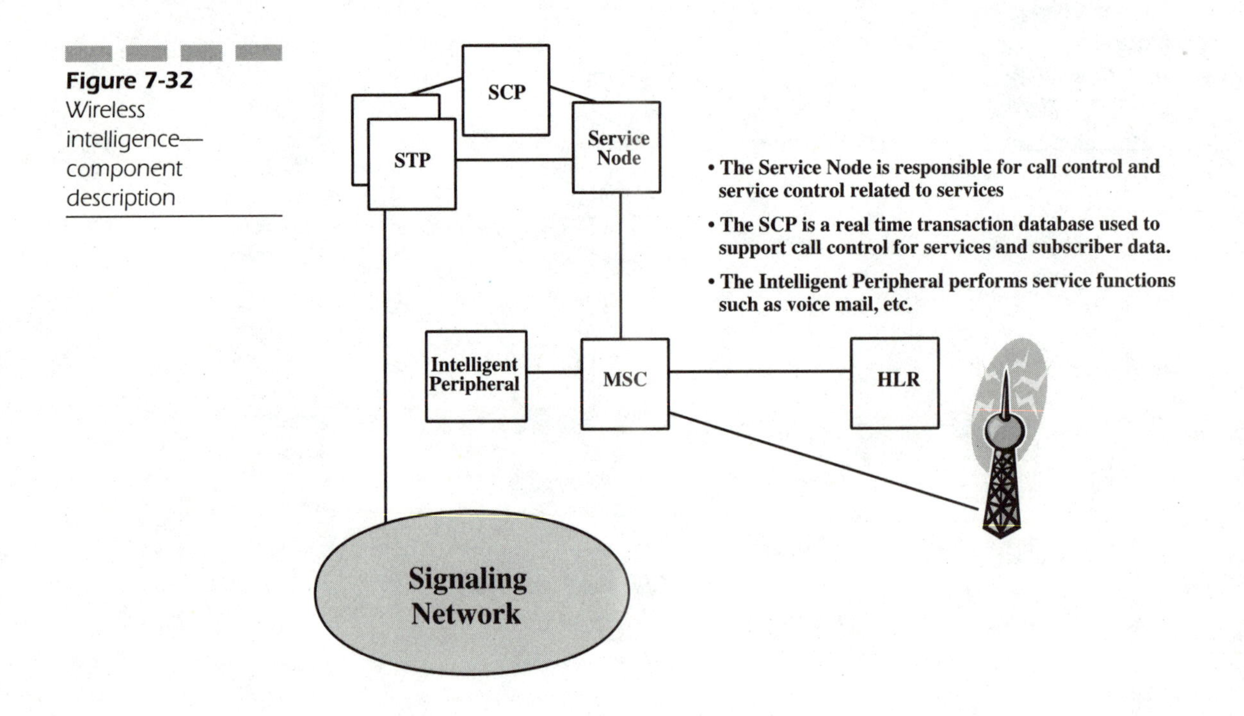

Figure 7-32 Wireless intelligence—component description

The Internet

The Internet is playing a key role in the convergence of the wireless network and other information businesses. The Internet is a high-speed information network that has grown throughout the globe. The Internet is the catalyst for a myriad of potential information businesses that the wireless carrier may enter. These businesses include

- *Multimedia services* Video, news services, entertainment, research and librarian services, image transmission, and so on
- E-mail
- Personal banking
- Electronic commerce
- *Database access* Corporate intranets will need to be accessed
- *World Wide Web browsing* Search engine accessibility and usage

The challenges faced by the wireless carriers involve supporting the following:

- *Access protocols* The wireless network is governed by a set of technical signaling standards not supported by the Internet. The Internet is a wireline originated information network. The wireless carrier must be capable of interoperating with the Internet (more on Internet protocols later in this book). The Wireless Access Protocol (WAP) is the wireless industry's industry effort to support the Internet and the various 3G efforts. Note that there will be more on 3G later in this chapter.
- *Handoff* Typical wireless handoff involves a series of actions and network messages that enable the mobile handset to be handed off from one site to another cell site. The Internet poses a problem. The switch or server must be involved in the handoff to the extent that the mobile handset's communication link is undisturbed and that the site to which it is handed off is capable of supporting an Internet connection at the time of handoff.
- *Roaming* The primary characteristic of a wireless network is the mobile nature of the user. The user can roam from the Home system to a Visited system. The network must still be able to maintain all connectivity with the Internet. Today, there are carriers who provide

Web browsing services and email to their subscribers. However, the users can only obtain the services on the particular carrier's network. In order for the user to find any usefulness in a service, the user must be able to use the service anywhere at any time.

- *Routing* The whole concept of traffic routing would have to be reviewed in order to ensure that routing concepts used in both the wireless carrier networks and the Internet networks are compatible or at least harmonize with one another.
- *Coverage* Coverage must be looked at a realistically. At times, coverage will be "spotty," noise will dominate, or weather may affect the quality of the coverage.
- *Transmitting or translating images and text between the Internet and the wireless mobile handset* The Internet and wireless carrier industry must find some way to adapt the large amounts of data (audio, video, and text) to appear on small handheld devices (the size of the current mobile handset). The process of adapting the different bandwidth types of data to the large variety of handsets that will be marketed by the wireless carriers is called *transcoding*.

Figure 7-33 illustrates the Internet in the wireless carrier environment.

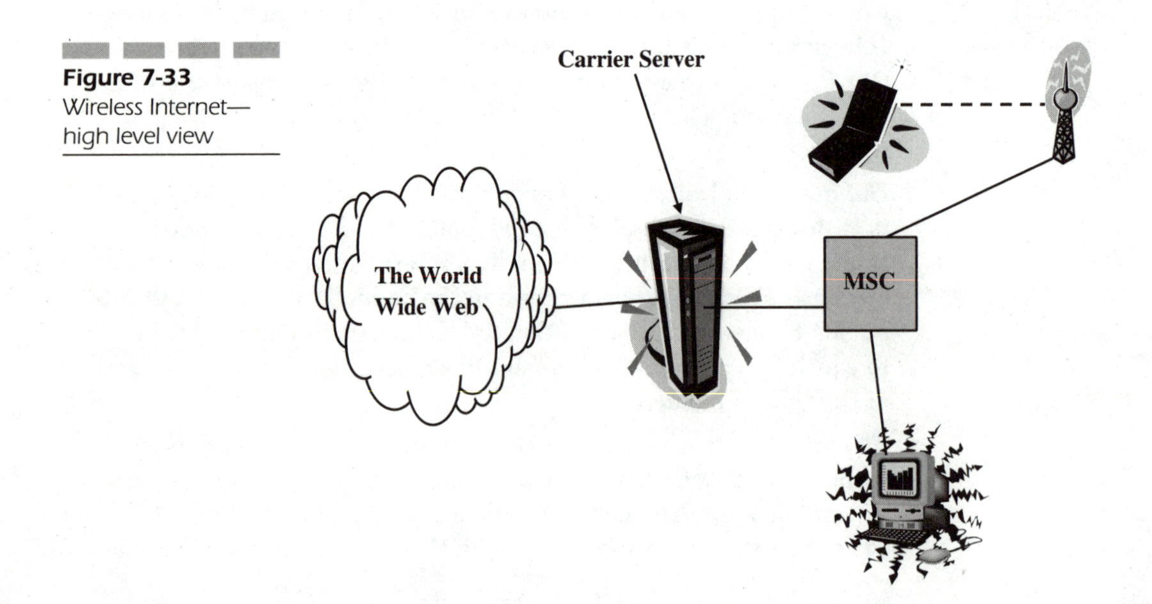

Figure 7-33 Wireless Internet—high level view

The Wireless Local Loop

The *Wireless Local Loop* (WLL) is wireless access to the home and business (the whole local loop). Wireless carriers envision this as a way of replacing the fixed phone on the wall or the office telephone set. Wireless carriers seeking to compete with the wireline carriers are looking for ways for their wireless service to replace the fixed telephone. Wireless carriers are using various rate plans and combinations of intelligent call forwarding services, voice mail, and other services to penetrate the loop. The wireline carriers are also seeking ways of using wireless technology to maintain their control of the loop. Wireline carriers that have "sister"/ "sibling" wireless carriers are working with their "sibling" companies to find ways of keeping the subscribers within the "family of companies." See Figure 7-34 for an illustration of the wireless local loop concept. The technical challenges of WLL are

- *Antenna placement* Antenna placement is a challenge because now the carrier needs to provide adequate coverage in a neighborhood to eliminate dead spots and provide fairly clear, noise-free transmission.

Figure 7-34 Wireless local loop

Wireless Local Loop - The objective is to replace the wireline telephone company. Coverage and Quality are the principal objectives.

Achieving the optimal antenna height is a challenge.

Carriers face uphill battles placing new 200-foot towers in communities. WLL poses a new challenge: placing antennas in residential neighborhoods without the benefit of towers. Rooftops can be used as well as perhaps telephone poles. It is possible that the WLL carrier can establish a network of rooftop antennas throughout a community in order to achieve coverage. Telephone poles or even home rooftops are not optimal locations for antennas. Coverage is influenced by antenna height. Telephone poles are normally approximately 30 feet tall. Many communities have telephone poles dating back to the 1920s. Some telephone poles from that era are only about 20–25 feet tall. Many of the newer telephone poles are 35–40 feet tall. In many communities, telephone poles are surrounded by tall foliage. Rooftops of homes are much lower than telephone poles. Also, homes are typically surrounded by foliage.

- *Coverage* Coverage in this context refers to interference avoidance. This can be a challenge in a suburban community that has hundreds of homes in a one square mile neighborhood. Each home must be capable of obtaining control and traffic channels with minimal blocking, yet it must be done with no noise on the channel. A good WLL carrier realizes that the customer will not accept poor quality of service.
- *Frequency Availability* A wireline carrier needs spectrum to provide wireless access. Unlicensed spectrum is unacceptable.
- *Cell site capacity engineering* Without good capacity engineering, the WLL carrier will face the possibility of cell site congestion.
- *Radio technology selection* The carrier, wireless, or wireline hoping to deploy a WLL network will need to select a radio technology. The radio technology must meet the following requirements:
  - High channel capacity
  - Subscriber service evolution
  - Privacy
  - Perceived wireline voice quality
  - Cells the size of less than one acre—these would be considered picocells
  - Data services
  - Internet access

Carriers deploying WLL systems face challenges unique to their particular wireless industry segment. Whereas typical wireless carriers such as cellular and PCS carriers work to provide full coverage but realize that the subscribers expect to run into dead zones, the WLL carriers must provide 100% coverage around the home and neighborhood. If a WLL carrier does not meet this criteria, it will have a enormously difficult time penetrating the marketplace. Unlike typical wireless, WLL will be measured against the perceived quality of the wireline service. By holding itself up against the wireline industry, the WLL carrier faces an engineering challenge both technically and financially. Figure 7-35 is a pictoral depiction of the major challenge facing the wireless local loop.

Growing into new industry segments is not the only way the wireless carriers are seeking to grow. The wireless carriers are seeking ways to leverage their assets in order to grow their businesses and reduce capital expenditures.

Leveraging Assets

PCS and cellular networks use the same infrastructure equipment. If the PCS or cellular service providers were to provide infrastructure support to other service providers, the following network components could be tapped

Figure 7-35 Wireless local loop—customer and network challenges

Figure 7-36
Leveraging assets

for "double duty": Figure 7-36 illustrates how the assets of the carrier can be leveraged.

- *Databases* SCPs, HLRs, and so on
- *Radio licenses* This isn't exactly a piece of equipment, but it is a tangible asset. Wireless carriers can resell their radio spectrum so other carriers can provide service to subscribers using the license of the other carrier.
- Radio infrastructure
  - Antenna
  - Towers
  - Backhaul
- Real estate
  - Rooftops of company buildings
  - Long-term deals with other types of carriers or companies in order to gain access to real estate. This has brought the railroad industry into the picture. The railroad industry's right-of-way through communities has enabled wireless carriers to extend their coverage while simultaneously creating a new business for the railroad industry.

- The wireless carriers are also leasing tower space to other carriers. However, the majority of the towers today are under the ownership of tower companies.

- *Transmission facilities* Wireless carriers are leasing spare facilities to other types of carriers. Colocated carriers are leasing transmission facilities to one another. These carriers may be satellite companies, paging companies, or even cable television companies.
- *Utilities/facilities* The need for transmission facilities has created a new line of business for the electric power and natural gas utility industries. Both industry segments maintain their own internal communications networks. Both industry segments are regional and national in nature. Wireless carriers can take advantage not only of the utility industry's spare capacity, but also their right-of-way.
- *Electric power* Colocated carriers purchase power from the wireless carrier landlord.
- *Heating* Another basic utility you might take for granted
- *Air conditioning* Another basic utility needed for warm weather
- Spectrum leasing

The above is fairly obvious. Many wireless carriers are pursuing the above; unfortunately, it is not because of aggressive market programs. Rising operating costs are forcing many to look at equipment sharing as a way of managing costs.

The cost of operating a wireless network may eventually force some carriers to even consider switching matrix sharing. In this case, a switch would be partitioned for at least two different types of call processing. For example, one may logically partition a switch to process calls for a wireline service provider and a wireless service provider. Partitioning brings its own problems. Switch security becomes problematic. Your tenant may even gain access to your subscriber profiles. Routing tables will probably be a challenge. The cotenants of the switch will have to learn establish joint maintenance and operating procedures. The ideal situation is that only one provider be given sole responsibility for running the switch. Figure 7-37 highlights the concept of multi-functional switching.

Leveraging assets does not need to be an exercise in reducing costs. Leveraging assets may in fact be viewed as a way of developing new business relationships.

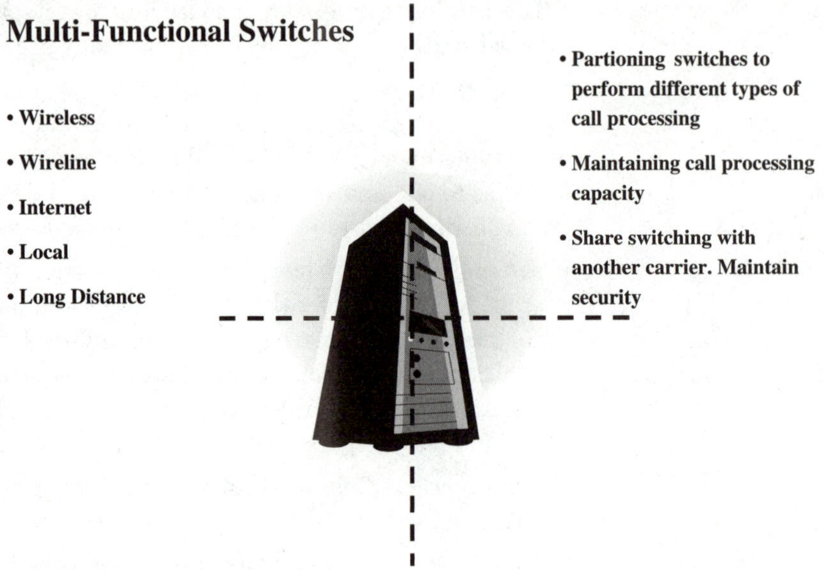

Figure 7-37 Multi-functional switches

Convergence: The Network and Business Interconnect

In the late 1990s, many wireless carriers were concerned about the lack of differentiation between cellular and PCS carriers. Price was one way of differentiating one carrier from another. However, at some point, price alone was and still is not enough. Even the leading wireless carrier can eventually inflict damage to its own revenue stream if it continues to rely on pricing as the differentiator. New services must be deployed. These services may be brand-new types of services, bundled existing services, or access to other platforms. Wireless carriers are concentrating on their strengths, which are wireless access and mobility. Focusing on wireless access and mobility has enabled the wireless carriers to concentrate on bringing services that would be of interest to a mobile user and reject those services that would not. This focused approach has facilitated the convergence of technologies and the creation of new business relationships. See Figure 7-38.

The growth of the wireless industry has caused many individuals to expand their definition of network interconnection. Although the physical network interconnection has not really changed, the way many perceive the importance of network interconnection has. Originally, wireless network interconnection was viewed as a way for the wireless subscriber to reach

Figure 7-37
Convergence—a wireless perspective

the wireline PSTN subscriber base. Traffic once flowed primarily from the wireless into the wireline subscriber base; the percentage splits were 70% wireless to wireline and 30% wireline to wireless. These numbers affected the way everyone in the telecommunications industry perceived wireless. Today, the operating paradigm has changed—these numbers are meaningless. As I write this chapter, the total number of wireless subscribers in the United States alone is over 65 million. The growth of new types of telecommunications industry segments, namely the Internet, has enabled the wireless industry to greatly increase its perceived value to the subscriber.

The network (radio and fixed portions) is not the only challenge. In addition to converging technologies at a network signaling level, the wireless carriers must deploy new types of handsets or terminal devices. These subscriber terminal devices must be capable of displaying video and high-resolution graphics. Further, the devices must be capable of accessing the Internet and other network services. The subscriber/user interface must be capable of supporting user interaction with services such as Internet access, multimedia services, and banking services. Human factors will figure prominently in the design of this new type of terminal device. Handsets are small and will continue to be small; therefore, multifunctional keys must be designed into the handset. Handset keys and buttons are multifunctional today and will need to manage even more functions in the future. Marketing departments may even wish to change the handset paradigm and sell a handset that is in fact bigger than the current pocket-sized devices.

The convergence effort underway in the wireless industry has taken on a life of its own. The current work in 3G, also known as 3rd Generation Wireless, has assumed all of the characteristics of what we have come to expect of convergence. 3G work includes data, Internet, video, music, smart handsets, intelligent networking, wireline replacement, entertainment, etc. For now, the list is as endless as one's imagination.

SUMMARY

The line between wireless carriers and other types of carriers is blurring. The need to stay competitive is driving the changes in the wireless industry. The other wireless technology, paging, will be discussed in the next chapter.

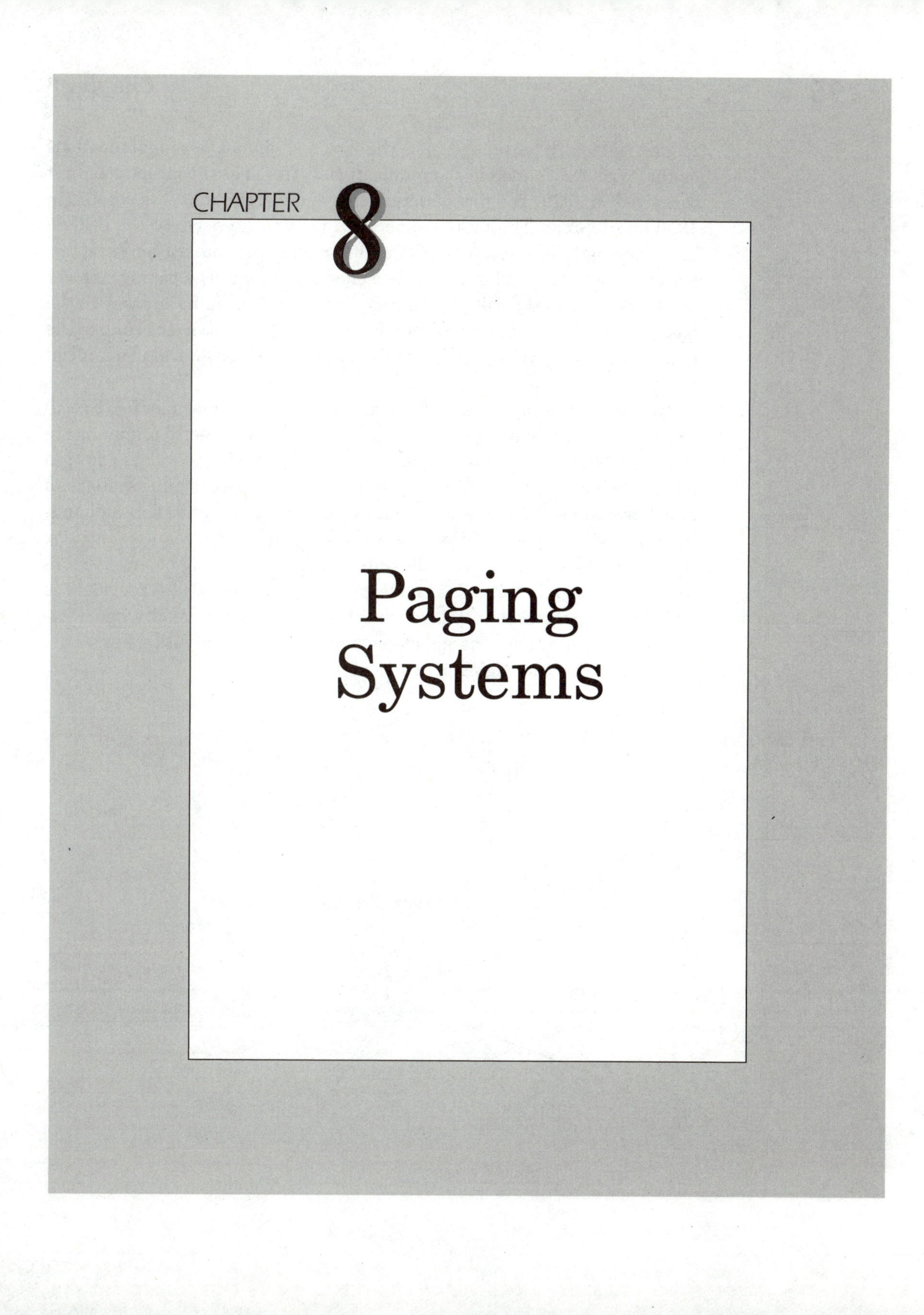

CHAPTER 8

Paging Systems

Paging is usually considered to be the "low end" of mobile communications or the "poor man's mobile communications." The paging concept was first conceived in 1939. The first practical and commercial unit was created in 1950 by Al Gross. The first commercial unit was licensed by the FCC in 1952. Paging is less expensive than other mobile communications systems because it was and still is primarily a one-way system. The paging receiver alerts the user to the call but does not verify or respond in any way to the base station. The cost and bulk of a typical mobile transceiver is due to the transmit portion, which is missing from a paging receiver; therefore, it can be small and cheap.

Paging has become a part of the mainstream consumer market. Executives and teenagers alike now use pagers that allow them to communicate wherever they are. This trend has given paging service providers cause to feverishly increase their subscriber base at a faster pace. Paging systems to date have satisfied most requirements for tone, numeric, and short alphanumeric messaging with sufficient subscriber capacity for the service providers. Cellular and PCS service providers are already integrating paging capabilities into their handsets. The challenge facing paging service providers is differentiating themselves. Pagers can either integrate low-quality voice messaging or expand their existing messaging and data capabilities. Figure 8-1 is a rendering of the Paging network.

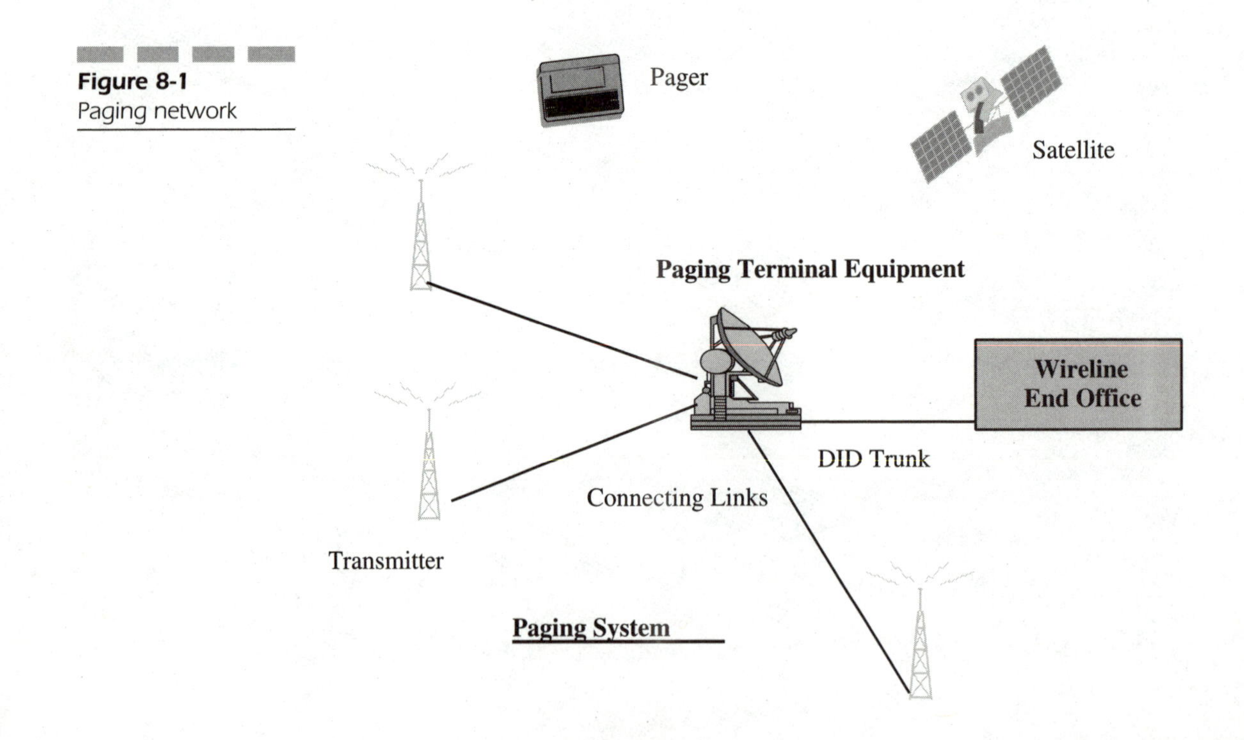

Figure 8-1 Paging network

Figure 8-2
Paging and the Internet

New types of paging devices and enhanced protocols have led to a dazzling array of new paging services involving messaging and low speed data. The services themselves may involve partnering with another provider of telecommunications services. For example, a paging provider decides that they will partner with an Internet service provider. The partnership results in a small handheld paging device/computer that serves as both a messaging device and an Internet access device. Figure 8-2 is a rendering of the integration of paging and the Internet.

Fewer pagers exist today than cellular and PCS devices, but they are still widespread.

Paging Radio Frequency Spectrum Requirements

The FCC has allocated multiple frequency bands to paging service providers. These bands are

- 35 MHz
- 43 MHz

- 150 MHz
- 450 MHz
- 931 MHz

These frequencies are licensed as paging frequencies. These bands were allocated over a period of 50 years starting in 1952. The 931 MHz band was allocated specifically for nationwide use only. The FCC has licensed only three nationwide carriers for this purpose. The frequencies list in Table 8-1 can also be used for paging services.

Wireless carriers, especially PCS carriers, have taken a broader interpretation of paging and are providing paging services in the narrowband PCS frequencies. These PCS carriers are marketing their services as two-way messaging services. Although these service providers are not FCC-licensed paging service providers, they are marketing themselves as such. There is no prohibition on selling paging services. These types of carriers are in fact selling and marketing the short message service capability as paging. The subscribers do not know the technical difference, regulatory difference, or marketing difference between a paging company and a cellular carrier or between a paging company and a PCS carrier. The subscriber attaches labels and names to things and businesses all the time. The word paging brings to mind images of little devices that beep or vibrate and are stored on a belt or in a pocket or purse. These devices light up or display a alphanumeric message. If a PCS license holder attacks the paging market by announcing that they will provide a new kind of paging to the subscriber base, that paging company has found a market differentiator. As competition in the traditional cellular and PCS markets heats up, they will be entering the paging marketplace. The narrowband PCS carriers have an advantage because they have more spectrum with which to work. The frequency blocks allotted to the traditional paging carriers are only one or two

Table 8-1

PCS Band

PCS Band	Spectrum Block	Frequency Range (MHz)	Coverage Area
A	30 MHz	1850–1865/1930 –1945	Major trading area
B	30 MHz	1870–1885/1950 –1965	Major trading area
C	30 MHz	1895–1910/1975 –1990	Basic trading area
D	10 MHz	1865–1870/1945 –1950	Basic trading area
E	10 MHz	1885–1895/1965 –1970	Basic trading area
F	10 MHz	1890–1910/1975 –1990	Basic trading area

MHz in each of the previous traditional bands. This does not mean that the traditional paging service providers cannot compete. The various advancements in device design and radio design are enabling the traditional players to remain competitive.

For illustrative purposes, I will explicitly differentiate the PCS carriers in the paging market and the traditional paging service providers.

Paging Architecture

The network architecture of a paging system is not unlike that of any other type of wireless telecommunications network. A paging network has a switching element, transmission facility element, and a radio component. In reality there is far more complexity to what I have described. However, what is important to remember is that one should not get intimidated with telecommunications network technology. All networks can be dissected in a way to better understand the technology.

Paging Network Elements

To understand traditional paging, you must gain a basic understanding of what comprises a paging system. A typical paging system is comprised of the following:

- *Paging control terminal* The paging service provider's switch
- Transmitter
- Paging device

Paging Control Terminal The paging control terminal is typically interconnected to the ILEC (that is, the PSTN). The paging control terminal is similar to a large piece of customer premise equipment. The paging control terminal receives the call and associates the telephone number or *personal identification number* (PIN) with a specific paging device. The paging control terminal is interconnected to the ILEC/PSTN via a *Direct Inward Dialing* (DID) trunk or a common two-way Type 2A trunk. The control terminal's interconnect point in the PSTN is a Class 5 end office.

Unlike the cellular, wireline, or PCS industry segments, the paging industry's switching technology is not as complex. The typical small controller is a 6,000 line controller with two RF channels and eight trunks. The control terminals vary in size and may even have a voice messaging system. Larger control terminals will support millions of subscribers and hundreds of RF channels. The paging systems are not as complex. The complexity is in the feature set offered by the typical paging service provider, not the size of the system. As I write this chapter, the paging industry is undergoing changes that will keep it competitive to cellular and PCS. Figure 8-3a is an illustration of how the paging network is currently configured and interconnected. Figure 8-3b is an illustration of how a Type 2A transmission facility is used in the interconnection of a paging terminal and the PSTN.

Transmitter The transmitter is comprised of two components: modulator and carrier. A radio frequency generator generates the radio energy that will carry the signal. This generally consists of an oscillator (which produces the initial signal) and a number of amplifier stages (which amplify the level to that required at the antenna). A modulator mixes the signal to be transmitted with the radio frequency signal (called the carrier) in such a way that the signal can be decoded at a distant receiver. Figure 8-4a is a block diagram of a paging transmitter. The antennas used in the base station are typically simple high gain omnidirectional dipole antennas. As noted in Figure 8-4b, omnidirectional antennas can assume a variety of different shapes. The thing to remember is that the omnidirectional antennas radiate radio energy in a uniform pattern outward from the antenna.

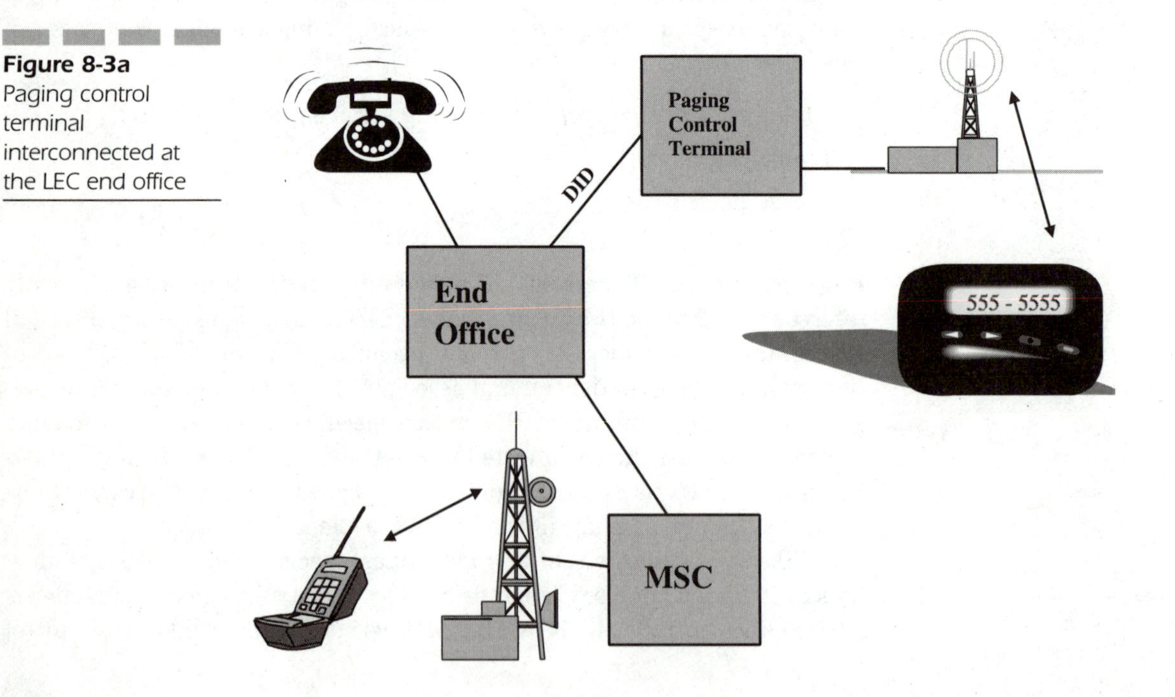

Figure 8-3a Paging control terminal interconnected at the LEC end office

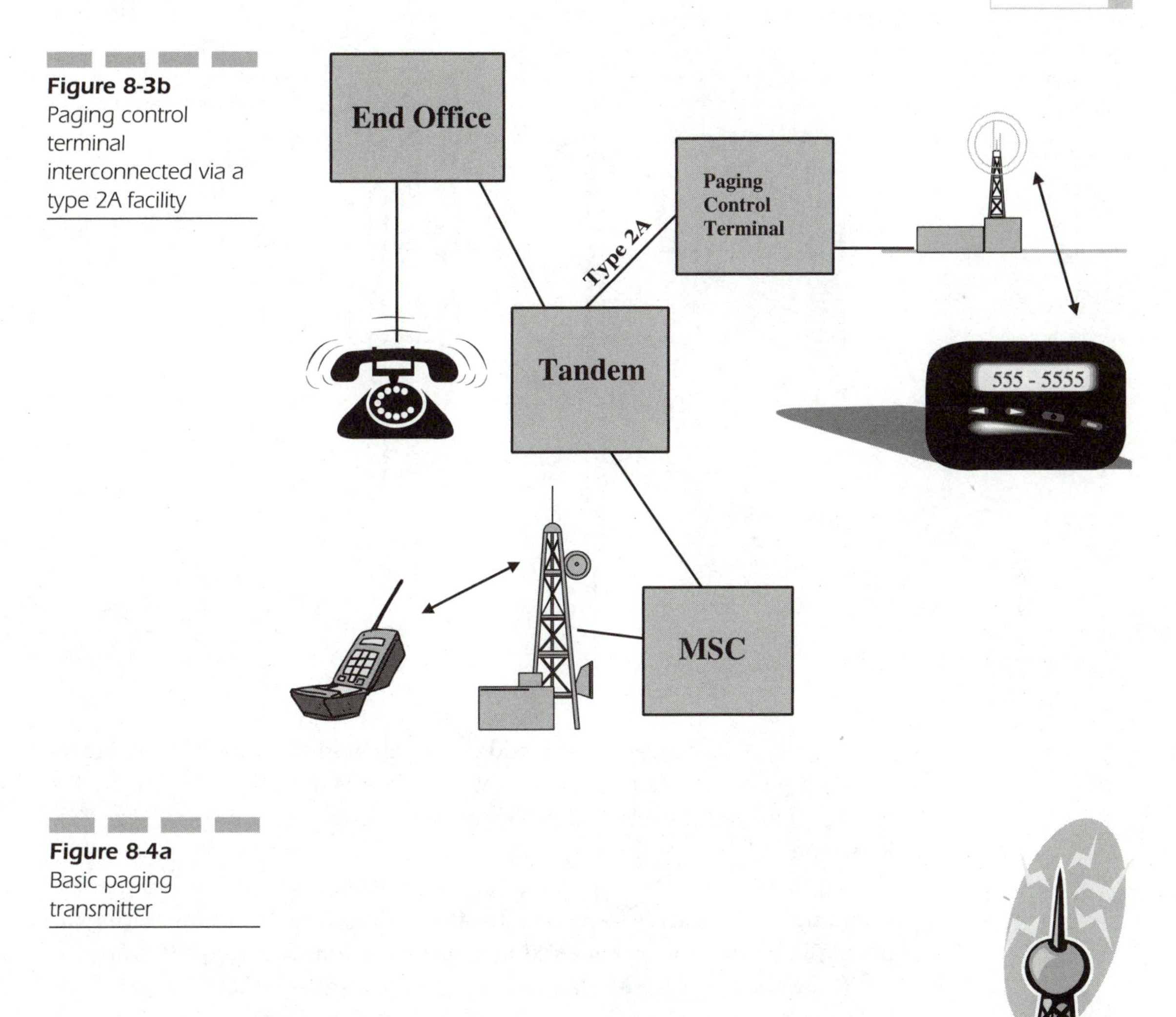

Figure 8-3b Paging control terminal interconnected via a type 2A facility

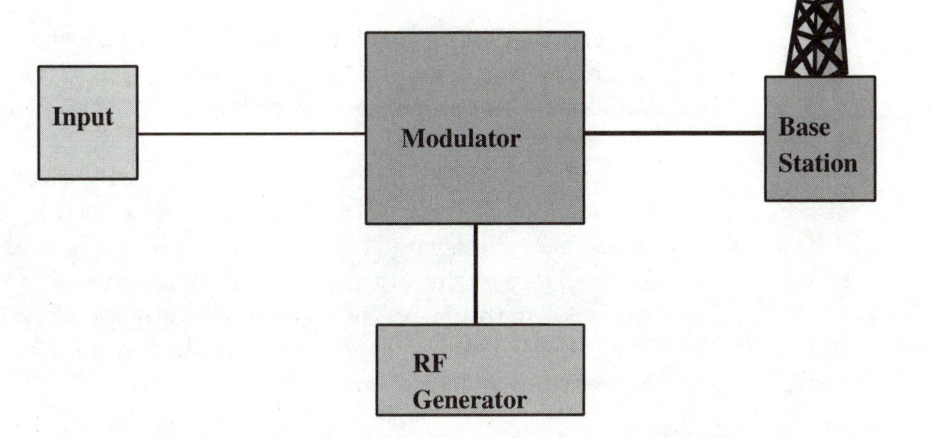

Figure 8-4a Basic paging transmitter

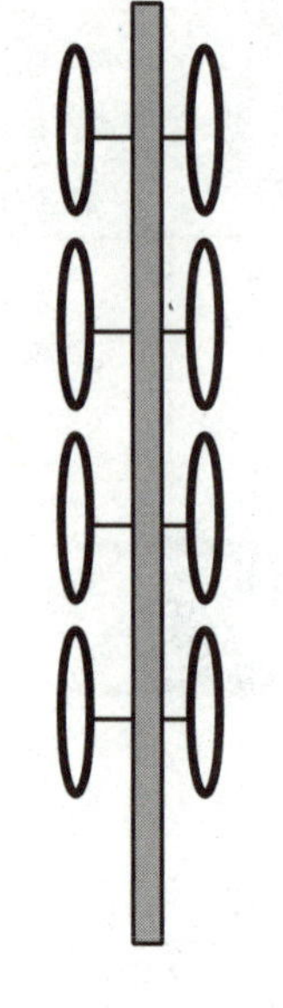

Figure 8-4b Omnidirectional antennas

Omnidirectional antennas can assume a variety of configurations.

Traditional paging service providers are allowed to transmit at effective radiated power levels of up to 500 watts. This is a transmission level far greater than cellular or PCS, which do not transmit above 100 watts. Cellular and PCS are prohibited from transmitting at such power levels. The reader should note the various antennas described are utilized in the cellular and PCS industry segments. Unlike cellular and PCS, tower siting for paging is not as a laborious and arduous experience. Given, the transmit power levels, lower subscriber density, and message packet size, transmit antenna design in the paging industry presents less tasking challenges.

Receiver—Paging Device The receiver gets a signal from its antenna, which also receives a number of unwanted signals. The tuned circuit tunes out all but the wanted signal, which is then demodulated (decoded) by the demodulator.

The very simple receiver (which consists only of a tuned section and a demodulator) is known as a *tuned radio frequency* (TRF) receiver. The term *tuned radio frequency* (TRF) is out of date today. Several years ago the TRF was replaced by the superheterodyne. In the case of the paging device, the receiver is in the pager. The pager is a small, pocket-sized, two ounce device. The receiver antenna needs to be installed in the aforementioned device. Figure 8-5 is a block diagram of a paging receiver.

Figure 8-5
Paging receiver

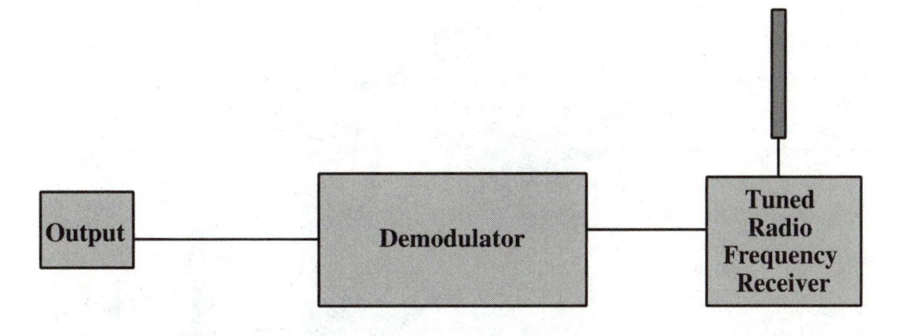

Pagers normally use a single loop antenna. Pagers come in a variety of different flavors:

- *Tone only* The original pagers supported only tone, which was a single "soft beep tone." Today pagers will beep, chirp, or even blip.
- *Alphanumeric* The alphanumeric pagers today display telephone numbers, stock reports, and the news. The alphanumeric paging systems utilize digital radio technology. More on this later in the chapter.
- *Vibrating* The paging system emits a signal that instructs the pager to vibrate instead of emitting a tone or lighting up. The pager has a mechanism that causes the paging device to vibrate.
- *Visual* The visual indication is a *light emitting diode* (LED) or a flashing icon.
- Combinations of the above

The original pagers were tone only and were fixed to react to a specific frequency. Most pagers today are still designed to operate on one frequency. However, newer pagers are being manufactured so the user can enable the pager to operate on different frequencies. Given the enhancements being made to the pagers, they are not just simple paging devices, but more like messaging devices capable of communicating more than just a number or a name. Pagers can be leased from a paging service provider or purchased through various retailers.

Typical Architectural Configurations

There are two basic configurations used in the paging network. One supports the original and still dominant one-way paging system. The second configuration supports two-way paging. Figure 8-6 is a rendering of a one way paging network configuration. Figure 8-7 is an illustration of a two way paging network configuration.

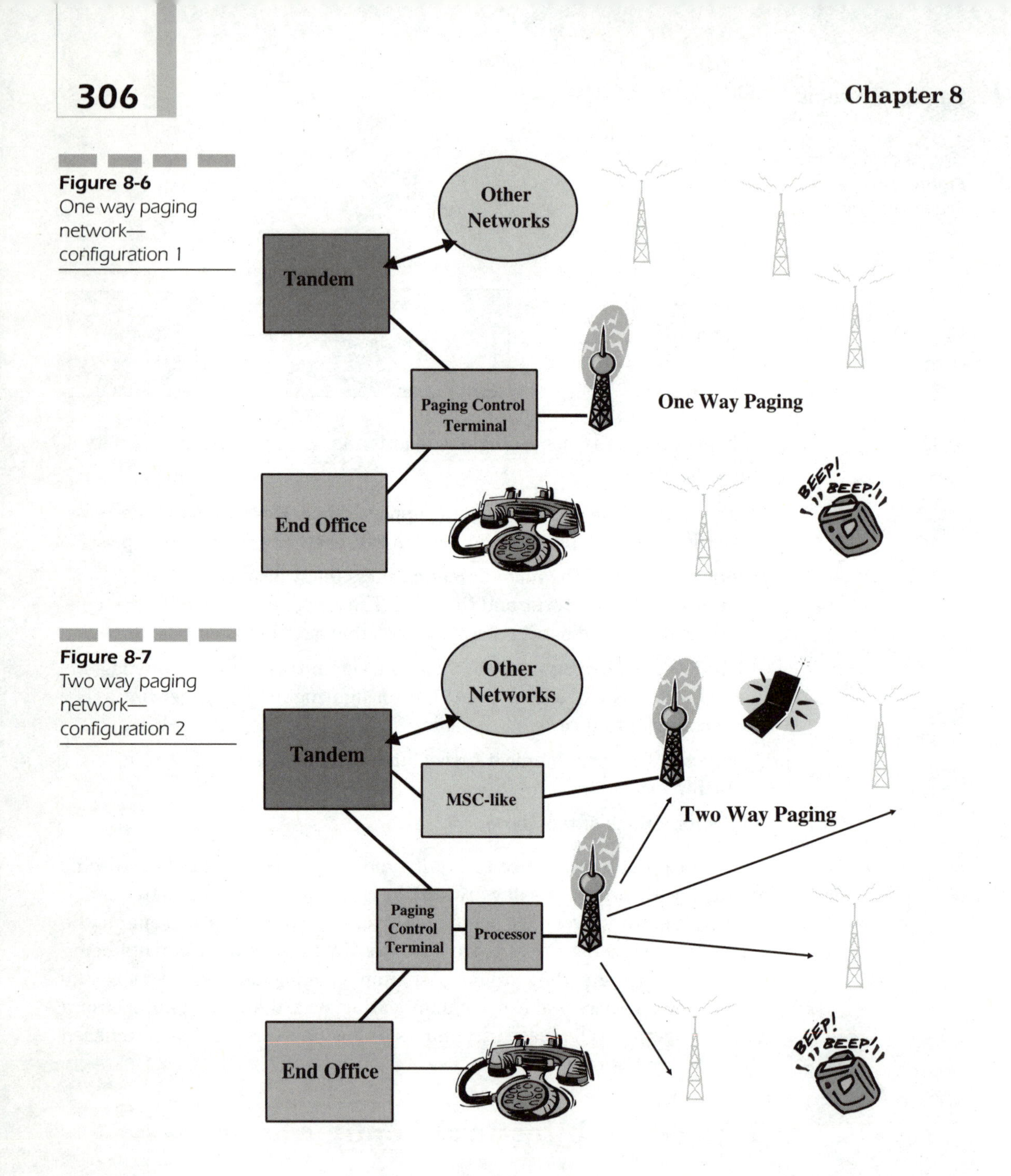

Figure 8-6 One way paging network—configuration 1

Figure 8-7 Two way paging network—configuration 2

In the first configuration, the paging device is a simple receiver. In the second configuration, the paging device is a transceiver (receiving and transmitting data). Two-way paging or, more accurately, the two-way messaging device is capable of supporting alphanumeric responses to incoming messages, sending voice messages to the paging subscriber, voice between parties, stock reports, the latest news, and so on. The second configuration requires the use of a paging switch and base station that can accept transmission from the paging device. The two-way paging system can assume one of two types of network switching architectures. The first type utilizes the same paging control terminal with an adjunct device for processing messages from the paging devices. The second type of architecture uses a MSC class switch that is capable of processing messages from mobile devices. In both configurations, satellite transmission may be used to support nationwide broadcasting. The satellite transmits the paging information to a terrestrial network for the actual paging broadcast.

From the perspective of the ILEC/PSTN, the traditional paging terminal is like a cellular switch or a piece of customer premise equipment. The PSTN interconnect is critical because without the PSTN/ILEC interface, the paging service provider does not have visibility to its own customers. At this time, traditional paging service providers do not enable their subscribers to dial out telephone numbers from their paging devices.

The traditional paging terminal is responsible for receiving, processing, storing, and forwarding information from the caller. The paging terminal validates the type of call, determines the authenticity of the subscriber, and serves as the interface to the *radio frequency* (RF) network or to other paging terminals within a multi-city paging network. The RF network accepts the data from the paging terminal via telephone lines, RF link, or satellite and decodes the data streams containing the paging data. Upon decoding

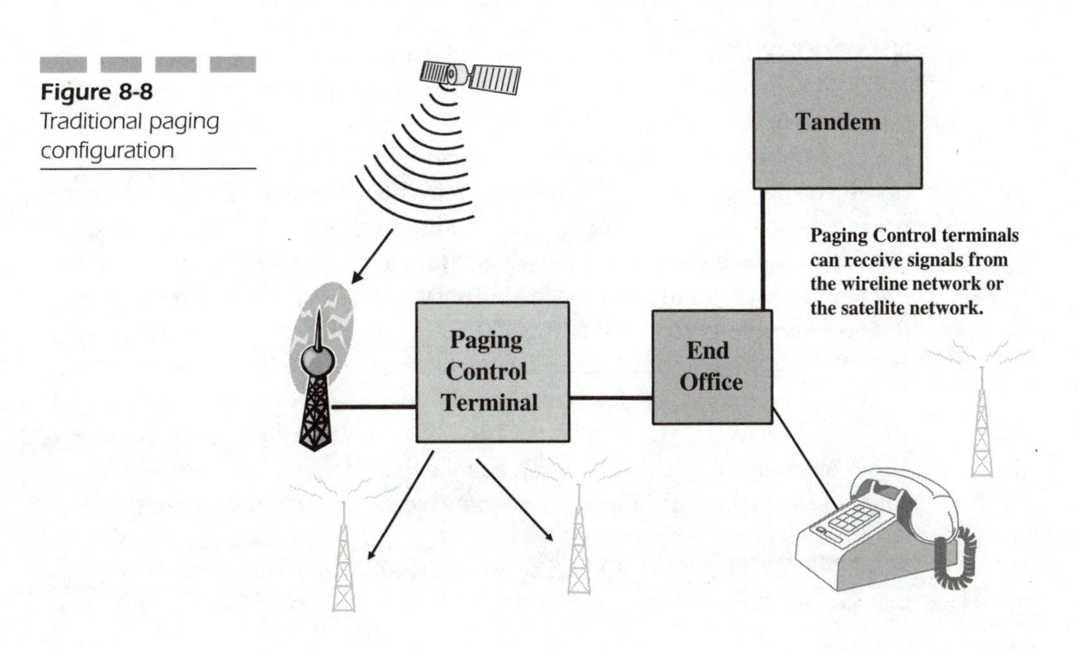

Figure 8-8 Traditional paging configuration

the data, the transmitter translates the paging data into signals that modulate the RF paging signal at the desired transmit frequency. Figure 8-8 is an illustration of the traditional paging network configuration.

Paging Company Types and Numbering

This section will cover numbering and its application in the paging industry. This section will also describe the types of traditional paging companies in operation. There is a relationship between the numbering schemas used and the types of traditional paging companies. The relationship is partially technically- and business-driven.

Addressing in Traditional Paging

Traditional paging companies are either assigned a single telephone number or a block of telephone numbers. The telephone numbers conform to the *North American Numbering Plan* (NANP). These telephone numbers are used by calling parties to send a page or message to a paging subscriber. The telephone numbers are the method of access to the paging company. The telephone numbers may be of the following format types:

NPA-NXX-XXXX The NPA (area code) may be a local area code. The NXX-XXXX part of the number may be assigned specifically to the pager or may be assigned to the entire paging company. In the case where the telephone number is assigned to the paging device, a calling party will be able to call the paging carrier. Because the telephone number is assigned to the paging device, the calling party will be able to simply enter their message.

In the case where a single telephone number (NXX-XXXX) is assigned to the paging carrier only, the paging subscriber has a unique *personal identification number* (PIN) assigned to them. In the PIN case, the calling party calls the paging carrier and is asked to enter a PIN before sending the alphanumeric message. This case usually means that the paging carrier is sharing the NXX (exchange) with the ILEC's subscribers. Remember the ILEC assigns the local telephone number in their areas. Figure 8-9 is a chart of the typical numbering schema used by the paging carriers.

888-NXX-XXXX, 800-NXX-XXX, 877-NXX-XXXX This numbering format is typically used by the large nationwide carriers to provide toll free (no

NPA-NXX - XXXX

- NPA = Area Code
- NXX = Exchange usually shared with other subscribers or carriers
- XXXX = Assigned to either the Paging carrier or subscriber

❖ PIN = Personal Identification Number sometimes assigned

Figure 8-9 Numbering scheme

888-NXX-XXXX

800-NXX-XXXX

877-NXX-XXXX

Figure 8-10 Numbering scheme—toll free

telephone access charges) service to calling parties wishing to page the carrier's subscribers. Nationwide paging carriers utilize PINs to identify individual paging units. Nationwide paging carriers are normally given a single NXX-XXXX to identify itself—this is why the use of PINs is so important. PINs can range from 6 to 20 numerical characters. Figure 8-10 illustrates the free charge numbering scheme.

Both nationwide paging carriers and local paging carriers support autonomous communication and service bureau communication to the paging device. In autonomous communication, the calling party simply enters the telephone number associated with the paging device. In the service bureau scenario, the calling party calls a service bureau and identifies the paging subscriber by name and sometimes by company. The service bureau alerts the paging subscriber via a tone and the subscriber calls the service bureau for the message. The service bureau was the predecessor of the alphanumeric paging device. The paging service bureau is still very much active within the paging industry. Figure 8-11 is a rendering of how a paging service bureau can be used.

Traditional nationwide paging carriers and local paging carriers are very similar. The only apparent difference is in the coverage of the carrier. Like cellular and PCS, even nationwide paging carriers do not have complete coverage. Paging carriers have the advantages of greater transmission power levels and minimal overhead setup messaging. The perceived differences between cellular, PCS, and paging are rapidly disappearing. The following section will take a more in-depth look at network deployment and operations. Figure 8-12 is a rendering depicting the benefits/advantages of Paging.

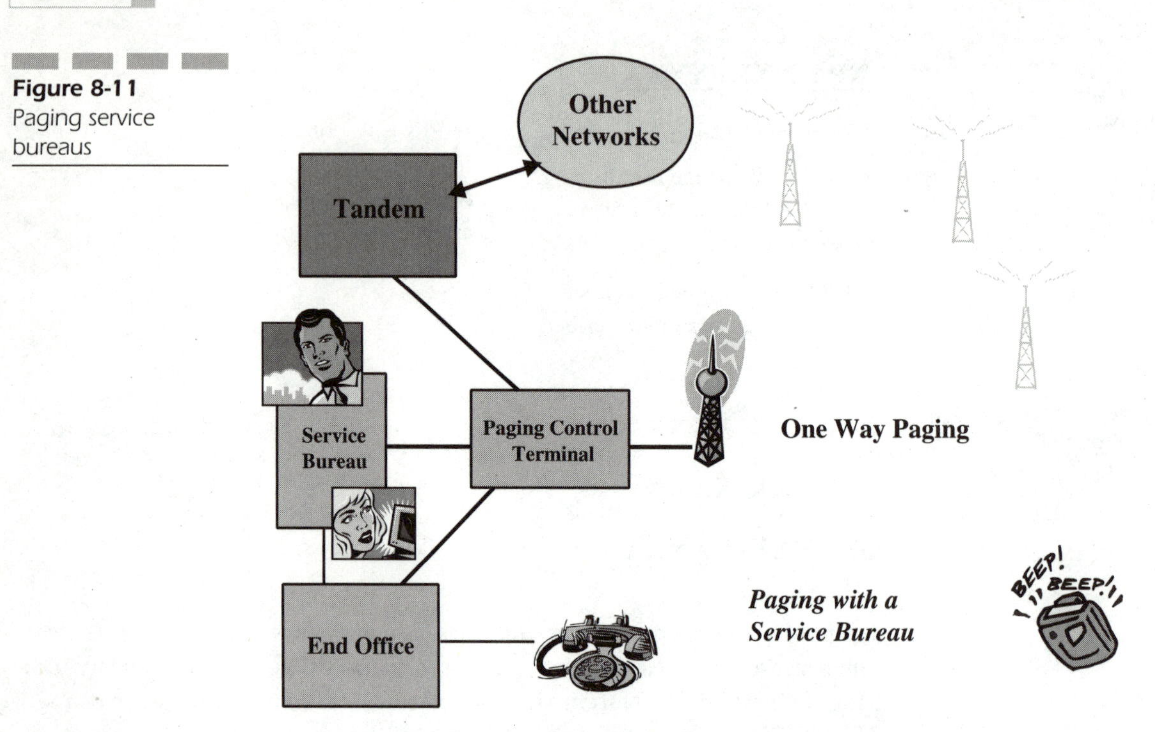

Figure 8-11 Paging service bureaus

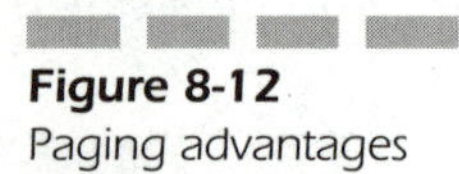

Figure 8-12 Paging advantages

- *Greater Power Transmission levels*
- *Less Overhead Messaging than Cellular or PCS*
- *Coverage is better than Cellular or PCS*

Traditional Paging Network and Interconnection

When an individual contacts a paging device, the device is called with a telephone number and probably some type of *personal identification number* (PIN). You can use almost any kind of terminal device or even service bureau provided it has access to the paging network. The calling device can be a personal computer, telephone, or an operator dispatch (service bureau) where someone takes and enters a message or sends and alerting tone.

Paging is different than either cellular or PCS service. In a cellular and PCS system, a calling party dials a number, the call is processed through the a service provider's network, and then probably through the PSTN to the wireless service provider's network. As soon as the call gets to the wireless service provider's network, the calling party is connected to the wireless called party. A real-time conversation connection is established. In the case of paging, communication does not occur in real-time. Communication is stored and forwarded. Up until recently, paging was strictly a one-way service: calling party to called party only. The reader should note that the stored-and-forward nature of a paging call tends to have an effect on the types of services that are offered. Two-way paging/messaging is entering the marketplace. I will discuss this more later. Figure 8-13 is a rendering of paging's strengths.

A sender uses one of the previously mentioned input sources to send the message or page through the local phone system, or PSTN. The PSTN "switches" the page to a carrier paging terminal. After the paging control

Figure 8-13 Paging—information storage is a strength.

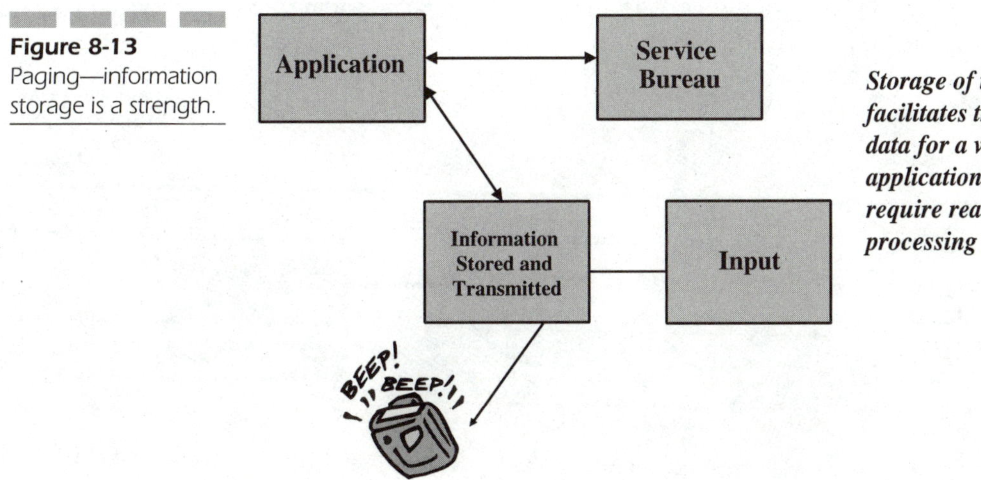

terminal receives the page, the terminal processes, stores, and forwards information from the caller. Additionally, it encodes the page for transmission through the carrier paging system. Typically, an encoder accepts the incoming page, validates the pager address, and "encodes" the address and page into the appropriate paging signaling protocol. After the page is encoded, it is sent to the RF link system, which includes the link transmitter and link receiver. The link transmitter sends the page to the link receiver, which is located at the various paging transmitter sites along the channel. The transmitter then broadcasts the page across the coverage area on the specified carrier frequency.

The interconnection through the PSTN (that is, ILEC) is critical in the paging carrier's operations. Without the ILEC interconnection, the paging carrier cannot be reached. The paging carrier obtains its PSTN interconnection from the wireline *local exchange carrier* (LEC). The dominant interconnection used by the paging carrier is the *Direct Inward Dialing* (DID) connection. The *Direct Inward Dialing* (DID) connection is a trunk-side wireline carrier end office connection. The DID connection is a two-wire circuit limited to one-way incoming service (LEC to other carrier). The DID connection gives the wireline local exchange carrier the perception that the paging control terminal is a customer premise equipment terminal. DID connections support DTMF address pulsing. Although the DID is a one-way interconnection type, it can support signaling back to the paging control terminal. The reader should note that the one-way nature of the DID connection prohibits outbound (from the paging carrier) calls to the LEC. The DID interconnection also does not support power ringing, also known as alerting ringing. The DID connection was one of the earliest interconnection types offered by the PSTN; hence, the DID is the dominant connection types used by the paging carrier. Figure 8-14 depicts paging's basic interconnection type with the PSTN.

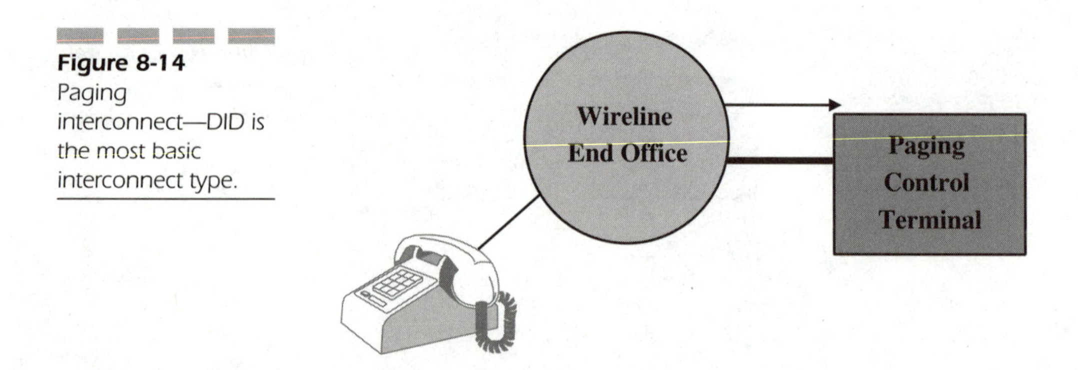

Figure 8-14 Paging interconnect—DID is the most basic interconnect type.

Other interconnection types supported by paging carriers include

- Type 1
- Type 2A
- Type S

The Type 1 connection is a trunk-side connection to an end office. The end office uses a trunk-side signaling protocol in conjunction with a feature known as *Trunk With Line Treatment* (TWLT). The TWLT feature allows the end office to combine some line-side and trunk-side features. TWLT enables the LEC to provide billing and interexchange carrier presubscription. Note that presubscription is a process in which the subscriber selects their primary long distance carrier; the subscriber then dials "1" followed by the 10-digit number and is automatically connected to their long distance carrier of choice.

The paging service provider will be given a telephone number(s) that reside only in the interconnected LEC end office. Type 1 connections also permits the mobile user to reach directory assistance, N11 codes (for example, 911, etc.), and service access Codes (for example, 700, 800, 888, 877, 900, and so on). The typical one-way paging subscriber will not be able to dial any telephone numbers. However, paging carriers are seeking ways of incorporating narrowband voice transmission in its product line for the two-way paging subscriber; therefore, the ability to dial out will be important. The wireline carriers seeking to maintain interconnection as a business will find ways of enhancing the appeal of the Type 1. Wireline carrier efforts to maintain the marketability of Type 1 have reduced the price of this interconnection to one below the DID rate. Figure 8-15 is an illustration of the Type 1 Interconnect in a paging network.

The Type 2A connection is a trunk-side connection to the LEC's access tandem. This connection allows the paging control terminal or the paging carrier's switch to interface with the access tandem as if it were an LEC end office. If the paging carrier is providing two-way messaging with dial-out capabilities, it will have access to any set of telephone numbers within the LEC network. The Type 2A comes in two flavors: a multifrequency version and a *Signaling System 7* (SS7) version. The Type 2A increases the visibility of the carrier to the overall network of networks. Many paging carriers are bypassing the ILEC's access tandems and end offices (and therefore the access charges) by interconnecting to the interexchange carrier *Point of Presence* (POP) using Type 2A-like interconnections (standard two-way MF trunks). The interexchange carrier can charge the paging carriers less than the ILECs in order to capture the paging carrier as its own access customer. If the ILEC's reduce the rates of the Type 2A, it is possible that the ILEC

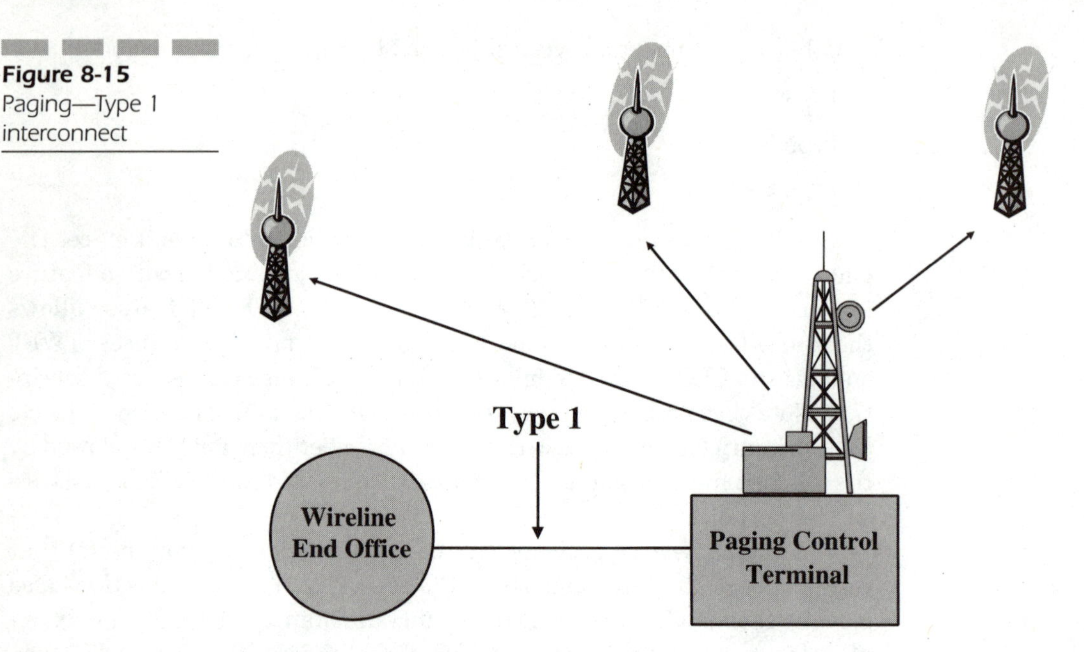

Figure 8-15 Paging—Type 1 interconnect

supported Type 2A may replace the Type 1 as the predominant interconnection for traditional paging carriers. Figure 8-16 is an illustration of how the Type 2A Interconnect is used. The Type 1 Interconnect is also shown for comparative purposes.

The Type S connection is not a voice path connection; it is a SS7 signaling link from the wireless carrier to the LEC. The Type S supports call setup via the *ISDN User Part* (ISUP) portion of the SS7 signaling protocol and TCAP querying. The Type S is used in conjunction with the Type 2A, 2B, and 2D. At this time, paging carriers have not installed switching systems to support *Signaling System 7* (SS7). However, Class 5 or MSC-type switching systems could open up opportunities for *common channel signaling* (CCS) based services. Figure 8-17 is an illustration of the Type S Interconnect.

Figure 8-18 is a complete illustration of paging carrier and wireline carrier interconnection. As the figure illustrates, the traditional paging industry is poised to evolve. Figure 8-18 is an illustration of the Paging network with all of the interconnects depicted.

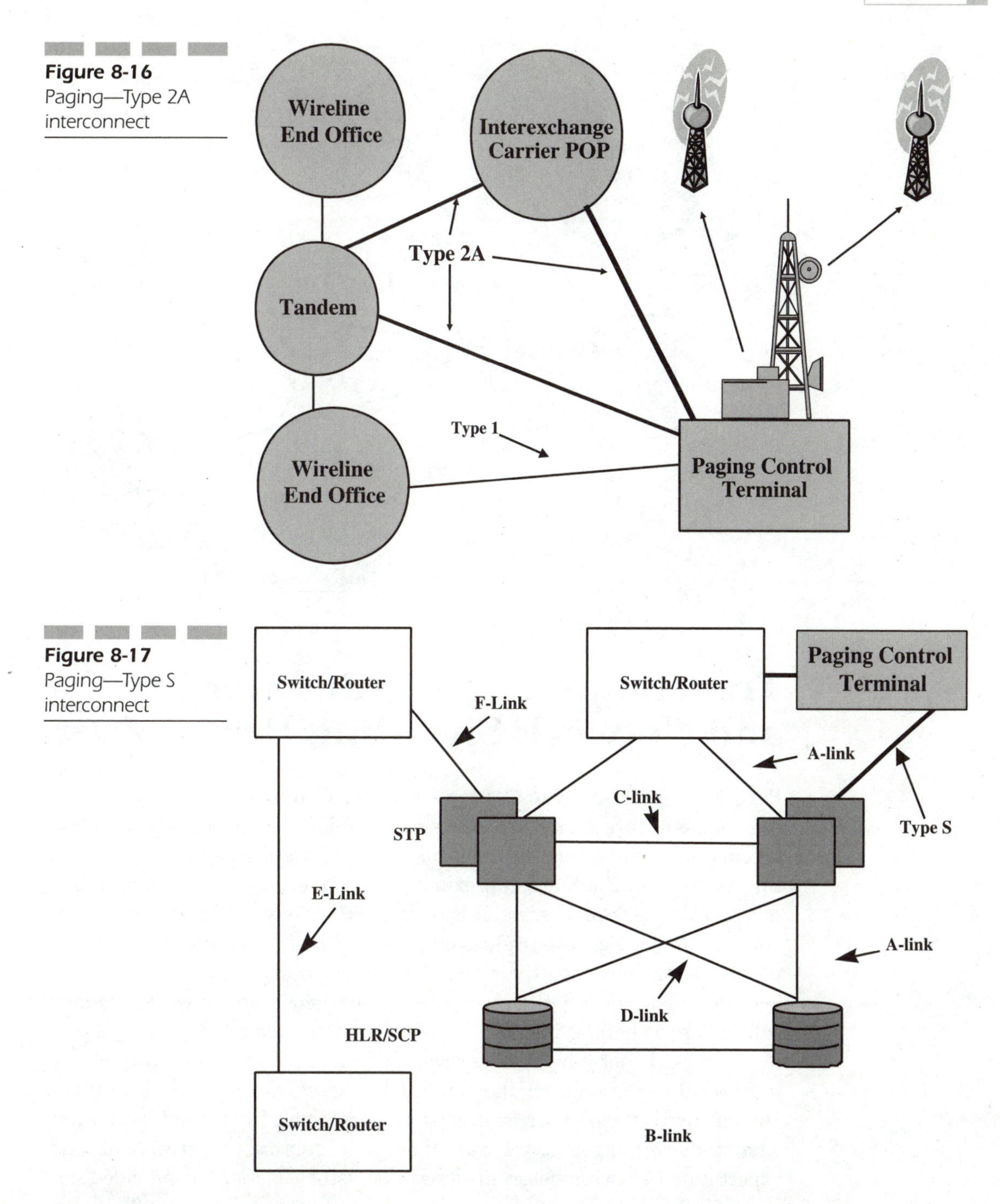

Figure 8-16 Paging—Type 2A interconnect

Figure 8-17 Paging—Type S interconnect

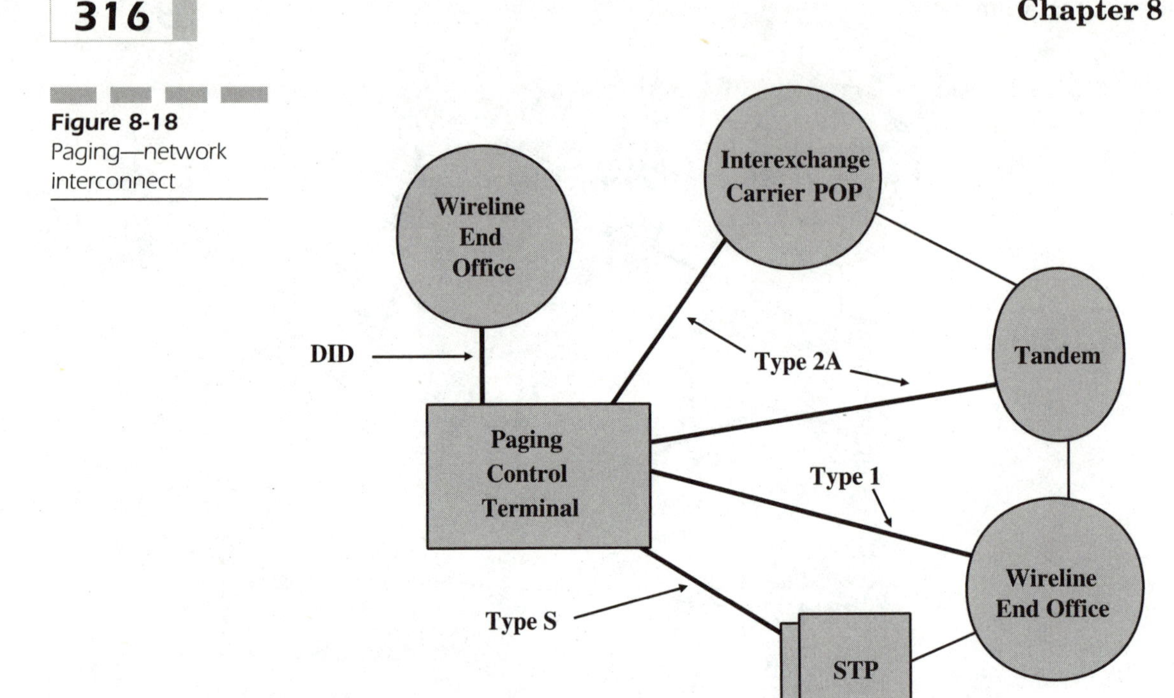

Figure 8-18 Paging—network interconnect

Traditional Paging Air Interface and Network Signaling Protocols

Paging air interface protocols have the benefit of not being a "standards development" target. In other words, the standards community has not had an opportunity to attempt to develop paging protocol standards. Of course, this means that there are approximately half a dozen major paging protocols and all are *de facto* standards in and of themselves. Unlike the cellular and PCS industries, the traditional paging industry does not need to support the same mobility attributes. In other words, traditional paging carriers provide a service that works only on their own networks. Paging subscribers do not need to roam from one provider to another. National paging carriers do not have service coverage in all areas of the country due to topological conditions, yet the subscribers do not raise a ruckus with the paging carriers' customer care departments. Note: In the end, no matter what industry segment we are addressing, subscriber perceptions and expectations are what determine how successful a carrier has provided service. Mobility is supported by paging carriers, but roaming is not. This simplifies the air interface and (fixed) network signaling protocols issues.

Fixed Network Signaling Protocols

The following fixed network signaling protocols have supported access to paging systems for decades. *Signaling System 7* (SS7) is new to the paging industry.

- Multifrequency
- *Direct Inward Dialing* (DID)

As I had indicated, the paging carrier network is essentially a separate network, but it requires access to the larger and more established PSTN. Typically, cellular carriers and PCS carriers interconnect to paging carriers via the PSTN local access tandem. Access to the PSTN by most carriers is principally via *multi-frequency* (MF) signaling. However, many ILECs have reduced and are reducing the number of Type 1 connections in their switches. This leaves the ILECs with a choice and therefore the flexibility to either grow the Type 2A base or the Type S interconnection base.

The simplest method of encoding a paging message is to use the DTMF keypad of a telephone or mobile handset to send a numeric message (usually the telephone number the called party is supposed to call back). The calling party dials the pager number desired and is then requested to input some type of message. This method requires no human service bureau operators and is the cheapest to implement. The message I refer to may be the number to call back. Sending alpha messages in the common one-way paging system requires a human operator. Figure 8-19 is a rendering that stresses the operator's role in a paging business.

Figure 8-19
Paging—operators

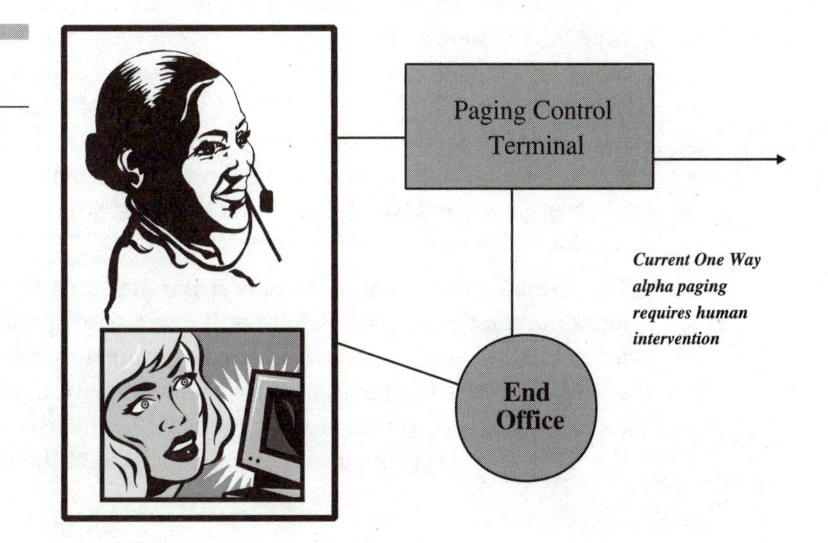

Paging Air Interface Protocols

The radio interface protocols supporting the paging industry are numerous and are all standardized. In this case, *standardized* does not mean ANSI or even ITU standardized; it means de facto standardized as a result of marketplace presence. As de facto standards, the air interface protocols do not support interoperability or service provider roaming. Given the way the traditional paging operates as a business, the incompatibility of the formats will continue until the traditional paging industry needs to change. The following are paging air interface protocols that have been in use for a number of years:

- *Audio Tone Signal (selected channels)* Tone only, no alphanumeric characters at all. The tone meant the individual either called an operator or simply called a pre-determined point. There are three types of tone-only protocols: two-tone, three-tone, and five-tone.
  - The two-tone signaling protocol was one of the earliest radio formats used. The paging carrier would alert the pager that a message awaited the subscriber. The subscriber would then call an operator for the message. The original answering services used such paging devices. Hospitals and doctors were among the first users of paging devices. The reader should note that even though the earliest paging systems used two-tone signaling, each paging company used its own two-tone signaling air interface.
  - The three-tone signaling protocol was used by the old Bell system in its Bellboy® service. This signaling format had a faster transmission time for the tones. The three-tone system converted the last four digits of the paging device's telephone number into a set of three discrete tones. The Bellboy system used a set of 32 audio tones. The key attribute of the three tone system was that a single company (with national coverage) supported a single format. No matter where the customer went, the Bell system offered a single product face for paging.
  - The five-tone signaling protocol is still in use today and is in use for alphanumeric paging as well. Like the two-tone and three-tone formats, the five-tone format is an analog air interface. The five-tone system converts the paging device's telephone number into a five-tone format. The system uses five out of 12 supported tones. Each of the 12 tones represents a numerical digit. Ten of the tones represents

the numerical digits 0 through 9. The eleventh tone represents an action, which is that a digit is repeated. The twelfth tone is used to instruct the paging device to emit a different audio alert pattern for tone-only paging.

- *Golay Sequential Coding* The Golay format is a digital format that can support tone, numeric, alphacharacter, and voice paging. The Golay format supports information transmission speeds up to 600 bits per second. Motorola introduced Golay.
- *Post Office Code Standardization Advisory Group* (POCSAG) is a high-speed paging protocol. It can handle up to two million addresses per carrier and supports tone-only, numeric, and alphanumeric pagers. POCSAG operates at 512, 1200, and 2,400 *bits per second* (bps) and is the most widely used format today.
- *European Radio Message System* (ERMES [1990]) is a 6,250 bps paging protocol used in Europe. The protocol supports message delivery for alphanumeric, numeric, tone, and data.
- *FLEX*™ Flex is a high-speed paging protocol introduced by Motorola a few years ago. Briefly, FLEX was designed to maximize channel capacity and speed, the pager's battery life, and data integrity, all key ingredients for a service provider evaluating a paging protocol. The FLEX protocol runs at four different speeds, allowing service providers a choice in matching the potential capacity of a FLEX protocol-based paging system to their individual requirements: 1,600, 3,200, 4,800, and 6,400 bps. FLEX allows lower latency for potential messaging as well as increased subscribers per channel. More importantly, FLEX supports two-way messaging. The ability to provide two-way messaging is a step towards maintaining the traditional paging industry's competitiveness. This is not an advertisement for any particular paging system or product. The intent of this information is to highlight the changes taking place in paging.

Figure 8-20 is an illustration charting the various radio protocols used in the Paging industry.

In the next few years, paging devices will support data transmission speeds of up to 160 kbps. By taking a broader perspective of data terminals and paging, we can envision a time when small card-sized devices will be considered to be mobile high-speed messaging and information terminals. I will discuss applications further later in this chapter.

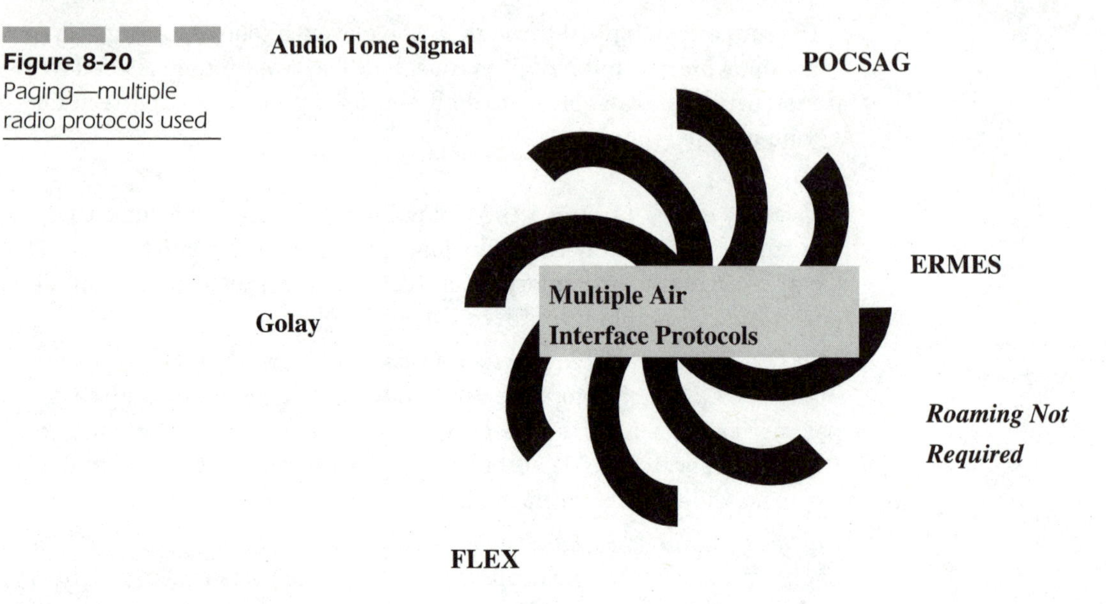

Figure 8-20 Paging—multiple radio protocols used

Paging Broadcast Operations

Unlike cellular carriers and PCS carriers, the paging carrier does not worry as much about co-channel interference or tightly designed RF transmission system coverage. Paging service providers look for broadcast sites that allow for wide area coverage. Cellular carriers note the last location of the mobile user transmits an incoming call indicator to the mobile handset and waits for a page response. If a response is not received, the network broadcasts the call indication throughout the local cell site network. Paging carriers contact the paging device differently.

Paging service providers broadcast the alert throughout the paging network. A paging device does not register in a network. The traditional device is designed to hear the paging carrier's broadcast system ID. Sometimes the device will provide a visual indication whether the device is in a coverage zone. Paging carriers are concerned about coverage and capacity but, because of the bursty nature and high *Effective Radiated Power* (ERP) nature of a page the concerns, are not nearly as large as that of a cellular or PCS carrier. Paging carriers also are not as concerned about paging device interference. Pages in a traditional paging carrier have the following characteristics:

- *Bursty* Data (telephone numbers) transmission speeds are on the order of a couple of seconds.

- High *effective radiated power* (ERP) of the data transmission
- No indication to the system if the page was received
- No requirement for the paging device to register or even indicate it has been turned on
- No requirement for the paging carrier subscriber to register their location. Therefore, the paging network must simulcast (simultaneous broadcast) the message throughout the network.

The most unique characteristic of a traditional paging network is the ability of the network to simulcast the page. This reduces the paging carrier's administrative overhead as it relates managing information regarding the paging device's location. Simulcast (a simultaneous broadcast of the page) is more of an issue for the nationwide paging carriers. In reality, the broadcast is coordinated amongst multiple controllers. Further, the system is designed so multiple transmitters do not interfere with each other. Unlike the cellular carrier or PCS carrier, the paging carrier is more like a commercial broadcast radio network. One similarity of the paging network is the system's ability to direct the page to specific paging device via its own device identification method. Figure 8-21 is a depiction of the nationwide paging carrier's characteristics.

Figure 8-21 Nationwide paging carriers

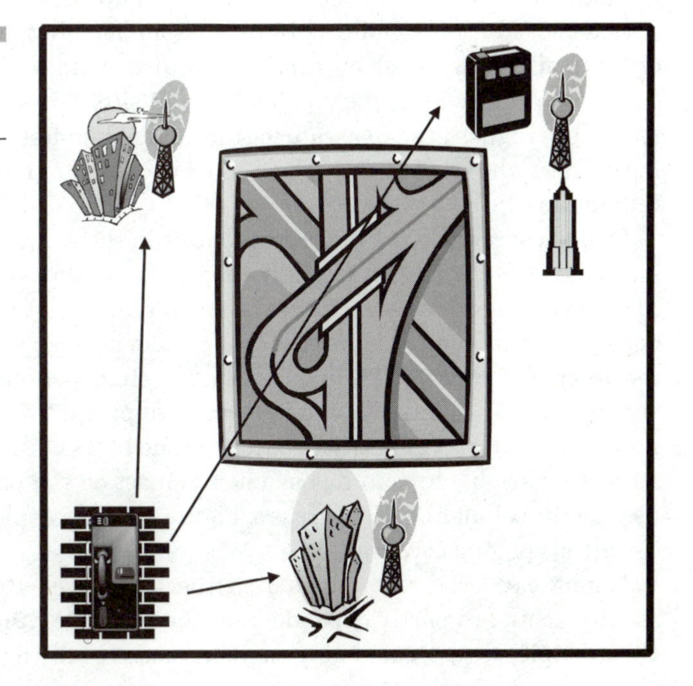

Do not maintain a location profile of the paging subscriber.

System wide broadcasting is necessary.

Nationwide paging carriers simply broadcast throughout the network, starting in one part of the nation and broadcasting until all sites have transmitted. This is why it can take upwards of 20 minutes before a paging device actually receives a page. Because real-time alerting is not a requirement, pages can take several minutes to be transmitted in either a local or national carrier.

Differences between Paging and Cellular/PCS Operations

Paging carriers deploy RF and fixed networks as well as the cellular carriers and the PCS carriers. However, there are differences in design and deployment requirements. The differences can be further defined to include coverage, base station design, antenna design, data transmission quality, and operational support systems.

Coverage

Cellular and PCS carriers deploy cell sites in pattern that usually results in sites deployed from 1/4 mile to 15 miles from one another. The distance or cell site size is affected by capacity requirements and traffic load. The denser the traffic load, the smaller the cell site. The smaller the cell site means the higher the reuse rate for the channel groups. In the case of paging carriers, the traffic load and subscriber counts are much lower than the cellular and PCS carriers.

Paging carriers do not reuse frequencies in the same manner as the cellular and PCS carriers. In fact, in local paging systems channels are reused in a less complex manner. The reader should note that the paging carriers deploy their antenna in a manner that results in large areas of coverage on the order of 25 to 100 miles in diameter. The channel group reuse pattern is simple. Topological studies are important to paging carriers because they need to qualify and quantify coverage on the basis of radio frequency field strength. Variable terrain has as much impact on the paging system as it does on the cellular or PCS carriers. Figure 8-22 is a depiction of how topology affects paging coverage.

Paging carriers are numerous on the local level. Broadcast coverage requirements are partly dependent on the type of customer being served. For example, some local paging companies serve only a single city or even

Figure 8-22
Paging—topology impacts coverage

one or two counties. The fact that many users can still receive their pages outside of the contracted service area is a non-issue for those subscribers because no one would complain about getting more for their money. National paging carriers seek approximately 90% coverage in their networks. Further paging companies seek call completion/message transmission rates of 90%.

Nationwide paging carriers use satellites to transmit the page from an originating Earth station to other Earth stations. The receiving Earth stations then re-broadcast the page through their respective local radio paging networks. Figure 8-23 is a high level illustration of how paging and satellite networking work together.

Antenna Design

The objective for a wireless carrier is to install antennas with the highest possible gain, yet still operating within the FCC's ERP restrictions. The RF gain is directly related to size and tilt. There is a practical limit to the size of the antenna; one does not want to install an antenna or tower that is obtrusive, presents an air traffic hazard, or cannot be properly secured

Figure 8-23
Paging and satellites

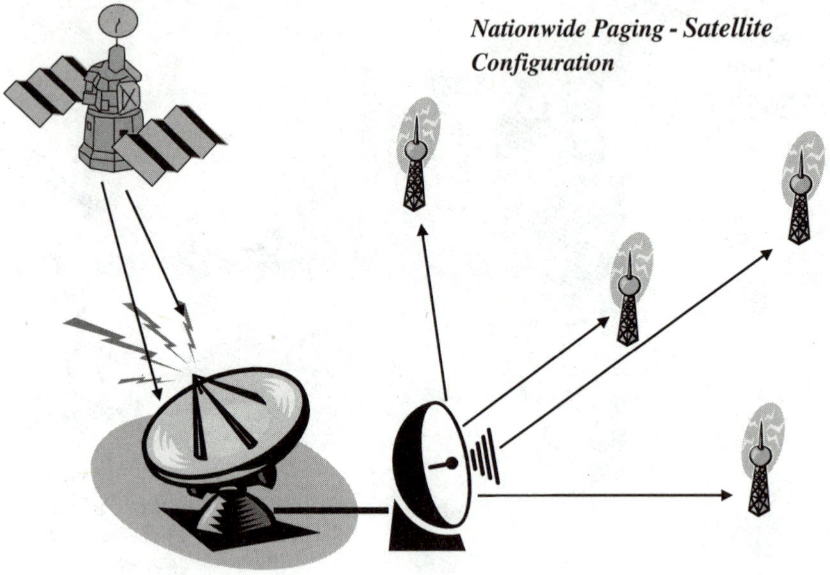

Nationwide Paging - Satellite Configuration

against collapse under high winds. The antennas used by the paging industry are similar to the antennas used in the cellular and PCS industries. Note that leaky coaxial cable is used in the cellular and PCS industries is used to reach hard to "get to" areas.

- Collinear antenna
- Omni-directional antenna
- Isotropic antenna

The collinear antenna has dual oval shape transmission field pattern, whereas the omni-directional antenna has a circular pattern. The isotropic antenna radiates radio energy equally in all directions. Sectorization is not performed and has no value in the paging industry. However, the field patterns can be shaped to optimize coverage. Field patterns can be shaped by taking the previous antenna types and arranging them in an array. The antenna array is comprised of two or more antennas that are arranged so the gains of the antennas add constructively and subtract constructively in certain directions. This results in field patterns that have higher gains in certain directions than if one antenna were used. Figure 8-24 is a view of a collinear radiation pattern.

Figure 8-25 is an illustration of the configuration of an omnidirectional antenna and its radiation pattern.

Figure 8-26 is a rendering of the radiation pattern of an Isotropic antenna.

Figure 8-24
Top view of collinear antenna radiation pattern

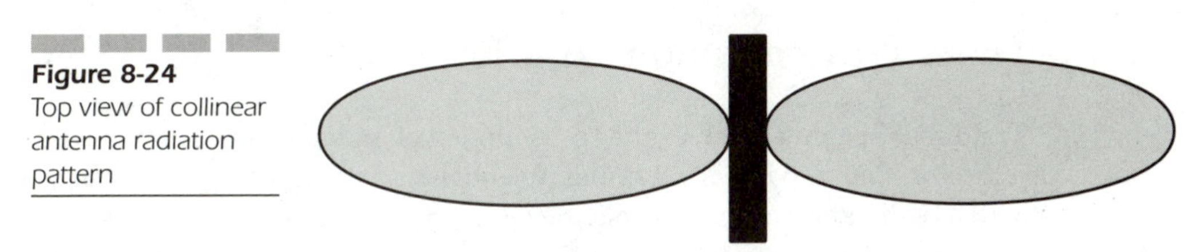

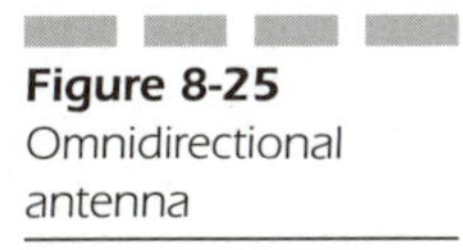

Figure 8-25
Omnidirectional antenna

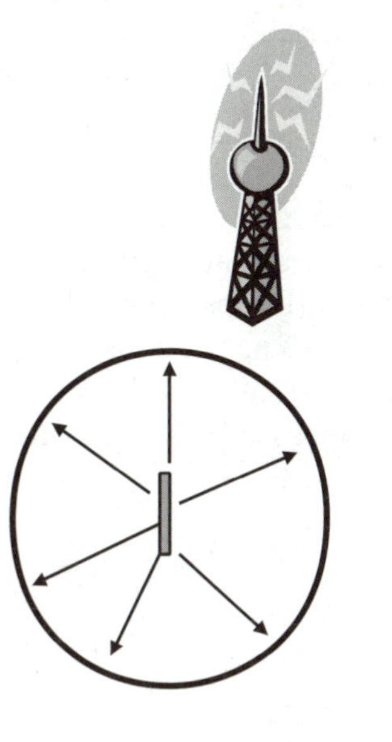

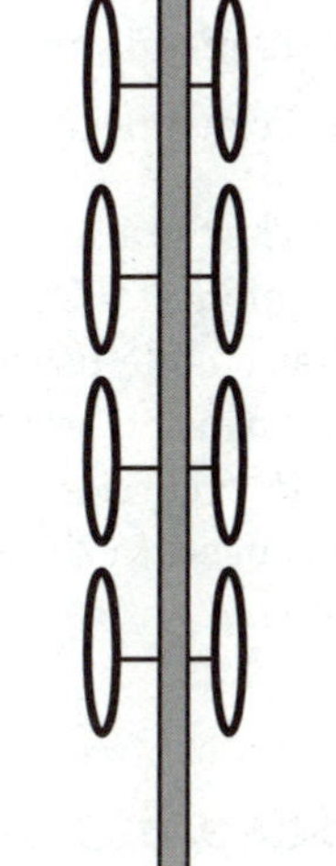

Omnidirectional antennas radiate radio energy in all directions.

Radiation pattern is not shaped.

Figure 8-26
Isotropic antenna

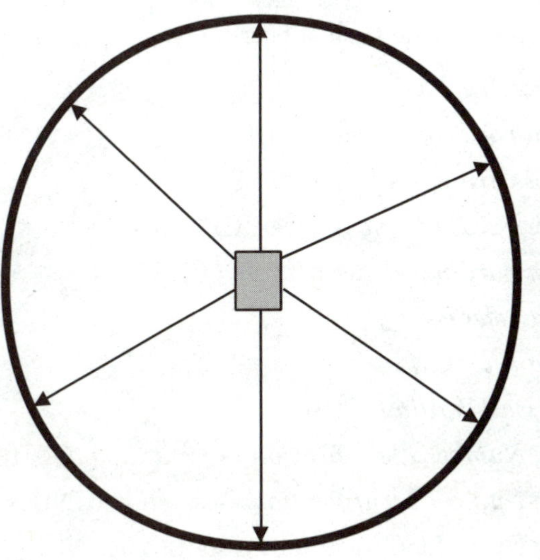

Isotropic antennas are engineered to radiate radio energy at equal ERP levels in all directions

Data Transmission Quality

Traditional paging involves the transmission of an indicator to emit a tone indicator that a page is incoming, telephone number, or alpha message. These really amount to a few paging device instructions:

- Indicator that there is a page or message. The device can be locally programmed to emit a tone, turn on a light, some other type of visual indicator, or vibrating action.
- Transmit numerical messages.
- Transmit alpha messages.

The data is bursty in nature and only requires a few seconds to transmit. Cellular and PCS carriers must maintain network transmission standards that enable users to understand the spoken word. Transmission quality is less of a concern to many paging companies than it is with cellular carriers or PCS carriers. As paging carriers enhance their systems to support voice, the need to design systems with low data transmission error rates will grow.

Base Station Design

The following is the list of subscriber services that are supported by a cellular carrier and a PCS carrier. The cellular or PCS base station needs to support all of the radio interface messaging to enable these services. The paging system has no such subscriber service requirements.

- *Call Delivery* (CD)
- *Call Forwarding—Busy* (CFB)
- *Call Forwarding—Default* (CFD)
- *Call Forwarding—No Answer* (CFNA)
- *Call Forwarding—Unconditional* (CFU)
- *Call Transfer* (CT)
- *Call Waiting* (CW)
- *Cancel Call Waiting* (CCW)
- *Calling Number Identification Presentation* (CNIP)
- *Calling Number Identification Restriction* (CNIR)
- Calling Name Identification Presentation

- Calling Name Identification Restriction
- *Conference Calling* (CC)
- *Three-Way Calling* (3WC)
- *Message Waiting Notification* (MWN)
- *Mobile Access Hunting* (MAH)
- *Password Call Acceptance* (PCA)
- *Remote Call Forwarding* (RCF)
- *Remote Feature Control* (RFC)
- *Screen List Editing* (SLE)
- *Selective Call Acceptance* (SCA)
- *Selective Call Forwarding* (SCF)
- *Selective Call Rejection* (SCR)
- *Messaging Delivery Service* (MDS)
- Short Message Service
- Voice Mail
- Fax Mail
- *Single Number Service* (SNS)

The base station not only needs to support subscriber service message signaling, but also needs to communicate with the central switch coordinating activities such as handoff. Paging system base stations do not need to support any of the previous subscriber services. Paging system base stations do not need to communicate with the paging control terminal to manage handoff. However, as traditional paging carriers enter new markets such as the Internet, the complexity of the paging control terminals will increase. At some point, the paging control terminals will be more aptly described as paging switching centers.

Operational Support Systems

Operational support systems used in the paging industry perform similar functions as those used in the wireline, cellular, and PCS worlds. These functions are

- Network management
- Customer care
- Billing

Network management as I have described it up until this point includes the following:

- Network Element Management
- Network Systems Management
- Service Management

Network management requires one to coordinate functions between each of the paging carrier's systems. Managing a paging network involves scheduling of activities and resources, routine maintenance, trouble reporting, and diagnostics. A generic description of network management would be "network maintenance and health." A paging network is no different in its requirements for management of the network. The difference between the paging carrier and the wireline, cellular, or PCS carrier is the level of network management sophistication needed. Network management typically includes the ability to dynamically route traffic. The current paging carriers do not need to route traffic to different points in their network for loading purposes. Most paging carriers are local and tend not to need a complex network management system. The nationwide paging carriers would benefit from having a sophisticated network management system.

As the paging industry enters new areas of telecommunications or seeks to create synergistic relationships with other types of carriers, network management systems will evolve into more complex integrated systems. Figure 8-27 is a rendering that highlights the ever growing role of network management in paging.

Customer Care and Billing

Customer care in the traditional paging industry amounts to calling in a trouble during daytime work hours. In the wireline carrier, cellular carrier, and PCS carrier worlds, customer complaint desks are open 24 hours a day. The paging carriers will be forced to open up 24-hour help desks as they deploy new services. For many of these new types of CLECs, customer care may be the difference between staying in business and going out of business. Figure 8-28 highlights the importance of customer care. Customer care centers have the following concerns and issues to manage:

- Customer service representatives' training
- Customer interaction
- Customer complaint resolution process

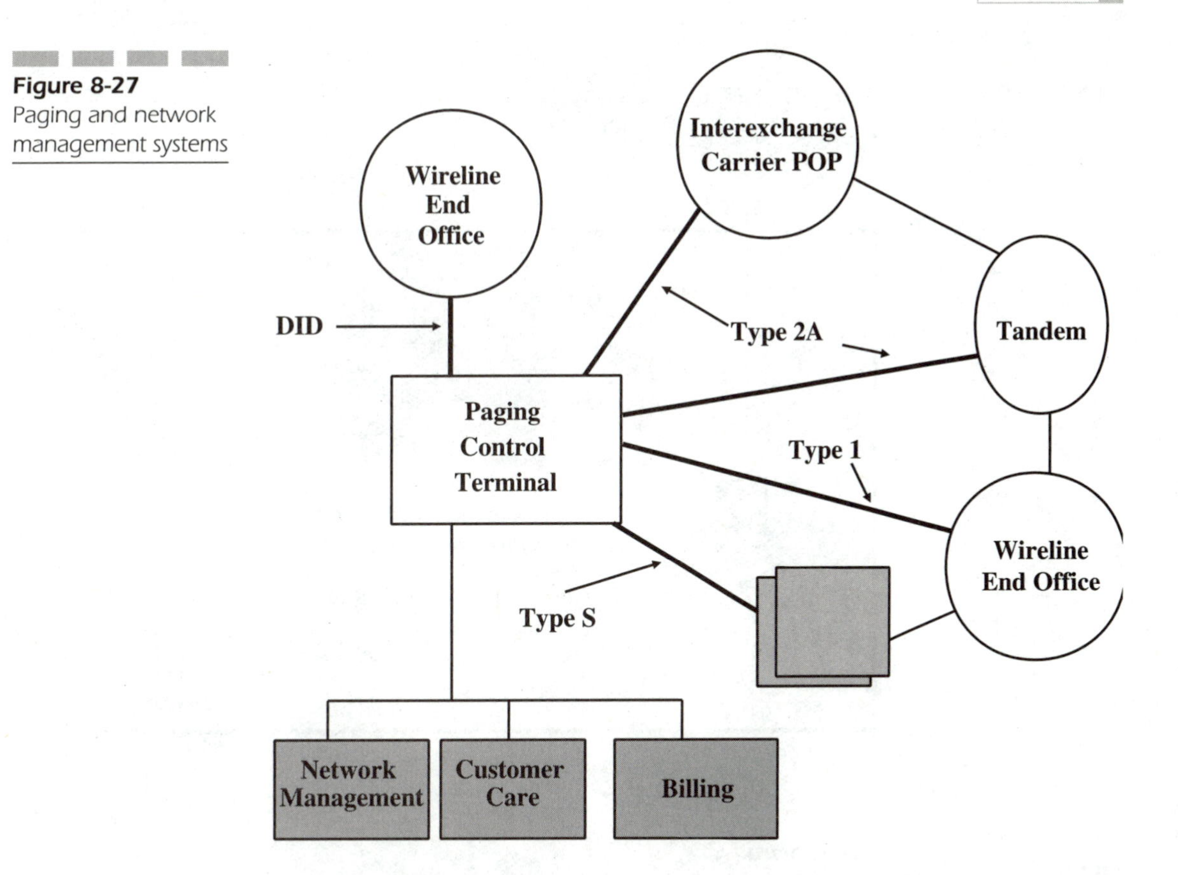

Figure 8-27 Paging and network management systems

- *Customer support* This is not limited to just Internet companies. Many would be surprised to know that a customer can ask even the traditional wireline LECs a technical question and have it addressed.
- Installation and Repair departmental coordination

Billing in the paging world today is simple flat rate. In order to stay competitive, the billing must become a usage-sensitive one. The billing system must become flexible enough to manage the billing of new services. Paging companies must be able to offer and manage a variety of rate plans. The following will become part of a paging carrier's billing concerns and managing those concerns:

- Types of billing plans
- Customer payment options
- Customer bill formats
- Customer billing periods

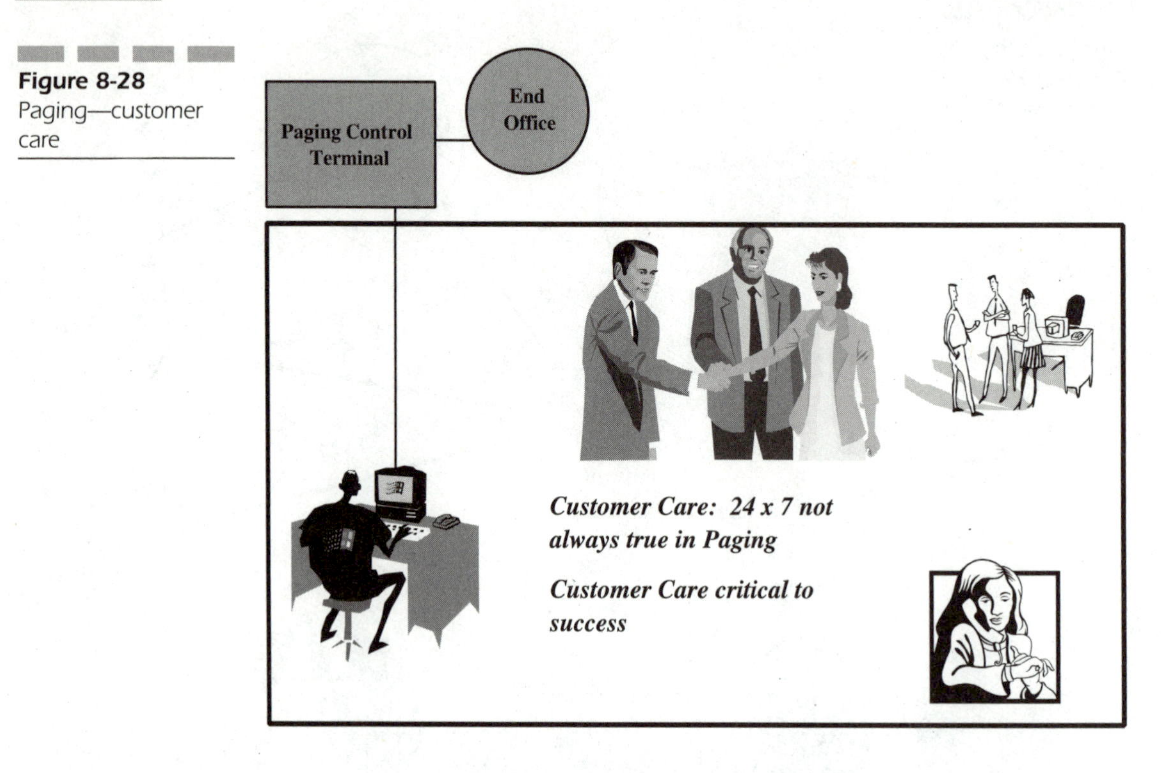

Figure 8-28 Paging—customer care

Figure 8-29 Paging—rate plans

Usage Sensitive normally refers to fee per # of pages.

Fee will be based on single page transactions and value added service packages

Flat Rate

Fee Per # of Pages

Usage Sensitive

Majority of Paging fees are either Flat Rate or a Fee based on the number of pages.

Figure 8-29 is a rendering that illustrates the changing nature of rate plans in paging.

Paging Business Paradigm/Environment

Traditional paging carriers use FCC-authorized frequencies intended for paging. The new paging carriers (using PCS frequencies to support paging and voice services) can utilize their PCS frequencies to provide the kinds of services subscribers have come to expect from cellular and PCS carriers. It is important to note that the paging business paradigm is changing. Paging was once restricted to the FCC-assigned paging frequencies; however, the reality is that cellular and PCS carriers have added device vibrating, tone emitting, 50 different types of ringing, and even auto-answer functions to the wireless handset. From the perspective of the customer, the differences between paging and typical wireless are

- *Price* Traditional paging carriers normally charge a flat fee per month. The low single rate plans being offered by the cellular and PCS carriers are piquing the interest of paging subscribers.
- *Performance* Customers are victims of more dropped calls than lost pages. The traditional paging carrier's right (by virtue of the FCC) to transmit at ERPs of 500 watts is an overwhelming advantage over cellular and PCS carriers. This means that paging devices will be able to receive pages in places where a cellular or PCS handset loses signal.
- *Coverage* Given the paging carrier's power transmission levels, the subscriber typically can receive a page in almost any location, in-building or out of building. Therefore, the perception from cellular and PCS subscribers is that paging subscribers can get their pages all of the time and in any location.
- *Applications and Services* Cellular and PCS service offers more different types of voice and data services. Furthermore, paging does not offer voice services of a quality high enough to equal voice quality offered by cellular and PCS.

As traditional paging companies deploy new technology to enhance their applications and services sets, the differences between paging and cellular and PCS will be minimized. Paging carriers will seek business relationships with other types of telecommunications service providers in order to create new business opportunities. New technology will be deployed. Paging carriers will need to make investments in backoffice operations systems such as network management, customer care, and billing. As the competitive nature of the paging business environment increases, the need to make capital investments in infrastructure and support systems increases.

Convergence: Applications and Services

The stored-and-forward nature of paging messaging is an attribute that can be further exploited. The fact that the information being sent can be stored someplace and then later forwarded means that information can

- Be stored for use by a third-party application provider, for example, intelligent network systems. Information can be used as a trigger.
- Be stored for polling at will by the recipient.
- Be stored for use by multiple parties.

Figure 8-30 is a rendering highlighting the importance in which the store-and-forward nature of paging can be exploited to further paging applications.

In effect, the information can be stored for database manipulation. Most paging systems today do not store called or calling party information for extended periods of time. However, as paging encounters more and more competition, information storage will become a factor in success.

Another way of looking at paging applications is by looking at terminal access. In the early 1990s, paging terminals were nothing more than very small handheld devices that hung from a person's belt or sat in a pocket. In a few years, paging devices got a bigger digital display and started playing

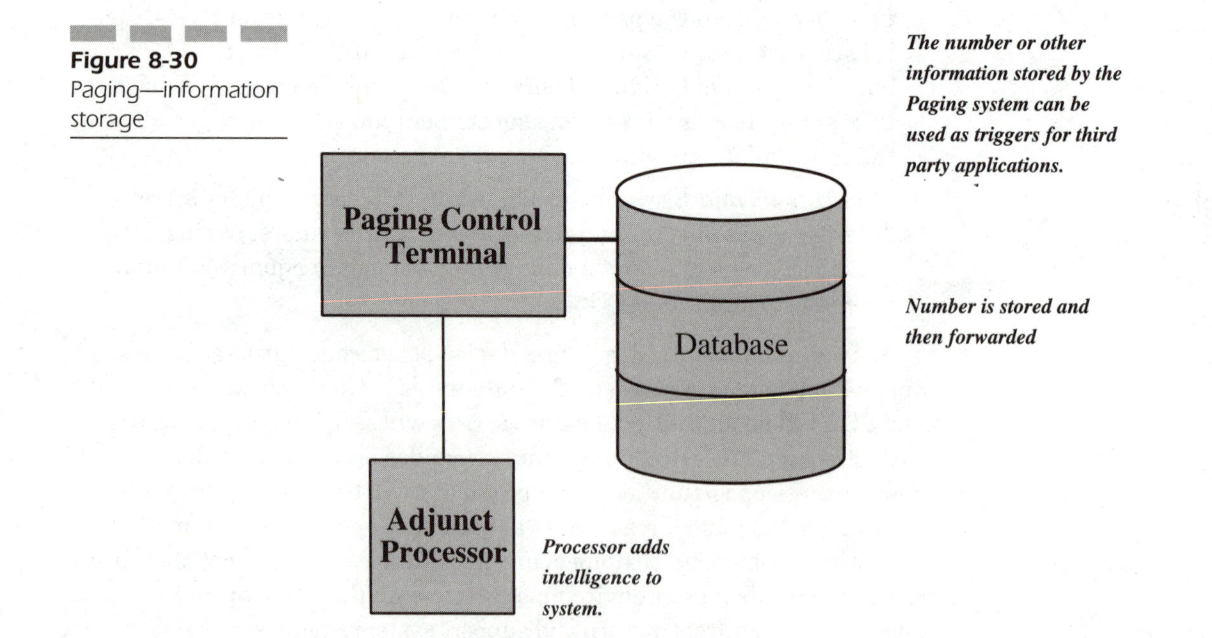

Figure 8-30 Paging—information storage

news highlights. In the last couple of years, paging went from a one-way service to a two-way service. The current two-way applications are mostly limited to acknowledging receipt of a page either via tone or alphanumeric response.

Paging can be integrated with voice mail systems today and even email systems. Paging even offers unique opportunities for mass advertising because paging devices tend to be left on all the time. Note that unlike most current mobile handsets, paging devices operate in an intelligent sleep-like/wake-like mode waiting for a page. Therefore, power demands from a paging device are much less than a mobile handset. As two-way paging devices are called upon to provide for more complex services requiring more transmission power, battery life will become an issue. Figure 8-31 is an illustration that highlights the power draw difference between a traditional paging device and a two way paging device.

The size of the device is an important and integral part of the paging industry's market plans. People do not want to have to carry large portable devices. A simple example is the laptop computer. In the last few years, the

Figure 8-31 One way paging device and two way paging device — power

Figure 8-32 The evolution of the paging device

laptop computer has gotten smaller. Of course, there is a limit to how small they can get because of keypad size issues (human factors engineering issues), but you can make it lighter in weight. Size and weight are utility factors that play major roles in a product's success. Today we have wristwatches that incorporate a pager and use the strap as an antenna; these have been on the market for some time. Figure 8-32 illustrates how the paging device is evolving.

Paging services themselves are still largely one-way services that simply tell the paged party who should be called back. What is interesting about paging is how this one capability is put to use across different customer segments. Who uses paging? Can these segments be served differently by the paging industry?

- *Hospitals* Doctors and medical staff
- Sales forces
- *Restaurants* Some restaurants issue pagers to customers awaiting tables

- Resorts
- *Amusement Parks* Paging devices are used by staff and can be rented out to park customers
- Teenagers
- Parents
- Construction
- Limousine drivers
- Business executives
- Any business or market segment where a person wants to be contacted at any given time

Figure 8-33 is an illustration of how paging is currently employed in the marketplace.

The traditional paging industry players need to undergo changes in order to stay competitive. As I had indicated, the paging function is already a feature in new cellular and PCS services. Traditional paging needs to offer something new. The answer is email and Internet access. The technology is already being deployed and marketed to enable the traditional paging carriers to offer the subscriber email and Internet access. A proprietary paging air interface has been created to enable this new application for the paging subscriber. By offering Internet access and email service, the paging carriers

Figure 8-33 Paging—uses in the marketplace

are positioned to combat the growing cellular and PCS threat. In order for paging carriers to offer Internet access, issues such as graphical and video transmission and display must be addressed. Paging currently supports a number of proprietary analog and narrow band digital interfaces. Assuming the paging industry wishes to support intersystem roaming, there will be a need to address the standardization of interface messaging between the paging air and network interfaces. The next challenge for the PCS carriers is to find an application or new feature capability to entice new subscribers. Figure 8-34 is a rendering that highlights the growing relationship between the paging and Internet industries. The Internet represents the next stage of growth for the paging industry.

SUMMARY

The key to finding that new application may be found by going back and reexamining paging's attributes. As I noted, the paging carrier stores a number for transmission to the paged subscriber. The paging carrier markets to professionals that are also users of the Internet and multimedia. The paging carrier must consider adding a intelligent database management system that will leverage its key attributes and its current subscriber base. A

Figure 8-34 Paging and the Internet

database contains information about the call and the subscribers involved in the call. This information can be used to enhance the call being set up between the parties. The information entered by the calling party may be used to support billing, Internet, data warehouse marketing information, or to create new routing methodologies to support time of day and day of week routing. The page can be used as a IN trigger.

As for the radio spectrum, higher data rates are needed. The higher data rate may enable the paging industry to converge its technologies in the direction of the multimedia industry. Convergence in the paging industry is as necessary as it is in the wireline, cellular, and PCS industries.

Satellite Communications Systems

Satellite communications is an outgrowth of the quasi-scientific, political space race. The first objects that were launched by mankind were satellites that emitted a simple, repeating radio tone. The United States launched itself into a space race with the Soviet Union in 1957 upon the launching of the Soviet satellite called Sputnik. Sputnik was the first artificial satellite launched into space.

Satellite communications was a natural outgrowth of this race. If man were to travel through space, work in orbit around the Earth, and land on the moon, a means of communicating between those on Earth and in space would be necessary. Essentially, satellite communications enable people to communicate over large distances. World War II drove the development of satellite communications. First, the war caused the rapid development and growth of rocket/missile technology, which enabled the launching of artificial satellites. Second, the war heralded an incredible growth in radio communications technology. Instead of carrying entertainment programs, radio during the war was used to transmit encoded messages either via special manually developed codes or via specially encrypted signals. The war also drove the development and growth of microwave communications. Figure 9-1 is an illustration of a satellite network.

By combining both technologies—and with the space race as the driving force—the telecommunications industry could facilitate the development

Figure 9-1
Satellite network

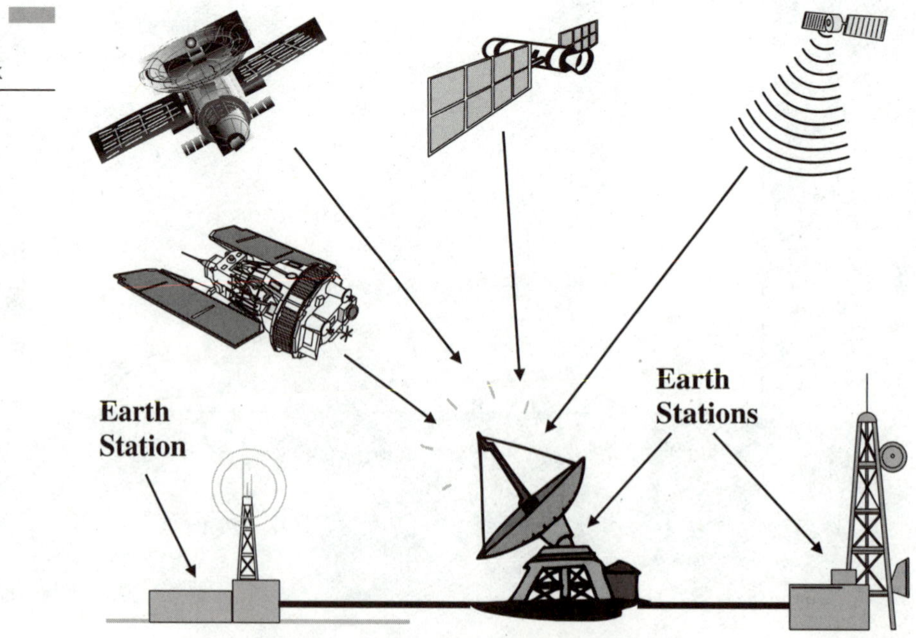

of the satellite communications industry. The objectives of satellite communications include achieving the greatest coverage and capacity at the lowest possible cost. Satellite communications systems can be broken into two parts:

- *Space segment* This part includes the satellite and the rockets that are used to launch the satellites. Essentially, the entire infrastructure that is used to create, maintain, and launch the satellites is considered part of the space segment. Unlike Earth-based communications systems (also known as terrestrial communications), satellite communications systems require support in a number of non-communication-related areas (such as rocket launchers, power supply in outer space, orbital propulsion motors, etc.).
- *Earth segment* This part includes Earth transmission systems and satellite communications receiving stations. The Earth segment also includes all satellite fixed-network interfacing equipment and switching systems that are needed to convert and process satellite radio transmissions into meaningful data and calls.

The Earth and Space segments are depicted in Figure 9-2.

As I had noted, the space age and space race began in 1957 with the launching of the first artificial satellite called Sputnik. Sputnik's success

Figure 9-2 Earth and space segments

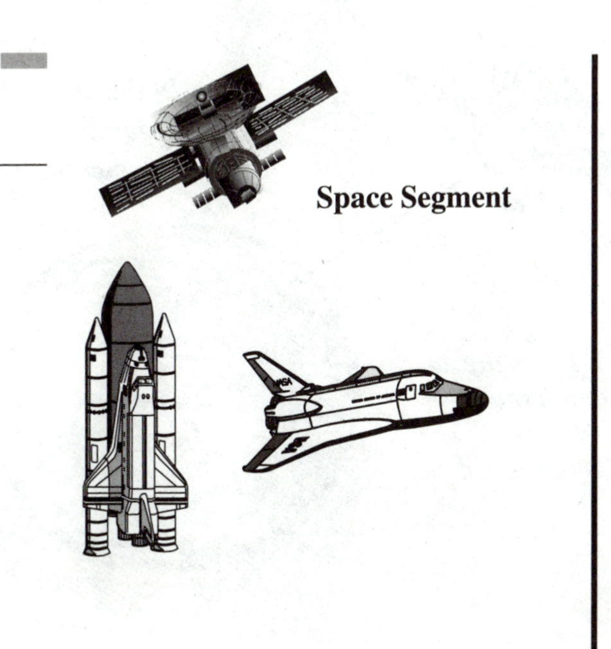

Earth Segment

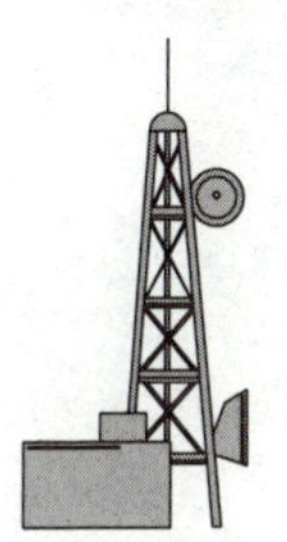

led to a flurry of activity by both the United States and the former Soviet Union. In 1963, the first geostationary satellite (called SYNCOM) was launched. In 1965, the first commercial geostationary satellite (called Intelsat I) was launched. Intelsat I was the first in a long series of Intelsat satellites. From a historical perspective, the Intelsats played a pivotal role in the commercial success of satellite communications. Satellite communications is an example of how technology transference from one industry to another can prove successful. Figure 9-3 is a rendering of Intelsat I.

What Is a Satellite?

Satellites only have a few basic parts: a satellite housing, a power system, an antenna system, a command and control system, a station-keeping system, and transponders. Satellite systems are complex technological systems, however. They orbit the Earth at varying altitudes, and they present challenges that Earth-based telecommunications systems do not face. Figure 9-4 is an artist rendering of these challenges. The satellite telecommunications industry faces the following challenges:

Figure 9-3
Intelsat I

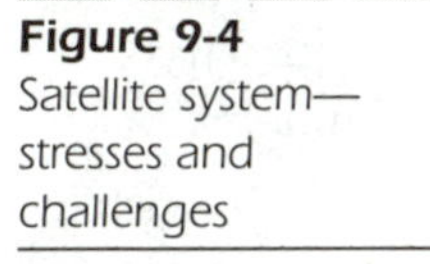

Figure 9-4
Satellite system—stresses and challenges

- **Physical stress**
- **Cold**
- **Heat**

- Satellites continuously operate in harsh environments with no air, no atmospheric pressure, and temperatures approaching absolute zero. In sunlight and in shadow, the satellites are subject to temperature extremes. Satellites are traveling through space at high rates of speed and therefore are exposed to the dangers of collisions with objects in space (such as micrometeorites and manmade garbage, such as lost tools, other satellites, and garbage).
- Satellites are subject to enormous physical stress during the launching and orbital placement phases.
- Satellites can only carry a limited power supply. The larger the power needs, the larger the power supply.
- Satellites need to be maintenance free. The space shuttle has enabled some satellites within the range of the shuttle to be repaired; however, satellites need to be designed and operated under the assumption that repair service will not be available.
- Satellites are expensive. Recently, there have been some satellite launch failures that resulted in hundreds of millions of dollars in losses to satellite companies.

Satellite telecommunications has the following characteristics:

- The signal must travel far. Distances are on the order of thousands of miles, which can result in signal degradation.

- The signal cannot be regenerated for enhancement, like the terrestrial systems can via signal repeaters. When the signal arrives on Earth, it must be enhanced for transmission through the terrestrial network.
- The signal is being transmitted through a medium that is filled with ambient noise and radiation. A ground-based system must be in place to interpret the signal that is being received.
- An Earth-based infrastructure must be maintained to ensure satellite space operation, the orbital status, and the system's health. This infrastructure is similar to the terrestrial telecommunications systems' network operations centers and is also known as a command and control center.

Figure 9-5 is a rendering of a satellite system's operating characteristics. As I noted, satellites have a few basic components: a satellite housing, a power system, an antenna system, a command and control system, a station-keeping system, and transponders. The systems are complex and expensive and require a high degree of technological sophistication to operate. Figure 9-6 is an illustration of a satellite's basic components.

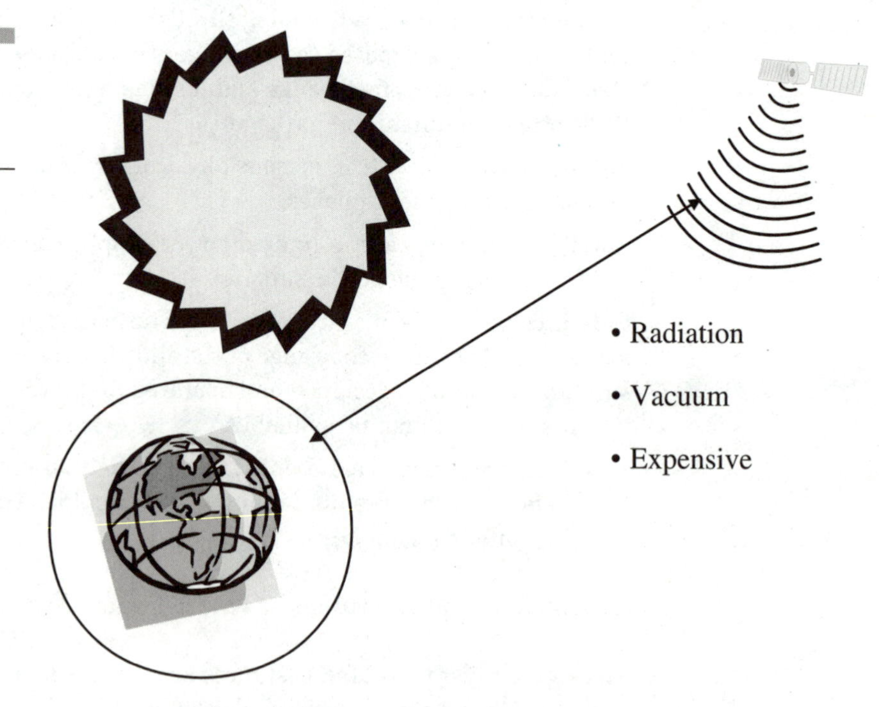

Figure 9-5 Satellite system—operating in hostile environments

Figure 9-6
Basic satellite components

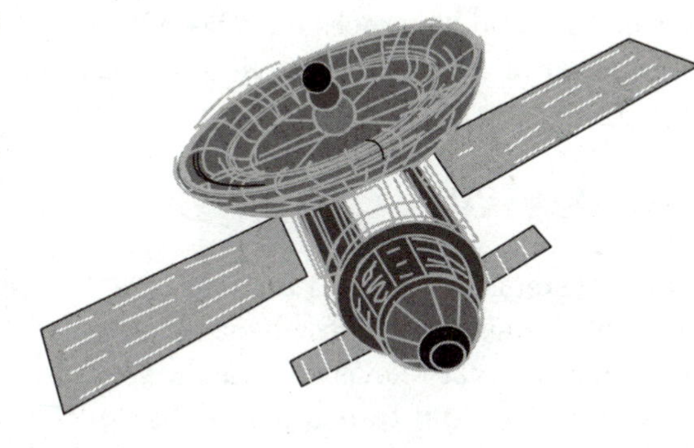

- **Satellite Housing**
- **Antenna**
- **Power System**
- **Station Keeping System**
- **Transponders**

Satellite Housing

The satellite housing configuration is determined by the system that is employed to stabilize the altitude of the satellite in its orbital slot. One method of stabilization is called three-axis-stabilization, which uses internal gyroscopes rotating at 4,000 to 6,000 *Revolutions per Minute* (RPM). Another method of stabilization is spin stabilization. In this case, the satellite's housing is cylindrical and rotates around its axis at 60 to 70 RPM to provide a gyroscopic effect.

Power System

Satellites must have a continuous source of electrical power, 24 hours a day, 365 days a year. The two most common power sources are high-performance batteries and solar cells. Solar cells are an excellent power source for satellites.

One problem exists with using solar energy, however. Twice a year, a satellite in a geosynchronous orbit will go into a series of eclipses where the sun is screened by the earth. If solar energy were the only source of power for the satellite, the satellite would not operate during these periods. To solve this problem, batteries are used as a supplemental on-board energy source. Initially, nickel-cadmium batteries were utilized, but more recently, nickel-hydrogen batteries have proven to provide higher power, greater durability, and the important capability of being charged and discharged many times over the lifetime of a satellite mission. Another source is nuclear power. Exploratory probes have been launched with nuclear power sources that power all telemetry and vital functions of the satellite.

Antenna System

The antenna systems' primary functions are to receive and transmit telecommunications signals to provide services to its users. Some of these services can be broadcast television, broadcast radio, public telecommunications, video imaging, and transmission of scientific data. The secondary function includes *Tracking, Telemetry, and Command* (TT&C) functions to maintain the operation of the satellite in orbit.

The command and control system includes TT&C systems for monitoring all of the vital operating parameters of the satellite, telemetry circuits for relaying this information to the Earth station, a system for receiving and interpreting commands that are sent to the satellite, and a command system for controlling the operation of the satellite.

Station-Keeping System

Minor disturbing forces exist that would cause a satellite to drift out of its orbital slot if the satellite is unguided. For example, the gravitational effect of the sun and moon exert enough significant force on the satellite to disturb its orbit. Interestingly enough, even land masses have a gravitational effect over and above the planetary pull. For instance, the continent of South America tends to pull satellites southward.

Station keeping is the maintenance of a satellite in its assigned orbital slot and in its proper orientation. The physical mechanism for station keeping is the controlled ejection of hydrazine gas from thruster nozzles that protrude from the satellite housing.

Transponders

A transponder is an electronic component of a satellite that shifts the frequency of an uplink signal and amplifies it for retransmission to the earth in a downlink. Transponders have a typical output of five to 10 watts. Communications satellites typically have between 12 and 24 on-board transponders. Each transponder transmits data at a rate of 36 MHz, which equates to about 60 Mbps with multiplexing techniques.

Satellite Spectrum Bands

Satellites operate in a number of high-frequency bands for a number of different applications. In the case of telecommunications, which includes radio, television, entertainment (satellite TV broadcasts), and public voice, satellites transmit in the following bands:

- L Band—1.5 GHz to 2.5 GHz
- S Band—2.5 GHz to 4 GHz
- C Band—3.4 GHz to 7 GHz
- X Band—8 GHz to 12 GHz
- Ku band—12 GHz to 18 GHz
- K Band—18 GHz to 26 GHz
- Ka Band—26 GHz to 40 GHz

The frequency bands support a multitude of communications services. Certain frequency ranges are shared by terrestrial telecommunications microwave systems. These microwave systems are used to support long-haul backbone telecommunications and other commercial applications. The sharing forces the satellite industry to deploy its Earth-based stations in a manner that minimizes interference. The following are the band classifications as managed in the United States by the *Federal Communications Commission* (FCC), with the L band supporting the following telecommuniations services:

- Mobile Satellite Service *(MSS)* *Low Earth Orbit* (LEO) mobile satellite services
- *Personal Communications Service* (PCS)
- *Ultra-High Frequency* (UHF) television
- *Microwave links* These links are used to support terrestrial communications.
- *Local and state government applications* This service might include law enforcement and public safety communications.
- Very High-Frequency *(VHF) television* Typical broadcast television
- Cellular service

The S band supports the following telecommunications services:

- Mobile satellite service
- Deep-space research

The C band supports the following telecommunications services:

- Fixed Satellite Service *(FSS)* This service includes deep-space research and other types of government applications
- *Fixed service terrestrial microwave* This service supports government applications.
- New types of terrestrial-based telecommunications services, such as location systems

The X band supports the following telecommunications services:

- Military communications
- *FSS* Earth exploration and research, including meteorological studies and research

The Ku band supports the following telecommunications services:

- *Satellite television* Direct broadcast to the home
- Broadcast satellite service
- *Terrestrial microwave transmission* Long-haul applications and terrestrial linking

The K band supports the following telecommunications services:

- Broadcast satellite service
- Fixed satellite service
- Terrestrial microwave transmission

The Ka band supports the following telecommunications services:

- Fixed satellite service
- Terrestrial microwave transmission
- New types of terrestrial mobile services, such as *Local Multipoint Distribution Service* (LMDS), use this band.

Satellite systems can support analog and digital transmission. Satellites multiplex signals using the following standard telecommunications techniques:

- Frequency Division Multiple Access *(FDMA)* A multiplexing technique in which a single radio transmission band is shared by multiple users. Each user is assigned his or her frequency within the band. FDMA is used in the analog cellular world. The users are effectively operating in a non-overlapping channel. FDMA enables the cellular carrier to divide the total spectrum band into discrete channels.

- Time Division Multiple Access *(TDMA)* TDMA is a narrow-band digital multiplexing technique in which the radio channel (not the band) is divided into time slots. Unlike FDMA, TDMA requires that the voice (or information) be digitized. Each channel is divided into three time slots.
- Code Division Multiple Access *(CDMA)* CDMA is a wide-band digital radio technology. The TIA standardized IS−95 CDMA technology requires 1.25 MHz of bandwidth, which is shared by multiple users. The theoretical capacity is 10 to 20 times the capacity of the same bandwidth if multiplexed by FDMA. CDMA digitizes the information and spreads the information stream across the 1.25 MHz bandwidth. The information is encoded so that unless one has the code, the intelligible information stream appears much like random noise spread across the bandwidth. The encryption key enables the service provider to assemble the information into an intelligible conversation or data packet.

Satellite Classifications

Several different types of satellites exist, with the general classes as follows and depicted in Figure 9-7:

- *Scientific* Weather, environmental, cartographical, and stellar observation
- *Communications* Commercial radio, broadcast television, direct satellite television, cable television support, public voice, data, and video
- *Military/government* Various applications of all of these classes

Each one of these classes can be subdivided into orbit types. The orbit classes impact the way in which satellite networks are configured. Applications in the realm of public telecommunications will be discussed later in this chapter. Figure 9-8 illustrates the following orbit classes. Figure 9-8 illustrates the orbit classifications.

- Low Earth Orbit *(LEO)* Orbital altitude of 400 to 900 kilometers
- Medium Earth Orbit *(MEO)* Orbital altitude of 5,000 to 12,000 kilometers
- Geosynchronous Orbit *(GEO)* Orbital altitude of 35,800 kilometers

Figure 9-7
Satellite classifications

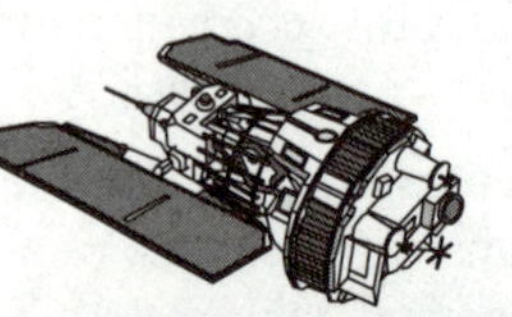

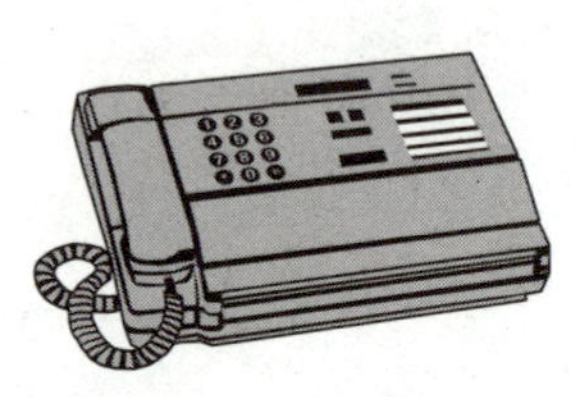

- Science
- Communications
- Military

Figure 9-8
Orbital classifications

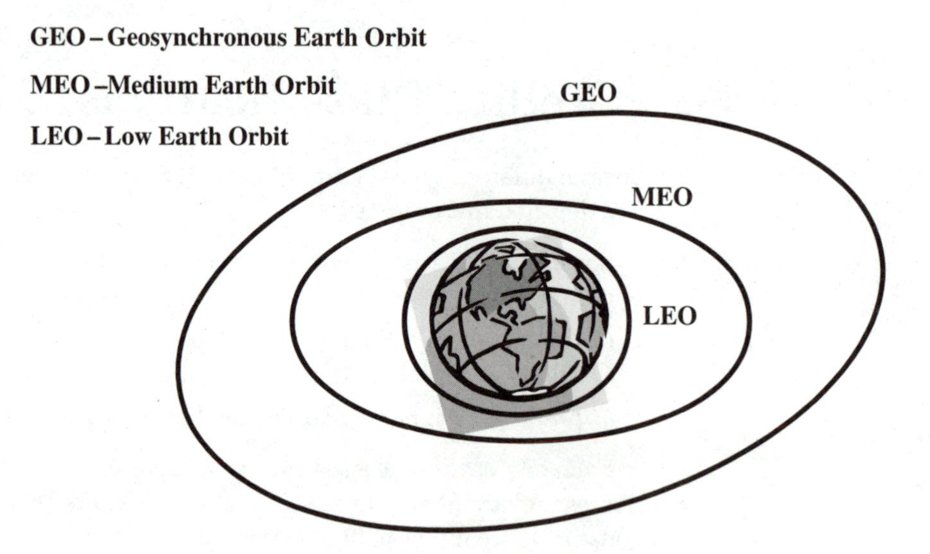

The FCC has created additional classifications for United States commercial and private satellite systems. These classifications are used for regulatory purposes. The satellite industry is categorized and labeled, as is the rest of the public telecommunications industry. The classifications are as follows:

- Fixed Satellite Service *(FSS)* FSS is a satellite industry term that is used to describe telecommunications services that are not mobile or broadcast. This classification applies to satellites that support scientific research endeavors.
- Mobile Satellite Service *(MSS)* MSS is a satellite industry term that is used to describe the struggling and new LEO-supported mobile telecommunications service.
- Broadcast Satellite Service *(BSS)* (BSS) is an industry term that is used to describe the direct broadcast of information to the home. In other words, this service is a direct satellite service. Satellite television is a common example.

Satellite Network Architecture

Terrestrial communications systems can be implemented by using a variety of network topologies: tree, ring, and star. Satellite communications systems are implemented in a fashion that is analogous to the star configuration, where the satellite forms the node of this star configuration. Figure 9-9

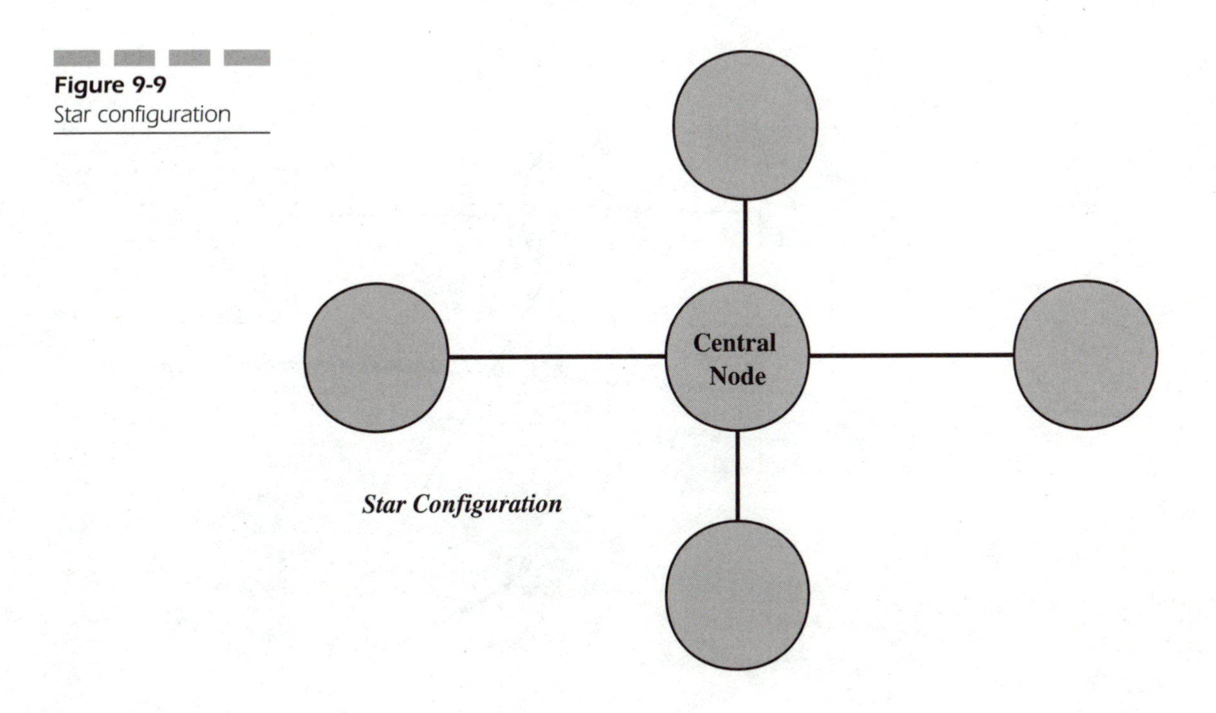

Figure 9-9 Star configuration

illustrates the star configuration, and Figure 9-10 describes the application of the star configuration to the satellite network.

As you can see, the satellite is the node of the star configuration. In a terrestrial network, the nodes of the network are interconnected. As Figure 9-11 indicates, the terrestrial network is interconnected so that the nodes (all

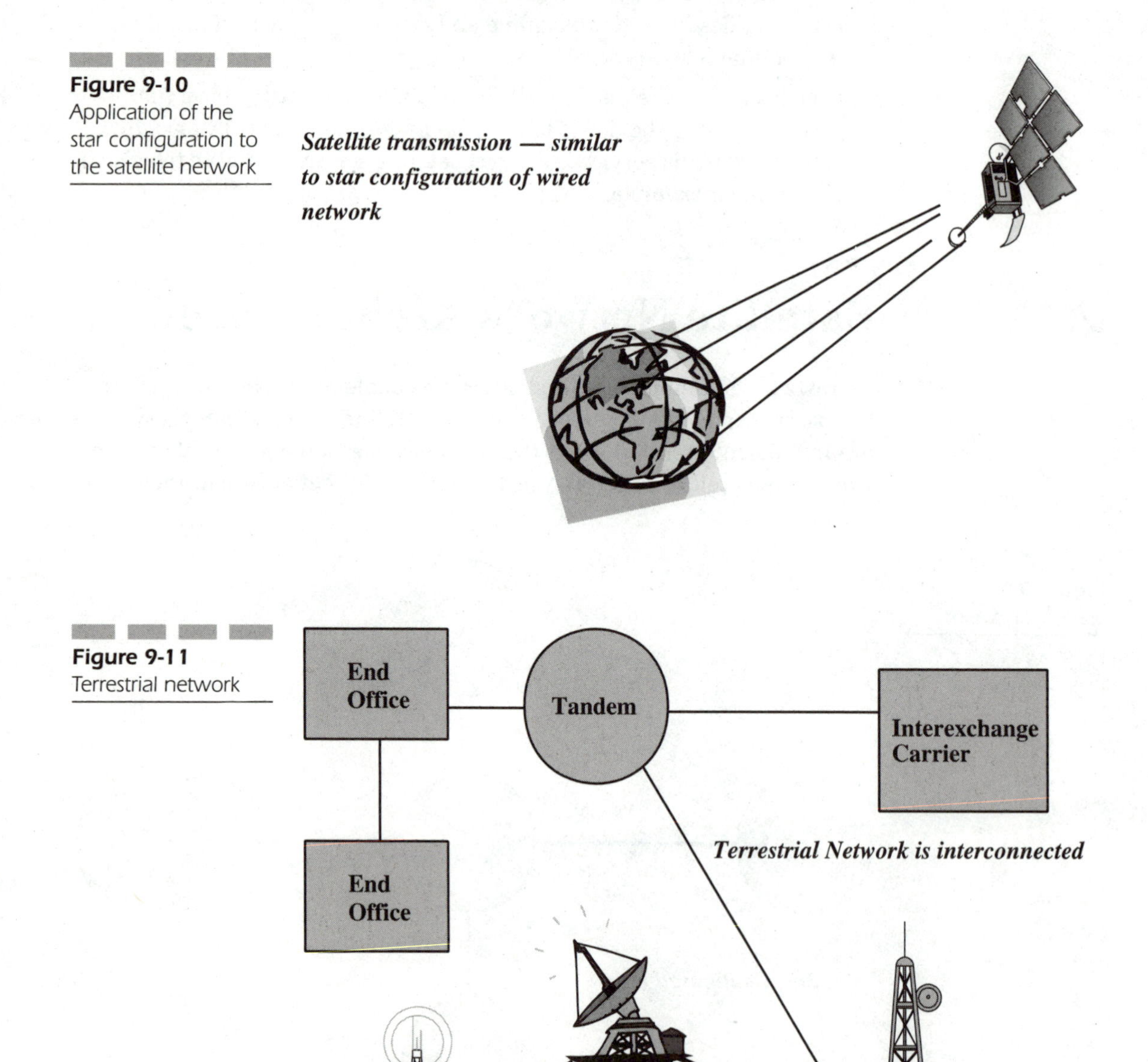

Figure 9-10 Application of the star configuration to the satellite network

Figure 9-11 Terrestrial network

switches) are interconnected. Essentially, the terrestrial network supports alternate routing schemes. The use of the word *terrestrial* applies to wireline, cellular, PCS, paging, Internet, and cable television networks. Satellite networks tend to be single, stand-alone networks. The satellites communicate with terrestrial networks for the purpose of receiving, translating, and transporting the satellite signals throughout the Earth-based networks. Satellites can communicate with other satellites; however, deployments tend to involve single satellites beaming information to the Earth.

The service or coverage areas of a satellite are called *footprints*. The cellular, PCS, and Paging industries also refer to coverage as footprints. These footprints can be quite large, covering whole sections of a country. By contrast, Earth stations uplink information from satellites on narrow radio beams. Figure 9-12 describes the typical relationship between the Earth station and the satellite.

As I noted, the terrestrial network is interconnected. There are few situations in which the terrestrial network has any single point of failure. Switches can fail, and a loss of total communications can result. The redundancy and multiple routes that are supported in the transmission facility network, however, greatly reduce the likelihood of routing failure. With regard to the switching system, terrestrial systems are by far easier to maintain than a satellite. As noted earlier in this chapter, satellites are expensive and are difficult to repair once they are in orbit. In other words, the satellite is the single point of failure of the entire satellite communication system.

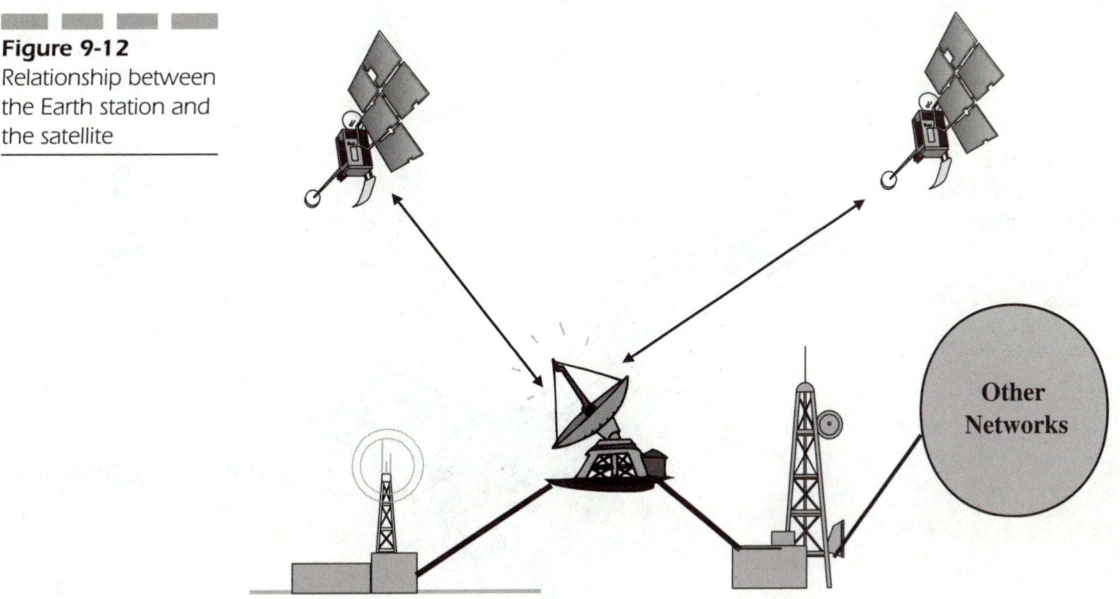

Figure 9-12 Relationship between the Earth station and the satellite

A satellite communications service provider can greatly alleviate this situation by deploying multiple satellites in the network. Possibly, in the event of a satellite failure another satellite can be repositioned to replace the lost satellite. The architectural configuration used by the satellite community is ideal for this operating environment; however, the configuration as deployed lends itself to coverage failure. Figure 9-13 illustrates how the satellite can be the single point of failure. Figure 9-14 further illustrates a configuration using multiple satellites.

Figure 9-13
The satellite can be the single point of failure.

Without the satellite, communication among the points on the globe is disrupted.

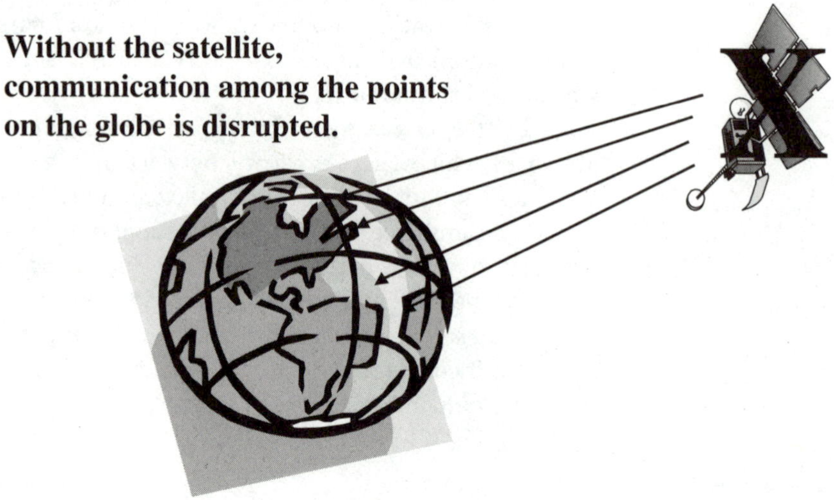

Figure 9-14
Configuration using multiple satellites

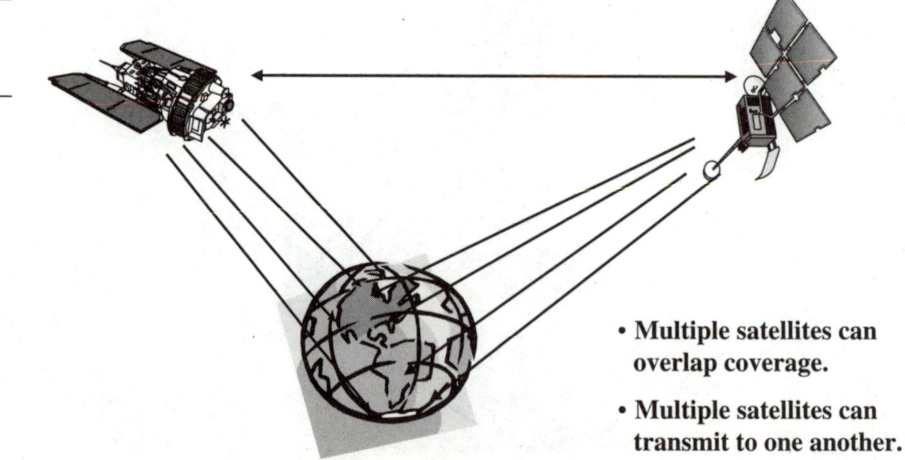

Multi-Beam Satellite Network Architecture

As the reader knows, satellites can beam information directly to one location or to multiple locations. Services such as direct television (i.e., BSS) require the satellite to beam information over a large expanse of land (refer to Figure 9-15). This service can be delivered by using shaped beams. A multi-beam system could be used by a commercial telecommunications company, in which the information streams are being directed to specific Earth stations. Multi-beam configurations are used in order to support frequency reuse and therefore increased channel capacities. Frequency reuse in this configuration is enabled, due to spatial separation. In other words, the various beams of information are physically separated to avoid interference. This technique is the way in which the cellular and PCS networks establish their reuse patterns (via physical separation). Refer to Figure 9-16.

You should note that even in the satellite industry, there are some basic traffic and engineering concepts that are applicable throughout the industry.

Earth Stations

A key component of the satellite network is the Earth station. The Earth station is the device, terminal, or switching center that receives the signal

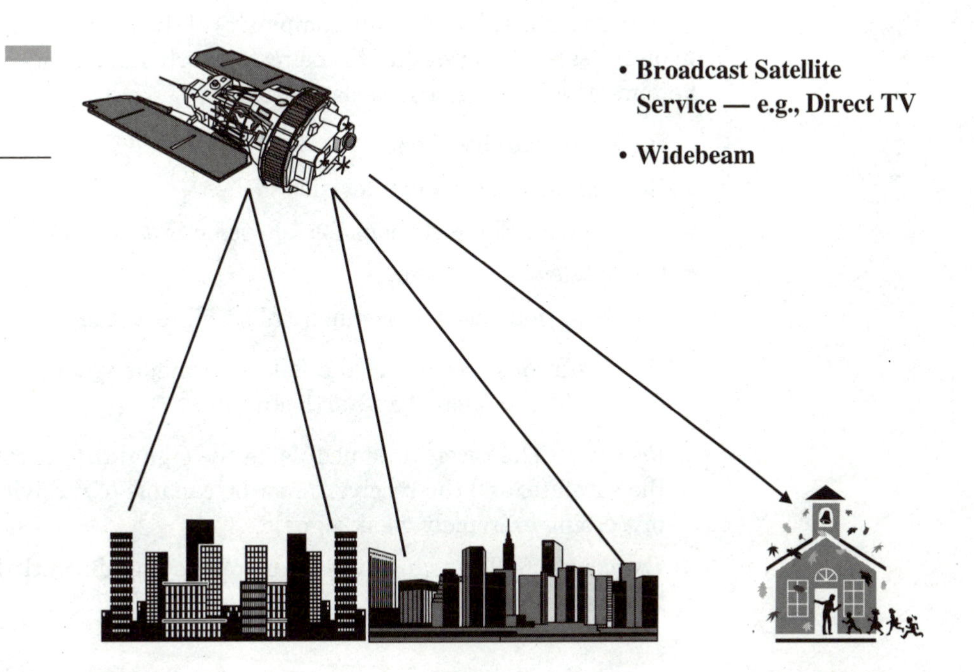

Figure 9-15 Direct television configuration

Figure 9-16 Physical separation in a multi-beam configuration

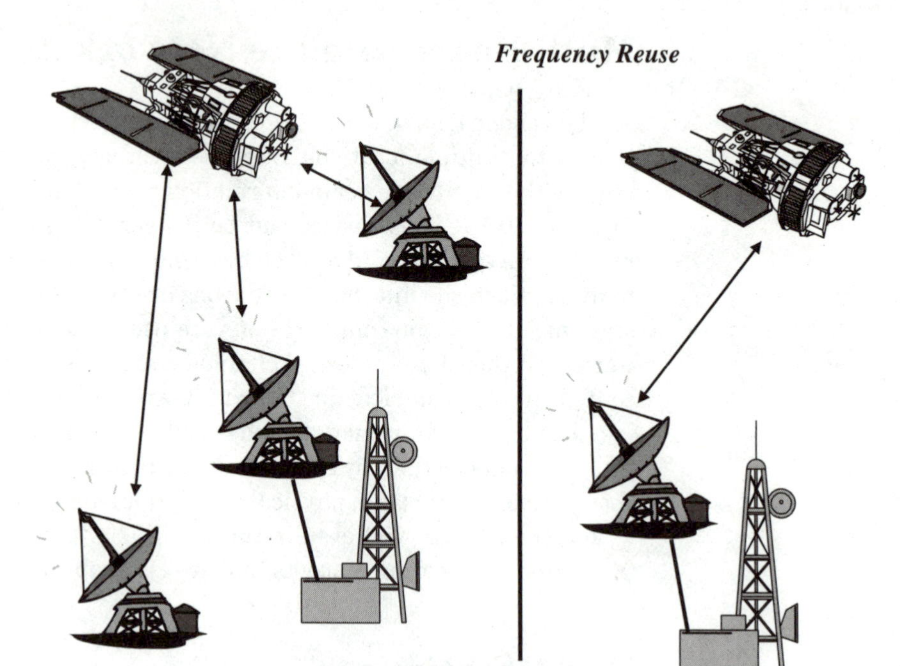

from the satellite. The size and complexity of the Earth station is dependent upon the satellite service that is being supported. Earth stations can be broken into the following usage classifications:

- Backyard satellite dishes
- Rooftop direct satellite television dishes
- Commercial Earth stations for television broadcasting
- Cable television networks
- Earth stations used by commercial satellite carriers

Earth stations cconsist of the following major systems (refer to Figure 9-17 for an illustration of an Earth station):

- *Receiver* The receiver demodulates the signal that is transmitted by the satellite, and the receiver must be capable of receiving and processing extremely weak signals.
- *Transmitter* The transmitter transmits at significantly higher power levels than the receiver.

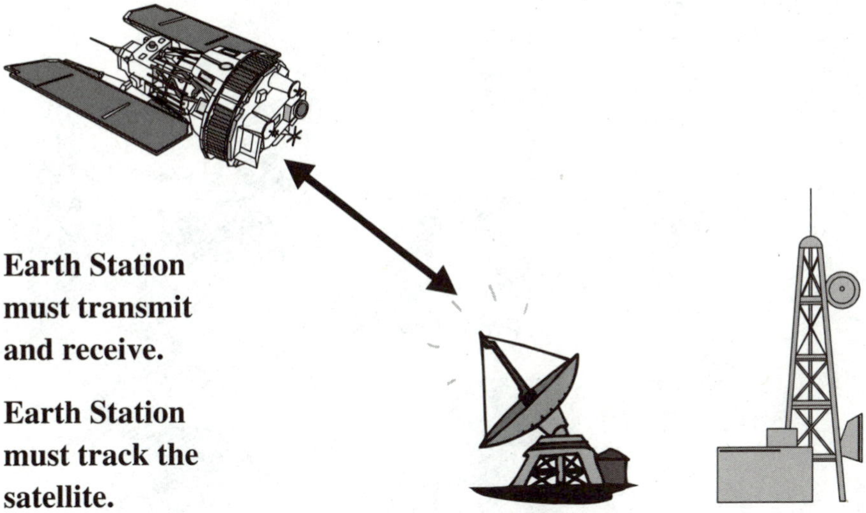

Figure 9-17 Illustration of an Earth station

Earth Station must transmit and receive.

Earth Station must track the satellite.

- *Antenna* The antenna is a critical component, because it supports both the transmitting and receiving functions.
- *Tracking equipment* Tracking the satellite from an Earth station is more appropriately broken into two functions: pointing and tracking. Using the Earth station's antenna requires you to know where to aim/point the antenna. Tracking involves maintaining the antenna's directional fix on the satellite. Tracking can be employed by stationary Earth stations or by moving vessels (e.g., ships, airplanes, trucks, etc.) in order to maintain the fix on the satellite—despite the motion of the vessel and the satellite. Pointing and tracking is dependent on a number of factors, including the following:
    - The antenna beam width
    - The motion of the satellite
    - The motion of the Earth station
    - *The type of Earth station* The type of Earth station determines the type of tracking equipment that is necessary. Motion means mobility; therefore, a mobile Earth station must have the capability to see the satellite—no matter what buildings or foliage might be in the way.

Figure 9-17 is an illustration of the basic Earth station concept. Figure 9-18 is a depiction of how far earth stations have evolved. Today, earth stations can be small satellite dishes sitting on the roof of a house, roof of a truck, or even the front lawn of a home.

Figure 9-18 Additional types of Earth stations

Satellite Access

As with any network, the issue with satellite networks is access to the network. Satellite networks can be comprised of a single link to an Earth station or comprised of multiple links. Satellite networks are either one-way link networks or multiple-link networks. The term *uplink* refers to the telecommunications from the Earth station to the satellite. The term *downlink* refers to the telecommunication from the satellite to the Earth station.

A one-way link is a simple network. In this scenario, the satellite has two antennae and one repeater, and there is one transmitter and receiver on Earth. The satellite does not provide the switching functionality that either an MSC or wireline switch offers. The satellite is similar to a smart relay station. Figure 9-19 illustrates the concept of the one way link.

Satellite broadcast networks involve multiple Earth stations. When a satellite is broadcasting to multiple receiving stations, the receiving stations must be within the downlink's coverage zone. Figure 9-20 represents a one-way link between a single Earth transmission station and multiple Earth reception stations.

Two-way communications links between Earth stations is simply a doubling of facilities that are used in a one-way system. A two-way communication system uses two uplinks, two downlinks, and four frequencies: one can have a single receiving antenna and a single transmitting antenna if

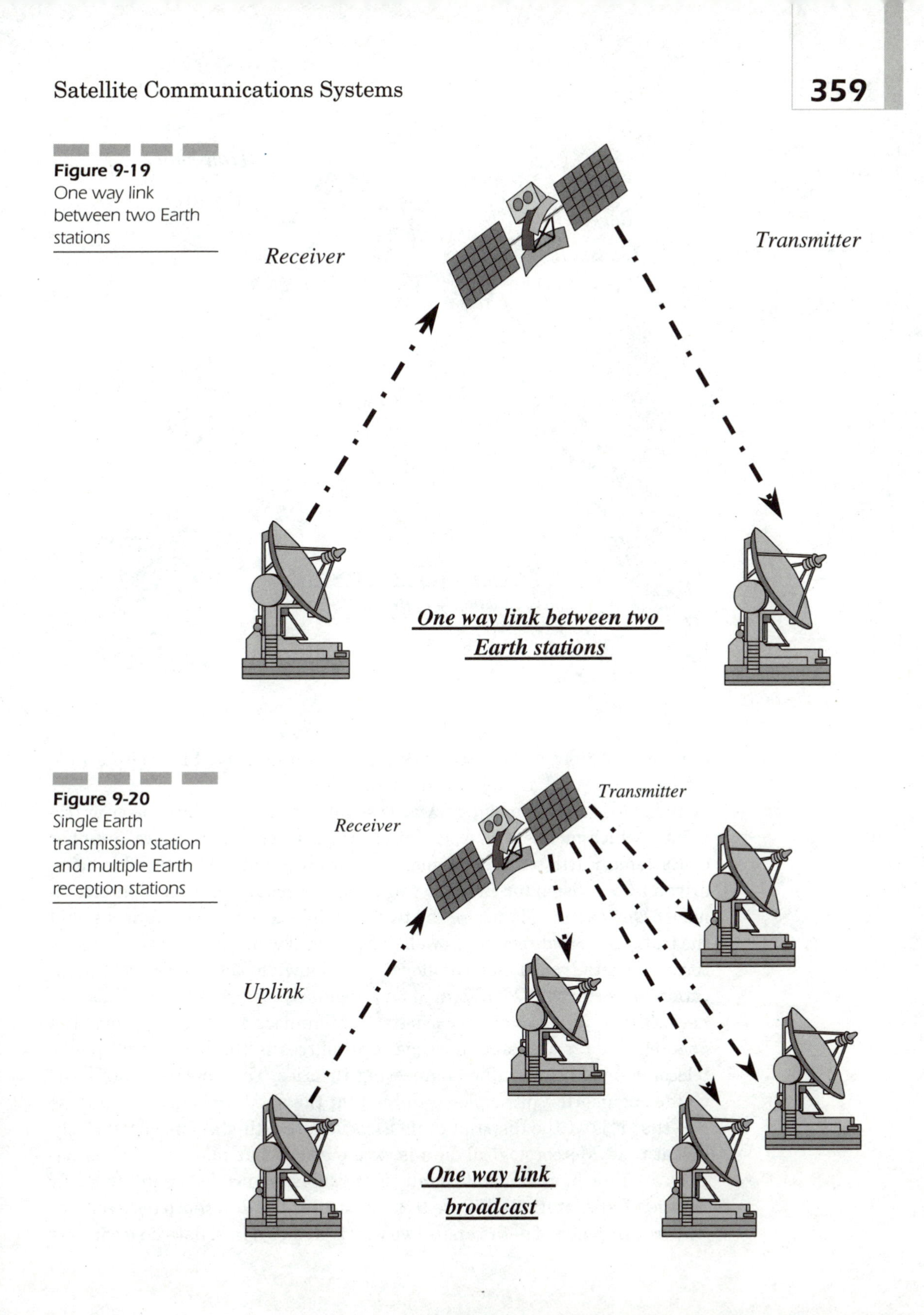

Figure 9-19
One way link between two Earth stations

Figure 9-20
Single Earth transmission station and multiple Earth reception stations

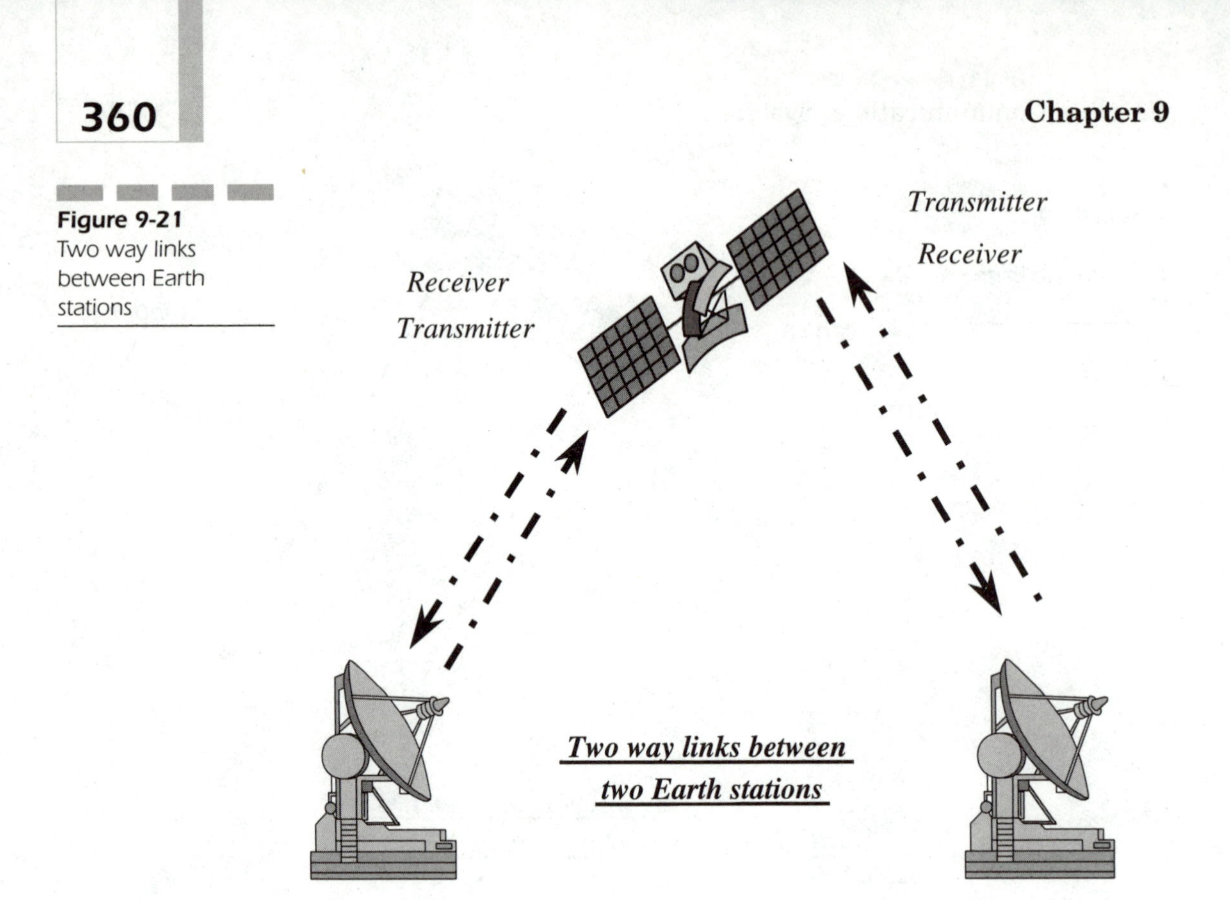

Figure 9-21 Two way links between Earth stations

the Earth stations are located in the same coverage area. Figure 9-21 illustrates two way link between two Earth stations.

Today, the trend is moving towards small, inexpensive, direct-to-user terminals, which means that a satellite needs to interconnect a large number of scattered Earth stations by means of two-way links. This trend is being driven by television (or, in other words, commercially accessible entertainment). One example is where the two-way links are not simultaneous, and the traffic in both directions is well defined (or if waiting time is acceptable, corporate data transactions or global or nationwide paging). The communication links can then be allocated on a regularly scheduled basis. This scenario is also an example of basic telecommunications traffic theory. The queuing of data and shared usage of resources is the basis of all public telecommunications traffic engineering theories. The application of basic traffic engineering principles ensures that the satellite is in constant use and that it is at the disposal of its associated Earth stations successively, two at time. Meteorological data is exchanged in this manner. Such situations are rare, however, and usually many links are needed simultaneously between Earth stations. Figure 9-22 illustrates the difference between one way transmission of data versus two way transmission of data on a satellite

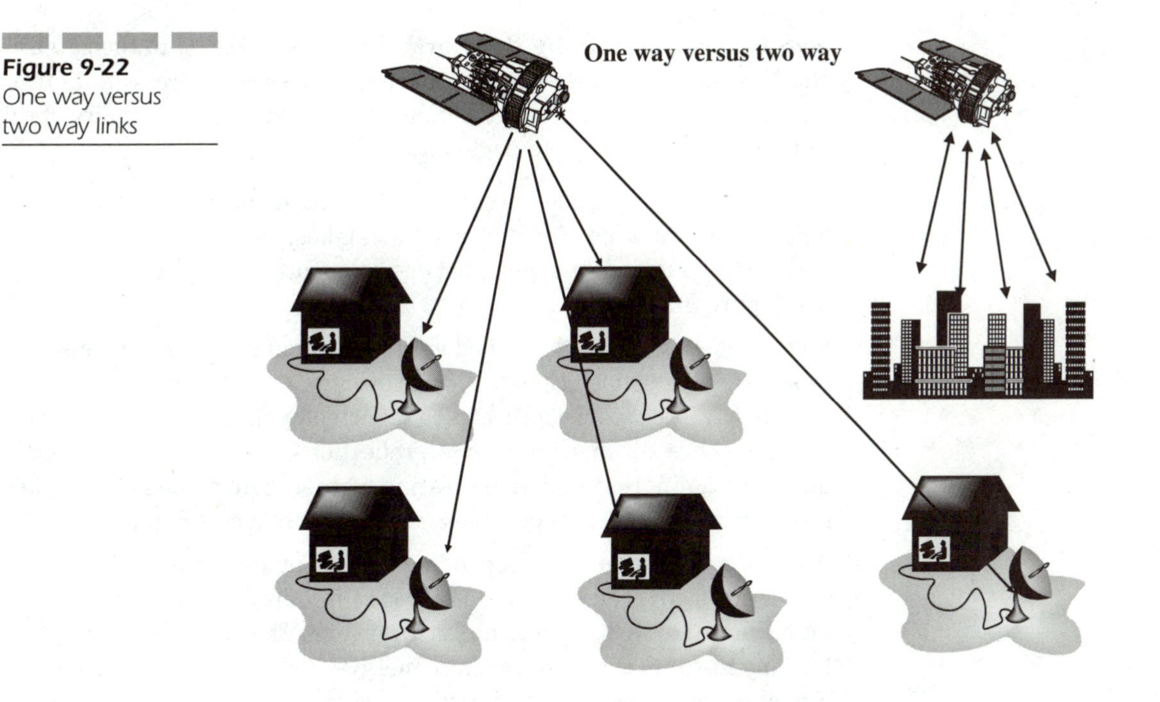

Figure 9-22 One way versus two way links

network. As I have indicated, the directionality is a factor in the traffic engineering of a satellite telecommunications system.

To deploy multiple two-way links, one could simply increase the number of transponders in a system. In a multiple-link scenario, it becomes advantageous to use a single wide-band transponder with appropriate multiple-access techniques. At some point, there is a physical limitation to the number of transponders that one can reasonably fit into a satellite. This technique is how the Intelsat II series of satellites has been engineered. Although one might think that the satellite can simply be made larger, the reality is that satellite payload size is dictated by economics and the economics of the launch vehicle. One can truly state that the money issue lies at the core of every telecommunications business. Regardless of whether the system is in space, one will find that the telecommunications business is driven by the same issues. More information about this topic will appear later in the chapter.

Satellite access has a number of concerns in order to make its capabilities accessible by users:

- Access to the satellite by the Earth stations
- Access to the Earth station by multiple users. For the sake of commercial and public communications, one should envision millions of subscribers seeking access to the Earth station.

Access to the Satellite by the Earth Stations Access to the satellite by the Earth stations does not refer to the permission to access data from the satellite. Access in this case refers to satellite visibility. Visibility of the satellite from the Earth is impacted by the following factors:

- *Objects* Objects include buildings, trees, manmade structures, and topological formations (mountains, hills, lakes, etc.) These objects can block signals or reflect signals between the satellite and Earth station. See Figure 9-23.
- *Signal reflection* Buildings, bodies of water, and manmade structures reflect radio and video signals. Copies of signals are made when reflected. These signal copies can be reflected by manmade objects. The reflections are known as multi-path reflections, and each reflection ends up taking a different path and creating further distortions as a result of interference from the other copies. See Figure 9-23 and Figure 9-24.
- *Atmospheric attenuation* Depending on the angle of arrival of the signal, the signal can be traveling through hundreds of miles of atmosphere before it reaches an Earth station. The atmosphere is just like any medium that can transmit energy—it can refract the signal. Signal refraction can result in degraded signal strength. See Figure 9-25.

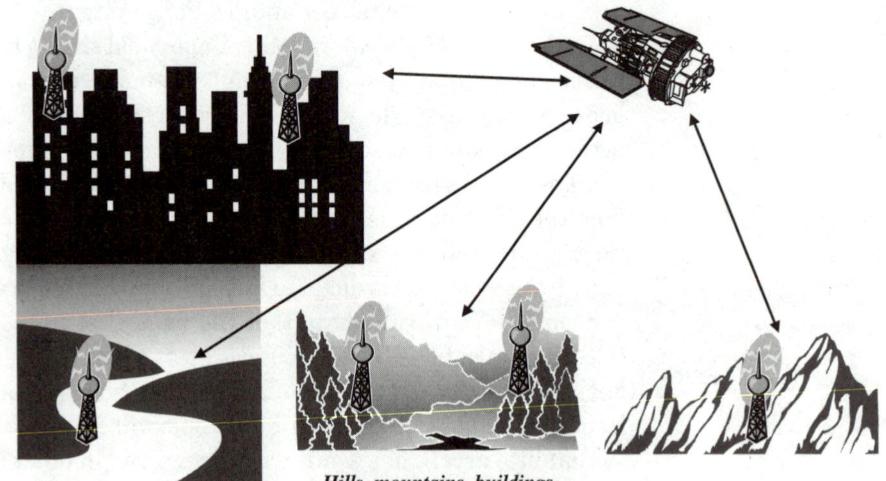

Figure 9-23 Objects blocking satellite signals

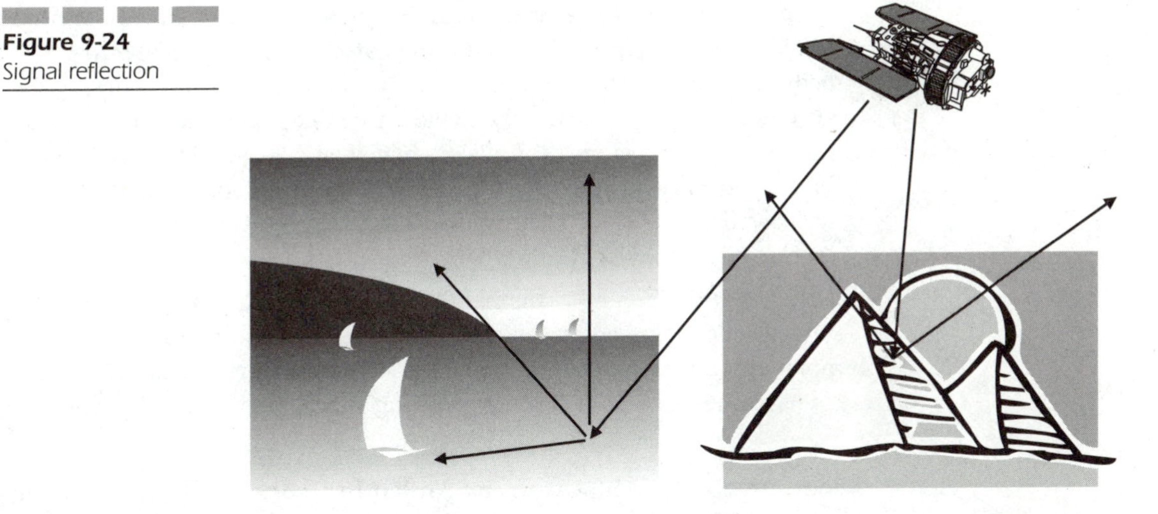

Figure 9-24
Signal reflection

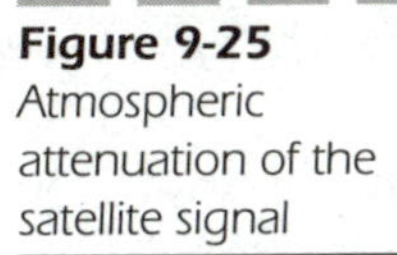

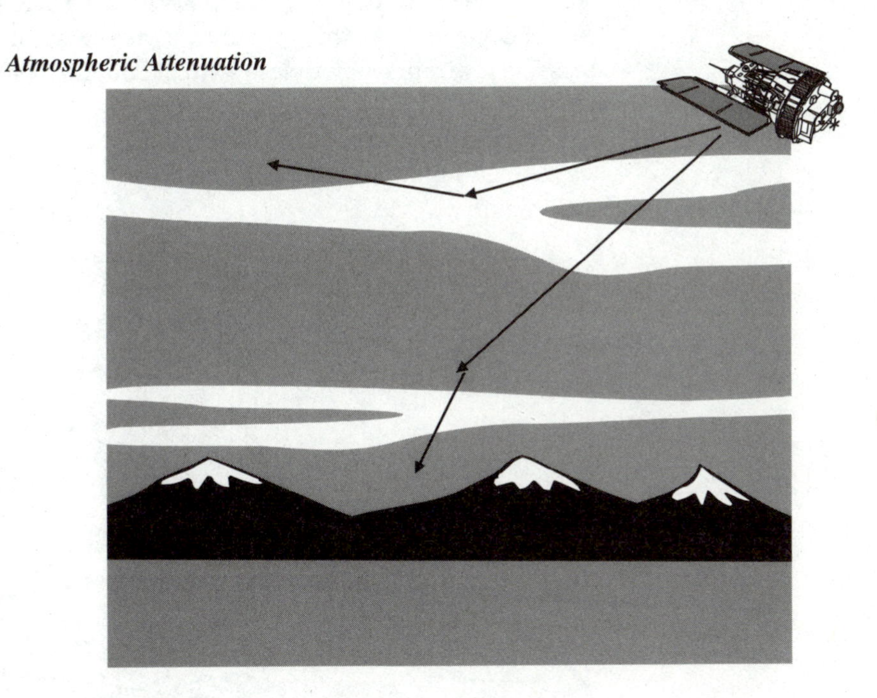

Figure 9-25
Atmospheric attenuation of the satellite signal

- *Orbital arcs* The orbital arc of the satellite impacts all of these factors. Satellites are launched into specific orbital paths around the Earth, and these paths are dictated by physics and governmental permission. The paths are not perfect circles around Earth, and for that matter, the Earth is not perfectly round. In fact, the Earth is ovoid in shape, so the orbital paths are elliptical in shape. The elliptical orbit is influenced by the Earth's gravity, and satellites are like planetary bodies. The orbits of satellites operate under the laws of physics, specifically Kepler's Laws. By global agreements between the various national space powers, orbital slots have been assigned to specific arcs in the geosynchronous orbit. Each government assigns slots in the arc for specific purposes. The arcs affect the Earth stations' pointing and tracking angles. See Figure 9-26.

- *Solar storms* Solar storms result in a level of solar magnetic interference. Sunspot activity is a result of solar storm activity. The magnetic interference is so high that it causes signal disruption for the satellite networks that are orbiting Earth. See Figure 9-27.

Access to the Earth station by Multiple Users As I had noted, one should envision millions of subscribers seeking access to the Earth station. The reader can view this situation as being similar to millions of users accessing a wireline or wireless switch. The Earth station can be any of the following:

- A direct broadcast television system, where each home has a satellite dish antenna. Figure 9-28 is an illustration of a direct broadcast TV system.
- A commercial Earth station that is linked to a wireline switching system. This wireline switching system converts the signals into usable information for subscribers. One example is the satellite

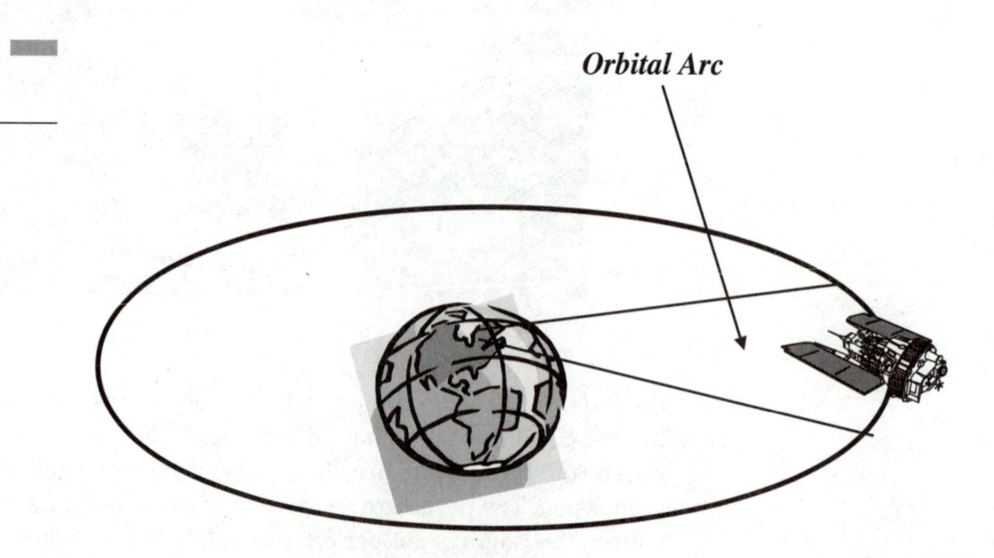

Figure 9-26
Orbital arc

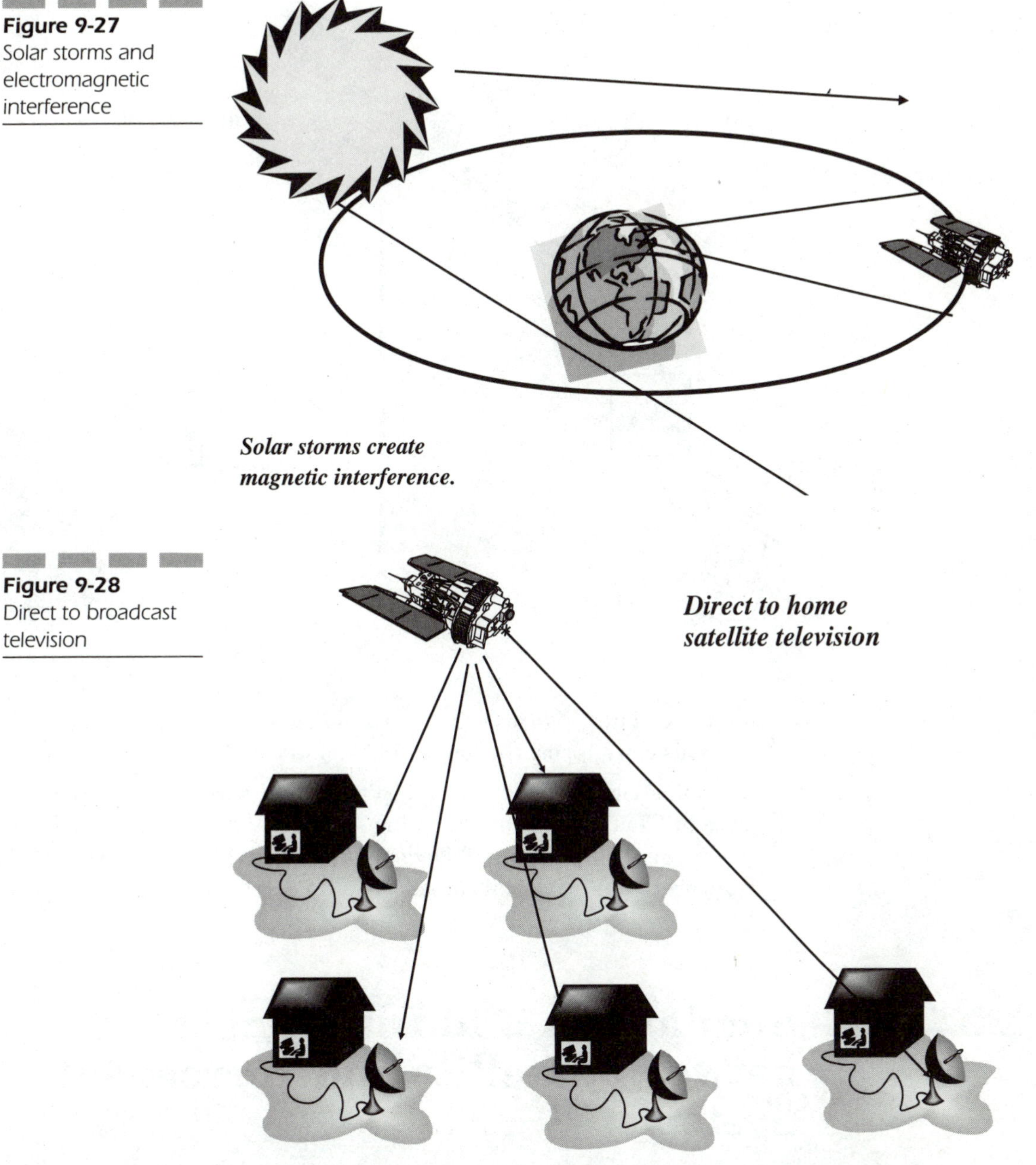

Figure 9-27 Solar storms and electromagnetic interference

Figure 9-28 Direct to broadcast television

gateway that is used by international long-haul telecommunications for the purpose of transporting a call or an information stream from one continent to another. This application of the satellite will be

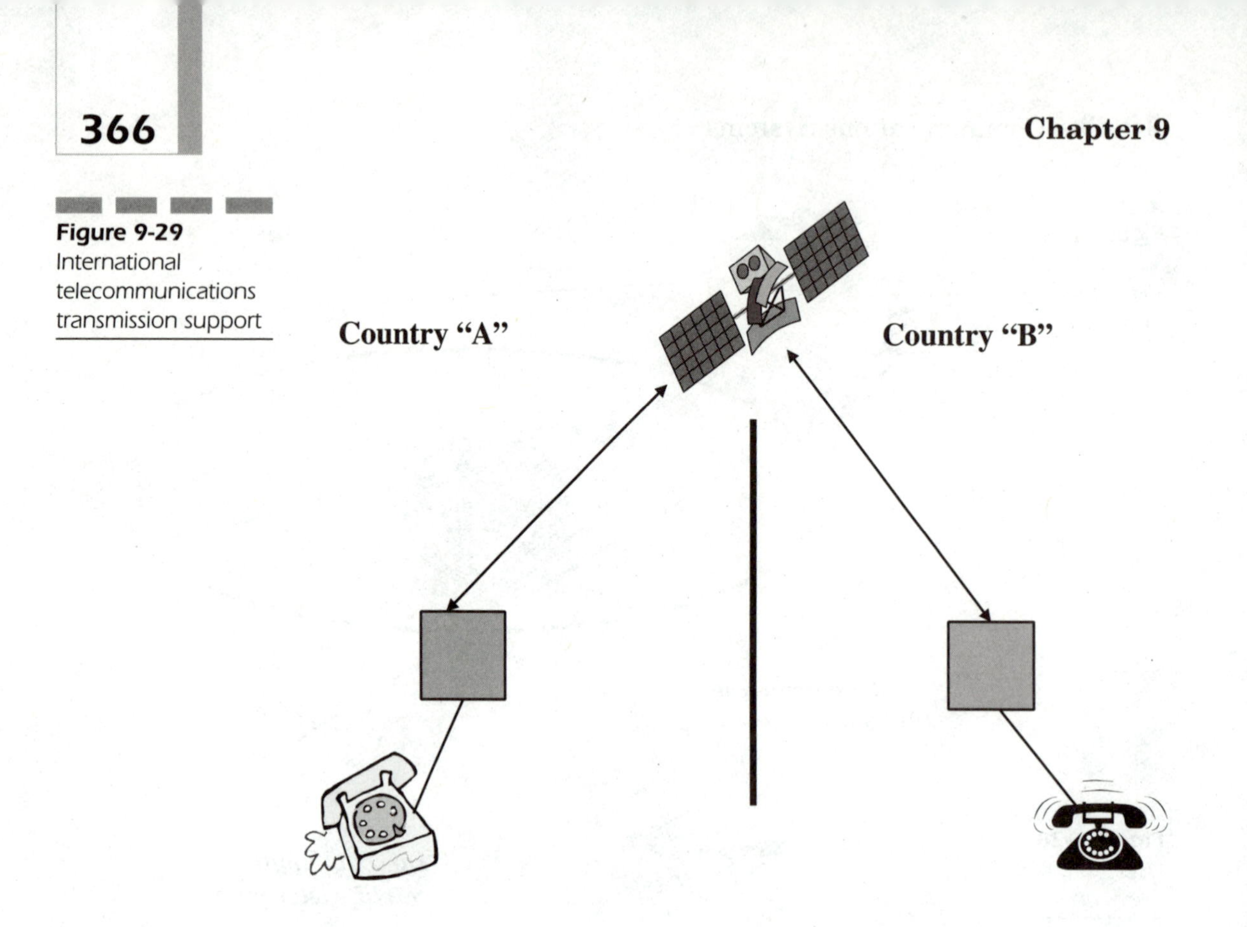

Figure 9-29 International telecommunications transmission support

discussed later. Figure 9-29 illustrates how satellite transmission systems can support the long-haul business.

- A commercial Earth station linked to another wireless broadcasting system. This wireless broadcasting system can be a radio station or a television station. Figure 9-30 depicts how satellite networks can support commercial radio broadcast.

Similarities and Differences between Satellite and Terrestrial Systems

Satellite networks have their own unique characteristics, while at the same time they share many similarities with networks in a wireline carrier and cellular/PCS carrier environment. Basic traffic engineering concepts are similarly applied in both space and terrestrial environments. We will start by describing the differences.

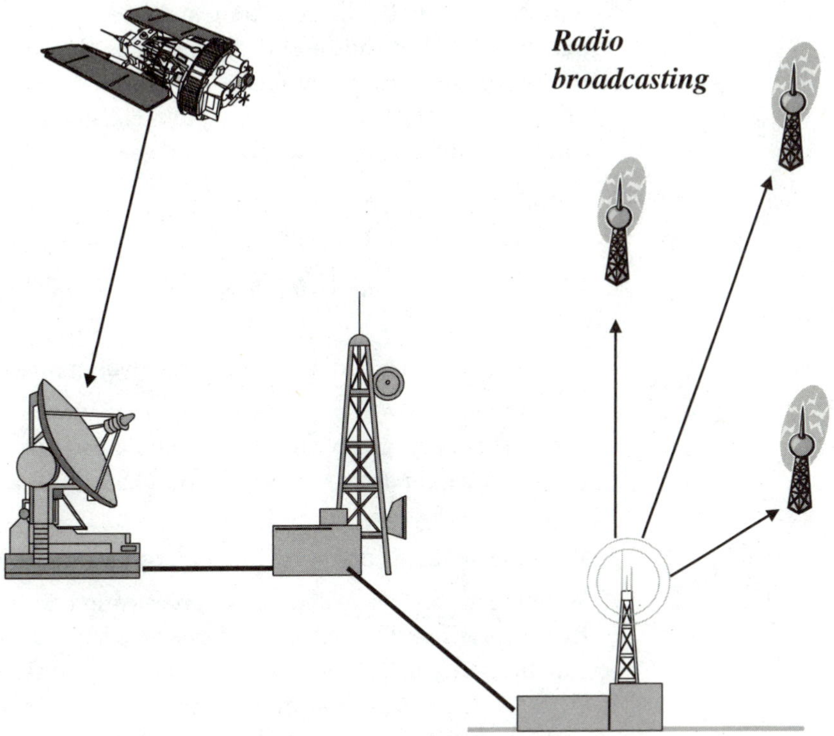

Figure 9-30
Commercial radio broadcast support

Radio broadcasting

Differences

Satellite carriers face the following unique operating challenges:

- Satellites continuously operate in harsh environments with no air and no atmospheric pressure. In sunlight and in shadow, the satellites are subject to temperature extremes.
- Satellites are traveling through space at high rates of speed. Space exposes the satellites to harsh operating conditions, including collisions with objects in space (such as micrometeorites and manmade garbage. Manmade garbage can include lost tools, other satellites, and garbage.
- Satellites are launched into space.
- Satellites can only carry a limited power supply. The larger the power needs, the larger the power supply.
- Satellite carriers cannot repair their satellites if there is any damage to them.
- Satellites do not process calls in the same manner as the terrestrial switches. Satellites are more akin to relay stations.

- Satellites are not call-processing switches in the sense that a user makes a call and the satellite determines the calling party's billing profile, service profile, or special routing considerations. Satellites do not apply vertical services to satellite calls (vertical services include call forwarding, caller identification, voice mail, etc.).

Wireline carriers, cellular carriers, and PCS carriers face the following unique operating challenges:

- Wireline switches and *Mobile Switching Centers* (MSCs) operate in the atmosphere.
- Wireline switches and MSCs do not travel in space.
- Wireline switches and MSCs are supplied power by external sources. Internal energy sources are used when main power is deactivated. These internal power sources are batteries and carrier-owned electric power generators.
- Wireline switches and MSCs can be repaired by humans.
- Wireline switches are capable of processing higher data speeds than a satellite. The biggest difference is the bandwidth transmission capability (in other words, throughput is the difference). Wireline switches process calls at higher speeds than even the MSCs. Today, wireline switches can process calls in the gigabit speed range. The wireline carriers are limited by the medium in which they transmit information. The largest capacity physical-transmission medium used by terrestrial switches is fiber optics. Although satellite transponders (when grouped together) can handle speeds in the gigabit range, the limits of fiber optics remain to be seen when coupled with compression and multiplexing techniques. The following list provides a quick review of the capabilities of fiber optics:
  - Fiber optics can support transmission speeds/bandwidth in the high GHz range. Fiber optics advantages are as follows:
    - Fiber optics can support the following data speeds: *Optical Carrier* (OC)−1, 3,12, 24, 48, and 192.
      - OC−1 equates to a data speed of 51.840Mbps.
      - OC−3 equates to a data speed of 155.520Mbps.
      - OC−12 equates to a data speed of 622.08Mbps.
      - OC−24 equates to a data speed of 1244.16Mbps.
      - OC−48 equates to a data speed of 2488.32Mbps.
      - OC−192 equates to a data speed of 9953.28Mbps.

- Large bandwidth applications (e.g., video, Internet video, and audio applications)
- It has limited access locations. Not all locations lend themselves to gouging out a large trench in order to lay outside plant.
- Transoceanic cable

Fiber optics advantages are as follows:

- Non-susceptibility to electromagnetic interference. Glass is a non-ferrous/non-magnetic material.
- Ground returns are eliminated. All electrical systems need to be grounded, and grounding establishes a difference in electrical potential between electrical conductors. Light does not need to be grounded.
- *Small size* One strand of fiber (the diameter of a human hair) can carry data at speeds in excess of 100 gigabits per second.
- *Light weight* A fiber strand is as light as a human hair.
- Low maintenance requirements
- Glass does not age as fast as metal.
- Glass does not corrode.

Similarities

The similarities between satellite and terrestrial systems are as follows:

- Both systems multiplex information.
- Both systems support traffic queuing. The underlying goal of queuing calls or other types of information is that a carrier cannot install a relay or circuit to support every single caller. The assumption by which all carriers design their systems is that a certain percentage of users will be seeking the resources of the system at any given time. If a carrier were to install a system that had the physical circuit configuration to support six billion simultaneous calls, that carrier would be out of business—because the cost of such a switch would break the company. Traffic engineering principles support both financial and technical objectives. The underlying financial assumptions are the primary drivers behind the technical consideration. One must remember that every system inside a carrier network is built around certain underlying cost assumptions. Without the financial structure in place, no system can ever be built.

- Both satellite carriers and terrestrial carriers are concerned about siting of their systems. In the case of the satellite carrier, the concern is orbital slot for the satellite and real estate for the Earth stations. In the case of the terrestrial carriers, the concern is real estate for switches, transmission facilities, and towers. In a sense, the issue is the same: real estate for siting.
- Both satellite carriers and terrestrial carriers must maintain skilled workforces to operate their respective systems.
- Both satellite carriers and terrestrial carriers interconnect their networks.
- Both satellites and terrestrial switches route calls.
- Many networking principles are similar, and this topic will be addressed in the next subsection.
- Satellite networks and terrestrial networks (specifically cellular, PCS, and paging) are concerned with coverage. Coverage will be addressed separately in the next subsection.

From a macro perspective, there are many similarities and differences between terrestrial and space networks. The similarities and differences have been leveraged in order to create synergies for the two types of telecommunications networks. The synergies have resulted in enhanced service for each industry segment while creating new opportunities that are combinations of both terrestrial and space systems. This subject will be further explored in the applications section of this chapter.

Common Networking Principles

Networking is a broad term. For the purposes of this chapter, networking refers to the interconnection of network elements in order to support the processing and transportation of a call or an information stream. Networking concepts are similar in both terrestrial and satellite networks. The basic configuration used by the satellite network is the star-shaped network, where the satellite is in the center of the star. Wireline networks and other terrestrial networks can assume several different basic configurations: ring, tree, or star. When one takes into account the interconnection of the Earth stations to either a wireless broadcast network or a wireline network, the satellite network takes on a layer of complexity that one would not see if they did not look beyond the satellite network. Figure 9-31 illustrates the

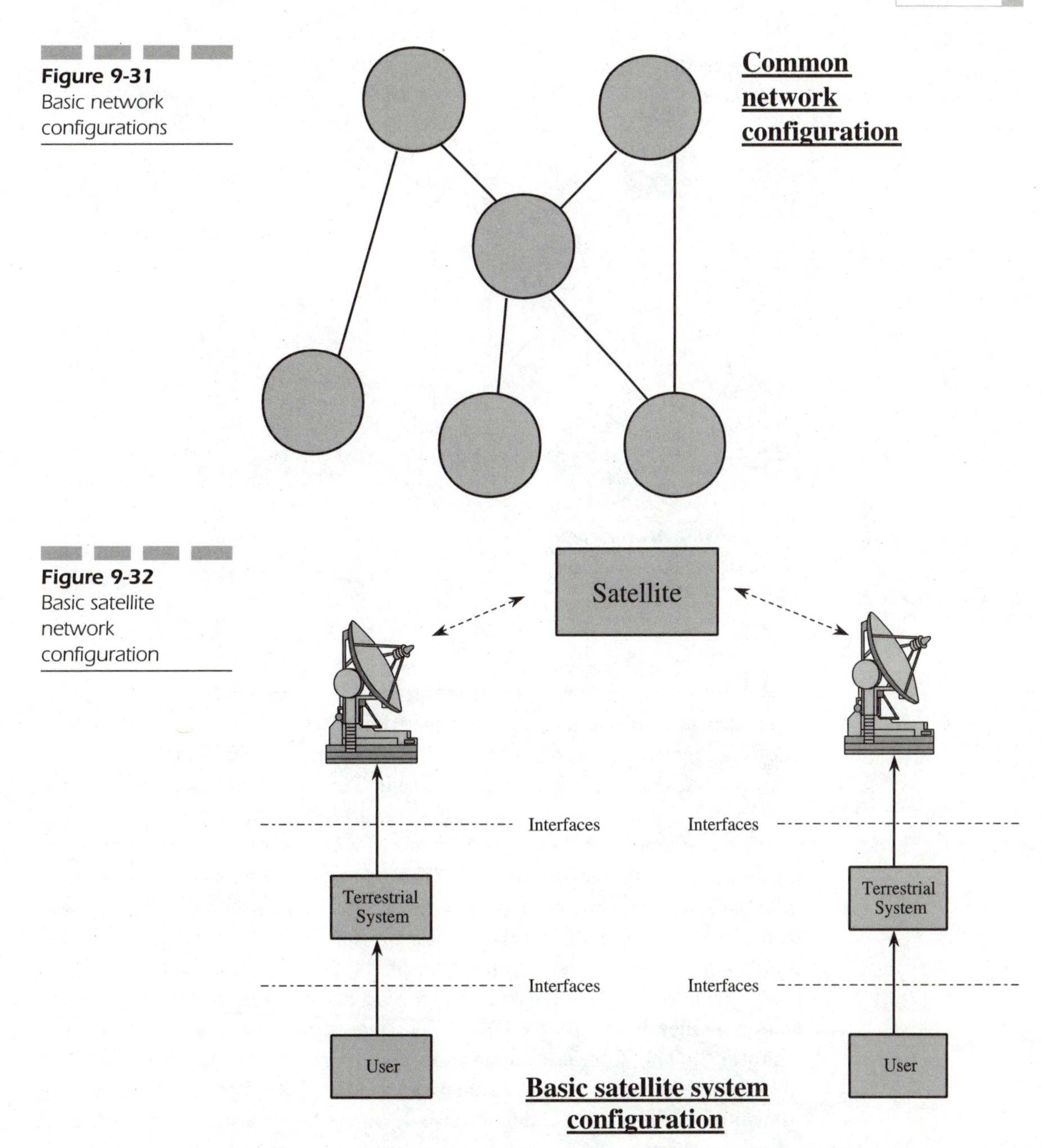

Figure 9-31 Basic network configurations

Figure 9-32 Basic satellite network configuration

basic network configurations, and Figure 9-32 highlights the basic satellite network configuration. Figure 9-33 takes an expanded view of satellite network interconnection.

Figure 9-33 Expanded view of satellite network interconnection

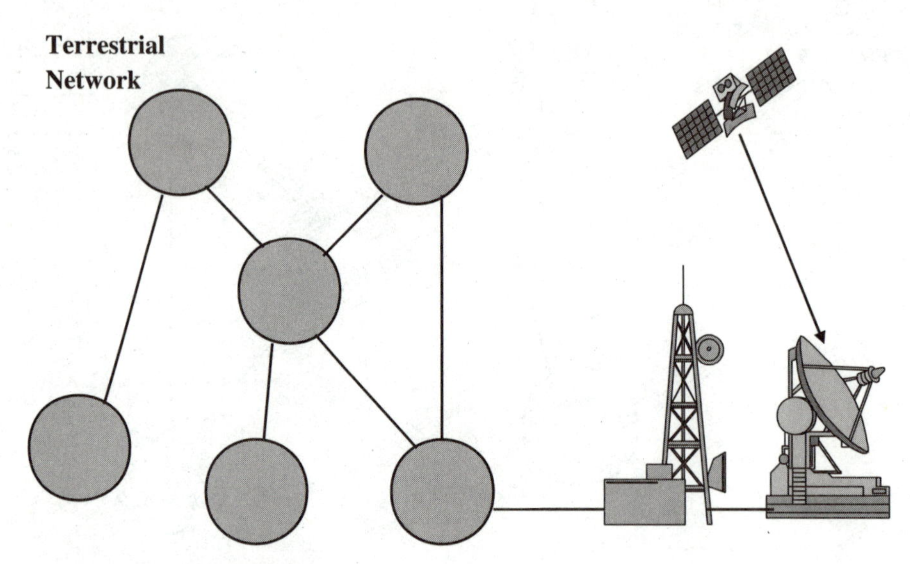

Satellite networks are not just stand-alone systems. In fact, satellites can communicate with one another, essentially by bouncing signals off one another. Signals can also be transmitted from the satellite to the Earth station and back to another satellite in orbit. This capability to pass information from point to point is not unlike how a terrestrial network will utilize wireline tandem switches and microwave relay points to long-haul information across the nation or globe. The terrestrial network will transport information from a switch to some type of regional switching center in order to begin the process of carrying information from the local market. The regional switching system then seeks some type of transmission network to carry the information across the region or country. Figure 9-34 illustrates this capability. In Chapter 1, "What Is a Telecommunications Network," and Chapter 2, "The Telecommunications Hub—Creating Value," we described networking as the interconnection of devices for the purpose of transporting information. This description is broad, but it is an accurate depiction of what networking performs.

The reader can view satellite networks as relay networks or as long-haul networks in space. A satellite network carries information from one point on the Earth to another point on the Earth. Despite the fact that the satellite throughput is limited by comparison to that of a fiber optic link, satellite transmission avoids many of the problems that physical transmission

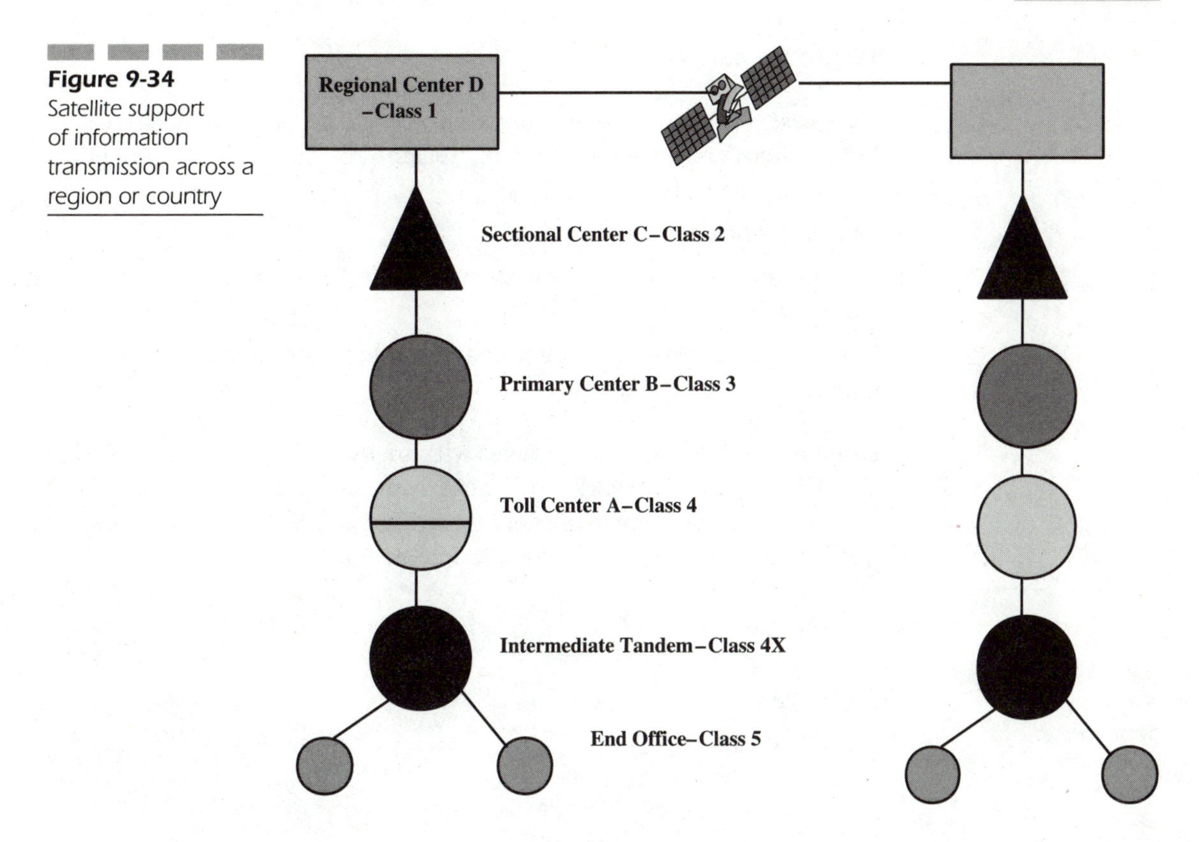

Figure 9-34 Satellite support of information transmission across a region or country

systems encounter (such as problems that are encountered by metallic, optical, and terrestrial microwave transmission media).

Metallic transmission media has a number of disadvantages:

- Because this form of media is made of metal, it is subject to electromagnetic interference of all kinds.
- Metal ages and will deteriorate.
- Metal is heavy to transport.
- Metal requires large storage areas.

Fiber optics disadvantages are as follows:

- Expensive for low-traffic load needs
- Requires special electronics for the light source, light detection system, encoders, and decoders
- Susceptible to signal attenuation (loss)
- Suffers from signal dispersion

Terrestrial microwave disadvantages are as follows:

- *High cost* The cost of deploying a microwave system can be described in two ways: political and financial (leasing real estate for towers is not an inexpensive proposition).
- Transmissions are not secure.
- Ambient environment (i.e., weather) can affect the quality of transmission.
- Line-of-sight transmission requires multiple towers to traverse a market.

Despite the obvious disadvantages with metallic or optical transmission media, the satellite networks still need some form of physical media to interconnect to the terrestrial networks. Figure 9-35 illustrates the role played by metallic or optical transmission media.

Figure 9-35 Terrestrial transmission media in a satellite network

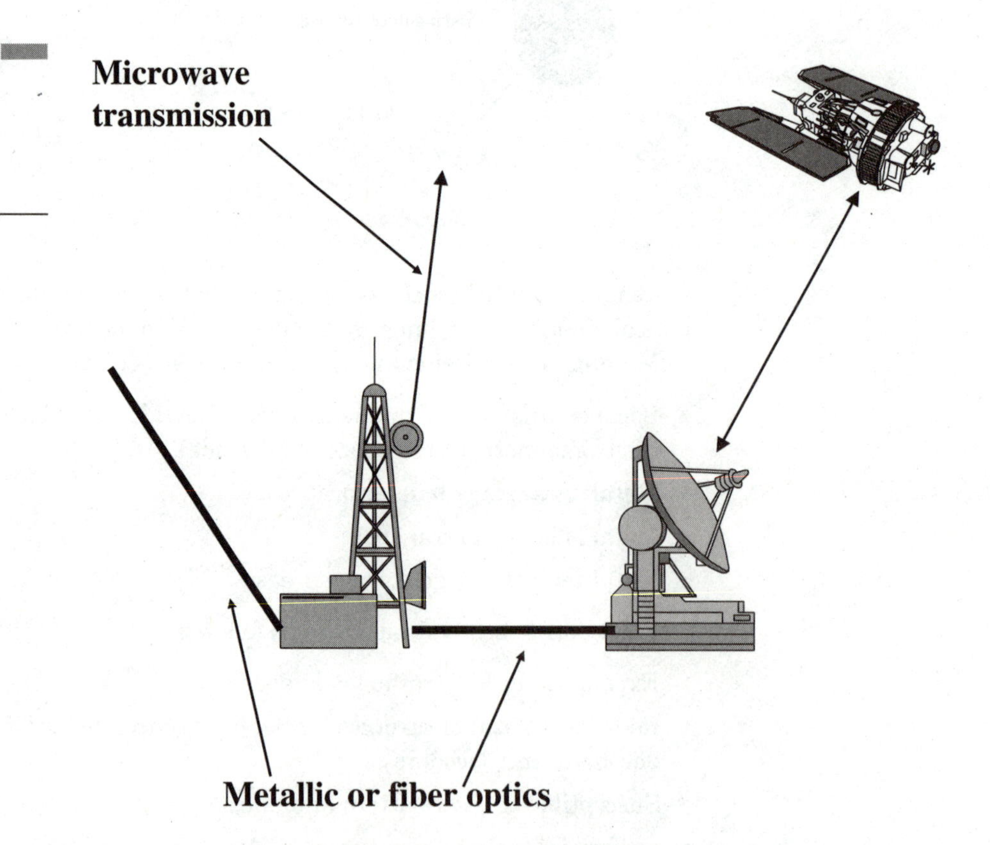

Figure 9-36 Satellite and terrestrial network interconnection

The satellite beams information to multiple Earth stations.

The Earth stations need to be interconnected to other networks to move information throughout the Earth.

Figure 9-36 highlights the point that terrestrial networks need to be interconnected to the satellite telecommunications networks.

Despite the difficulties of deploying and managing physical transmission facilities, these facility types are needed to interconnect the satellite carrier's Earth stations to the other public and private networks in operation. No matter what one might believe, as it relates to maintaining a stand-alone LEO telecommunications network (for consumer use), the LEO subscribers still need visibility to other networks. Without the capability to interconnect the Earth stations to other networks, the LEO subscribers are islands of communications. Therefore, the Earth stations must and do support standard terrestrial interconnection procedures and signaling protocols. Figure 9-37 depicts the LEO configuration's need for multiple satellites

If one takes the view of visibility one step farther, the satellite network can be likened to a type of long-haul network that is interconnected to another long-haul network (or even to a local network). In some sense, the satellite network fits into the network hierarchy as a Class 1 type of switching entity. The satellite can serve as both a nation-wide long-distance network element or as an international gateway network element. This application will be discussed in the next section of this chapter. Figure 9-38 stresses the role of the satellite network in an international long haul traffic network.

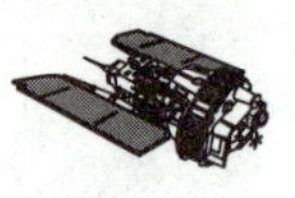

Figure 9-37 LEO networks—multiple satellites

The satellite beams information to handsets.

The Earth stations need to be interconnected to other networks to move information throughout the Earth.

LEO networks use numerous satellites.

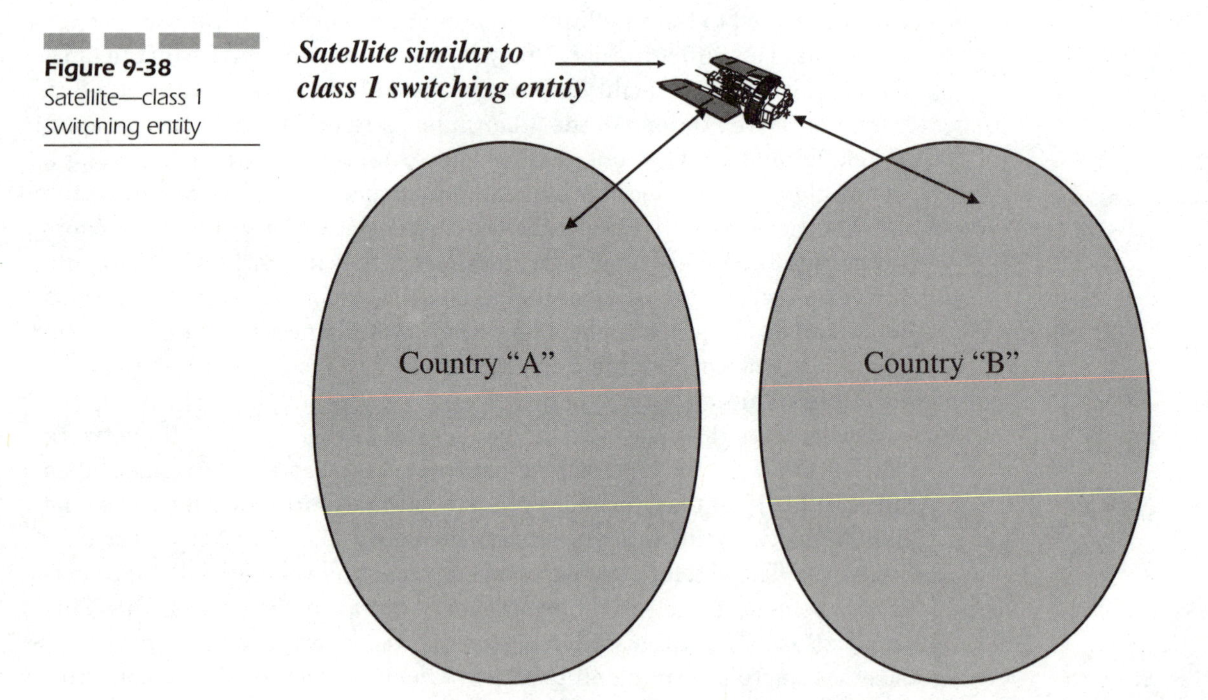

Figure 9-38 Satellite—class 1 switching entity

Applications

When one thinks of satellite services, one typically thinks of stand-alone satellite services such as direct broadcast satellite TV, weather satellites, satellite phones, satellites used for scientific research, and military communications. Except for direct broadcast satellite TV, none of the other services are explicitly observed or used by the everyday, average subscriber. The subscriber might be the beneficiary of these satellite capabilities, but these services are not subscriber services as I have defined them. Subscriber services would include services such as CCS-based services, voice mail, fax mail, call forwarding, caller identification, and call waiting.

Satellites are particularly well-suited for data that is graphics heavy. Satellites are also more cost-effective than T1 for running business applications that involve weekly or daily downloads of large files. The problem with satellite access is throughput, however. Rates vary but tend to top out around 400 Mbps. The rate is high, but it could be higher. Data compression helps. More importantly, most satellite technologies are broadcast-only—where users need a modem for uploads. The following list describes standardized satellite services/capabilities that are currently being provided by this segment of the telecommunications industry. These are standardized industry terms, and the principle space radio communications services are as follows:

- Fixed Satellite Service *(FSS)* This service addresses communication between Earth stations at specified, fixed points via one or more satellites (e.g., Intelsat).
- *Mobile satellite services* This service provides communication between mobile Earth stations and one or more space stations, or between mobile Earth stations via one or more space stations. Earth stations can be situated on board ships, on board aircraft, and on board terrestrial vehicles. This service might also be used to detect and locate emergency signals from people in distress. This service has been expanded to include LEO-mobile handset satellite services.
- Broadcasting satellite service enables sound and visuals to be received by individuals or by communications via satellite.
- Earth exploration satellite service enables observation of the Earth for various purposes, such as weather or geological.
- Space research service, where spacecraft are used for scientific or technical research

- Space operation service is concerned exclusively with the operation of spacecraft.
- Radio determination satellite service, for the purpose of determining the position and velocity of an object by using one or more space stations
- Amateur satellite service for radio amateur use, carrying out technical investigations, and learning about intercommunications
- Intersatellite service provides links between artificial Earth satellites.

These terms can be redefined from a consumer perspective. The reader can categorize all of these terms into basic marketplace categories. The reader should maintain the commercial focus of even space radio communications. We are no longer living in a world where money (billions of dollars) can be spent on pure research. The average consumer and voter will no longer tolerate vast expenditures unless there is a defined payback. The average investor will not gamble millions or even billions of dollars unless there is a defined product. The commercialization of space was long in coming and is not a bad thing, but rather a realistic approach to managing limited resources. The scientific pursuit of knowledge for the sake of knowledge is not a bad thing. The reality of the situation, however, is that the money has to come from somewhere, and in a world in which people are suffering, the words "return on investment" take on a more immediate sense of urgency. The United States has already won the space race. Now is the time to utilize the knowledge that has been gained for the benefit of mankind. For the purposes of this book, the benefit of mankind is defined as "meeting marketplace needs." The following categories define a marketplace view of satellite services:

- *Entertainment* Includes television programming and radio programming
- *Education* Includes distance learning and research
- *Data transmission* Includes data transmission to support businesses, financial transactions, and video
- *Public telecommunications* Includes wireline, cellular, PCS, paging, Internet access, and other types of wireless telecommunications

Entertainment

Entertainment includes all visual and audio forms of media, including the news, situation comedies, dramas, regularly scheduled broadcasts, one-time media events, concerts, sporting events, movies previously released in theaters, and made-for-TV movies. These forms of entertainment are primarily shown in broadcast television (commercial and free TV), cable television (pay TV), and radio. Satellite application of entertainment communications includes direct-to-home transmission of programming.

Broadcast (over-the-air) television and cable television distribute their programming in a wireline network manner. The distribution of television signals is done in a multipoint/multicast manner, where a few locations serve as a transmission center to multiple locations. In the wireline, cellular, PCS, and paging environments, a switch is responsible for communicating information to multiple handsets or to terminal devices. Radio programs are an excellent example of broadcasting to multiple customer-owned terminals. Broadcast television refers to the broadcasting of television signals from terrestrial-based television towers. Figure 9-39 is a rendering of how satellites play a role in entertainment.

Multicast/multipoint distribution of information is a cost-effective way of communicating information to multiple locations. Distributing information in this manner requires the distributor to know and understand the shared usage of network components. In other words, the distributor must optimize the effectiveness of a single element by using it for either multiple purposes or for multiple users who have the same purpose. Figure 9-40 illustrates the

Figure 9-39 Satellite and entertainment

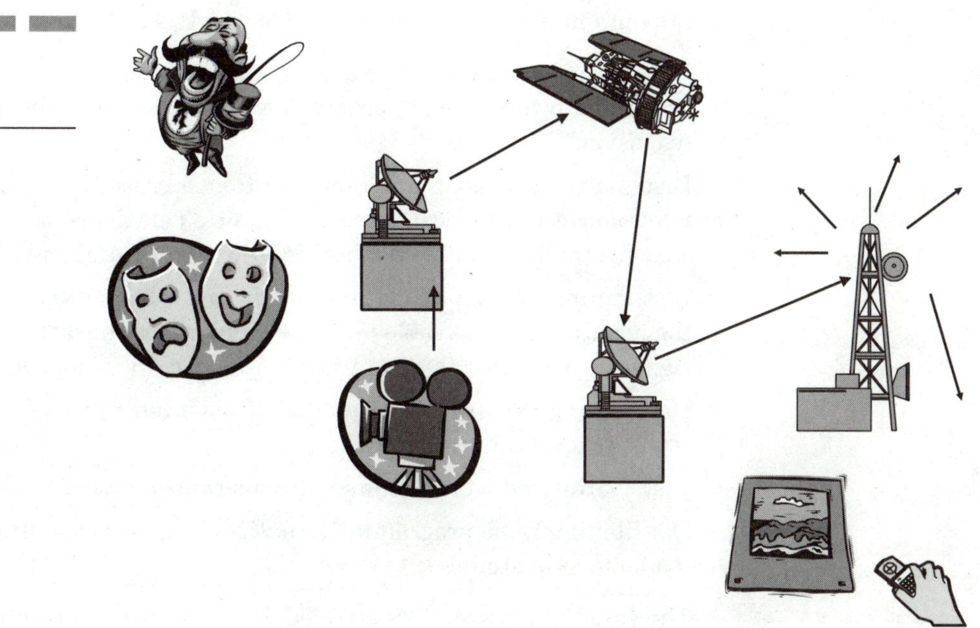

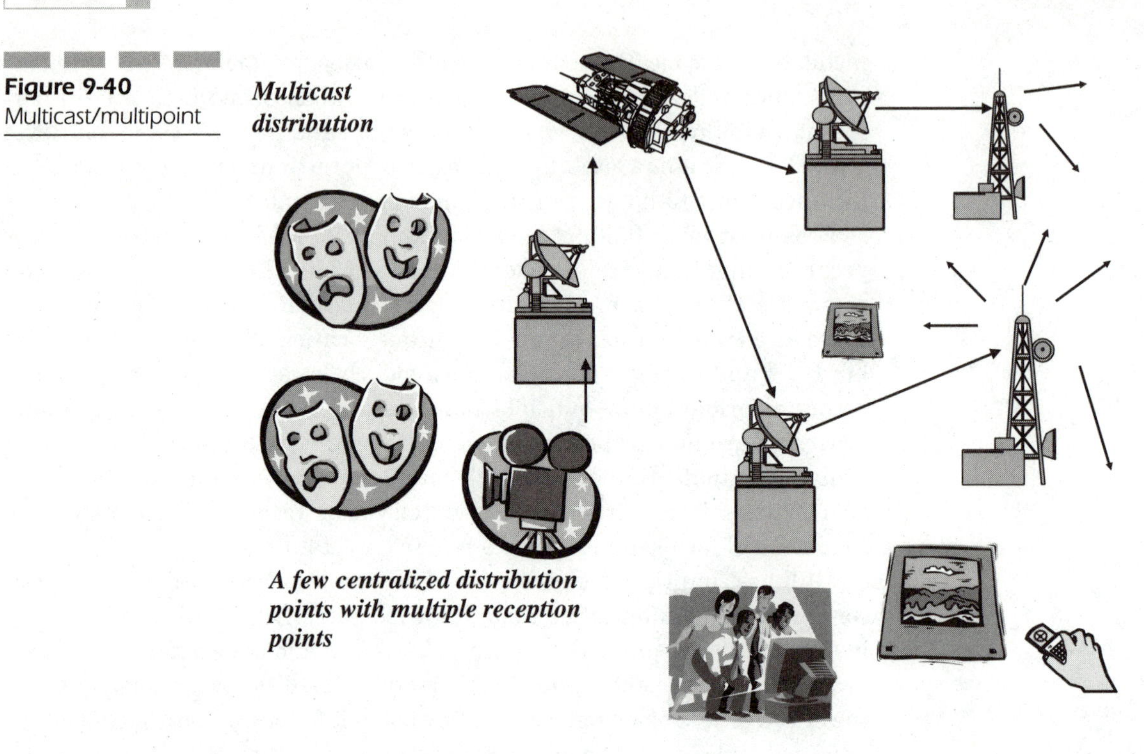

Figure 9-40 Multicast/multipoint

multicast/multipoint distribution system. In the world of television, the multipoint/multicast concept can be described as follows:

- Distributing television programming over the airwaves (television broadcast towers) from a local television studio to the customer-owned television
- Distributing television programming from a broadcast network-owned television studio to local television studios. This distribution is akin to a wireline tandem that distributes information to local end offices.
- Distributing television programming from a cable television studio to the home or business cable television subscriber. The distribution of such signals is done over physical transmission facilities (not over the air).
- Distributing television programming from a centralized cable television studio to local cable television studios
- Distributing radio programming from a central studio to listeners
- Distributing radio programming (national radio shows) from a central studio to local studios

The satellite network has enabled entertainment to be broadcast over a wide area in real time, thereby improving the efficiency of the transmission of television signals. Prior to the application of satellite technology to television, the transmission of television signals over the air via towers encountered the same problems that today's cellular and PCS carriers have encountered:

- *Terrain* Hills, mountains, rivers, lakes, deserts, etc. all impact the quality of the over-the-air broadcast signal.
- *Siting* Finding a site for a tower is subject to political, environmental, and financial constraints.
- *Towers require maintenance.* Maintenance is a cost factor in operating a terrestrial broadcast network. Nationwide television networks must bear this cost factor in mind.

The satellite network facilitates the broadcast of television signals in the following ways. The reader should not forget that no single technology is without its own unique concerns:

- Terrain is no longer a factor.
- *Siting* Hundreds of towers are no longer needed to transmit a signal across the country.
- Fewer towers mean reduced tower maintenance costs.

Education

The ability to reach classrooms across the nation and around the world has enabled educators to communicate and teach students from remote locations. Having a teacher in a classroom in person is best; however, the ability to invite guest speakers to classrooms has been made possible via satellite networks. For the adult who is working and is earning a living, yet wishes to further his or her education and therefore create new opportunities, distance learning has facilitated going to school and not losing any time traveling to a school or campus. Distance learning provides the following benefits:

- Broadens the pool of information that is made available to students
- Facilitates communication between schools of different regions. This situation is good in the sense that the students can interact with people of other cultures.
- Facilitates the working person's efforts to continue their education

Data Transmission

Data can be transmitted via physical facilities, such as copper or glass. Satellite networks facilitate rapid global and nationwide transmission of data, however. This data could be used to support scientific observations of astronomical phenomena, global environmental research, weather observations, or financial data. The data can be almost any kind of information.

A single satellite transponder can transmit at a rate of about 60 Mbps, which can support high-speed data transfers. Financial data, such as credit card transactions and bank transfers, are handled by satellite. Inventory control functions for large corporations are handled via satellite networks.

The Internet has caused the satellite industry to begin deploying high-capacity multimedia satellite networks. Large-capacity multimedia systems enable the satellite networks to carry high bandwidth-demand technologies. Work has begun on interconnecting satellite networks with ATM Earth stations and switches. Work is also underway for onboard processing of information that supports high-bandwidth applications. Onboard processing enables a telecommunications company to more easily process calls and to transport calls/information. Onboard processing also eliminates the need for signaling interworking.

Public Telecommunications

Public telecommunications encompasses wireline, cellular, PCS, paging, and the Internet. Satellite technology has served as a form of national and international glue by facilitating the interconnection of various types of networks. The satellite technology has facilitated the growth of public telecommunications throughout our planet, bringing people together in even the remotest corners of Earth. How public telecommunications has been affected will be further explored in the next section.

Access—An Application

We have just described categories of satellite applications. In this section, we will take a closer look at specific applications of satellite technology. As I have indicated, many of the problems that terrestrial-based systems encounter are not issues related to satellite technology. In fact, world-wide access is the satellite network's single greatest capability. Unlike the other telecommunications technologies, satellite networking is capable of reaching nations and people regardless of the lack of a wired/terrestrial infrastructure. Wider access is an attribute that can be translated into the following applications. Figure 9-41 depicts the concept of access in satellite telecommunications networking.

- Long distance voice and data services
- International voice and data services
- Global broadcast of television
- National broadcast of radio
- Global broadcast of radio
- Direct-to-home television
- Satellite telephony

Satellite transmissions cover huge expanses of land, and satellite networks use far fewer physical transmission media than terrestrial networks. Satellite networks are excellent transmission or distribution networks, but they were not meant to overlay vertical services for subscribers. Satellite

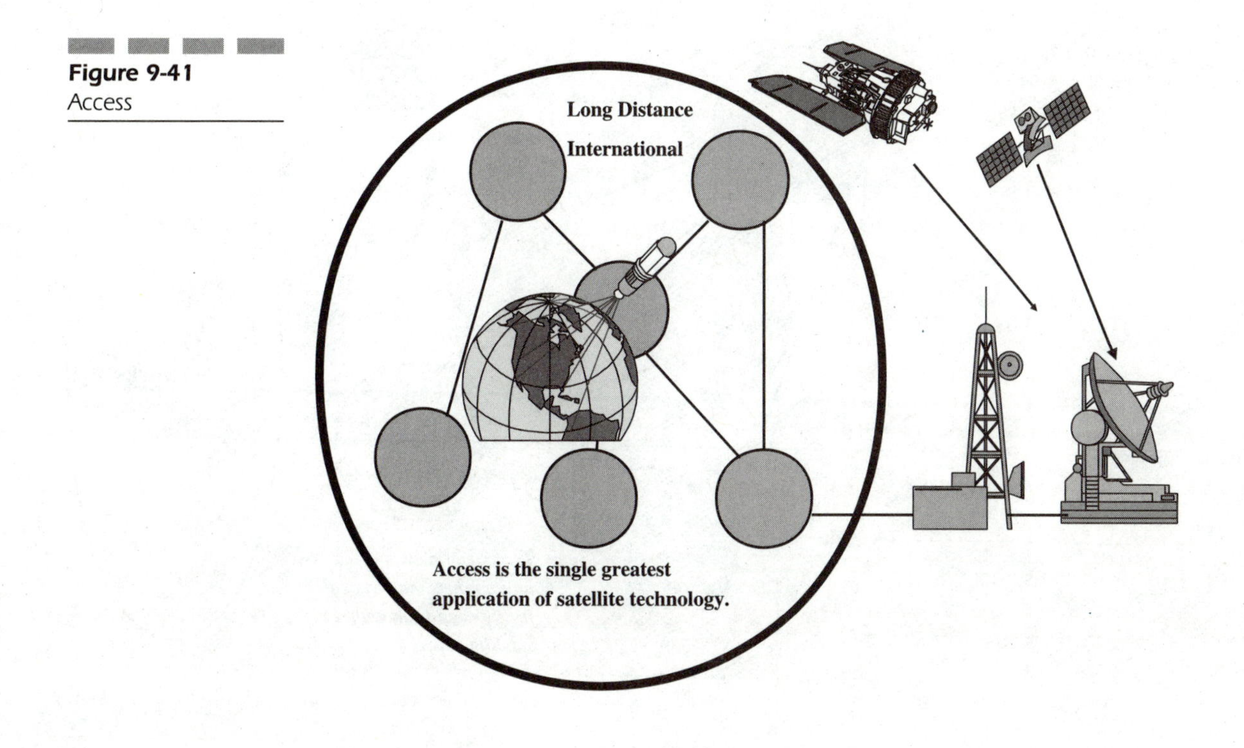

Figure 9-41
Access

networks are access networks that facilitate access to people, homes, businesses, and nations.

Long-Distance/International Voice and Data

Voice and data services that are provided by terrestrial network service providers are being supported by satellites, thereby facilitating nationwide transmission and network interconnect. Wireline, cellular, PCS, and paging carriers use satellites by either interconnecting their networks to one another or by interconnecting internal network elements. Figure 9-42 illustrates this point.

Wireline, cellular, PCS, and paging carriers need to interconnect if they are to ensure that their respective subscriber bases have visibility to other subscriber bases. Satellite networks can serve as backbone networks for these types of carriers, and this backbone network function can be used to support nation-wide and international access (refer to Figure 9-43). As the reader will recall, the satellite network element serves as a Class 1 type of hierarchical element.

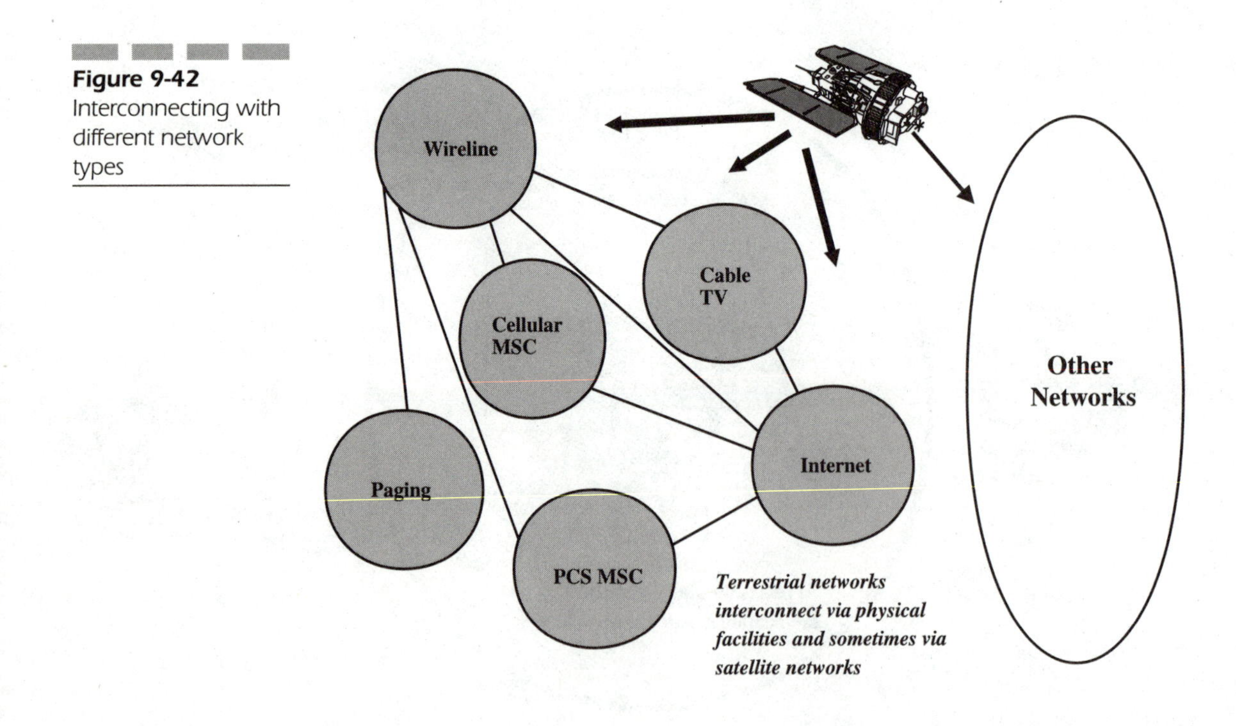

Figure 9-42 Interconnecting with different network types

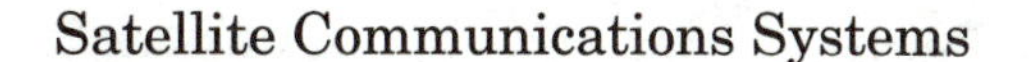

Figure 9-43
Backbone network function

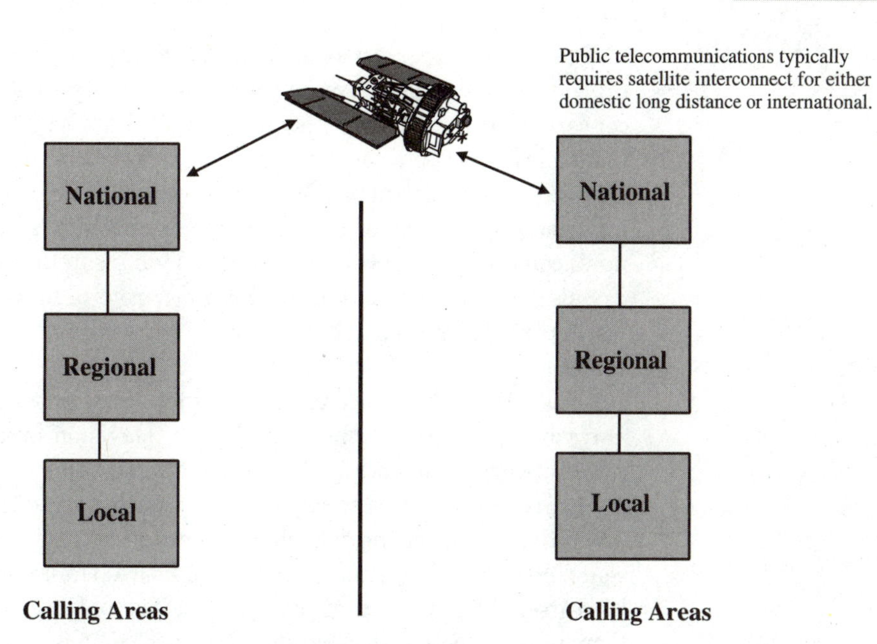

Television Broadcasting: Over the Air and Cable

Over-the-air broadcasting of television programs operates in the *Very-High Frequency* (VHF) and *Ultra-High Frequency* (UHF) bands. Note that this programming is the so-called "free TV" programming that most viewers watch. The programming is paid for by product commercials. The satellite network is used to support over-the-air broadcasting by transmitting the television signal that is received from a central television studio Earth station. The satellite transmits this signal to multiple Earth stations, which in turn transmit the signal to one or more local television transmitters. These television transmission distribution points transmit the signals to the television sets, and this transmission path can be reversed so that the local station can transmit back to the central television studio Earth station for the purposes of rebroadcasting to the nation or world. Baseball's World Series is an example of this type of local rebroadcast to the nation. Breaking news to the country is also handled in this manner.

The capability to broadcast around the world to several nations (simultaneously) is enabled through the use of satellites. There have been musical concerts that have been broadcast around the world in real time. A physical transmission system could not support a broadcasting requirement without the use of a satellite network. Unfortunately, some disasters that have been reported by news agencies in as close to real time as possible for breaking news could only have been transmitted via a satellite network.

Satellite systems have played a major role in linking the nations of the world through the reporting of news. Figure 9-44 is a rendering illustrating this role.

Unlike the public voice telecommunications carriers (e.g., cellular carrier and wireline carriers), the over-the-air television broadcasting companies (the networks and local television stations) tend to not rely on physical transmission facilities to carry their information signals. The fact is that given the nature of the television programming business, a high degree of real-time flexibility is necessary. The physical transmission facility does not lend itself to that kind of operation. Fiber optics can be used to support transmission between Earth stations and television transmission towers; however, the fact is that the kind of programming that one sees on television does not require megabits or gigabits of bandwidth. The only reason to install such fiber optics would be for operating-cost considerations.

Cable television programming is a subscription television programming service and was originally known as community-access television. Community-access broadcasting brought programming and local news to

Figure 9-44 Interconnecting the globe via real time news coverage

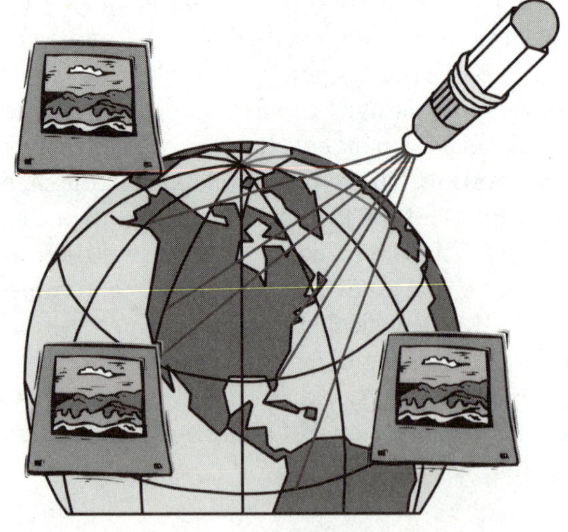

those areas that could not be reached by over-the-air broadcasting television stations. Popularly known today as cable TV, it has become a highly popular alternative program medium for various kinds of entertainment and news programs. Today, one can find programs that focus on original entertainment programs, news, sports, history, science, wildlife, cooking, home improvement, nature, comedy, theatrical movie releases, etc. The list is quite long.

Cable television uses physical transmission media to carry its signals. The transmission facilities might include metallic facilities, fiber optics, and wireless. Wireless cable TV was implemented by early cable TV stations in order to reduce transmission costs. Laying physical facilities is expensive. The wireless cable systems would ask subscribers to install small roof-mounted antennas that would receive the particular cable station's programming. The subscriber would have a cable box installed in the home to decrypt the encoded transmissions. Back then, the cable programs were supplying their own programming feeds to the home. These early systems would require line-of-sight transmission in order to optimize the quality of the received signal. Eventually, these early cable systems moved into pure programming and enabled other parties to carry their programs. This situation eventually led to cable television transmission companies that specialized in bringing programming created by others (in other words, today's cable television systems). Satellite systems are used to beam programs/content to Earth stations, which in turn retransmit the programs to the home via physical transmission facilities. Satellite proves to be a major factor in providing a backbone transmission function. Cable TV will be discussed in greater detail in the next chapter. Figure 9-45 illustrates how satellite networks support cable television networks.

Direct to Home Broadcasting

Direct to home broadcasting refers to satellite television programs beamed to the home. There are direct to home television services today. These highly popular television systems involve the satellite transmission of various kinds of programs to homes and businesses. The recipients of the programs (the customers/subscribers) have small satellite dishes installed on the rooftops of homes and businesses.

Direct broadcasting to the user enables the user to gain access to as many as 500 channels of programs, which sounds like a great deal. The fact of the matter is, however, that the subscribers want this access. One should

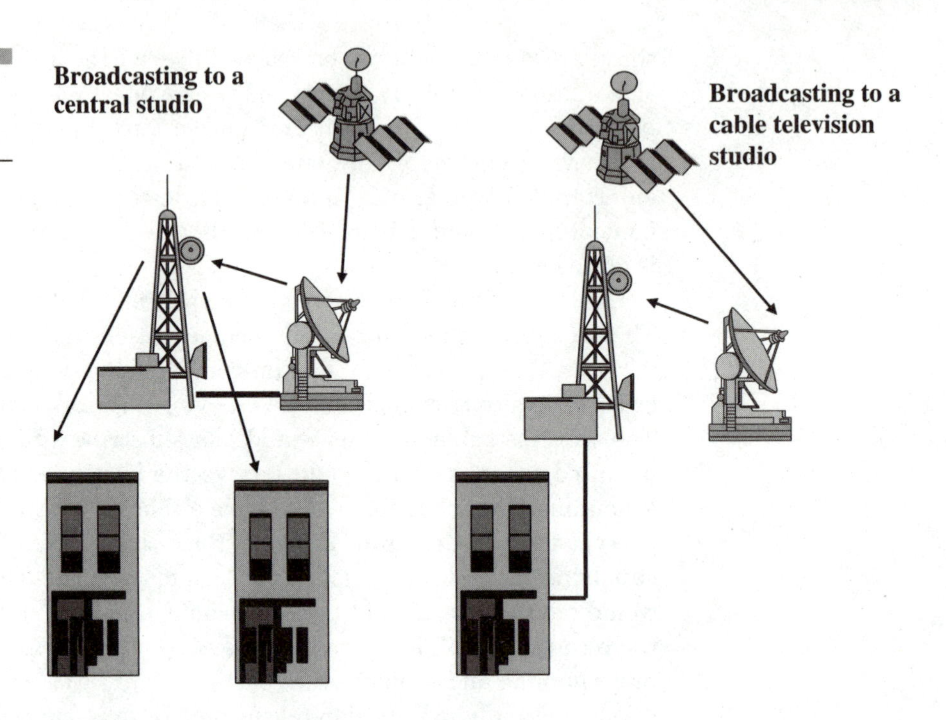

Figure 9-45 Satellite networking and cable television

be reminded that a rule of mass media is to give the people what they want. This perspective is an interesting twist on the concept of meeting the needs of the marketplace (but is true nevertheless). This type of television broadcasting offers additional outlets for new programs. Figure 9-46 is a rendering of direct broadcast to the home. As the illustration notes, the Earth stations in this model are the homes of the subscribers.

Radio Broadcasts

The broadcasting of national radio shows via satellite systems is popular today. Radio programs tend to be local affairs; however, there are a few popular national radio programs that are simulcast throughout the nation. There are times when newsworthy topics are broadcast around the nation and around the globe in real time. Holiday addresses by religious leaders such as the Pope are simultaneously broadcast via television and radio. Such a broadcast can only be accomplished via satellite networks.

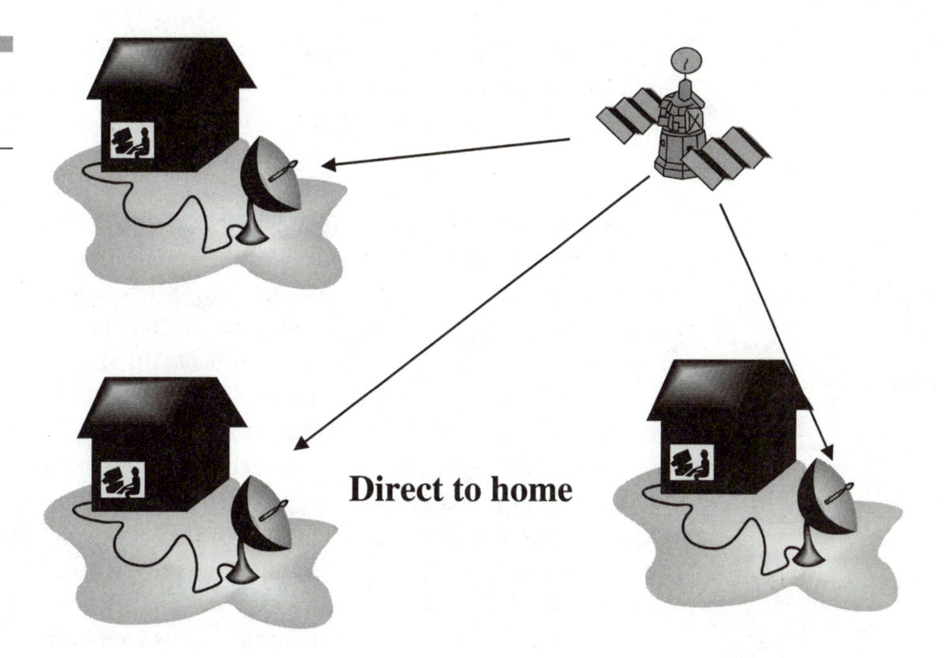

Figure 9-46 Direct to home broadcasting

Satellite Telephony

The reader might view satellite telephony as typical cellular telephone service. This view would be incorrect. Satellite telephony supports mobility, as do cellular and PCS. Satellite telephony does not require an infrastructure of towers or a land-based transmission facility infrastructure, however. Satellite telephony is satellite-to-handset transmission and vice versa. As an application, satellite telephony is useful for international travel, where cellular and PCS systems in various nations are totally incompatible with one another. A satellite radio telephone would enable a user to travel anywhere in the world and to make and receive calls regardless of whether there is a cellular carrier or PCS carrier in operation. To make this situation work, a specific satellite network configuration is necessary.

The *Low Earth Orbit* (LEO) satellite can be used in public telecommunications applications. The challenge for such satellite systems is that multiple satellites must be deployed in order to ensure that a satellite has coverage over the Earth. Satellites that are placed in LEOs move around the Earth with respect to the Earth's surface. A LEO satellite appears to be moving over a point on the Earth. In order to ensure that a LEO satellite system can provide a useful public communications function, it must maintain a communications connection during the entire length of a conversation. While the lengthy conversation is taking place, an LEO can move out of range of the subscriber. In order to maintain telecommunications connectivity, several LEOs must be placed in orbit about the Earth. This scenario boils down to coverage. A constellation of LEOs must be maintained to ensure full coverage over the Earth. The number of LEOs is so vast that the group is called a constellation of satellites. The reason for establishing a LEO network for public telecommunications is market viability. In order for the user on the ground to transmit useful signals to the satellite without a high-powered source, the satellite needs to be close enough to the Earth to receive the weak signal. If the telecommunications satellite were in a high orbit (on the order of thousands of miles), the handset would have to be of a commercially unwieldy size in order to support a powerful enough transmitter and power source. The handset size would defeat the commercial viability of the satellite telephone service. The small handset in use by commercial LEOs today is fairly high in cost. Such services are already having enormous difficulty establishing themselves in the marketplace, due to the high cost of the handset and the unfocused market thrusts.

The *Medium Earth Orbit* (MEO) satellite requires far fewer satellites to maintain full Earth coverage. Similar to the LEOs, the MEO satellites appear to be moving with respect to the Earth's surface. The handset's capability to communicate directly with the satellite dramatically decreases in comparison to an LEO handset. The mobile public telecommunications application is less viable; however, MEOs can support commercial broadcast of radio and visual entertainment. MEOs today support scientific applications.

The *Geosynchronous Earth Orbit* (GEO) satellites are orbiting at altitudes where they appear to be stationary to a point on the earth's surface. To an observer on the ground, a GEO satellite would appear to be stationary to the observer. A GEO satellite orbit is a shaped like a circle around the Earth's equatorial plane. A GEO satellite revolves around the Earth once a day in synchronous motion with the Earth's rotation. The geosynchronous orbital point is a point in which the satellite's centrifugal force is precisely balanced by the Earth's gravitational pull. GEO satellites are so high above the Earth's surface that it takes a quarter of a second for a round trip between the satellite and the Earth. Because of the delay, GEO satellites are not useful in real-time public telecommunications applications. There have been satellite telephony companies that have used geosynchronous satellite networks for data applications, however. GEO satellites are useful for commercial broadcasting of radio and video/television. Possibly, as mobile users

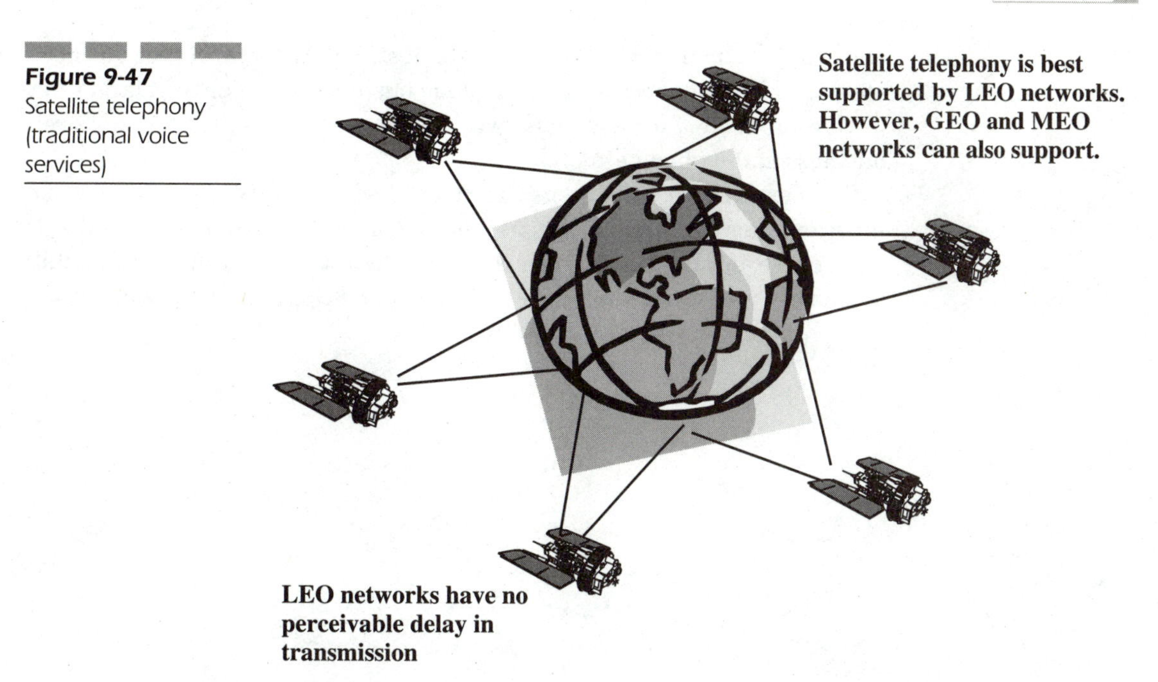

Figure 9-47 Satellite telephony (traditional voice services)

of all kinds seek Internet access, the GEO satellite configuration will find a home in broadcasting news events or data to next-generation handsets. Figure 9-47 is an illustration of satellite telephony.

SUMMARY

Satellites are playing a key role in interconnecting networks and even interconnecting users to carriers. A number of different applications of satellite technology exist, such as military, scientific, telecommunications, entertainment, and education. All of these applications are facilitated or are made possible through the use of a satellite network. The capability to reach or access billions of people around the globe is a tremendously positive feature. The expense of maintaining a satellite network is high. A satellite's life expectancy varies depending on usage, orbital conditions, design lifetime, power source, and reliability. The useful life expectancy is on the order of four to nine years, however. Therefore, replacements must be ready. Satellite networks are expensive (approximately $200 to $300 million for just one telecommunications satellite), but when one considers the scope and scale of the impact, the cost can be justified. Whether or not one operates a satellite network as a stand-alone business or as part of a larger network comprised of terrestrial networks is dependent on the company's core business and its business plan.

The next chapter will explore the cable television industry. Cable television programming has brought a mass-market appeal to satellite networks. In some sense, if it had not been for the desire for more entertainment, satellite technology would not have gained the consumer appeal that it has today.

Cable Television Networks

Cable television (originally known as *Community-Access Television*, or CATV) is primarily an entertainment medium.Originally, it was deployed to bring entertainment and local news to remote, rural areas where broadcast television signals could not be received. Community-access television deployment was highly popular during the late 1940s and throughout the 1950s. Early television broadcasting was not widespread, and there were many broadcast fringe areas during those days of television broadcasting. The fringe areas could be described as poorly covered areas in which a television viewer could see static-filled pictures or sometimes no picture at all. As recently as the late 1960s and early 1970s, even rural areas that were close to major metropolitan areas were still poorly covered. People (residences and small businesses) discovered that if they raised their rooftop reception antennas high enough, they could obtain a good-looking picture. These same people discovered that by using antenna splitters, they could share their signal with their neighbors. This type of signal distribution has its limits, however. One can only split the signal so much before the picture quality degrades below viewable levels. Figure 10-1 is an illustration of early cable television systems.

This situation led to the creation of companies that delivered television programming over coaxial cable. These coaxial cable television companies

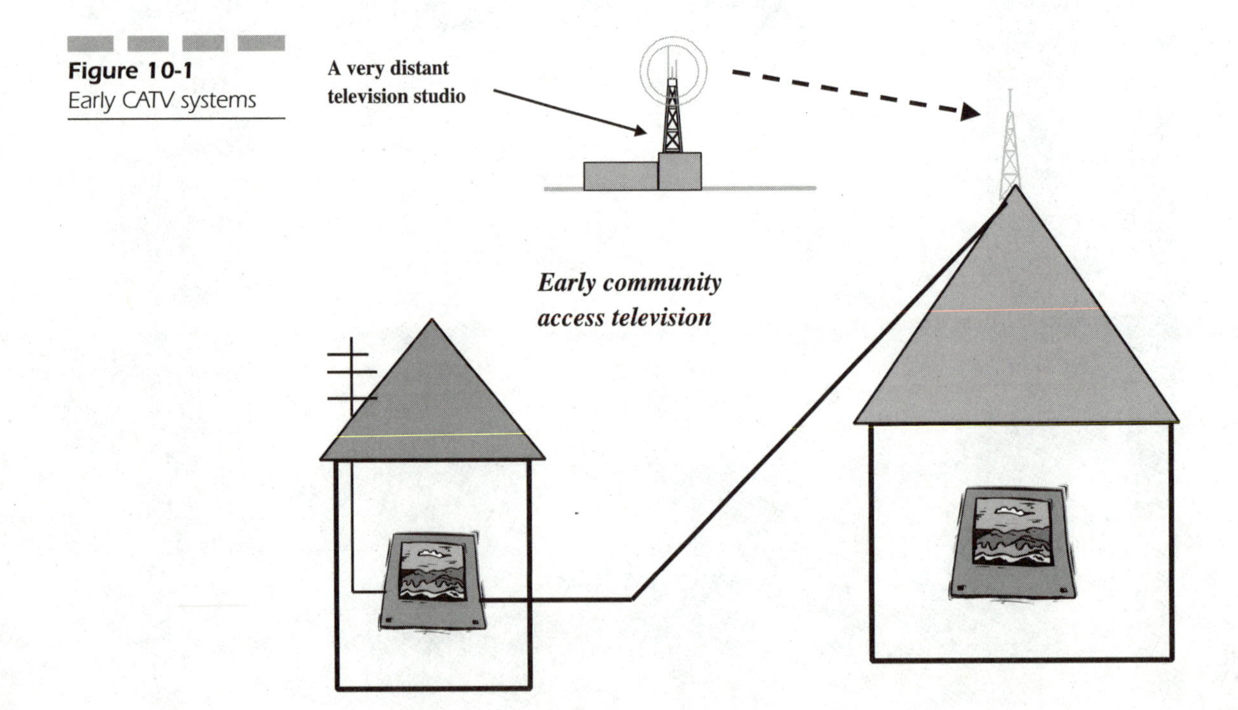

Figure 10-1 Early CATV systems

would construct a high antenna tower system that would receive broadcast television signals. The coaxial cable television companies would then redistribute the television signals to the community. CATV enabled rural communities to see the same commercial programming that the big cities viewed and also provided a way for local citizens to advertise local events and communicate information that affected the community. Today on cable television, one can still see messages from local school boards about library-run children's events and other community-specific events. Today, CATV is commonly known as cable television, rather than community-access television. Figure 10-2 illustrates the use of coaxial cable in CATV.

Cable Television Network Architecture

Cable television networks are currently designed to transmit multiple conventional-analog television signals to multiple subscriber locations. This

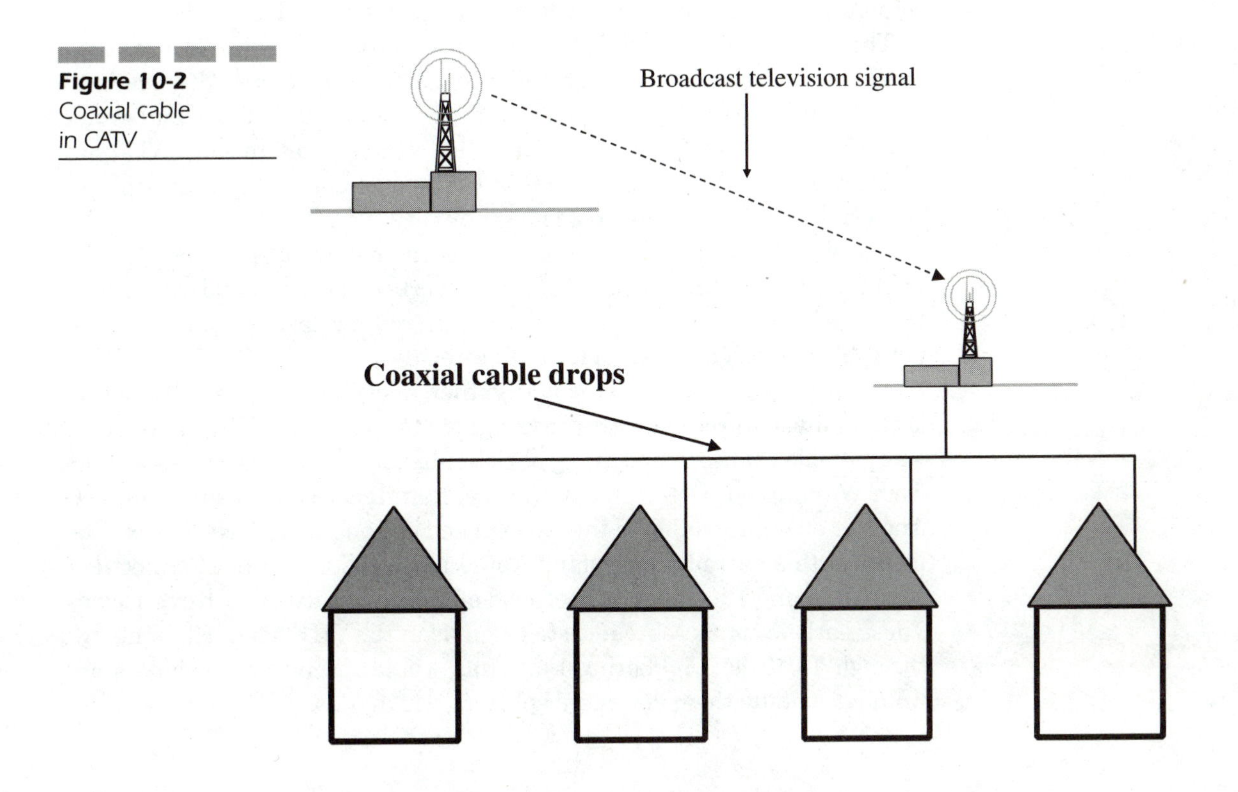

Figure 10-2 Coaxial cable in CATV

functionality is a one-way system for distributing the same set of signals to each subscriber location, but historically, these systems have had limited capability for return transmissions from designated subscriber locations. The advent of new types of equipment that enable a television to act as both a television and as a computer terminal will eventually convert the cable television network into broad-band multi-media information networks and Internet-access devices. Cable television has the capability to become a two-way telecommunications device.

Early CATV System Operation

The early cable television systems used coaxial cable as transmission facilities. The early systems consisted of the following components:

- Head end
- Service drops
- Subscriber premises equipment

Figure 10-3 depicts early the cable television network configuration. The configuration has remained largely unchanged to this day.

The early cable television providers would receive the broadcast signal from the television stations at a device called a *head end*. The head end received these television signals at a high tower. The head end would amplify the received signals and distribute the signals to the CATV subscribers. The reader should note that CATV customers paid (and still are) subscription fees to receive these high-quality signals.

The signals were distributed from the head end to subscribers via a distribution system that consisted of a neighborhood antenna and signal splitters. The initial CATV systems were, in fact, rebroadcast systems that were not cable television per se (refer to Figure 10-4).

The early CATV systems would gather television signals from nearby cities and would rebroadcast these signals to the community. Therefore, it was possible that someone living in Columbia, Maryland would see stations from Wilmington, Delaware; Baltimore, Maryland; and Washington, D.C. Another example would be the eastern end of Long Island, New York. Residents of this part of Long Island would receive signals from Connecticut, New York, and Rhode Island because of the proximity of all three locales. These early systems rebroadcasted signals in the VHF band (channels two through 13). The neighborhood antenna would receive these signals and would distribute them via signal splitters, which consisted of the device and

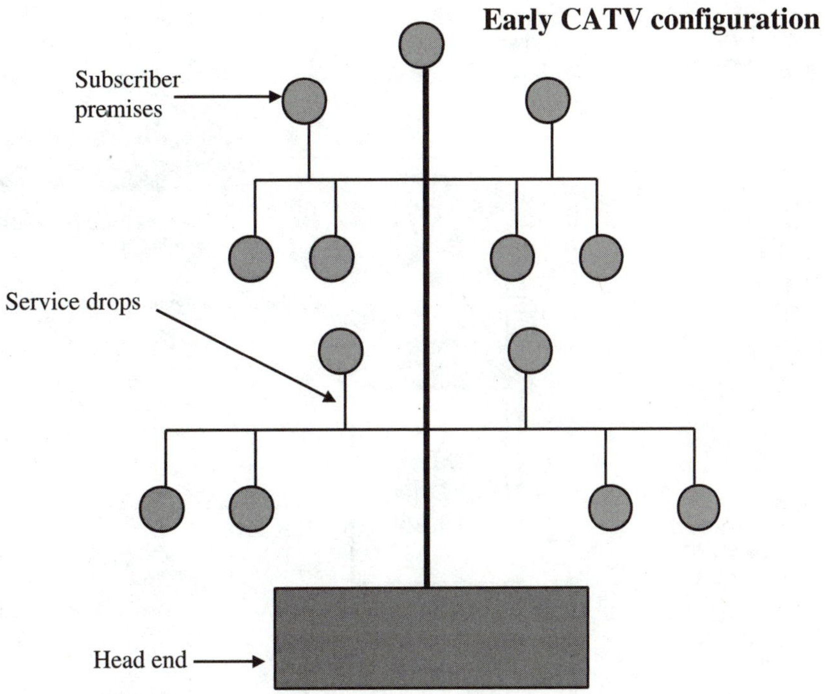

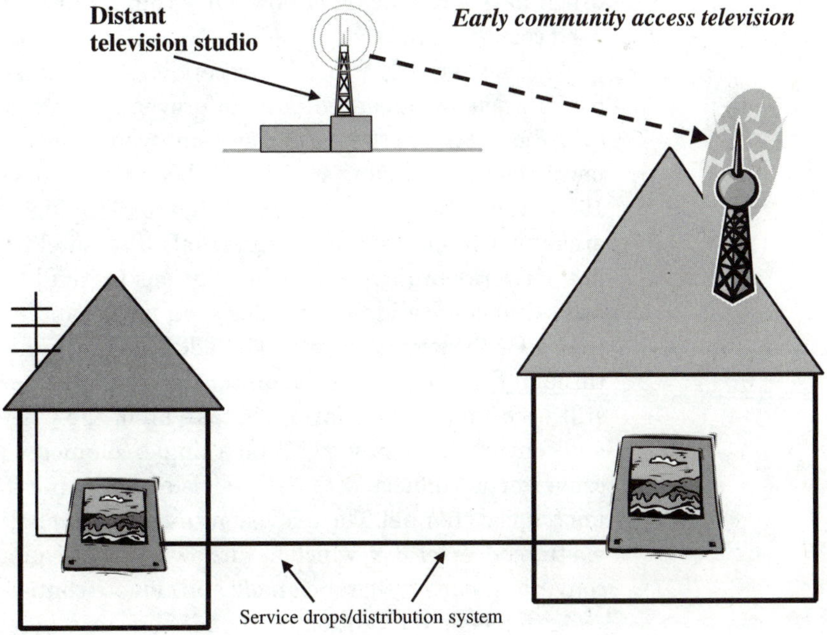

Early CATV configuration
Subscriber premises
Service drops
Head end

Figure 10-3
Cable television network configuration

Distant television studio
Early community access television
Service drops/distribution system

Figure 10-4
Initial CATV systems

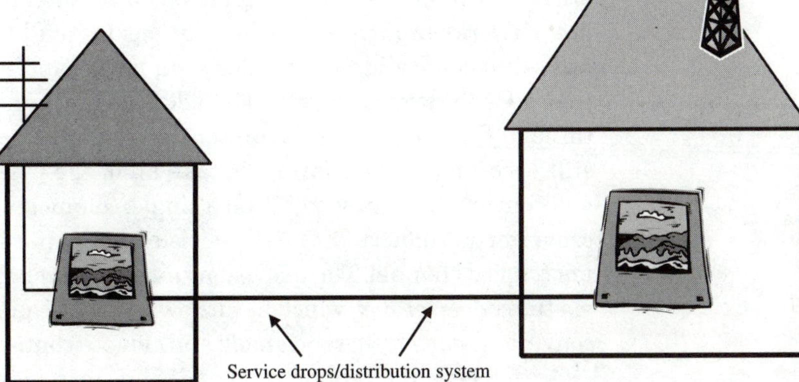

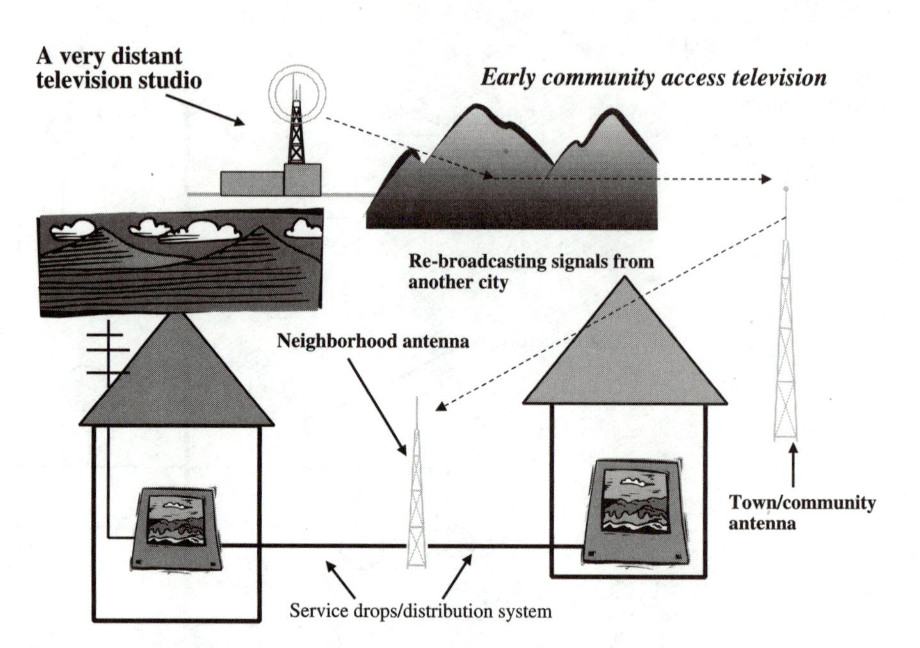

Figure 10-5 Coaxial cable in the early community access television systems

the coaxial cable. The drop wire to the home consisted of coaxial cable. Figure 10-5 is a rendering of how early community access television systems used coaxial cable.

As time passed, television reception and transmission improved, and fringe areas disappeared. This improvement to broadcast television, however, did not mean that CATV systems would go away; rather, it meant that new ways of improving service had been found. In the late 1970s and early 1980s, these improvements amounted to receiving more stations from other areas, which required receiving signals from satellite networks. These satellite networks brought television stations from all over the nation. The typical television could not then (or even now) handle the number of stations that a CATV system brings. The additional channels could only be shown through the use of a television set top converter box. These converters are still used today. The converters take all of the CATV-transmitted stations and convert them for viewing on a single, common, unoccupied channel. The converter is connected to the television set, while the set is tuned to this unoccupied channel. The television stations (channels) are viewed/controlled via the converter box, which has its own dial for channel selection. Today, the converter boxes have touch pads and remote controls for channel selection.

As television reception, transmission, and independent programming improved—and as the marketplace for new types of programming grew and airwaves became more crowded with wireless systems—antenna placement became more challenging. The need to find alternative ways of bringing the CATV signal to the home also grew. In the early 1980s, the CATV programming companies attempted to expand their customer base by moving into the metropolitan communities. These program companies were working with local CATV systems to deploy thousands of antennas in city communities. Small dipole antennas were mounted on posts on top of homes, on sides of buildings, in alleys, on top of flagpoles in backyards, and even in other people's backyards. Coverage and antenna visibility quickly became a problem. The challenge to the growing CATV companies was finding some way of bringing these new programs to the market. The answer was to deploy physical transmission systems directly to the home. During the early 1980s, these CATV companies lobbied the federal government for proactive regulation that eventually forced the telephone companies to enable these new programming companies to lay facilities underground and on top of telephone company-owned telephone poles. In other words, the telephone companies no longer had sole claim on the right of way for their facilities distribution networks.

These CATV companies eventually became more accurately known as cable television companies and began to string coaxial cable to the home. These cable television companies maintain their own programming switches that distribute programming from a centralized location. Microwave and the VHF bands are still used by some cable television companies for limited collection and distribution of programming. The most expensive part of a cable television operation in the 1980s and early 1990s was the construction of the facility transmission network. Today, the price of programming has surpassed the network. Many of these cable television companies have mature and established networks, and there is no new construction. These cable television companies are converting their networks to high-bandwidth networks and multimedia and multi-functional switches in order to process broad-band information services that go beyond simple entertainment programming. More information about this topic will appear later in this chapter.

Figure 10-6 is a rendering of how cable television programming became so important.

Cable Television Systems Layouts

Current cable television system layouts are comprised of the following components. These components are similar to those that were used by the original

Figure 10-6 Cable TV and programming

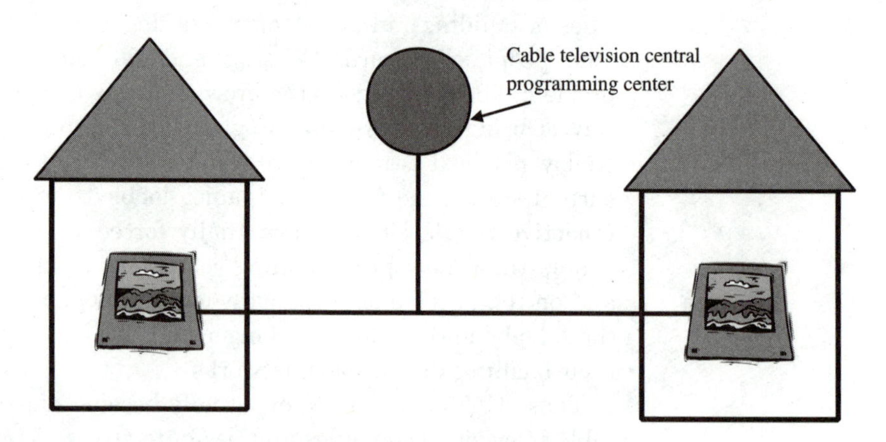

CATV systems; however, today there is a reliance on physical transmission facilities to carry the video and audio signals to the home.

- Head end
- Trunks
- Feeders
- Service drops
- Subscriber premises equipment

The basic network configuration is similar to the early CATV system layouts. Figure 10-7 provides greater detail than previous illustrations.

Head End The head end is the heart of a cable TV network. Television signals are collected at the head end by various methods. Head ends can receive signals from a variety of sources, and these sources can include over-the-air broadcasting stations, directed satellite transmissions, local television studios via physical facilities, microwave transmission, or remote pickups from a centrally located reception point. The video signals can be modulated in a television transmitter, or modulator, on the designated television channel. The signals from the various sources are assembled and are

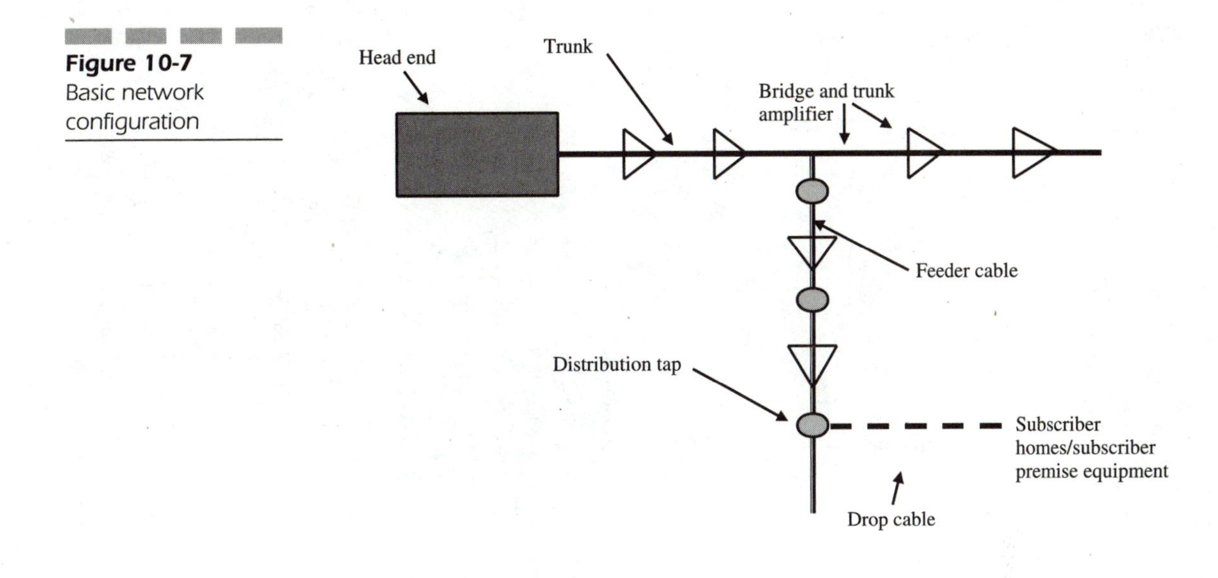

Figure 10-7 Basic network configuration

combined (multiplexed) in the desired configuration for distribution. The processing of the signals is done in such a manner to ensure that each subscriber sees the programming for which they have paid. Figure 10-8 highlights the functions of the headend.

Physically, a head end can be located in almost any structure but is typically located in an environmentally controlled space. This environmentally controlled space is similar to the environmentally controlled space used by a public telecommunications carrier. Generally, the head end has auxiliary power to run equipment and power-critical services in the event that commercial power is lost. Three to six satellite-receiving antennas can be situated adjacent to the head-end building but can be remotely located and connected to the head end by microwave or by a special coaxial or fiber optic link. The antenna sizes range from small antennas that are one meter in diameter to those that are 10 meters in diameter. Satellite systems are used for the large nation-wide cable television systems. The nationwide systems must ensure a ubiquity in programming. Satellite transmission ensures that everyone sees the same picture at the same time, which is especially important to advertisers who seek to have their message seen by the right audience at the right time. There is no point in showing an advertisement on hair-care products during a children's movie. Figure 10-9 depicts how a headend can be deployed.

Local cable television systems will use a land-based, computer-controlled multimedia switch as the media source. Local systems might simply run all

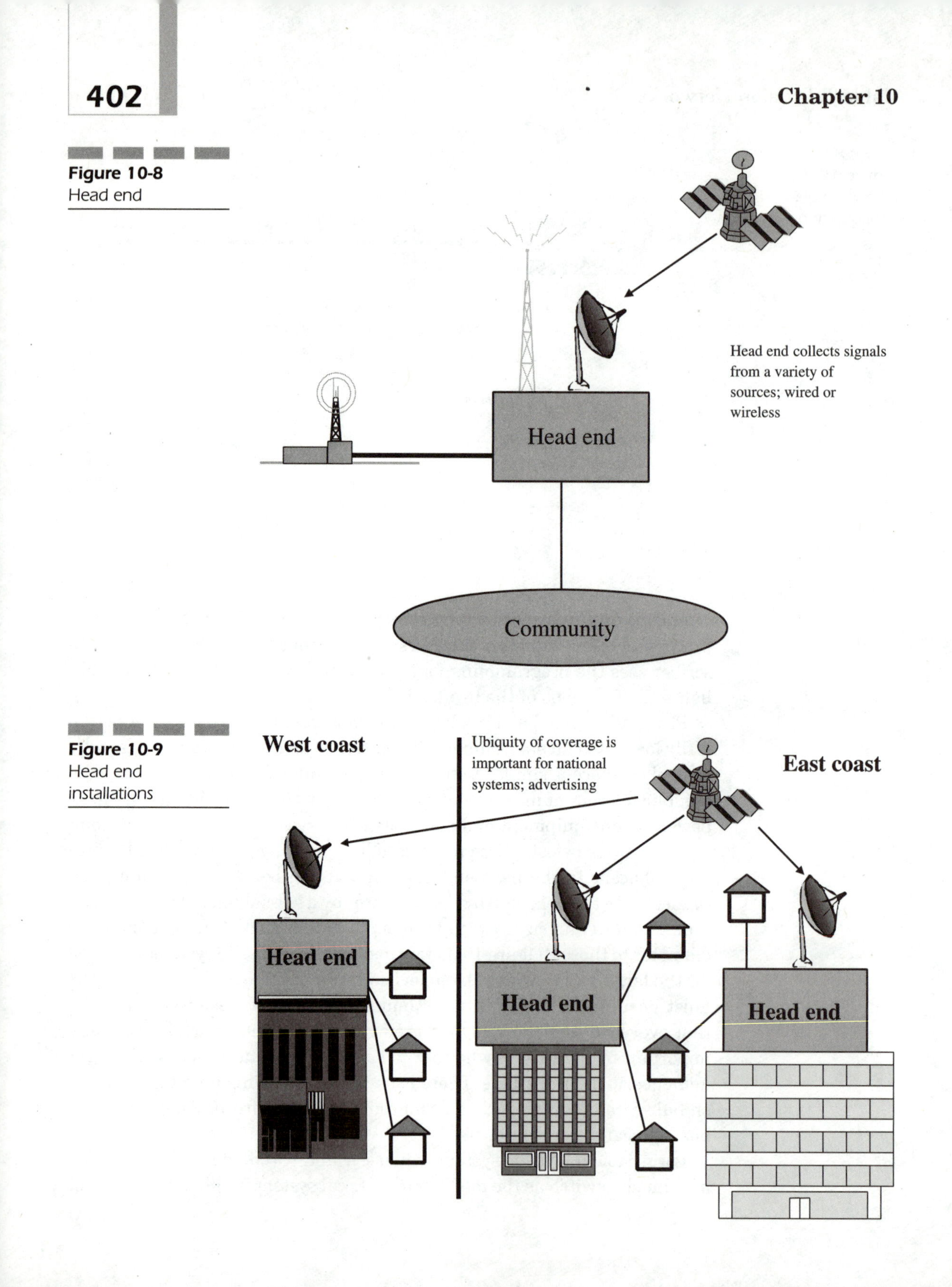

Figure 10-8
Head end

Figure 10-9
Head end installations

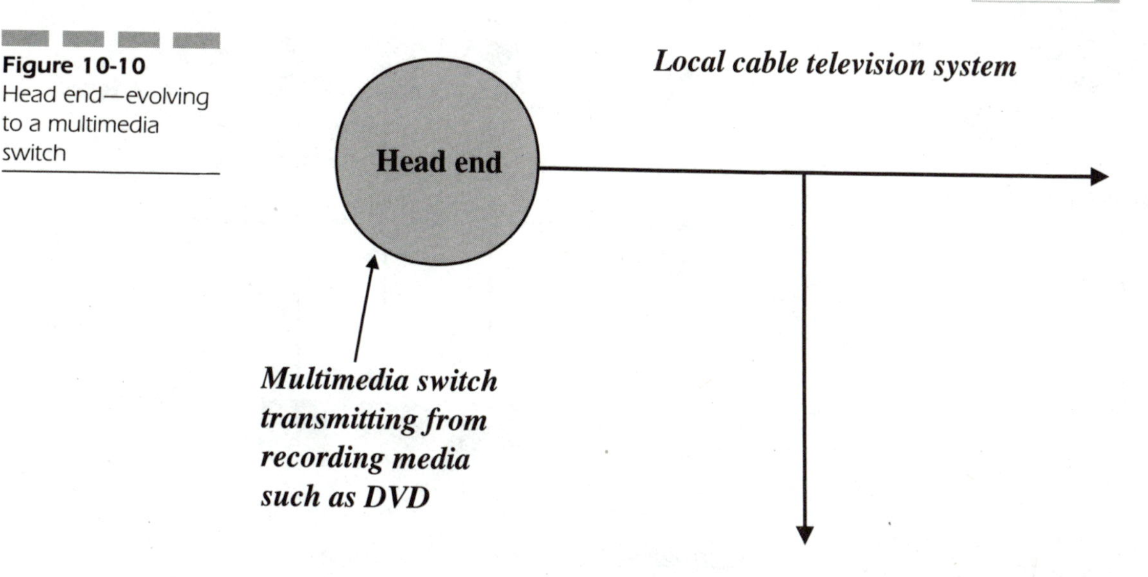

Figure 10-10 Head end—evolving to a multimedia switch

of the media from a physical recording media, because the need to maintain national ubiquity of programming is a moot point. Figure 10-10 highlights the evolution of the head end from simple programming distribution node to a multimedia switching network element.

Head ends can serve only one cable TV system or many cable TV systems. The choice depends on a number of factors, including ownership, geography, business requirements, politics, and demographics. In most large systems, one or more distribution hubs are used to limit the number of signal amplifiers in the service areas. These hubs can be supported via fiber optic transmission facilities.

Trunks Trunks represent the backbone of the cable television company's coaxial cable distribution network. The trunks are used to transport the multiplexed television stations/channels from the head end or hub to networked feeder cables (refer to Figure 10-11) that are part of the cable TV company's distribution network. Subscribers are not connected directly to the trunk cable, except in rare cases in which subscriber density is extremely low. The trunks in this type of network are similar to those that are used in public telecommunications networks. They carry large volumes of information to specific distribution areas. Refer to Figure 10-12 for a comparison of a cable television signal distribution network and a public telecommunications network.

Cable television networks are islands of one-way communication. Cable networks do not support transmission back from the home to the central distribution node. The basic configurations that are used in both the public

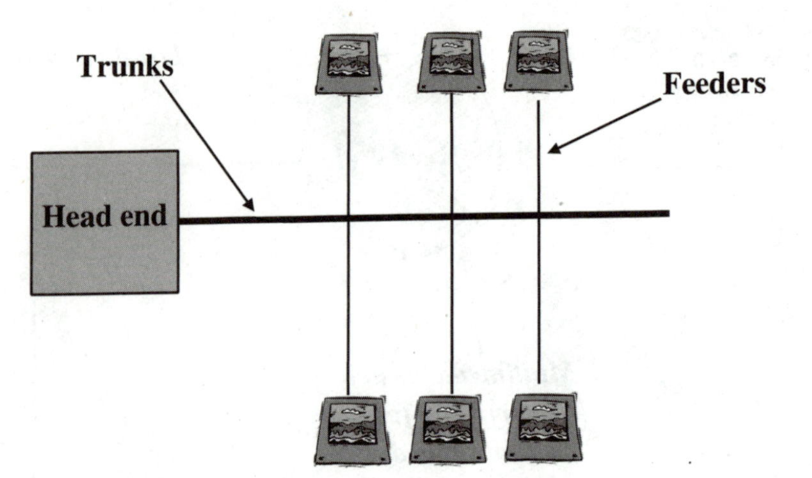

Figure 10-11 Trunks that are part of the cable TV company's distribution network

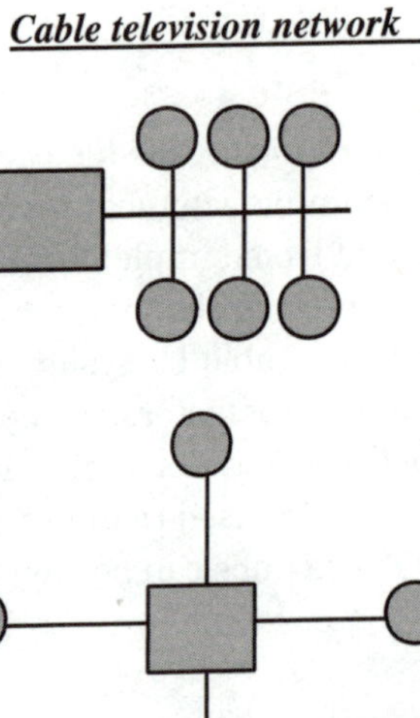

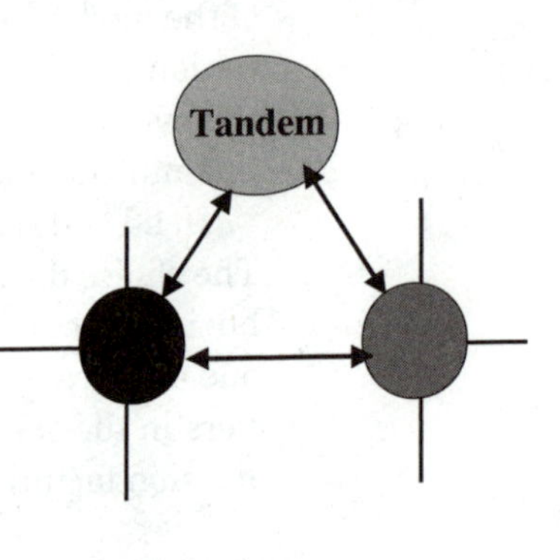

Figure 10-12 Cable TV signal distribution network versus a public telecommuniations network

telecommunications and cable television networks are similar. The similarity exists because the configurations support distribution of something—in both cases, the distribution of information. The network configurations are technology independent.

These trunks can be metallic or optical fiber. The type of transmission material used was originally simple coaxial cable; however, the demand for more channels and different types of broad-band services has forced many cable television systems to upgrade their networks. The original trunks that were used were classified as *N* Carrier facilities. This type of carrier is the historical predecessor of *T* Carrier. As the reader can imagine, *N* Carrier has bandwidth limitations in comparison to optical fiber. There will be a more detailed discussion of transmission networks later in this book.

Feeders Feeders, or distribution cables, take the signals from the trunk cables and extend these signals throughout the hub that is serving the area. The feeder cables are connected to the trunk through a coupler and a bridge amplifier (with four outputs) to provide the signal power that is delivered to the subscribers. Coaxial cable comprises the feeder cables. The subscriber access ports are called taps.

Service Drops Service drops connect the subscriber's home to the feeder cable taps. The drop cables are made of flexible coaxial cable. Drop cables (also known as service drops) are commonly used equipment for telephone companies. Drop wire is a term that is commonly used by the wireline telephony industry. Similar to trunk and feeder cables, drops can be placed underground or can be run aerially from a pole to the subscriber's premises.

Subscriber Premises Equipment The converter box described in the previous sections of this chapter have been largely unchanged since their inception 40 years ago. The converter box's basic operation is to serve as a television set top multiplexing device for multiple channels. Modifications to these boxes are envisioned in order to support new multimedia services. These services include the Internet, graphics, intelligent movie selection, and other types of services. There will be a more detailed discussion of these services later in this chapter.

Basic Network Configurations

Cable television companies use three basic architectural configurations. The topological configurations used are not much different than those that are used by wireline telecommunications service providers or wireless service providers. The two popular configurations are tree and star; however, for the sake of completeness, I should mention the ring configuration. The tree and star architectural configurations are used in order to maximize the utilization of the network.

Tree Architecture As indicated previously, the tree architecture looks similar to a tree. The characteristics of the tree architecture are the same as those of a typical LAN (in some wireline telephone company networks or party lines, or even in some cellular carrier networks). Since the inception of cable TV, nearly all cable television systems have been deployed by using the tree configuration. The tree architecture is the most efficient way of distributing the same set of communications signals to multiple terminals. Figure 10-13 illustrates how the tree architecture is deployed in a cable television environment.

Star Architecture The star architecture is the most typical wireline telephone company configuration. The star topology enables separate transmission paths to be established to each subscriber. Each path can be designed to carry the messages or different messages. For a cable television system, this method is also an efficient way of distributing programming services. Figure 10-14 is a depiction of how the star network topology is deployed in cable television.

Ring Architecture The ring architecture loops traffic so that it returns to its original starting point. The ring architectural configuration is a network configuration in which the same messages are transmitted simulta-

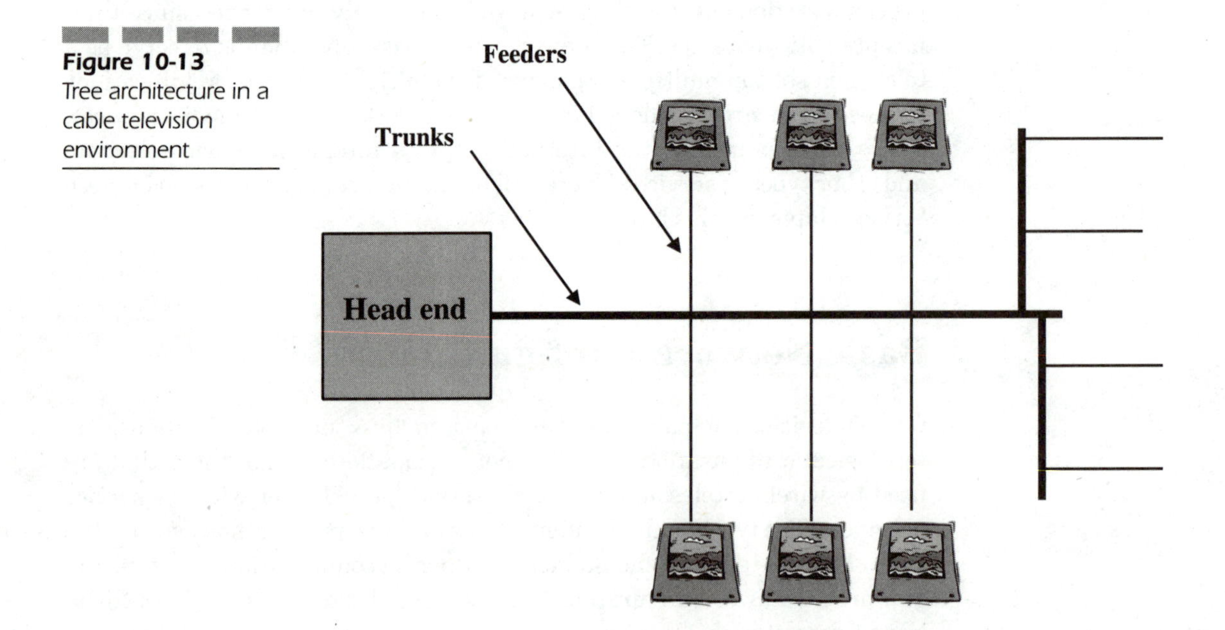

Figure 10-13 Tree architecture in a cable television environment

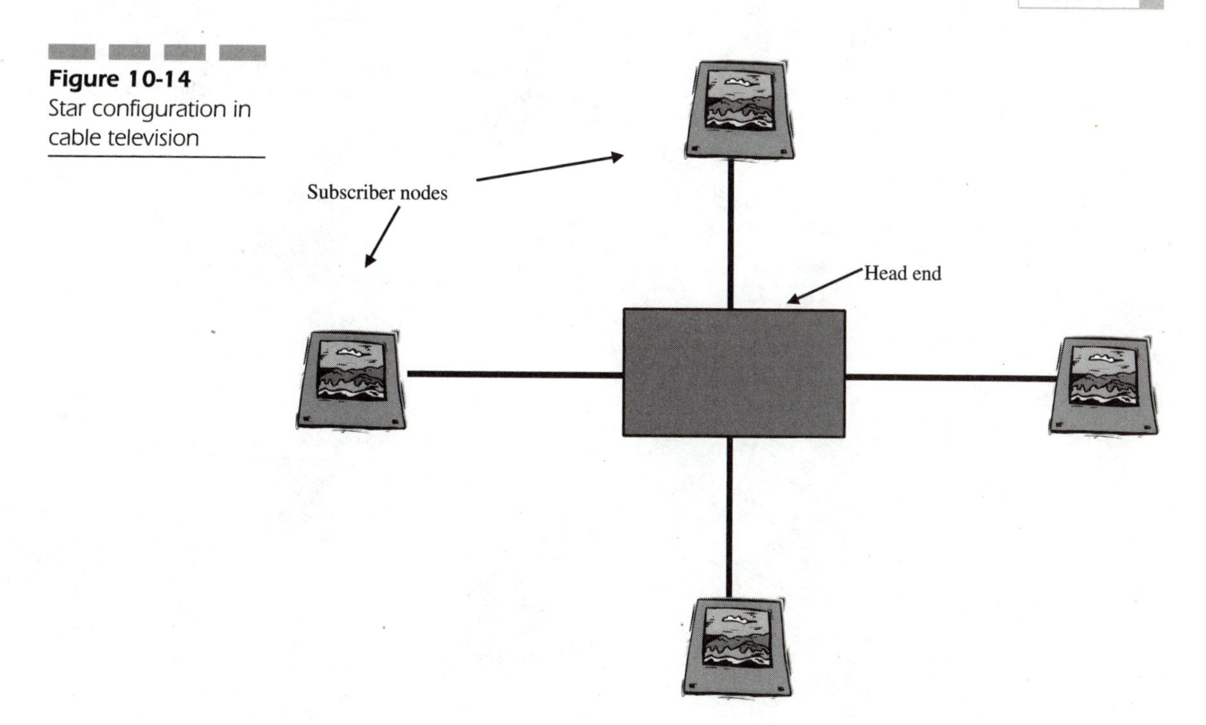

Figure 10-14 Star configuration in cable television

neously in opposite directions on parallel rings. In case of a service interruption on the ring, the messages are automatically transferred to the other ring. These rings operate in support of each other, providing redundancy. This architecture is an expensive but highly reliable configuration for maintaining data transmission.

Cable TV is an expensive network configuration for television entertainment. If the cable television companies are attempting to find new business opportunities for their industry, however, then supporting multimedia information services will require reliable and redundant systems. Figure 10-15 depicts the way in which ring networking can be deployed in cable television.

CATV Transmission Facilities

Coaxial Cable Coaxial cable was (and still is) the primary transmission facility used by CATV systems. For many years, coaxial cable was a medium that was heavily used to carry large-bandwidth information streams over

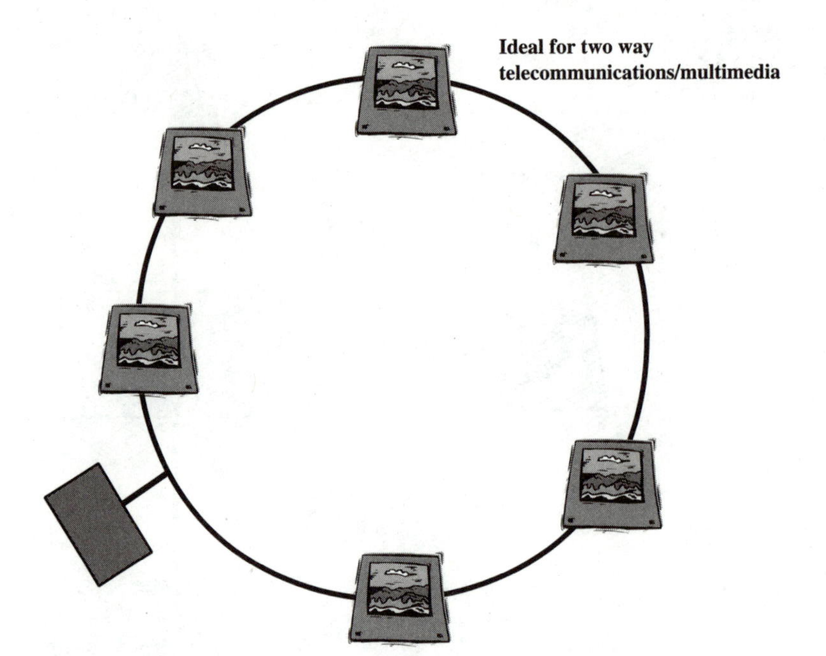

Figure 10-15 Ring Architecture in a cable television environment

distances greater than 200 meters. In the public telecommunications sector, the other (and principal) medium to transport information was (and still is) twisted-pair wire. This wire is being replaced by other media, such as optical fiber. There are two types of coaxial cable: broad-band and base-band. Broad-band is typically used for cable television, whereas base-band coaxial cable is used for data network connections.

As a transmission medium, coaxial cable is classified as an unbalanced medium. The information is sent through one conductor, while the other conductor is used as the electrical ground. Figure 10-16 illustrates the construction of coaxial (coax) cable. Coax consists of an inner conductor and an outer conductor. As the figure shows, coaxial cable has an inner conductor that is made from some type of metallic material (normally copper). The inner conductor is surrounded by a non-conductive material, which serves as a dielectric. The dielectric material is also surrounded by a sheath of conductive material, which is normally made of copper as well. The outer conductor is braided so that it also bends easily around the dielectric. The outer conductor is surrounded by an insulating material. Coaxial cable is used for a variety of construction purposes—normally, for fluorescent lighting. This cable is still used for communications purposes, but like twisted-pair wire, it is being replaced by optical fiber or by enhanced local loop access technologies. Figure 10-16 illustrates the structure of a piece of coaxial cable.

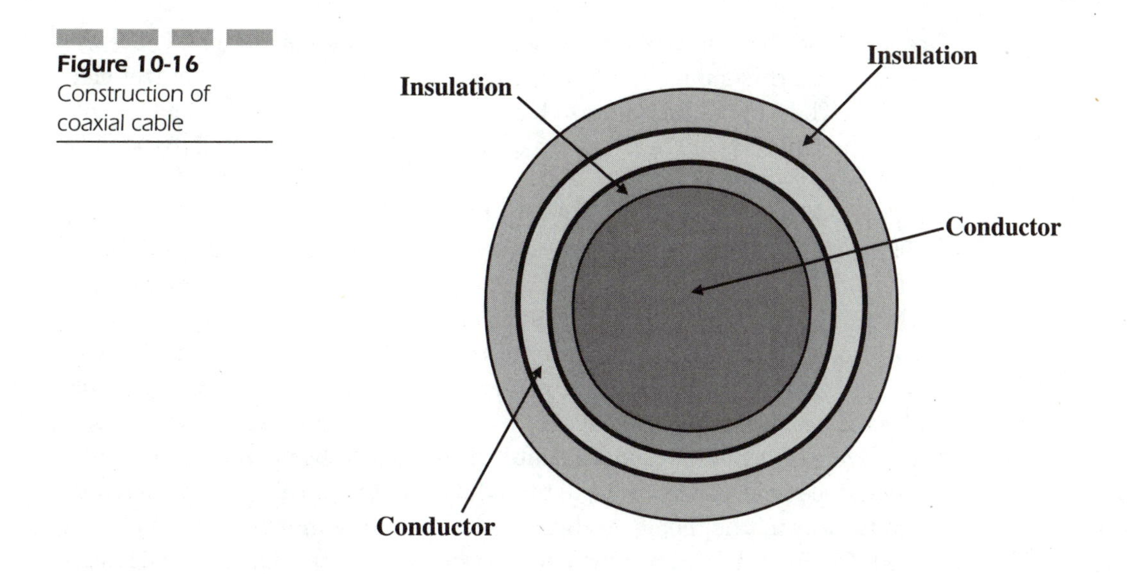

Figure 10-16 Construction of coaxial cable

Optical Fiber Optical fiber trunks are also used in cable television networks. In such networks, *Amplitude Modulated* (AM) laser transmitters are used. Fiber optic facilities are capable of transmitting high-quality signals up to 15 miles without amplification. Optical fibers have enormous bandwidth to utilize, and they are not affected by the electromagnetic noise (unlike metallic wire). The wireline telephone industry currently uses optical fiber facilities to provide telephone service to the home. At this time, telephone companies are preparing to use these facilities to provide (to the home) a variety of multimedia services. These facilities will also enable cable television companies to provide the same types of media services. Optical fiber is capable of transmitting large-bandwidth information at speeds in the gigabits per second range.

Optical fiber had been expensive not too long ago; however, fiber today—along with other transmission facilities—is becoming a commodities business. In other words, the prices are coming down. In the early to mid-1990s, fiber optics was expensive. The prices have come down so far, however, that the business of leasing dark fiber is no longer a large-margin business. To maintain profit margins at a high level, most companies that were not too long ago in the business of selling fiber capacity are now seeking partners or customers that will enable the lessor to leverage additional value from the fiber business. So, many companies (telecommunications, electric power utilities, water utilities, etc.) are now in the business of leasing the excess capacity that the transmission facility business is transitioning into Internet-related businesses, e-commerce-related businesses, Internet service providers, or wireline public telecommunications carriers. There will be more on this subject later. Dark fiber is fiber that has no control electronics. The fiber is just a glass facility with no optical equipment, amplifiers, multiplexers or transmission control equipment connected to the fiber.

Hybrid Fiber Coaxial Cable As I had indicated, in the early to mid-1990s, optical fiber was an expensive medium. Due to the cost of fiber, an alternative was sought. The solution to fiber was hybrid fiber coaxial cable, which is a combination of fiber optics and coaxial cable. This type of cable was created to facilitate the entry of the large, established wireline carriers and startups into the multimedia business. Similar to many product ideas, there was no single vision of what multimedia meant or what it was to provide (or even to whom this multimedia would be provided). What was understood was that in order to take the first step towards providing more bandwidth to the home or business, the telecommunications industry needed to first break down the transmission facility cost barrier. As a result, hybrid fiber coaxial cable was invented.

Hybrid fiber coaxial cable was built upon the strengths of coaxial cable, using optical fiber as the catalyst. Hybrid fiber coax is not coaxial cable with fiber wrapped around it. The term *hybrid* refers to the way in which the total CATV trunking and feeder cable system is constructed. In hybrid fiber coaxial cable, the trunking portion of the cable television system is replaced with fiber optic cable. The service drops are still coaxial cable. Hybrid fiber coaxial cable system advantages are as follows:

- Higher bandwidth
- Lower signal impedance
- Improved signal-to-noise ratio
- Less affected by thermal noise
- Much less affected by ambient electromagnetic fields from other surrounding transmission systems that are sharing the same telephone pole or transmission conduit

These advantages enable the hybrid fiber coaxial cable system to perform the following tasks:

- Reduce the number of signal repeaters in the field, which means fewer capital equipment to install and maintain
- Greater range for the transmission signal without having to use a signal repeater
- Less field maintenance on the trunking system

- Overall greater reliability by virtue of the fact that there are fewer individual pieces of equipment to maintain. In other words, having fewer tasks to complete means fewer items to fix.

As the reader will recall from previous chapters, optical fiber has a number of advantages overother types of transmission media.

Figure 10-17 depicts how a hybrid fiber optical coaxial cable system is configured.

A Broader View of CATV/Cable Television: Technology and Services

The cable television business is an entertainment medium that is poised to provide the kinds of public voice and data telecommunications services that people have come to expect from the wireline and wireless telecommunications carriers. The cable television business is a one-way signal distribution business that needs to become a two-way distribution business. The more appropriate names for the cable television industry in the near future are *distributed audio-video* or *distributed television*. For the sake of continuity, however, I will continue to use the terms cable television and

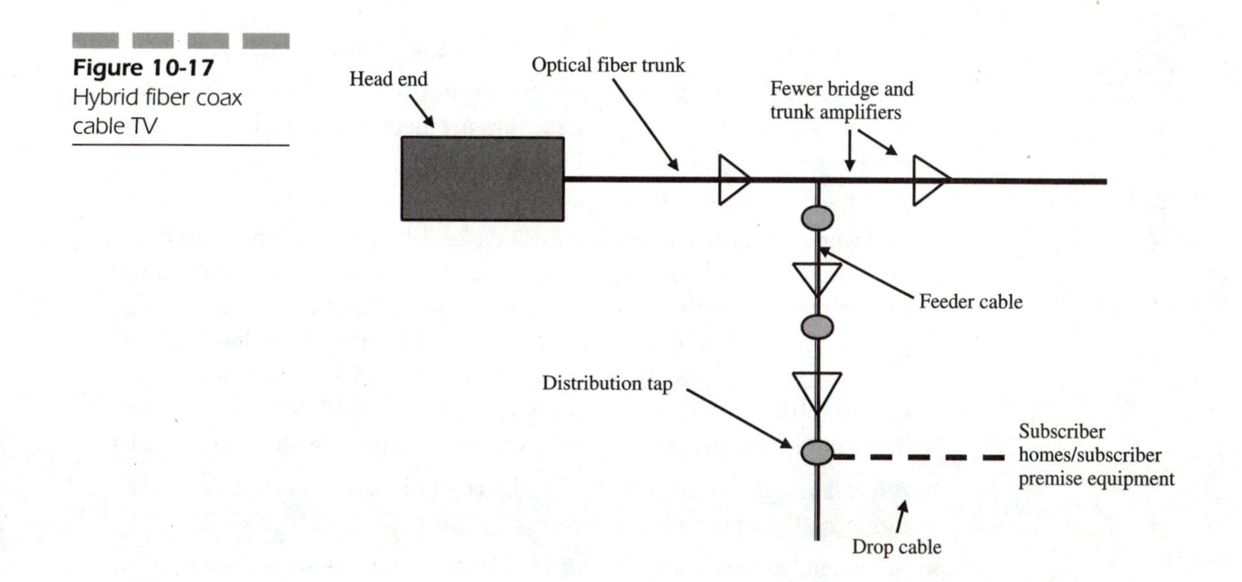

Figure 10-17 Hybrid fiber coax cable TV

CATV. The cable TV business can be broken into two basic pieces: content and network.

Content

Traditional content is simply entertainment programming. The network piece is the traditional programming distribution system; however, the cable television industry needs to change the business model. With the advent of direct-to-home satellite television, the cable television industry is meeting direct competition. Cable television tends to offer kinds of programming that broadcast television might consider too adult for the broadcast audience. As I write this chapter, however, the broadcast television networks are offering a variety of new kinds of innovative programming that the cable television has not offered yet. Today, cable television is primarily an entertainment medium that must pay a number of individuals in order to bring the subscriber the entertainment that they see. Remember, the cable television industry is a medium that presents programming and does not normally produce the programming. The companies involved in program development are independent companies that sell their programs to the cable television networks.

The cost components of programming and cable television have no analogues in the public voice and data telecommunications networks. These components are as follows:

- *Production costs* For example, film editing, sound, staging, etc.
- *Distributors* The distribution system in this case does not refer to a technology network. The distributors are people who own television stations, movie theaters, and cable television stations.
- *Residuals* After a telecommunications industry deploys a network and the activity is completed, all folks associated with the network are paid. In show business, actors continue to receive a percentage of the profits even after the program is no longer in production. Actors are paid for every time the show is repeated. The closest analogy in the business world might be the stock option; however, every employee shares in the wealth of the company. In the case of the residual, usually only the actors, producers, and directors continue to share the wealth.
- *Advertising and marketing* Unlike the public telecommunications industry, the entertainment industry has children's toys tie-ins.
- *Actors, producers, and directors* Highly paid workers who are essential to the development of entertainment programming. Given the

high salaries in the business world, many might disagree with this point. To some extent, a few of today's business leaders have become media stars; however, the salaries and incomes of actors, producers, and directors are associated with a one-time event. Once the job is completed, the actors, producers, and directors reap an enormous financial win. The scope and scale of the effort is beyond comparison.

- Entertainment production is a one-time effort. Public telecommunications is a 24-hour-per-day, 7-days-per-week, 365-days-per-year effort. Scope and scale are different.

Services The cable television industry is changing the business paradigm within which it is operating. Rather than simply presenting entertainment, the cable television industry is using the television set as a telecommunications device. There are efforts to offer two-way voice and data services and Internet access. The television set is a common piece of electronics, and there are more television sets in the home than there are computers. As the years go by, the difference will disappear, and the television set and the computer will be one and the same. To enable this scenario, there will be a need to change the network. In order to support two-way communications, the current cable television network has to be converted to a two configuration, which will result in the addition of more transmission facilities and new types of technology. More information about this subject will appear in the next section of this book. The competition is fierce within the entertainment business. The cable television companies control entertainment content, unlike the traditional public telecommunications carriers.

By combining entertainment and telecommunications, the cable television companies offer a new set of services that the current public telecommunications carriers do not offer. These services will consist of the following elements:

- Voice
- Data
- *Internet access* Web hosting, e-mail, and some form of information storage
- Information storage
- Video
- Music
- Distance learning
- Combinations of the above

The Network

The cable television network still primarily consists of coaxial cable. In some areas, hybrid fiber coaxial cable and optical fiber cable are used. In the early and mid-1990s, the public telecommunications industry view of the cable television network was negative. The cable television network was a hodgepodge of various network implementations that were characterized by poor transmission quality and frequent network failures. This situation was quite true at the time, because the cable television industry was still filled with mom-and-pop shops (independents, or companies that did not have the resources to maintain the operational expenses needed for high-quality service). If you will recall, you could pick up any newspaper and read about subscriber cable television service complaints and price gouging. These complaints and high fees resulted in cable television regulation by both the United States Congress, the President, and the FCC. The regulation of the industry forced many of the small mom-and-pop shops to close down and sell to larger cable television systems. The old cable television business suffered from the following problems:

- *Lack of technical standards* The lack of technical standards meant that system design was not standardized, which could result in frequent service outages.
- *Lack of operational standards* Installation practices varied within the cable television company. Unmanned studio facilities left many subscribers without service for hours on end in some areas, until an employee showed up for work the next morning.
- Lack of customer care
- *Unconditioned outside plant* Transmission facilities were not properly conditioned to carry video to an area due to poor installation, incorrect cabling type, lack of signal repeaters, etc.

The merging of small cable television systems with larger systems has been good for the marketplace. The marketplace no longer sees the problems that it once saw prior to regulation. The creation of the super/mega cable television companies has enabled financial resources to focus on fixing the problems of the past and positioning the cable television industry for the future. This situation has resulted in the following:

- Better and more uniform programming
- *Reducing the number of outages* Today, outages are unheard of from the large systems—unless a cable physically fails.

- High-quality customer care
- Satisfied customers

In order to support two-way communications, the cable television industry needs to add additional equipment for upstream and downstream communications. In the cable television world, transmission to the home is considered downstream, while transmission towards the head end is considered upstream. The current cable television systems are designed for downstream communications. One might think that the upstream is a simple reversal of the information stream, and this statement is not true. Some of the challenges that are faced by a one-way system being converted to a two-way capability are as follows:

- Establishing a reverse path for the upstream communications. Even the traditional wireline carriers have a reverse path back to the central office, which means more equipment in the field.
- Establishing signal repeaters on the upstream side of the network
- Establishing a baseline for the signal to noise on the upstream. An unusual aspect of the cable television network is noise that is generated by the television set. Unlike the public telecommunications industry, the FCC's rules concerning noise generation from a television set are not strictly enforced. In other words, the television sets generate so much electromagnetic interference that two-way communications would be interfered with by the television set itself.
- Establishing a signaling plan for the entire cable telecommunications network, which includes the type of signaling structure that is used and the type of network interconnection plan that is followed

Figure 10-18 is a depiction of the kinds of enhancements the cable television industry needs to perform in order to provide two way telecommunications services.

Network Interconnect In order for the cable television company to create a two-way telecommunications business, it would not necessarily have to follow every standard to which the current public telecommunications companies adhere. The cable television company would have to use network interconnection as one of the baselines to follow for conditioning the network for two-way use. If a cable television company were to become a two-way communications company, it would have to enable its subscribers to speak with subscribers of other networks.

The cable television company would have to establish points of interconnection with other carriers, and the cable network would have to consider

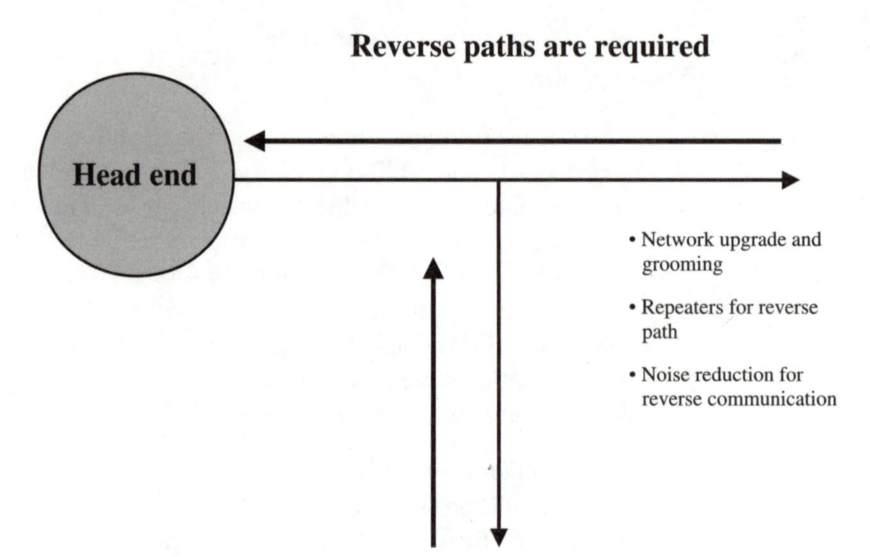

Figure 10-18 Upgrading the cable television network

moving away from the tree configuration that it currently employs and concurrently employing the star configuration in order to establish visibility with other service providers. A great deal of architecting will be required. Portability of service might be a desire, but the optimal scenario for initial deployment would be a Type 2A interconnection to a wireline tandem (the PSTN). The Type 2A will enable both subscriber bases to communicate with each other provide the broadest visibility possible for both companies. Figure 10-19 illustrates interconnect to the PSTN. Figure 10-20 illustrates a possible network configuration that the cable television companies could employ in order to support two-way communications.

The baseline for any telecommunications company that is attempting to become a carrier of voice and other services to the mass market is the capability to pass information between service providers. Essentially, a company that wishes to become a carrier must seek the lowest common denominator in order to meet the overall service needs of the subscriber base. This task requires interconnecting their network to another network.

The Terminal Device The television set will either have to be changed, or some other means of communicating with the network will have to be created. The most likely event will be the modification of the cable television set top converter box. The converter box is the smallest and the least-expensive alternative to making the subscriber pay a few hundred dollars for a new television set. That the fastest way to lose customers is to tell them to change their terminal equipment, because you will deny them new services

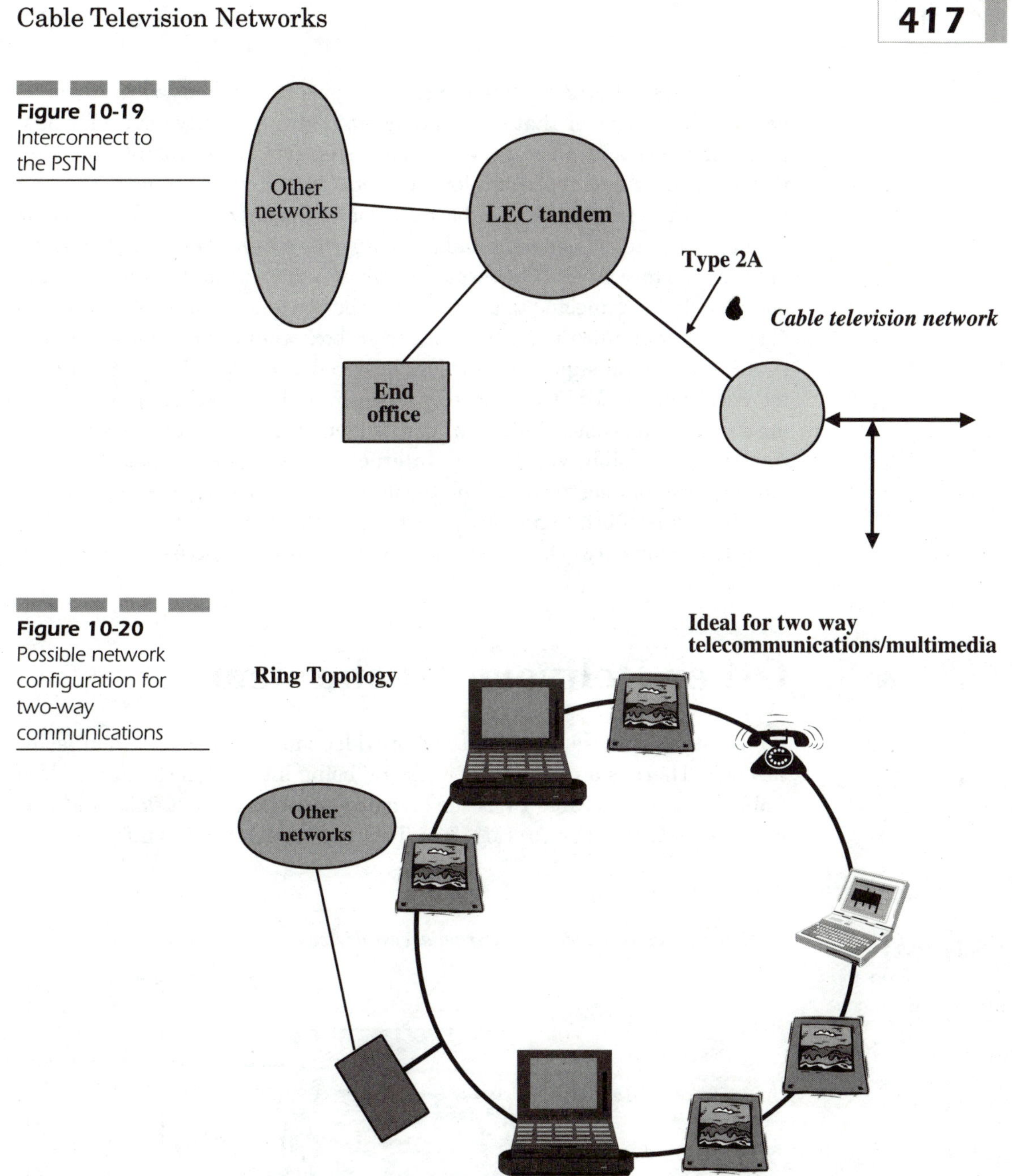

Figure 10-19 Interconnect to the PSTN

Figure 10-20 Possible network configuration for two-way communications

unless they make the changes. Carriers have been able to get subscribers to change terminal equipment through a combination of new services, cheap rate plans, and customer giveaways—all of which have required an intensive and extensive marketing effort.

The easiest alternative for the cable television company is to install a new type of converter that can serve as an interoperability device for new types of video and voice services. This step would be an interim step towards selling new types of television sets. The converter could serve as a modem for the Internet and as base station for a telephone device (wired or cordless). The television set could then serve as a dumb terminal for Internet access. The converter box would serve as a modem, and a separate keyboard could be connected to the converter box or television set. In regard to voice telecommunications, the converter box would serve as an access device that would support a cordless or wired telephone device. Finding a technical synergy for FCC-licensed wireless would be a long-range challenge for the cable television industry. At this moment, the technologies are not compatible. Possibly, via the new Internet access protocols that are being developed, a common technical platform can be found. Business imperatives will drive this effort. Figure 10-21 highlights the need for changes that the cable television network will need to undergo in order to access the Internet.

Other Delivery Mechanisms

Cable television does not need to be provided via fixed transmission facilities only. There is a new type of cable TV being introduced called Wireless Cable TV. Wireless cable TV is the layman's term for *Multi-Channel Multi-Point Distribution Service* (MMDS). This type of cable TV is a fascinating

Figure 10-21 Cable television and the Internet

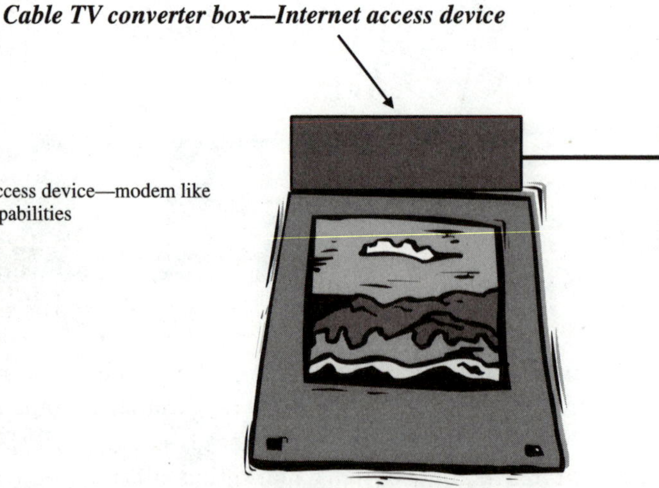

combination of satellite transmission/reception and fixed-transmission facilities. MMDS is a type of cable television system that offers its subscribers a mix of satellite channels by transmitting the programming over MMDS frequencies. Wireless cable uses *Super-High Frequency* (SHF) channels to transmit satellite cable programming over-the-air, instead of through overhead or underground wires. MMDS also offers bandwidth. Unfortunately, MMDS development in the area of interactive voice and video services has been limited. MMDS is a economical way of providing cable TV in high operating cost areas.

Scrambled satellite cable programming is received at a central location, where it is processed and is fed into special transmitters that are located in various Earth stations. The SHF transmitters distribute the programming throughout the coverage area. The signals are received by special antennas installed on subscribers' roofs, are combined with the existing VHF and UHF channels from the subscriber's existing antenna, and are distributed within the home or building through coaxial cable into a channel program selector located near the television set. This application combines satellite service in such a way that it embraces the benefits of satellite, rather than treating it as a competitive technology. This scenario is also an example of synergy between telecommunications business segments. Benefits of wireless cable TV are as follows:

- *Availability* Wireless cable can be made available in areas of scattered population and in other areas where it is too expensive to build a traditional cable station.
- *Affordability* Due to the lower costs of building a wireless cable station, savings can be passed on to the subscribers.

Figure 10-22 is a rendering of wireless cable television.

Wireless cable operations might have as many as 33 channels of broadcast and cable programming, depending of course on which channels are already used in your area. If you use digital compression, you can squeeze in another 150 to 300 channels. Wireless cable systems can carry any of the normal basic and premium cable television channels.

Wireless cable systems can optimally range up to 25 to 30 miles, which depends largely on the terrain, transmitting power, both the transmitting and receiving equipment, and many other factors. In order to receive the signal, the transmitting and receiving antennas must be line-of-site. Equipment requirements for the home are as follows:

- A *Multichannel Multipoint Distribution Service* (MMDS) antenna
- *Cable down converter* The cable television down converter is the cable television set top converter box. This box contains an addressable

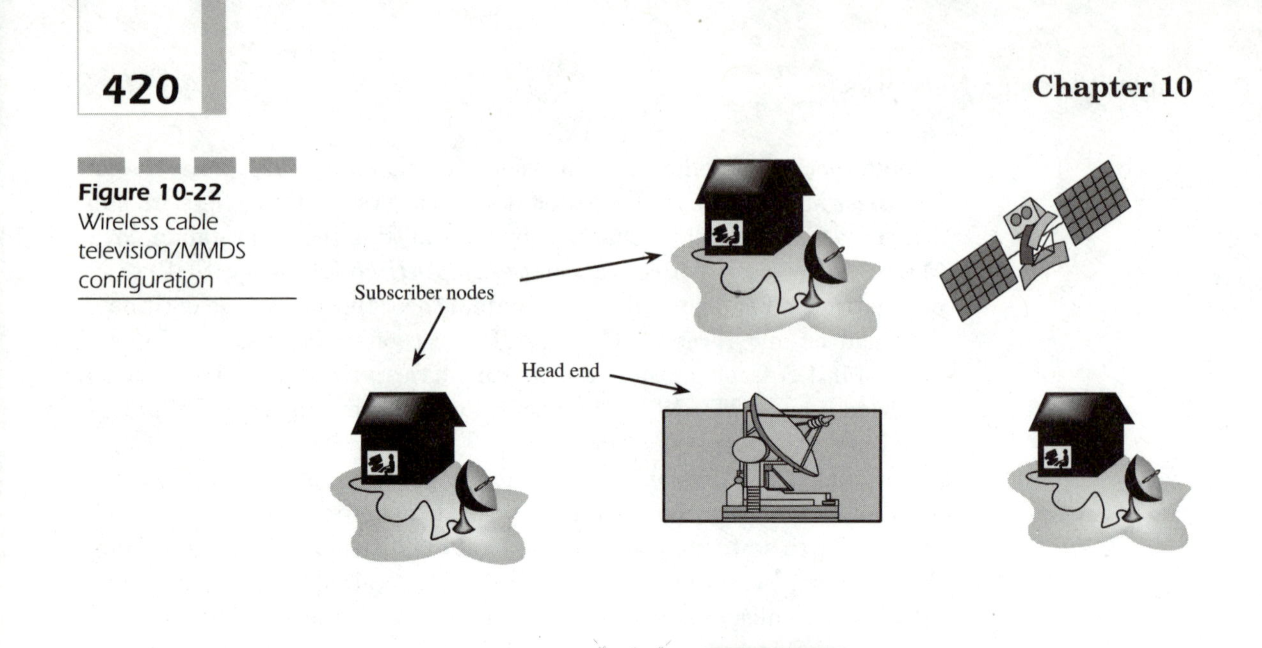

Figure 10-22 Wireless cable television/MMDS configuration

decoder and a VHF/UHF tuner that is built in, which gives the down converter the capability to tune in broadcast channels without having to use valuable MMDS channels.

- A UHF and VHF antenna for normal broadcast channels

Convergence

Convergence in the cable television industry involves creating new technologies and applying these technological strengths in a way that brings basic voice and data service to the mass market. All telecommunications companies are working towards the same goal: using their respective core strengths in order to gain a foothold in the mass marketplace. The cable television industry has access to entertainment programming, which has added a new dimension to the traditional telecommunications industry's view of content. Content is no longer just the customer's voice; rather, this concept now includes entertainment, history, education, music, and news. The cable television industry was once just an entertainment-delivery

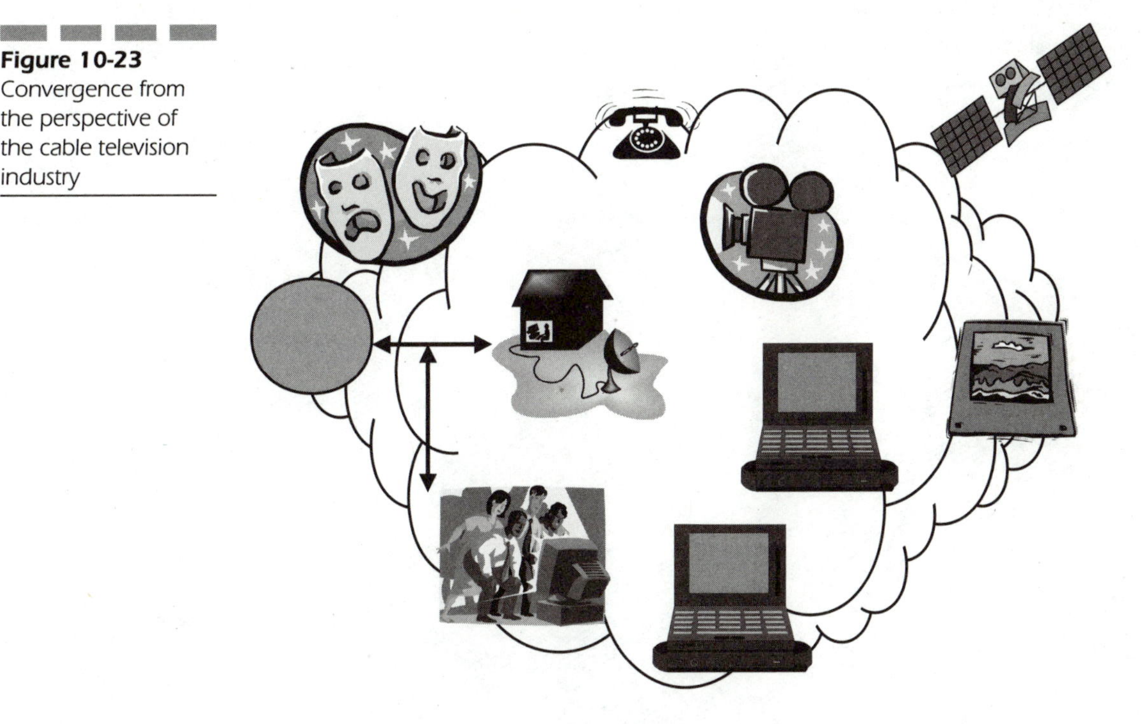

Figure 10-23 Convergence from the perspective of the cable television industry

mechanism. Today, the cable television industry is positioning itself to be a delivery mechanism and processor of information. Figure 10-23 is a rendering of the convergence of telecommunications from the perspective of a cable television operator. From the perspective of a cable television operator, entertainment/news media will be at the heart of the convergence.

SUMMARY

The cable television industry started as a delivery mechanism for broadcast television and has grown to become a potential player in the public telecommunications arena. Cable TV has the capability of easily supporting the Internet and wired voice business segments.

CHAPTER 11

The Internet

The Internet was conceived in the 1960s. A report had been written by the Rand Corporation for the United States government in the 1960s. This report outlined a vision of the future information network and described a view of information becoming a material good and the principal commodity of the 21st century. The U.S. government agency called the *Defense Advanced Research Project Agency* (DARPA) invested several billion dollars in developing packet-switching networks starting in the mid-1960s through the early 1970s. Initially, the customer for these switching networks was the U.S. Department of Defense; however, many of the researchers that had been contracted to carry out the research and development were scholars. These scholars became enamored with the test systems that they had built and realized that their work had applications beyond defense. Initially, the research network was called the *Advanced Research Project Agency Network* (ARPANET), but later it was called the Internet (which stands for Inter Network). The first e-mail was sent in 1972. Figure 11-1 is a rendering of the Internet.

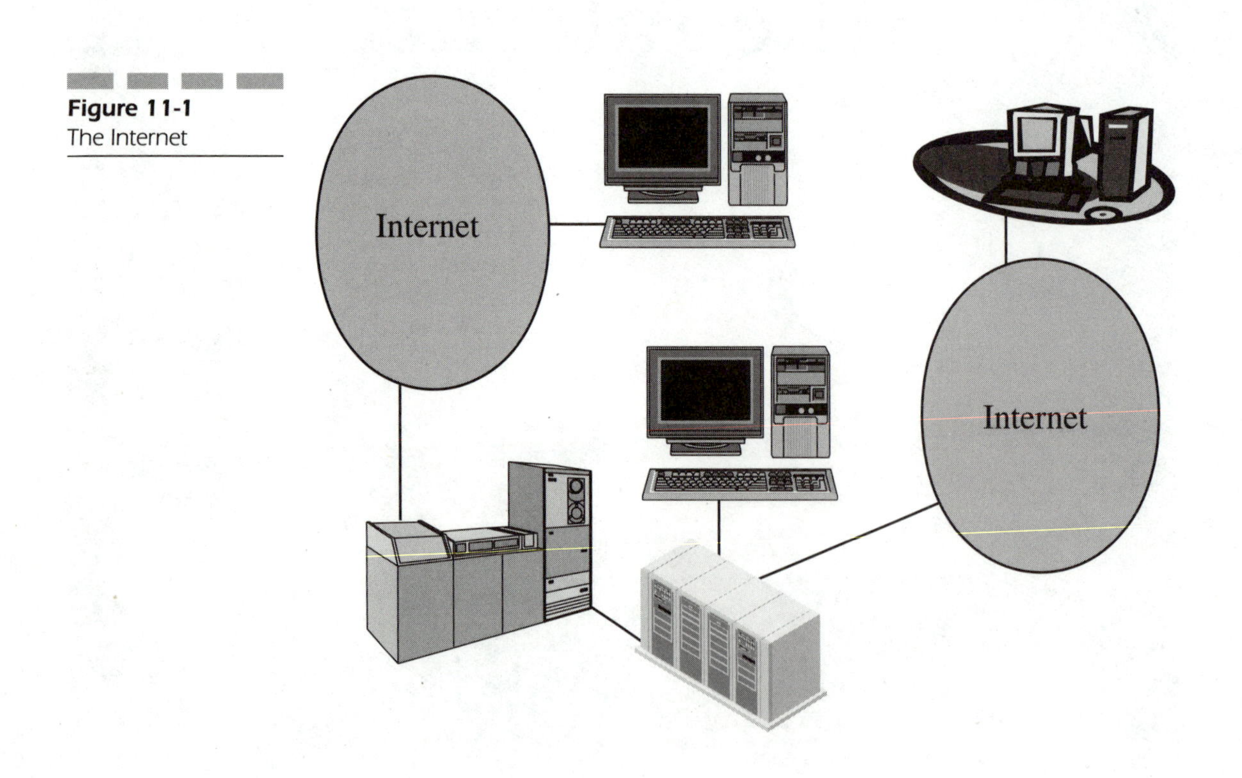

Figure 11-1
The Internet

What Is the Internet?

By the early 1980s, two other important pieces of technology in the United States had emerged. The first was the workstation/server system, which was emerging as a way to provide cost-effective computing to the desktop. The second was the Ethernet *Local Area Network* (LAN). The Ethernet was widely accepted as the way of providing communications between the desktop and server computers in the same organization (LAN). The computer and information technology researchers who were involved used all of these systems for computing, local, and national communications. The United States government had funded most of the initial implementations of the Internet technology on the basis that it would be made freely available to others.

By the mid-1980s, a large number of universities and research labs in Europe and the United States had access to the Internet through largely government-subsidized network links that were leased from the public network operators. This situation changed rapidly, however, so that now many Internet connections are paid for directly by the subscribing organization. Unlike most telecommunications standards that are developed under the auspices of two groups, the *American National Standards Institute* (ANSI) and the *International Telecommunications Union* (ITU), the Internet protocols were literally being developed out of sight. The Internet was and still is an information network structure that is not controlled by an single entity. As a result of this lack of centralized control, most people in the telecommunications sector ignored the Internet.

Internet users and companies that were interested in providing Internet equipment recognized the need to develop protocol standards. Hence, the Internet Society was created—an international not-for-profit professional body that was formed to enable individuals, government agencies, and companies to have a direct say in the direction that the protocols and the technology could move. The Internet Society is not controlled by ANSI or ITU. Figure 11-2 illustrates the apparent free flowing nature of the Internet.

Internet Network Architecture

The Internet is an excellent example of convergence. The Internet is an open pipe of information that enables access to anyone who uses a computer or some type of terminal-access device. The Internet currently carries voice, data, and video (real-time and stored).

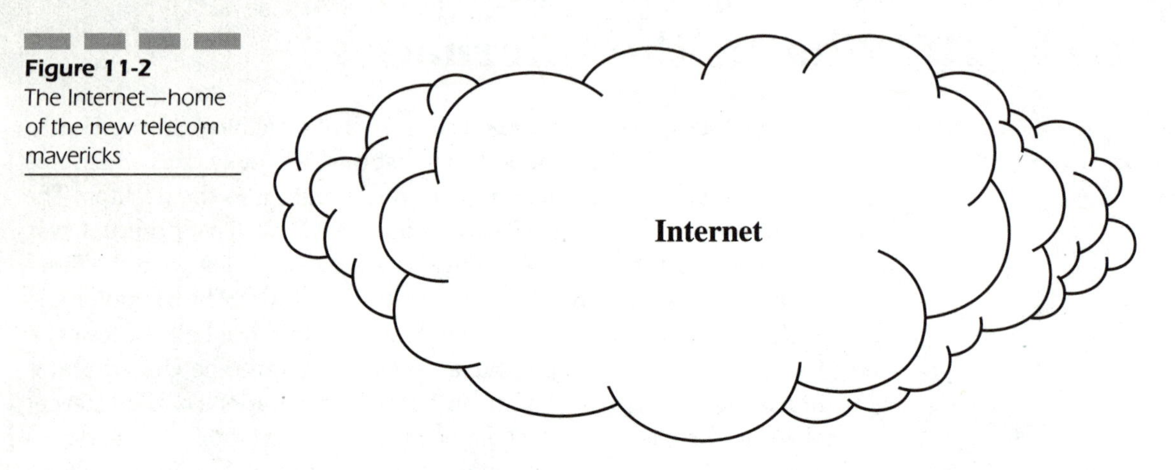

Figure 11-2 The Internet—home of the new telecom mavericks

The Internet is not controlled by a handful of companies or associations.

The Internet is the home of the new telecom mavericks.

The Internet is not a single network; rather, it is a web of networks (or a network of networks). Intelligence does not reside within a single component of the Internet; instead, it is embedded within the components of individual network elements. When you compare the Internet with the older and traditional voice networks, you will find that the Internet is decentralized in its intelligence and control. The power of the Internet is in the software that supports the applications and is in the protocol itself. No one controls the pipe. Figure 11-3 is an illustration highlighting the major differences between the Internet and the other telecommunications industry segments.

Another way of looking at the Internet is that the *public telecommunications networks* (PSTN and wireless voice networks) look at voice as their principal business. They also perceive data (non-voice information) to be a line of business that they want to enter heavily. The Internet, on the other hand, looks at data (non-voice information) as its principal line of business and looks at voice as a line of business it is beginning to heavily enter. The traditional public telecommunications carriers, satellite, and cable television are all seeking ways of entering the Internet business. The challenge for these companies is understanding what product they will be providing. The Internet is not a product; rather, it is an information medium that sup-

Figure 11-3 Anyone can enter the Internet business

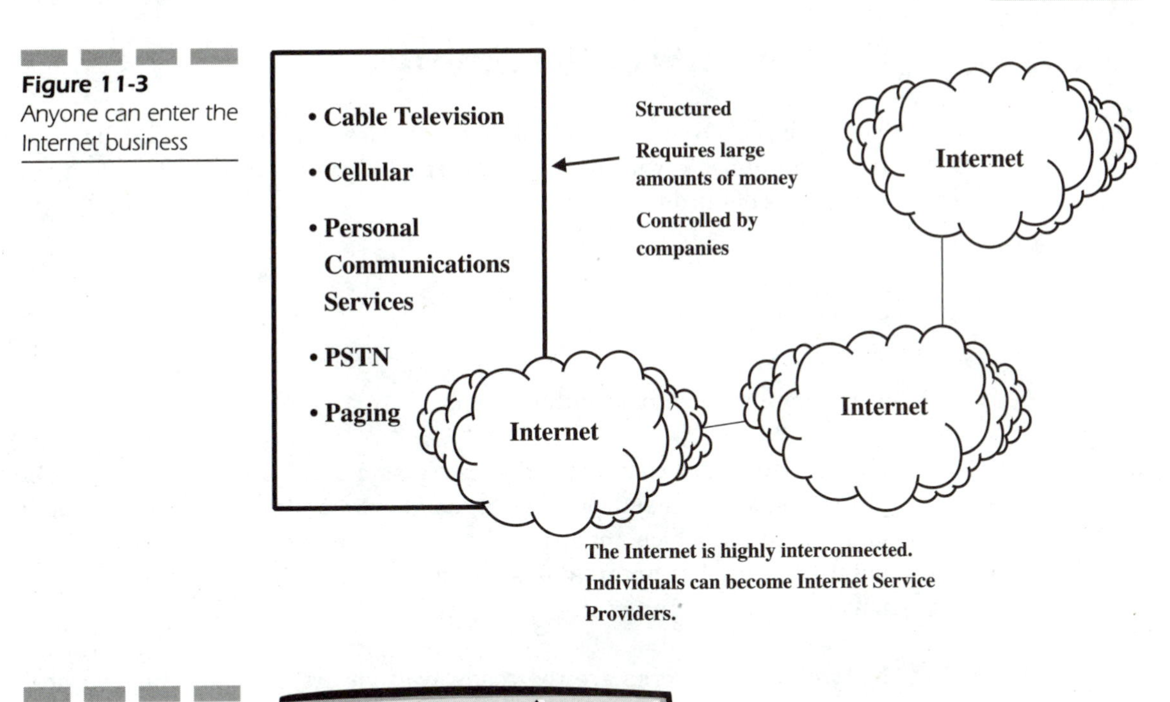

Figure 11-4 The Internet—the new information medium

An information highway

ports bidirectional transmission. Figure 11-4 illustrates the points just made. The Internet is the new information highway.

The differences go even deeper: network architecture is different, traffic engineering profiles are different, and even the pricing mechanisms are different. More information about this subject will appear later in this chapter.

Components of the Internet

The Internet is a collection of computing devices and transmission facilities that communicate with one another. The Internet is basically made up of four pieces of hardware:

- Hosts
- Networks
- Routers
- Computers/terminals with modems

Hosts Hosts are computers that serve as central nodes for information processing and transactions. Hosts can be workstations, PCs, servers, and mainframes on which applications are run. The hosts are the central computer systems that are used to support the users who wish to connect to a network or to the Internet.

Networks Networks are the roadways that are built to connect a host and its subscribers to other hosts and their subscribers. Networks in this case are conceptually not any different than the traditional voice networks in a wireline or wireless voice service provider business.

LANs (Ethernets), point-to-point leased lines, and dial-up links (telephone, ISDN, and X.25) are all transport mechanisms that carry traffic between one computer and another. These elements glue together all of the different network technologies to provide a ubiquitous service that is used to deliver information packets.

Routers Routers are special-purpose computers that are good at talking to network links. Some people use general-purpose computers as low-performance (low-cost) routers; e.g., PCs or Unix boxes with multiple LAN cards, serial-line cards, or modems. Routers are akin to wireline tandem switches. Routers direct traffic but are smart traffic switches that select the best route possible. Routers provide network-management capabilities such as load balancing, dynamic route selection, trouble alert, and trouble identification.

The typical network configuration of the Internet appears as follows. Figure 11-5 is a generic representation of the Internet. The Internet is a conglomeration of computers and LANs. All of these elements are designed to communicate with each other.

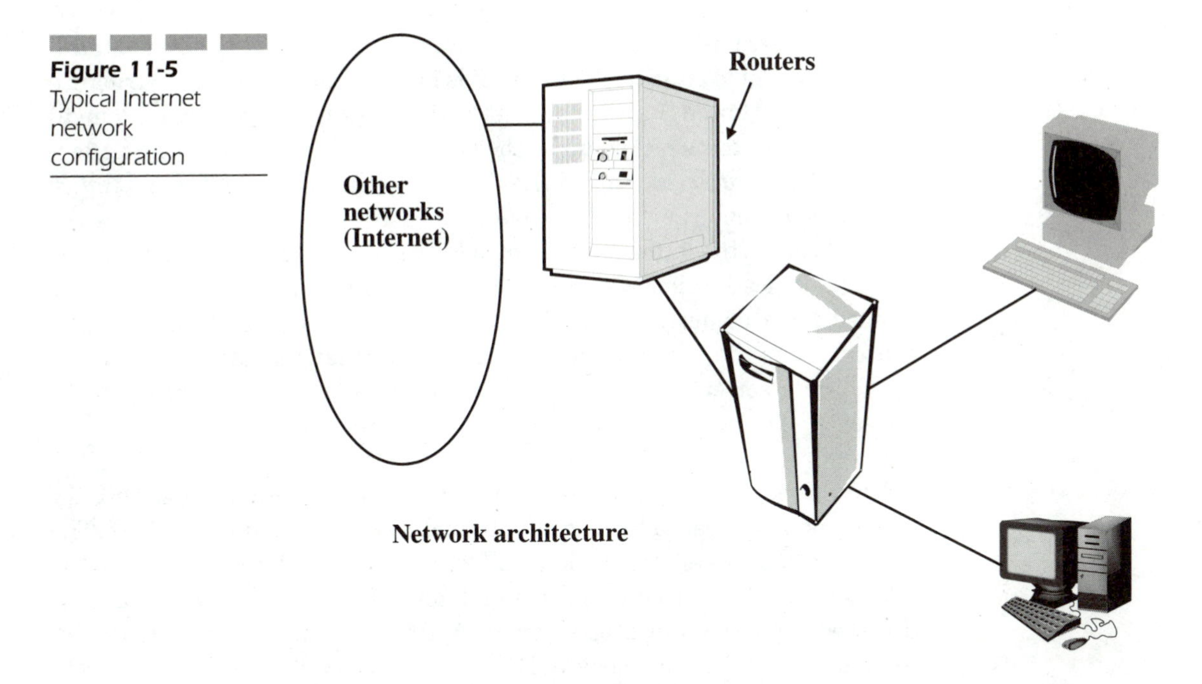

Figure 11-5 Typical Internet network configuration

Computer/Terminals with Modems Computers come in all shapes and sizes; desktops, laptops and palm sized. Computing devices coupled with a modem or some type of high speed network connection serve as the user interface to the Internet.

Internet Routing

Every computer (host or router) in the Internet has a name that is given by the host/router operator. Internet names are actually hierarchical and look similar to postal addresses. This name system is also known as the *Domain Name System* (DNS).

In order to be reached via the Internet, the called party must have an address that can be reached. The address describes to the Internet addressee the respective location and is assigned by the Internet Society or designated entity. The address tells the user where the location is topologically. An example of an Internet address is 237.22.9.74. The number given to me by the Internet Society is 237.22.q.r, and the q and r are free for me to choose. The digit place called q represents the LAN, while the digit r represents the host. q and r can be any number in the range of one to 256.

The Internet is similar to the highway system in the United States. Another way of looking at the Internet is by comparing the entire structure to that of the United States postal network. For example, you decide to mail a letter. The letter is picked up by the postal worker. The postal worker represents the *Local Area Network* (LAN) or the transport piece. The post office sorters represent the Internet router. This router (i.e., the postal letter sorter) looks at the letter destination address (i.e., the data package's called address or destination address). The sorter sees that the letter is either headed for a domestic or an international address. The sorter (i.e., the router) sends the letter to the appropriate destination post office (i.e., the appropriate sorter/router). The letter continually gets moved to the next-lowest level router/sorter until it reaches its destination.

Addressing DNS is designed to be an internal information service for use by computer-based tools, rather than directly by users/subscribers. The names used by a system are called addresses of hosts. These names are the addresses that are used to route e-mail and to identify the system as it is accessed for other types of services. A DNS maintains the database of addresses, and there are several DNSs in operation (none of which are controlled by a single company).

DNSs can be owned by anyone. The naming system is maintained by a single organization and is mutually agreed upon by the members of the Internet Society. The naming servers function as a centrally controlled and orderly way of reaching Internet Web sites. This setup is similar to the 800/888/877 databases that are maintained by the wireline telecommunications industry. The 800/888/877 number assignments are maintained by a single company and are supported by the telecommunications industry (besides being authorized by the FCC). Prior to the divestiture of the AT&T Bell System, there was a single database that stored and maintained control of all 800 numbers (888 and 877 were not in use). The 800 database was controlled by AT&T. Today, the 800/888/877 database is several databases that are owned and operated by many carriers. Each carrier updates his or her databases based on the public announcements that are issued by the FCC and by the industry-authorized group called the North American Numbering Plan Administrator. The DNSs are owned by several companies and are accessed by multiple hosts.

Although it is not obvious, the Internet is a hierarchical structure that is designed to interact with hosts and routers in much the same way that commercial voice telecommunications networks switches interact. The voice-switching analogy would include local switches and the way they relate to tandem switches. The tandem switch is a higher-level switch. The

Figure 11-6
Internet addressing

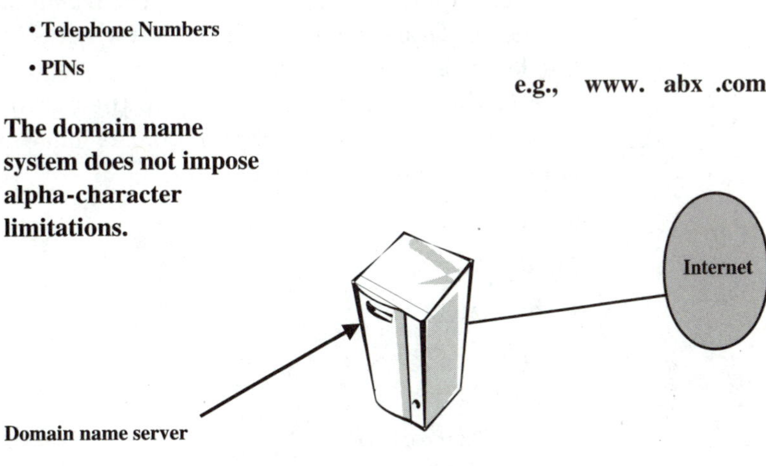

switching addresses used by wireline carriers and wireless carriers are called *Common Language Location Identification* (CLLI) codes or *Signaling System 7* (SS7) point codes. Figure 11-6 highlights some of the key characteristics of the Internet addressing schema.

The Internet hierarchy is evident in the way that Internet network elements are named. For instance, the address `book.computerstuff.california.bd.com` shows that the address is a United States code (i.e., .com). The rest of the address shows who owns the host, which is then followed by the owner of the address. The address works it way down to the ownership and destination chain, providing more information about the host site and sometimes even providing information about the routers that are involved in the process. Finally, the address points directly to a name or alias that the user/subscriber would like to be known by to others. The last address extension,.com for example, is assigned based on user type. The following domain name extensions are currently used:

- *.com* Commercial (United States)
- *.edu* Educational (United States)
- *.org* Non-profit organizations
- *.net* *Internet Service Providers* (ISPs) who typically own their own network (or a user who has his or her own network)

- *.mil* United States military
- *.gov* United States government
- *.us* United States (used when international boundaries are crossed)

International extensions are used when boundaries are crossed by Internet users. Think about sending a letter across international boundaries. Senders need to identify the country and the locale in some manner. In the case of the Internet, countries have their own designations (which are agreed upon by the Internet Society). The following list describes examples of international designations:

- *.ar* Argentina
- *.at* Austria
- *.au* Australia
- *.bm* Bermuda
- *.bo* Bolivia
- *.ch* Switzerland
- *.co* Colombia
- *.dk* Denmark
- *.fi* Finland
- *.fr* France
- *.ie* Ireland
- *.nz* New Zealand
- *.sg* Singapore
- *.th* Thailand
- *.uk* United Kingdom
- *.us* United States

The Language of the Internet

The language of the Internet is actually a combination of two protocols. The protocol suite is called *Transmission Control Protocol / Internet Protocol* (TCP/IP). TCP/IP is responsible for the communication portion of the language, while IP is responsible for the work of routing the message. As I had indicated in Chapter 4, "Network Signaling and Applications," the Internet layer (also known as the Network layer in the OSI model) is responsible for managing the *Internet Protocol* (IP). The IP provides the Internet addressing for routing and is a connectionless protocol that provides a method of transmitting information. The information packet is broken into sections and is transmitted in packets across the network.

The Transport layer, which corresponds to the same transport layer in the OSI model, transports the data. The *Transmission Control Protocol* (TCP) is run at the Transport layer. The TCP/IP protocol is designed to be capable of supporting a variety of different application protocols. This flexibility has enabled Internet Web designers to create Web pages with various capabilities. For instance, programming languages such as Java have been applied to the Internet in such a way that a user can see and perform the following functions from a Web page:

- Animation
- Audio
- Animated banner advertising
- *Interactive applets* Downloadable programs for immediate use
- Interactive icons and images

The Internet protocol has also assumed another function. TCP/IP has become the common language of all *Operational Support System* (OSS) network elements. Wireline and wireless public telecommunications networks tend to use OSSs from a variety of different vendors. As a result of the different vendor equipment, one will find that a base, common language is needed not only for intrasystem communication but also for operating purposes (one cannot expect carrier personnel to learn two or three dozen different operating languages). Digressing for a moment, the OSSs of the public telecommunications networks have taken on a more hierarchical structure. Not too long ago, the OSS environment was characterized by multiple vendor monitoring and control equipment—none of which was interoperable. Today, compatibility and interoperability are goals of the OSS environment. The use of TCP/IP is a reflection of this more structured and more uniform environment. Figure 11-7 is a rendering that highlights the importance and role of TCP/IP.

Internet Access and Network Interconnection

The Internet is accessed and is used by corporate users and home users. The corporate user typically accesses the Internet via the corporate intranet, which interconnects the company's LAN via a host. There will be more information about the intranet later in this chapter. The home user accesses

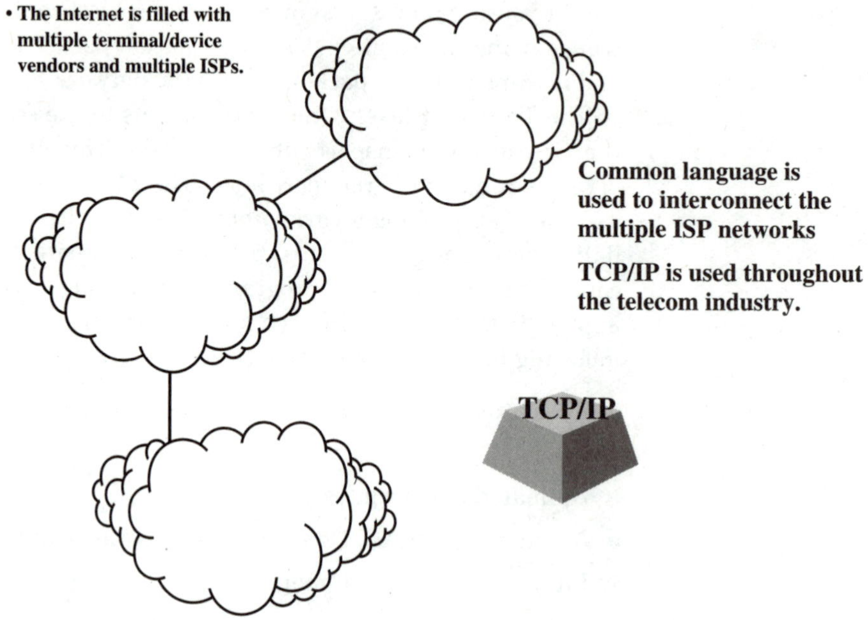

Figure 11-7
The Internet—TCP/IP

• The Internet is filled with multiple terminal/device vendors and multiple ISPs.

Common language is used to interconnect the multiple ISP networks

TCP/IP is used throughout the telecom industry.

TCP/IP

the Internet via an *Internet Service Provider* (ISP). The host uses TCP/IP to access the Internet, and the home user accesses the Internet via a modem in the home computer.

The Internet is not a single network that is owned by a single company or person; rather, it is a network of networks. Each network communicates with the other by using computers as the principal terminal device. The corporate user is using a data network to communicate with the Internet, which is also known as the World Wide Web (the Web). The data network is typically connected to the PSTN, and the home user purchases access from an ISP. The ISP serves as the host to subscribers of their service and is reached by home computers via modems that dial telephone numbers that are associated with the ISP. The modem makes these ISP calls on twisted-pair wire. The ISP is connected to several DNSs, as is the corporate intranet. The DNSs enable the users to find the destination/requested Web site (another host). Figure 11-8 illustrates the standard network interconnect.

You should note that we are looking at two distinct ways to connect the various Web sites, yet there is still a common node called the PSTN. The corporate intranet is a LAN that is effectively an internal company Internet. The intranet supports everything that the Internet user has access to, but instead, its principal function is to support corporate communications. The

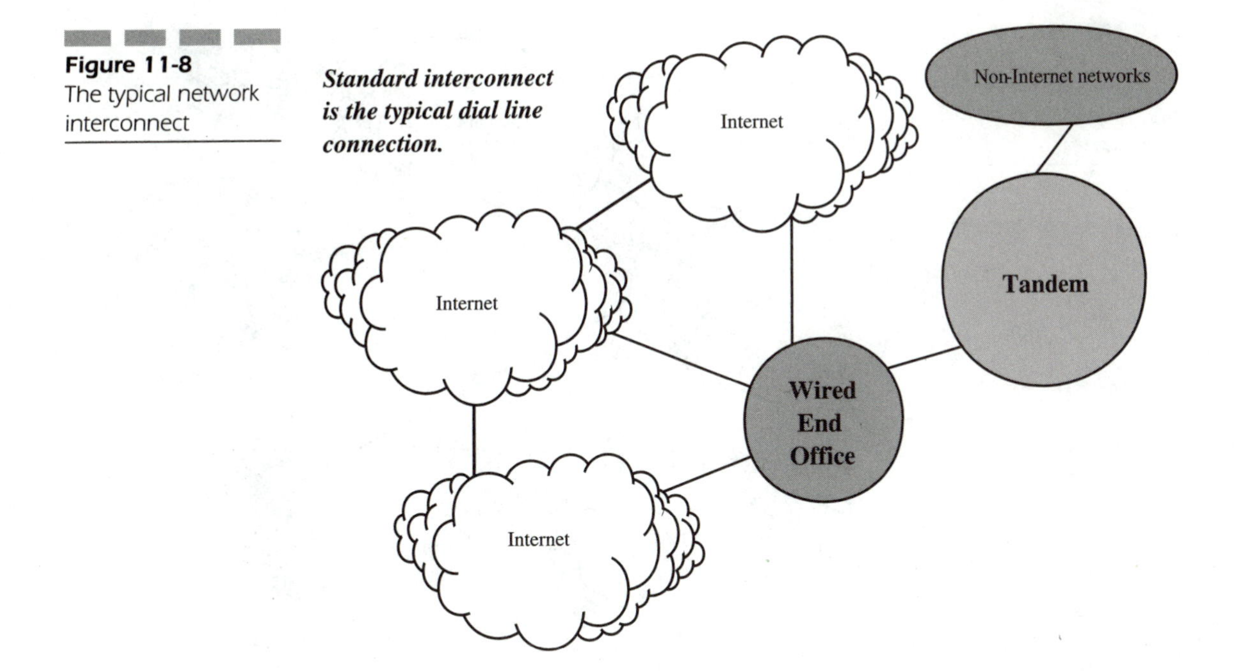

Figure 11-8 The typical network interconnect

intranet obtains a data line, probably a facility at T1 or T3 rates, from the wireline telephone company. The intranet is most likely interconnected to an ILEC end office. The wireline telephone company assigns a dialable number to the company, and the dialable number is associated with a domain name and address that hosts can access via their TCP/IP connection.

The home user accesses the ISP via dialable numbers and essentially calls the ISP via the home computer, then reaches other Web sites through this ISP. The standard interconnect to the Internet for an ISP is via a typical dial line from the end office. Today, telephone companies have installed routers in their networks that do nothing but interconnect intranets and ISPs. Network interconnect in the realm of the Internet is no different than that of any other type of carrier. Figure 11-9 is an illustration of a generic network interconnect involving the ISP.

ISPs are interested in network interconnection as much as any other carrier. ISPs are required to pay for long-distance access. Every time an ISP reaches out to the Internet outside of the LATA boundaries of the ILEC, the ISP is required to pay an access charge to the ILEC. There are some ISPs that usually pass on the cost of the long-distance portion via some type of long-distance transport charge. This charge would be over any hourly fee that the ISP would levy on the subscriber/user for use of the ISP services.

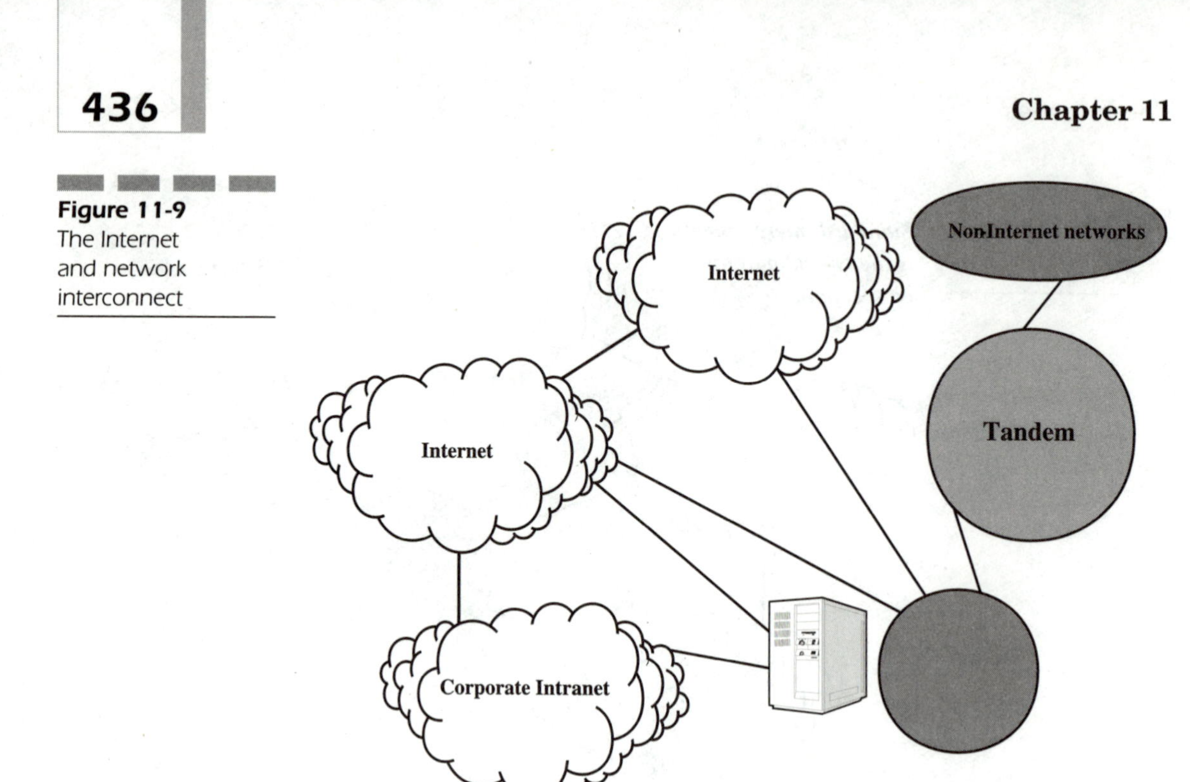

Figure 11-9 The Internet and network interconnect

Note that there are some ISPs that simply charge the subscriber/user for access to the ISP services and that make the subscriber/user responsible for the cost of the local telephone call via the normal monthly telephone bill.

As one might have assumed, anyone can be an ISP. Many understand this fact, and for this reason, the Internet has been flooded with a plethora of ISPs. There are ISPs that serve local communities, regions of the country and the nation and that provide international access. I use the word *Web* to describe the Internet. The thousands of ISPs and intranets—if interconnected—would form a picture of a spider web. Many people speak of the Internet backbone. This scenario is difficult to imagine, because there is no single company that is in control of the Internet. There are long-distance carriers, however, that own nation-wide transmission networks. In this manner of speaking, there is a backbone to the Internet. Figure 11-10 is an illustration of the "loose" nature of the Internet business. This situation may change as content monitoring becomes a "must" in order to protect children from various unsavory influences.

Figure 11-11 is an illustration of the need for a nationwide backbone network in order to support nationwide Internet connectivity.

The Internet is accessible by everyone and is controlled by no one. Although the ISPs must seek ILEC interconnection, the fact is that the

Figure 11-10
The Internet is not a complicated business to enter.

growing CLEC community is slowly taking away the ISP interconnect business. Therefore, network visibility is maintained through the CLECs. Figure 11-12 illustrates the role of the CLECs.

Applications

The Internet is a medium that enables communication between computer users. The Internet is not an application, per se; rather, it is a medium that is a blank slate as it relates to subscriber value. No one individual can say what product the Internet provides. To many, it seems to be the principal form of communication. The Internet supports a variety of capabilities, but it does not support a single product. For example, the following functions describe the Internet's services, not its products. These functions are not purchased by subscribers:

- *Research* Library access
- *Advertising* Advertising space is sold as banners on a Web page.

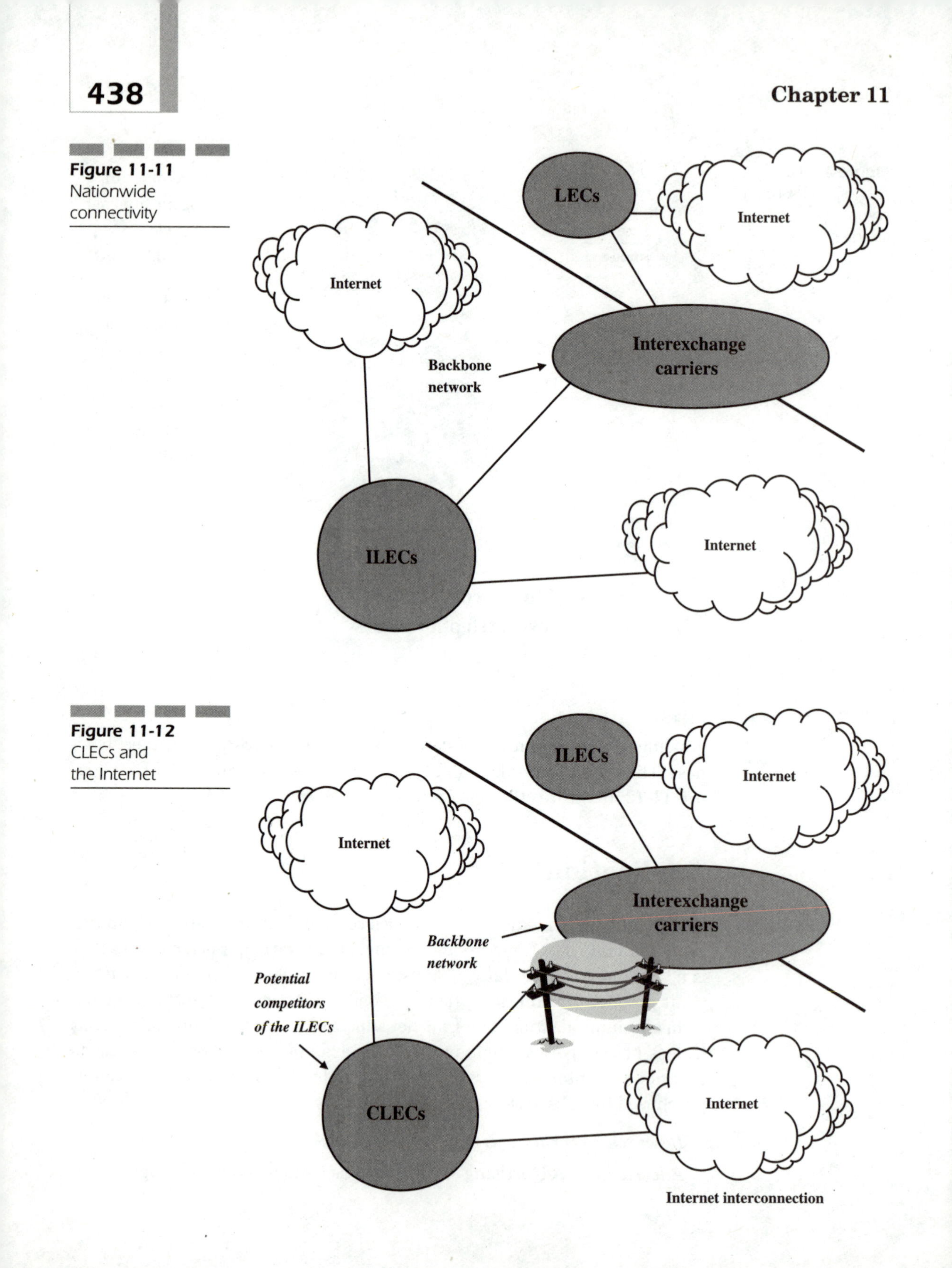

Figure 11-11 Nationwide connectivity

Figure 11-12 CLECs and the Internet

- *Web page publishing* This service concerns itself with any type of text material that can be found on a Web page.
- *Applets* This service includes software applications that are either found on the computer or are downloaded from a Web site for use at that moment.
- *Web browsing* Search engines are used to search, in an organized fashion, for other Web sites or topics (includes URL support).
- Communication between people and computers (i.e., e-mail)

Figure 11-13 is an illustration highlighting the various applications of the Internet.

Although these services are not purchased by Internet users, they are value-added capabilities that make the Internet attractive to users. The question still remains, "What products and services do the users buy from the Internet?" The answer is that the Internet itself is not a product; rather, it provides a service through access to people and information (databases). This access is what makes the Internet a valuable tool. Today, a child or adult can quickly obtain information from a library on the availability of a book, find out airline flight schedules, train schedules, movie time schedules, and information about almost any topic. The information obtained can used for a variety of purposes, including purchasing products from the database holder. The user is limited by what the targeted database will allow

Figure 11-13 Applications

and by the software on either the desktop computer or the targeted host site. The purchasing of products is an activity that has come to be known as *electronic commerce* (e-commerce). E-commerce includes catalog purchasing off the Internet, where the selling party displays pictures of its products, displays interactive (point-and-click) icons, and provides online purchase order forms. These sellers of products even have their databases tied into credit card validation databases. E-commerce is an activity/application and is not a product to be sold to an Internet user. Figure 11-14 highlights how the Internet has become a medium for retail shopping

Another application of the Internet is the corporate intranet. Corporate intranets were established to support internal corporate communications and to provide access to external organizations. The intranet has led to improved communications between parties of a company and between companies. The value of the intranet can be described as a series of activities and functions that are listed as follows:

- Internet access
- Internal e-mail
- Corporate directory services
- *Archival storage of files* Corporate servers are now being used to store electronic copies of corporate documents.
- Fax service

Figure 11-14 Online shopping and e-commerce

The Internet is a medium—shopping is one application

- **E-commerce**
- **Online shopping**

- Video conferencing
- *External communications* Includes e-mail
- Corporate advertising for external communications

Figure 11-15 depicts how Internet technology is used in the corporate world.

Another activity that the Internet can support is multimedia service (broad-band service) transmission. The Internet already has the bandwidth. The desktop computer or laptop computer is the limiting factor, however, and not the transmission facilities. The processing speed, the type of video driver, the type of audio driver, the type of audio speakers, the type of media reader (disk or CD-ROM), the type of monitor, and the software that is supported by the computer are what dictate the kinds of multimedia services that the Internet user can access. Today, one can see an entire movie or hear a song on a computer.

Up to this point, I have described a number of Internet applications—but none describe a specific Internet product. The answer is that there is no commercial Internet product. The Internet does not sell voice like the wireline or wireless carrier does. The Internet does not sell entertainment programming, as does cable television. The Internet also does not sell paging

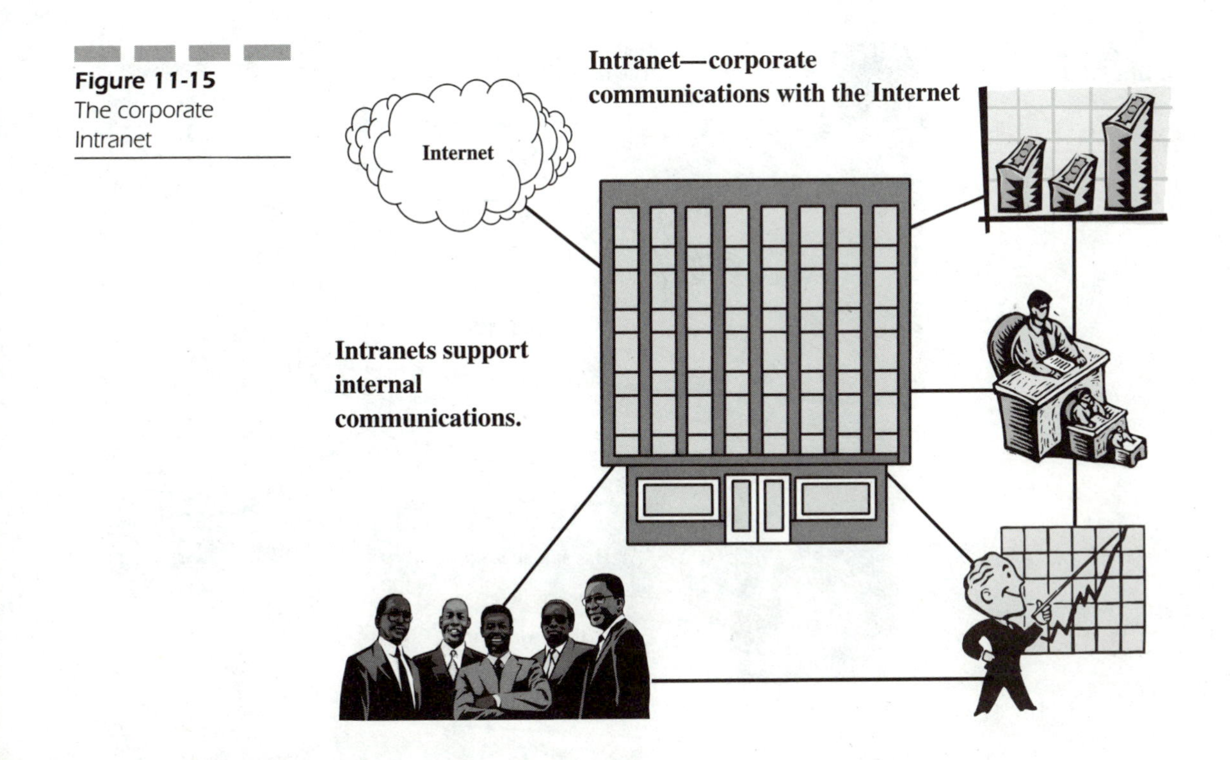

Figure 11-15 The corporate Intranet

like a paging carrier does. The telecommunications industry is seeking ways to exploit the Internet but still has not hit upon the *killer app*. The term "killer app" refers to the "killer application," which will capture the imagination of the customer, galvanize an industry, and become the underpinnings of the business. There are really no "killer apps" per se but rather "a convergence of opportunities" that occurs at the right time in the marketplace. The Internet is a medium through which companies and carriers can transact business and is a blank slate or blank medium that has not yet been fully exploited by the telecommunications industry. The Internet is in its business infancy, and its principal service is access. Figure 11-16 reiterates the key attributes of the Internet.

Convergence—Business of the Internet

The Internet is representative of technology convergence, in that it is a network of networks and a collection of LANs and computers where no single network or computer controls the network of networks. The Internet is highly decentralized and has an embedded intelligence that resides within the individual networks and the network elements. The Internet offers new

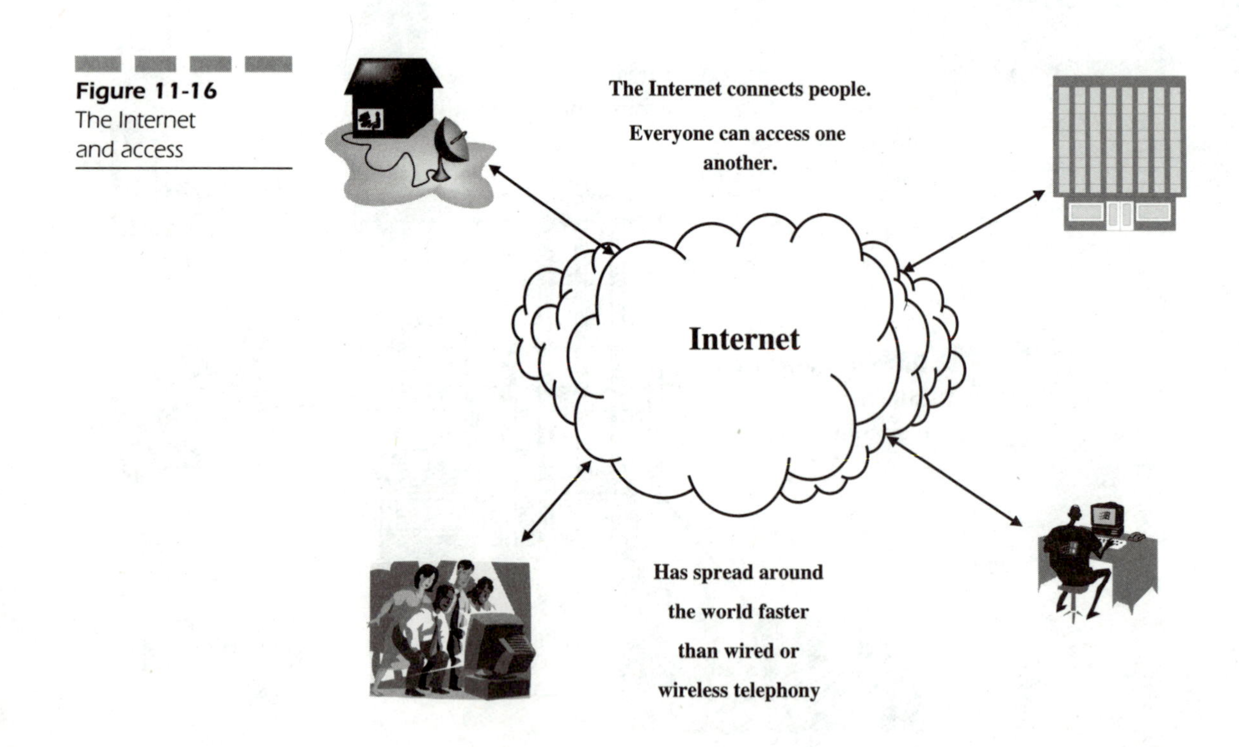

Figure 11-16 The Internet and access

business opportunities through its capability to access people and databases. How the access is used is the key to how the telecommunications industry and the business community will grow. This access capability has yet to be fully explored.

The Internet has brought communications to the subscriber/user in ways that even the common household telephone, cellular handset, or PCS handset could not. The Internet is similar to the household telephone and is something to which nearly everyone will have access. The Internet today is used by many people for research, shopping, advertising, looking for employment, playing video games, sending e-mail, meeting people, reading the news, and so on. The Internet is also capable of becoming all things to everyone. Figure 11-17 is a depiction of how deeply embedded the Internet has become within the marketplace in just a few short years.

The attention that the Internet is receiving from both the telecommunications industry and the business community is causing the Internet to become a focal point for convergence. I noted previously that the telecommunications industry is seeking ways to exploit the Internet by offering Web browsing services, e-mail, fax mail, news, and voice over the Internet. The wireline carriers are working hard to provide voice over the Internet, which is also known as *Voice-Over IP* (VoIP). By offering voice over the

Figure 11-17 The Internet and the marketplace

Figure 11-18 Voice Over IP—voice the common service of the marketplace

Internet, the wireline carriers are seeking ways to keep subscribers on their network by offering access to the most basic of telecommunications services: voice. Figure 11-18 highlights the ever growing importance of voice over the Internet.

Convergence is not just about merging wireless, wireline telephony, satellite, cable TV, data, the Internet, and entertainment to form information networks. Convergence is also about leveraging strengths to create new opportunities.

SUMMARY

The Internet can be likened to a collection of LANs, which are themselves comprised of desktop computers with graphics monitors. The Internet is not just one network but a huge network of networks (or possibly even a virtual network). Accessing the Internet can be done via a variety of media. The Internet cannot be characterized as being either strictly terrestrial or strictly wireless; rather, it is almost like a media in itself. The limitations of the Internet are those of the desktop computer and the transmission facility. The Internet is much like a living organism, because it is growing and changing—and no single individual or company has absolute control over it.

Figure 11-19 The Internet—a medium for convergence

The Internet a medium for convergence and creation

Figure 11-19 is a rendering of the Internet and its role in the convergence of the telecommunications industry.

Accessing the Internet via cable TV modems is one way of enhancing the home shopping business. Cable TV transmission facilities offer the bandwidth needed for real-time interaction between the user/subscriber and the home shopping channel. The Internet is even affecting the types of retail selling methods, however. The brick-and-mortar stores are giving way to Internet online shopping. You will be able to look through a catalog of goods while viewing a live advertising transmission for a specific product. Accessing the Internet can be conducted via satellite transmission, the wireline telephone network, the wireless network, a corporate local area computer network, a two-way paging network, etc. The only limitation will be in the terminal device and in the transmission network.

The Economics of and Requirements for Becoming a Telecommunications Carrier

Convergence is as good a word as any to describe what is occurring in the telecommunications industry. Corporations are finding synergies in their business plans, and by using the synergies found in their technologies, they are either building new businesses or enhancing existing businesses. To understand what it costs to become a public telecommunications carrier, let's briefly review the impact of the Telecommunications Act of 1996.

The Telecommunications Act of 1996 and Other Regulatory Commitments

February 1, 1996, Congress passed the Telecommunications Act of 1996, also known as "the Act." The Act rewrote the rules of the Communications Act of 1934 and changed the rules for competition and regulation in all major sectors of the telecommunications industry. The areas that were impacted ranged from local and long-distance telephone services to cable television, broadcasting, and equipment manufacturing. The provisions of the Act can be placed into five major areas:

- Telephone service
- Telecommunications equipment manufacturing
- Cable television
- Radio and television broadcasting
- The Internet and online computer services

Figure 12-1 is a chart of the five major areas addressed by the Act.

The Act declared invalid all state rules that restricted entry or limited competition in telephone service—both local and long distance—and established provisions that dismantled the AT&T and GTE antitrust consent decrees (modification to the final judgment), including federal prohibitions on entry by the baby *Bell Operating Companies* (BOCs) into the inter-LATA telephone market (known to the layman as long-distance service). Competitive safeguards were established to protect the industry against anticompetitive behavior by the incumbent local exchange carriers. These safeguards included the requirement to establish separate affiliates to enter long-distance and other telecommunications businesses and a prohibition of cross-subsidization. The Act was designed to break down both local and long-distance market barriers not only for the established carriers, but also for the new startups.

Figure 12-1 The Telecommunications Act of 1996

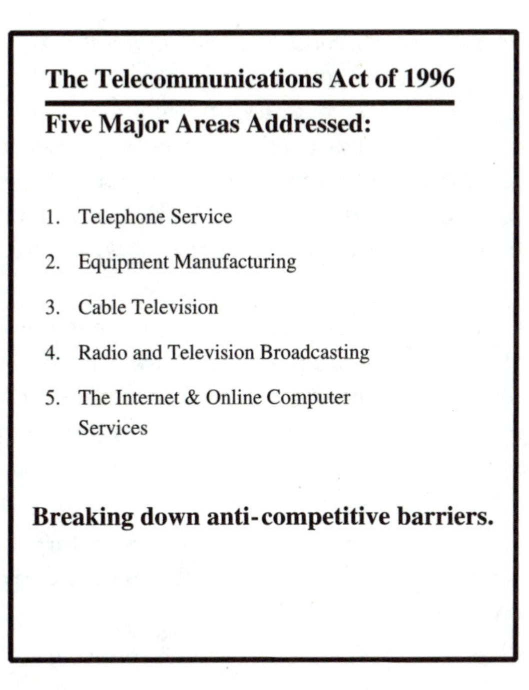

The ILECs will be permitted to offer inter-LATA/long-distance service once they have successfully completed steps to remove entry barriers to competition for local telephone service. At the time of writing this chapter, Bell Atlantic Corporation became the first ILEC to be given permission to enter the long distance market. The Act requires the BOCs and other ILECs to implement a series of reforms that are known as the Competitive Checklist in order to qualify for providing long-distance service outside their regions. The Act also requires all local exchange carriers to interconnect with new entrants, to unbundle their networks, to enable resale by competitors, to provide number portability so that subscribers can keep their telephone numbers when switching local providers, and to take other steps to promote an effectively competitive local exchange market. The 14-point checklist established by TA 1996 can be summarized as follows:

- *Interconnection* The ILEC must enable other carriers to link networks to its network for the mutual exchange of traffic.
- *Access to unbundled network elements* The telecommunications network is comprised of individual network elements. In order to provide access to an unbundled network element, the ILEC must provide a connection to the network element (at any technically feasible point) under rates, terms, and conditions that are just, reasonable, and non-discriminatory.

- *Access to poles, ducts, conduits, and rights-of-way* In order to serve customers, telephone company wires must be attached to, or pass through, poles, ducts, conduits, and rights-of-way. The ILEC must demonstrate that other carriers can obtain access to its poles, ducts, conduits, and rights-of-way within reasonable time frames and on reasonable terms and conditions—with a minimum of administrative costs—and that this access is consistent with fair and efficient practices.
- *Unbundled local loops* Local loops are the wires, poles, and conduits which the telephone company end office uses to connect to the customer's home or business. Non-discriminatory access to unbundled local loops ensures that new entrants can provide quality telephone service promptly to new customers, without constructing new loops to each customer's home or business.
- *Unbundled local transport* Non-discriminatory access to the ILEC's transport facilities ensures that calls that are carried over competitors' lines are completed properly.
- *Unbundled local switching* A switch connects end-user lines and connects trunks that are used for transporting calls. The ILEC is required to provide non-discriminatory pricing for the switching of calls for competing carriers and for non-competing carriers. As an illustration, this switching refers to switching calls from CLEC "A" to CLEC "B" or from CLEC "A" to the ILEC itself. Current pricing ranges from .0025 cents to .0045 cents for local switching and .0015 cents for tandem (regional) switching.
- *9-1-1 and Enhanced 9-1-1 services, operator services, and directory assistance* The ILECs must provide competing carriers with accurate and non-discriminatory access to 9-1-1/E9-1-1 services so that the competing carriers' customers can reach emergency assistance.
- *White Pages directory listings* These listings are the listings of customers' telephone numbers in a particular area. The ILEC shall provide competing carriers with access to all White Pages listings in the ILEC's operating area. Access shall include updated subscriber listings and permission to copy all White Pages listings.
- *Numbering administration* The ILEC must demonstrate compliance with industry guidelines and FCC requirements to ensure that its competitors have the same access to new telephone numbers in a given area code as provided to the ILEC.
- *Databases and associated signaling* New entrants must have the same access as the ILEC to these databases and signaling systems in

order to have the same capability as the ILEC to transmit, route, complete, and bill for telephone calls.

- *Number portability* Number portability enables customers to take their telephone number with them when they change local telephone companies. The ILEC shall ensure that all ILEC subscribers are unimpeded in the transference of service and associated telephone number from the ILEC to the competing carrier. The ILEC will be facilitate the transfer of service and number to the competing carrier.
- *Local dialing parity* Local dialing parity ensures that end customers are not inconvenienced simply because they subscribe to a new entrant for local telephone service.
- *Reciprocal compensation* The ILEC must compensate other carriers for the cost of transporting and terminating its local calls, unless a mutually agreeable alternate arrangement is established.
- *Resale* The ILEC must offer other carriers all of its retail services at wholesale rates, without unreasonable or discriminatory conditions or limitations, so that other carriers can resell those services to customers. This policy shall also include non-discriminatory access to all OSSs.

Figure 12-2 is a rendering that illustrates the requirements of the ILEC.

Figure 12-2 The competitive checklist

This list is a checklist and a road map for CLECs that are seeking to enter the marketplace. Many have discovered the challenging nature of launching such an endeavor, however. The nature of the business represents a regulatory and financial challenge for new entrants. More information about this subject will appear later in this chapter.

As the reader will note, the FCC has an increased role in defining the minimum thresholds of the 14 obligations. The FCC essentially judges whether the ILECs have met the 14-point checklist of requirements in order to offer inter-LATA services and to enter new telecommunications businesses. Under the Act, the state regulatory commissions are charged with the responsibility of overseeing the implementation of local telephone competition. Despite the one-sided nature of the Act, the fact is that the other LECs, current independents, and even wireline CLECs have obligations that they must meet. These obligations remind the CLECs and smaller LECs that they have an obligation to support competition and the public good. These obligations are largely the same as those that are applied to the ILECs. The biggest cost to the wireline CLECs is the obligation to provide 9-1-1 connectivity, to offer network interconnection, and to offer resale to potential competitors. The CLECs face a challenge in this regard, because while they are working to launch a company that will compete with the ILEC, they must simultaneously offer these services to potential CLEC competitors. For example, all LECs (not just ILECs) must make available all of their telecommunications services for resale to requesting telecommunications carriers. The FCC did not specify a detailed list of services that must be made available for resale; rather, the Act noted that state commissions, LECs, and resellers can determine the services that a LEC must provide at wholesale rates by examining that LEC's retail tariffs. Fortunately for the CLECs, they are neither required to resell exchange access services nor are they required to resell services that are available at wholesale rates to parties who are not telecommunications carriers (or even to carriers who are purchasing the service for their own internal use). In general, CLECs cannot place restrictions and conditions on the use of resale services without making a credible case to a state regulatory commission that such restrictions are not unreasonable.

As noted, the FCC effectively has the authorization to determine whether or not baby BOCs and GTE have lowered barriers to entry in the local market. The process has explicit parameters to measure the anti-competitive nature of the local telephone company. These parameters, however, are measured in a subjective manner. In this matter, the success of the baby BOC or GTE is whether or not a CLEC is successful. To a great extent, the long-distance future (and possibly other ventures as well) will be determined by the long-term success of the CLEC. The fallacy in the current way the baby

BOCs and GTE are being measured for success requires that the CLECs actually stay in business long enough during the FCC review. Unfortunately for the baby BOCs and GTE, many of the CLECs have been fly-by-night operations that are intended to make a quick profit or are intended to be sold to a much larger CLEC. The question facing the ILECs is, "How do you demonstrate success in lower barriers to local market entry if all of your competitors are being sold or are going out of business?" The ILEC cannot be held accountable for every action of the CLEC.

All aspects and obligations of the Act apply to all wireline CLECs that have more than two percent of the subscriber lines nationwide (as determined by the FCC). Under Section 251(f)(2) of the Act, a small LEC meeting this subscriber line count is exempted from Section 251 (b) and (c) requirements for network interconnection and resale. The small LEC must request an exemption from its state regulatory commission once it has received a bona fide request for network interconnection. In order for a carrier to receive an exemption from providing network interconnect, the requesting carrier must demonstrate that the interconnection request would cause "undue economic burdens beyond the economic burdens typically associated with efficient competitive entry." State regulatory commissions will determine, on a case-by-case basis, whether to grant an exemption to small LECs.

The Telecommunications Act of 1996 is a road map for any carrier that wishes to enter the business of public telecommunications. The CLECs have an opportunity to enter the marketplace while the barriers to entry have been artificially lowered by the federal government.

Other Regulatory Commitments

The new-entrant wireline and wireless carriers, including all other carrier types that are seeking CLEC status, must also meet other federal regulatory commitments. These commitments are laws that were passed outside the Telecommunications Act of 1996 but that still carry the same/equal weight of legal obligation as the Act of 1996. These acts include the Communications Assistance for Law Enforcement Act (CALEA), which was signed into law in 1994; the FCC rules for Ensuring Wireless Carrier Compatibility with Enhanced 9-1-1 Emergency Calling Systems (FCC Docket No. 94-102); and the Universal Service Fund. The CALEA law and the FCC rules on wireless carrier locations for 9-1-1 are requirements for all carriers that are entering the marketplace. The Universal Service Fund, as redefined by the Act of 1996, has the potential of having the most far-reaching impact at every income level in the United States.

CALEA In October 1994, Congress passed (and the President signed) the Communications Assistance for Law Enforcement Act (CALEA). The law was designed to position law enforcement to better respond to the rapid advances in telecommunications technology. The law was also designed to eliminate obstacles faced by law enforcement agencies in conducting electronic surveillance. Electronic surveillance is defined as "both the interception of communications content (wiretapping) and the acquisition of call-identifying information (dialed-number information) through the use of pen register devices and through traps and traces." Since 1970, telecommunications carriers have been required to cooperate with law enforcement personnel when conducting electronic surveillance. CALEA took this requirement one step farther, because it required telecommunications carriers to modify and design their switching systems and other support equipment, facilities, and services to ensure that authorized electronic surveillance could be performed. All wireline, cellular, and PCS carriers are affected by the CALEA law.

One must understand the challenges that are faced by law enforcement agencies today. The rapid advancement of communications technology, the growth of new types of services, and the increase in the number of carriers has complicated the work of law enforcement to the extent that surveillance is hampered.

CALEA requires telecommunications carriers to ensure that their equipment, facilities, and services will meet four functional (or assistance-capability) requirements that enable law enforcement to conduct authorized electronic surveillance. These requirements are as follows:

- Telecommunications carriers must be capable of expeditiously isolating and enabling the government to intercept all wire and electronic communications within that carrier's network, to or from a specific subscriber of such carrier.
- Telecommunications carriers must be capable of rapidly isolating and enabling the government to access call-identifying information that is reasonably available to the carrier (not the information that is acquired solely through pen registers or trap-and-trace devices). The call-identifying information cannot include any information that could disclose the physical location of the subscriber, except to the extent that the location might be determined by the telephone number alone.
- Telecommunications carriers must be capable of delivering intercepted communications and call-identifying information to a location specified by the government, other than the premises of the carrier.
- Telecommunications carriers must be capable of conducting interceptions and providing access to call-identifying information

unobtrusively. Furthermore, telecommunications carriers are required to protect the privacy and security of communications and call-identifying information not authorized to be intercepted, as well as information concerning the government's interception of the content of communications and access to call-identifying information.

Wireless 9-1-1 The FCC rules for wireless telecommunications calling party location in support of 9-1-1 emergency calls were passed in 1994 and were reaffirmed in 1996 and 1999. The FCC rules concerning wireless 9-1-1 require wireless carriers to provide the location of the party that is calling for help in a two-stage process. Stage 1, which is also known as Phase 1, requires the wireless carrier to provide the calling party's telephone number, cell site, and sector location (the cell site and sector from which the emergency call is being made). Phase 2 requires the wireless carrier to provide the latitude and longitude of the calling party upon initiation of a 9-1-1 call. Issues regarding liability and cost recovery (for the installation and operation of these systems) have occurred for a number of years. The deadline to comply with the FCC's rules is October 1, 2001.

Universal Service Fund The Universal Service Fund was established under the direction of the FCC in support of the Telecommunications Act of 1996. The Act redefined the universal service policy as it had been defined by the Communications Act of 1934, which was the forbearer of the Telecommunications Act of 1996. The Act of 1934 had created the FCC and noted the following:

> The purpose of the Act of 1934 was " . . . to make available, so far as possible, to all the people of the United States, without discrimination on the basis of race, color, religion, national origin, or sex, a rapid, efficient, Nation-wide, and world-wide wire and radio communication service with adequate facilities at reasonable charges, for the purpose of the national defense, for the purpose of promoting safety of life and property through the use of wire and radio communication, and for the purpose of securing a more effective execution of this policy by centralizing authority heretofore granted by law to several agencies and by granting additional authority with respect to interstate and foreign commerce in wire and radio communication, there is hereby created a commission to be known as the "Federal Communications Commission, ..."
>
> The Act of 1934 defined *universal* to consist of the following principles, as quoted in the Act of 1934:
>
> " . . . and the Commission shall base policies for the preservation and advancement of universal service on the following principles:
>
> (1) Quality and rates.—Quality services should be available at just, reasonable, and affordable rates.

(2) Access to advanced services.—Access to advanced telecommunications and information services should be provided in all regions of the Nation.

(3) Access in rural and high cost areas.—Consumers in all regions of the Nation, including low-income consumers and those in rural, insular, and high cost areas, should have access to telecommunications and information services, including interexchange services and advanced telecommunications and information services, that are reasonably comparable to those services provided in urban areas and that are available at rates that are reasonably comparable to rates charged for similar services in urban areas.

(4) Equitable and nondiscriminatory contributions.—All providers of telecommunications services should make an equitable and nondiscriminatory contribution to the preservation and advancement of universal service.

(5) Specific and predictable support mechanisms.—There should be specific, predictable and sufficient Federal and State mechanisms to preserve and advance universal service.

(6) Access to advanced telecommunications services for schools, health care, and libraries.—Elementary and secondary schools and classrooms, health care providers, and libraries should have access to advanced telecommunications services as described in subsection (h).

(7) Additional principles.—Such other principles as the Joint Board and the Commission determine are necessary and appropriate for the protection of the public interest, convenience, and necessity and are consistent with this Act."

The Telecommunications Act of 1996 redefined the concept of universal service and expanded this concept to include access to advanced telecommunications services and all kinds of information services. All regions of the United States (and all income levels) are required to be included. Regardless of the economic level of the region, every person, child, school, healthcare facility, and library shall be provided (to the same level and type) with telecommunication and information services in areas that have the economic infrastructure to support the most advanced information services.

In order to support this government initiative, a fund was established by the FCC and was administered by the *National Exchange Carrier Association* (NECA) to pay for the installation of computers in classrooms, to subsidize carriers in low-income rural or urban areas to enhance the telecommunications infrastructure, etc. This fund is paid into by all wireline LECs, including CLECs. Some might even call this fund a government entitlement; however, this fund is not a simple government welfare program. Rather, this fund is a necessity in order to ensure the education of the nation's young and to ensure the economic viability of the nation itself. Payments into the fund is based on a fixed formula that requires every carrier to pay a portion of its gross revenue to the fund.

Cost Components and Pricing Approaches

The business of being a telecommunications carrier does not just involve installing technology to make a call, track billable minutes, and render a bill. The business of being a carrier involves meeting regulatory commitments, which does not necessarily support the business itself but meets a government-ordered activity to support the public good. All of these benefits come at a cost, however. We have already described some of the basics. The opportunities that are presented by the telecommunications industry cost money to implement. But let's take a look at the costs that are involved with implementing the basic network infrastructure, as opposed to the new services within a network. Breaking the network into basic components, we find that a network can consist of the following elements:

- Network local access
- Network long-haul/long-distance access (i.e., connecting local networks)
- Hardware, which can include the following items:
  - Transport
  - Switching/routing
  - Database management (hardware)
  - Information management (non-hardware)
  - Radio antennae
  - Application boxes (e.g., voice mail)
  - Terminal devices (e.g., mobile handsets and computing terminals)
  - Radio towers
  - Network management
- Software
- Labor
- Customer care
- Operational expenses other than labor
- Real estate
- Government licenses

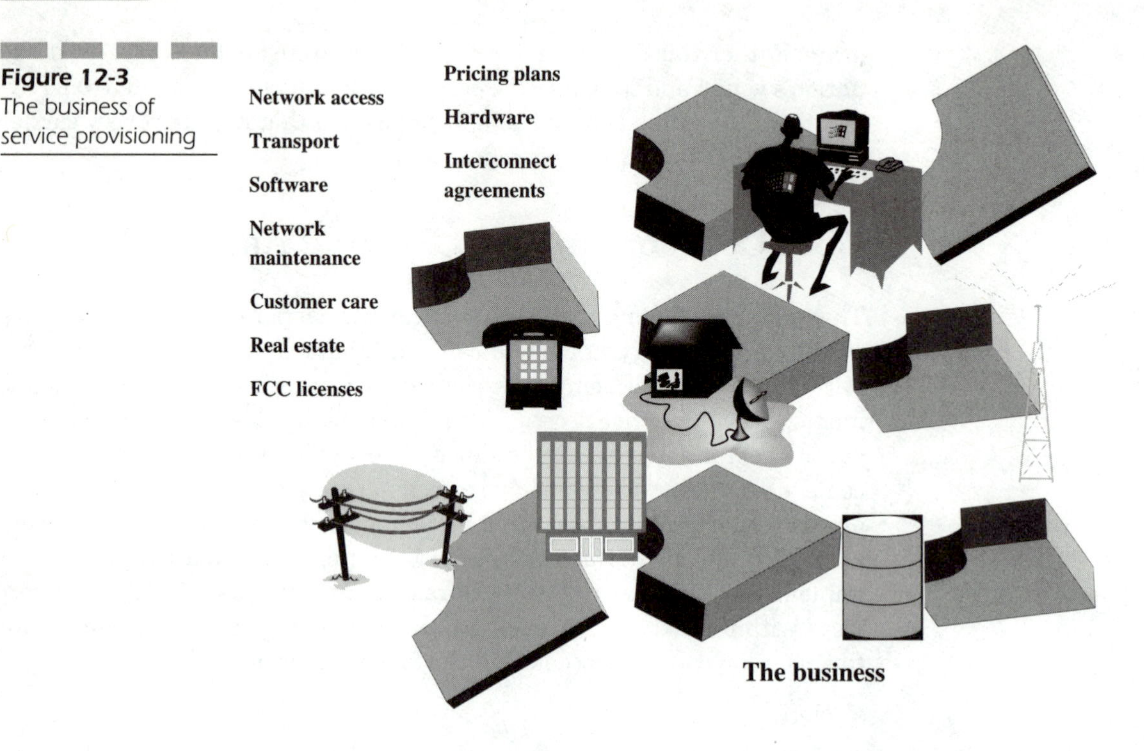

Figure 12-3 The business of service provisioning

Figure 12-3 illustrates the piece parts and requirements one must compile/meet in order to enter the telecommunications service provisioning business. As the figure depicts, there is a great deal to do enter the marketplace as a service provider.

All of these elements require outlays of money or some degree of financing. The initial startup costs for an operation can easily run into the tens of millions of dollars. Outsourcing is one way to reduce capital expenditures. Financial institutions need to review business plans of CLECs and other types of carriers in order to determine not only the cost structure, but also the revenue potential.

The pricing plan that is implemented by the telecommunications carriers impacts the way in which subscribers perceive the service, the value of the service, the perceived expense of the service, and the level of service usage. Pricing plans can assume a number of different forms:

- Peering agreements between carriers
- Usage-based pricing
- Flat-rate pricing
- Regulated rate of return

Peering Agreements Peer agreements refer to the business agreements reached between carriers. The agreements address the following factors:

- How each party will interconnect
- What types of interconnect will be available
- How much the interconnect will cost
- Where the parties will interconnect
- Responsibilities of each party
- Restrictions on use
- How communications traffic will be passed from one party to the other
- How the carriers will address failures
- General processes and procedures of doing business

The overall cost to each carrier might be close to or equal to zero. The overriding concept behind peering agreements is that the carriers who are involved in the agreement agree to treat each other as equals. Peering agreements usually do not involve usage-sensitive pricing. Remember, the operative word is *peer*.

Usage-Based Pricing Usage-based pricing enables the carrier to charge the subscriber or carrier based on the amount of time that is spent using the service or the number of times that the service is used. The usage parameter can be time-based or frequency-of-use based and can be applied to almost any type of service.

Usage-based pricing is difficult to apply between service providers if the usage parameter is based on time. Frequency of use is much easier to handle. One example of frequency of use is dipping into a database. Each database dip might be one cent. This figure is easy to track and maintain a record for, because this figure is similar to the peg count in the wireline telephony world.

Flat-Rate Pricing Flat-rate pricing is great for the subscriber. Typically, flat-rate pricing means one fee per month for as many calls or as much Internet service provider network time as you want (per month, of course). Flat-rate pricing can choke a service provider's network because it encourages users to use the network for long periods of time for a single fee. Today, the fees are normally pretty small; the competitive marketplace has forced prices way down from where they used to be. In the past, flat-rate pricing was equated to premium charges for the privilege of obtaining the service. Today, however, the market is a buyer's market. If you want to stay in business, then flat-rate pricing might be the way to go —but the provider will have to work on a volume basis.

Regulated Rate of Return Regulated rate of return can be defined as the old phone company system—before the AT&T breakup. Rate of return is still governed by state public utility commissions; however, the point I am making is that the environment is less regulated.

In a regulated rate of return schema, the service provider can charge some fee for services that will enable the provider to cover expenses and to provide some additional operating capital/profit. The formula for determining the rate of return varies from state to state (each state has its own requirements).

Construction and Operational Goals

Service providers that sell their network infrastructure as a product to other carriers can position their assets as a tool that enables these other carriers to operate. The aspects and elements of the telecommunications carrier's business can be turned into assets and profit centers. Figure 12-4 illustrates these goals. The goals of every telecommunications carrier are as follows:

- To save on operating costs
- To reduce the time to market
- To reduce capital outlays
- To increase flexibility
- To ensure availability and reliability

Savings Savings are fairly obvious. A startup carrier that is seeking to reduce the overall cost of running a network can do so by simply turning over network operations to someone else. This situation equates to outsourcing the following components:

- Field technicians
- Leasing building space (i.e., real estate)
- Outsourcing maintenance
- Outsourcing network management
- Outsourcing billing

Figure 12-4 Construction and operational costs

Network access
Transport
Software
Network maintenance
Customer care
Real estate
FCC licenses

Pricing plans
Hardware
Interconnect agreements

Construction and operational costs

$$$

- **Time to market**
- **Flexibility**
- **Availability**
- **Reliability**

Savings can also take the form of workforce management, which involves managing the time schedules of all employees to the point that overtime is eliminated and salary increases/bonuses are eliminated. A favorite tool for startups is the stock option. The telecommunications carrier pays less cash up front for the services of the employee and instead compensates the employee with stock options. Stock options in a startup have no value until the company goes public on the stock market.

Reducing the Time To Market Startup service providers that are seeking to reduce the time to market can do so by using the wholesale network infrastructure services of an existing carrier. The startup has to bear in mind that the carrier from which it is seeking help will not help a competitor in its own market segment. Why spend the time and energy building all of the infrastructure that you need? The answer is that you do not have to build all of this infrastructure. There are carriers that lease their network capabilities and provide outsourcing work, which would include the following:

- Network management systems
- Network operations
- Billing
- Call detail recording
- Subscriber service profile storage

- Database querying
- Voice mail storage
- E-mail access
- E-mail storage
- Service creation services
- Customer care

Components of a business plan that are used to generate savings are also used to reduce the amount of time spent trying to commercialize the company and to provide services to the subscriber.

Reducing Capital Outlays A service provider that is purchasing infrastructure support will find that one can quite literally be a service provider without spending a penny on equipment. In such a scenario, the provider will become the ultimate reseller. Another way to reduce capital expenditures is by financing all main infrastructure systems through the infrastructure vendor.

Increasing Flexibility Service providers (especially startups) that need to react quickly to the dynamic marketplace will find that leasing capabilities offers a lows cost and fast way to respond. This flexibility refers to meeting the needs of a fast-changing market.

To the carrier that is providing the infrastructure services, the challenge lies in the capacity to provide diversity in capability. Diversity means potentially higher costs. The capability to be flexible will probably require a switching platform that enables service-creation capabilities. Flexibility does not necessarily mean higher costs. The new multi-functional switching platforms will enable CLECs and other carrier types to maximize the value of their dollars.

Ensuring Availability and Reliability To the purchaser of infrastructure services, buying support means buying availability and reliability. To the supplier of infrastructure services, this concept means spending the money that is needed to ensure a high degree of network availability and reliability. The supplier of such services will be required to sign an agreement with the recipient of these services that will detail the following policies:

- Guarantee of service availability some percentage of time per day. The supplier of infrastructure services (the carrier's carrier) will not want to guarantee anything 100 percent of the time. Networks are not engineered for 100 percent availability, so do not make an agreement that you cannot possibly hope to meet.

- Performance metrics by which the supplier can be measured
- Time to repair non-functioning equipment or services
- Time between failures of equipment
- Punitive measures in the event that the supplier fails in its obligations and commitments

The burden of performance is on the supplier of services, and these aforementioned points are what comprises a service-level agreement between the customer and the supplier.

Telecommunications Carriers: Interconnecting for the Future

When one thinks of network interconnection, the following elements should come to mind. These attributes of a network are also necessities of a telecommunications carrier. Figure 12-5 illustrates these attributes. If a

- **To the user, the type of technology is not relevant.**
- **The customer wants to use their services regardless of time and place.**
- **Price and performance**

Figure 12-5 Attributes sought by the carrier and the customer

telecommunications carrier or startup carrier wishes to provide service, it must address these broad issues:

- Connectivity
- Access
- Seamlessness
- Convergence

Connectivity Interconnection enables people to connect to different types of networks. Service providers are capable of connecting to one another, and connectivity is becoming more and more important. Concepts such as convergence describe what is occurring in the areas of technology and business. Convergence is the merging of wireless technology; wireline telephony, satellite, cable TV, data, the Internet, and entertainment are merging to form information networks. Convergence accounts for both the technical and business aspects of integration of technology and business. A better term might be connectivity. The industry and the marketplace want to connect people, places, and things.

Standards must be established and followed to govern the manner in which multiple networks and service providers connect and communicate. Appendix B in this book describes in detail the types of network interconnect that are available. In order to ensure connectivity between networks, the following plans must be created from a planning concept:

- *Architectural switching plan* This plan would include hand-off methodology, anchor-switching design, switching selection points, and a routing plan, not to mention efficient routing of calls or data packets based on network efficiency or business arrangements. Route diversity is a major concern for many service providers, but there is a dollar cost.
- *Transmission plan* Types of facilities used and transmission design requirements
- Addressing plan, numbering plan, point codes, and the like
- Network signaling plan
- Subscriber services plan
- Subscriber billing/customer support plan
- Network management and operational support systems plan

Access Connectivity enables the service providers to not only technically connect their networks but to also access one another's network elements.

Access in this case refers to a service provider's capability to use another service provider's network infrastructure to support their network operations. Access enables database entry and is a broad term that includes freedom of movement through another service provider's network. Access enables network management system use and calls between networks.

Service providers are technically capable of accessing one another's network elements (of course, with the necessary permissions). Access is a doorway to the service and network infrastructure.

Seamlessness Interconnection supports seamlessness, which is generally viewed from the perspective of the user. The user sees nothing; in other words, the user does not see any difference in the quality of service, the manner in which service is provided, the types of services provided, the capability to receive calls, and the way calls can be made when roaming between service providers. Seamless is the ultimate goal for service provisioning.

From the perspective of the service provider, seamlessness is viewed differently. The service provider does want the user to see a difference, because differentiation is what makes one service provider different from another. To some extent, the user even wants to see differences, because differentiation is part of the competitive landscape. Seamlessness from the perspective of the provider generally refers to how the user uses services that are provided on different vendor-switching platforms. In other words, a provider has four different providers of computers. The provider does not want any of its subscribers to see any difference in the way they invoke/use the service as the subscriber travels from one point in their network to another point in their network. Competitors want differences, and allies and partners do not want differences.

Roaming agreements in the future will assume additional attributes. In the future, service providers that sign roaming agreements (with each other) will agree on the types of services and quality of service to be provided. Roaming agreements will become an additional vehicle through which strategic partners will support one another.

Convergence Convergence is occurring as you read this book. Convergence is the merging of systems and services in such a manner that service providers seem to belong together. Convergence is about corporations and carriers finding synergies in their businesses and using the synergies that are found in their technologies by either building new businesses or enhancing existing businesses. Convergence is the breaking down of barriers that have separated service providers for years.

Convergence is about creating new opportunities by building bridges between technology and businesses.

Moving Forward

As the industry moves forward, the need to be able to interconnect will become more and more important. Despite competitive "chestbeating," service providers understand that they will not have the entire market cornered. In a perfect world (from the provider's viewpoint), perfect service might exist, but this world is not perfect. Service providers will need to ensure technical interconnection, or else subscribers will not have the capability to make or receive calls. Service providers are working towards creating business opportunities and finding new customers while continuing to support existing customers and existing core products.

In the near future, we will see networks that will support voice, video, and data. These networks will support seamlessness, intelligent switching, mobility, and access to one another. The network of the future will be a network of networks that will (in reality) be made up of a multitude of providers but will look and feel like one big network (a virtual network). These providers will be wireline carriers, wireless carriers, satellite carriers, Internet, resellers, wholesalers, and every imaginable kind of carrier that has not yet been envisioned by regulatory bodies or by technology.

The growth of the telecommunications industry will only be limited by one's imagination. The network interconnect will serve as an enabler of telecommunications service provisioning as well as a tool in business combat. Interconnect will continue to play a role in the growth and evolution of the telecommunications industry. Carriers will use interconnect, regulation interpretation, and technology to create new business paradigms from which to operate and provide service.

APPENDIX A

Acronyms, Definitions, and Terminology

Access Link (A-link)—Used for switch-STP signaling connections and STP-HLR

Access Network—Portion of a public switched network that connects access nodes to individual subscribers

Access Service Area—A geographic area established for the provision and administration of communications services. An access service area encompasses one or more exchanges, where an exchange is a unit of the communications network that consists of the distribution facilities within the area served by one or more end offices, together with the associated facilities used in furnishing communications service within the area.

Access Tandem—A switching system that concentrates and distributes traffic for inter-LATA traffic originating or terminating within an *Local Access and Transport Area* (LATA). An access tandem also can provide equal access for non-conforming end offices.

Access Tandem (AT)—An EC switching system that provides a traffic concentration and distribution function for interexchange traffic originating or terminating within an access service area

Address Signals—These signals convey call destination information or the digits that are dialed by the calling party. There are several types of address signaling: dial pulse, *Dual-Tone Multi-Frequency* (DTMF), and *Multi-Frequency* (MF).

Alerting Signals—These signals alert the user to an incoming call. In other words, alerting signals are the ringing that the user hears (also known as power ringing in the wireline world).

Amateur Satellite Service—This satellite service is for radio amateur use to carry out technical investigations and to learn about intercommunications.

American National Standards Institute (ANSI)—A non-profit organization that coordinates voluntary standards activities in the United States. The institute represents the United States in two major telecommunication organizations: the *International Standards Organization* (ISO) and the *International Electrotechnical Commission* (IEC).

American Standard Code for Information Interchange (ASCII)—A widely accepted standard for data communications that uses a seven-bit digital character code to represent text and numeric characters. When companies use ASCII as a standard, they have the capability to transfer text messages between computers and to display devices (regardless of the device manufacturer).

Analog Signal—A signal that is modified in a constant fashion, such as voice or data

Answer Supervision/Answer Signal—The signaling state of a circuit that indicates idle or busy or on-hook or off-hook status. The signaling state can be indicated in various ways, depending on the signaling system.

Area Code—A three-digit number that identifies the home area of a telephone. In North America, the *Numbering Plan Area* (NPA) represents the area code.

Asymmetric Digital Subscriber Line (ADSL)—Modems that are attached to twisted-pair copper wiring that transmit from 1.5 Mbps to 9 Mbps downstream (to the subscriber) and from 16 Kbps to 800 Kbps upstream, depending on line distance

Asynchronous Transfer Mode (ATM)—A multiplexing and switching technique that organizes information into 53 byte cells. Each cell of data is transmitted asychronously.

Attenuation—The decrease in power of a signal, energy, light, or a radio signal. This decrease can occur either as a fractional value or as a complete attenuation.

Authentication—A process during which information is exchanged between a communications device (typically a mobile phone) and a communications network, which enables the carrier or network operator to confirm the true identity of the unit. This feature inhibits fraudulent use of the mobile unit. Another way of viewing authentication is by the process of user identity confirmation. Identity confirmation can involve checking the handset/terminal device identity by interpreting secret keys/data messages. If the data keys/data messages have been altered or do not show a specific format, the call will not be completed.

Automatic Number Identification (ANI)—The number that provides the billing number of the line or trunk that originated a call. This number is a charge number that is used to support exchange access and billing. This number might or might not be identical to the calling party number.

Backbone—A common distribution channel that carries anaolog or digital telecommunications signals for many users (also known as the central distribution cable from an interface)

Bandwidth—The width of a radio channel (in Hertz) that can be modulated to transfer information

Basic Rate Interface (BRI)—In ISDN, the network interface that provides 144 Kb/s information transfer, as defined in ANSI Standard T1.607

Bearer—In the communications industry, this term refers to a transmission channel that is used to carry data. In ISDN, there are 64 Kbps bearer channels used. These channels are used to carry data.

Billing System—A system that records the occurrence of a call or some event, the identity of the originating party, the identity of the destination party, and the length of the call. This system must also process the data for rendering a bill to the subscriber.

Bit—The smallest part of a digital signal, typically called a data bit. A bit typically can assume two levels: a zero (0) or a one (1).

Bit Error Rate (BER)—A measurement that is used to determine the quality of a digital transmission channel. BER measures the ratio of bits received in error compared to the total number of bits transmitted.

Bridge—A data communications device that connects two or more networks and transmits information between the networks

Broadband—Typically refers to voice, data, and/or video communications at rates greater than wideband communications rates (1.544 Mbit/s)

Broadband Integrated Services Digital Network (B-ISDN)—A digital network with ATM switching that operates at data rates in excess of 1.544 or 2.048 Mbps. ATM enables the transport and switching of voice, data, images, and video over the same infrastructure.

Broadcast Messaging—Messaging between users of telecommunications networks. Typically, broadcast messaging is between one user and many users.

Broadcasting Satellite Service—Enables sound and visuals to be received by individuals or communications via satellite

Broadcasting Service—A service in which the transmissions are intended for direct reception by some consumer group (generally, the public at large)

Bursty Data—Describes data rates that fluctuate widely with no predictable pattern

Busy Hour—A time-consistent hour in a specific measurement period when the total load offered to a group of trunks, a network of trunks, or a switching system is greater than at any other time-consistent hour during the same measurement period

Cable Television (CATV)—See *Community-Access Television* (CATV).

Call Delivery (CD)—A call routing process that permits a subscriber to receive calls to his or her directory number while roaming

Call Detail Recording (CDR)—A telecommunications system's capability to collect and record detailed information about all outgoing and incoming calls

Call Forwarding-Busy (CFB)—A call routing service that permits a called subscriber to have the system send incoming calls addressed to the called subscriber's directory number to another directory number (forward-to number) or to the called subscriber's designated voice mail mailbox when the subscriber is engaged in a call or service

Call Forwarding-Default (CFD)—A call routing service that permits a called subscriber to send incoming calls addressed to the called subscriber's directory number to the subscriber's designated voice mail mailbox or to another directory number (forward-to number) when the subscriber is engaged in a call, does not respond to paging, does not answer the call within a specified period after being alerted, or is otherwise inaccessible (including no paging response, the subscriber's location is not known, or the subscriber is reported as inactive).

Call Forwarding-No Answer (CFNA)—A call routing service that permits a called subscriber to have the system send incoming calls addressed to the called subscriber's directory number to another directory number (forward-to number) or to the called subscriber's designated voice mail mailbox when the subscriber fails to answer or is otherwise inaccessible (including no paging response, the subscriber's location is not known, the subscriber is reported as inactive, call delivery is not active for a roaming subscriber, do not disturb is active, etc.)

Call Forwarding-Unconditional (CFU)—A call routing service that permits a called subscriber to send incoming calls addressed to the called subscriber's directory number to another directory number (forward-to number) or to the called subscriber's designated voice mail mailbox. If this feature is active, calls are forwarded regardless of the condition of the termination.

Call Processing—Steps that occur during the duration of a call. These steps are typically associated with the routing and control of the call.

Call Progress Signals—Voice band tones and announcements that are used to inform a calling end-user or operator of the progress or disposition of a call. Examples include a busy tone, voice announcement, special information tones, and audible ringing. The tones or announcements can be used after a number has been dialed and until the called telephone is answered or the attempt is abandoned.

Call Routing—In circuit switching, the process of determining the path of a call from the point of origination to the point of destination

Call Transfer—A call-handling feature that transfers a call from one station or extension to another

Call Transfer (CT)—Call transfer service enables a subscriber to transfer an in-progress established call to a third party. The call to be transferred might be an incoming or outgoing call.

Call Waiting—A telephone call processing feature that notifies a telephone user that another incoming call is waiting to be answered. This signal is typically provided by a brief tone that is not heard by the calling party. Some advanced telephones (such as GSM mobile telephones) are capable of displaying the incoming phone number of the waiting call.

Call Waiting (CW)—A call routing service that provides notification to a controlling subscriber of an incoming call while the subscriber's call is in the two-way state. Subsequently, the controlling subscriber can either answer or ignore the incoming call. If the controlling subscriber answers the second call, the system might alternate between the two calls.

Calling Name Identification Presentation (CNaIP)—A call processing service that provides the name identification of the calling party to the called subscriber

Calling Name Identification Restriction (CNaIR)—A call processing service that restricts the presentation of the calling party's name to the called subscriber

Calling Number Identification (CNI)—A feature that enables a telephone customer to view the telephone number of the person who is calling. A related service called *Calling Number Identification*

Restriction (CNIR) enables the caller to inhibit the display of his or her telephone number when placing a call.

Calling Number Identification Presentation (CNIP)—A call routing service that provides the number identification of the calling party to the called subscriber. One or two numbers can be presented to identify the calling party.

Calling Number Identification Restriction (CNIR)—A call processing service that restricts presentation of that subscriber's *Calling Number Identification* (CNI) to the called party.

Calling Party Number—In the *Integrated Services Digital Network* (ISDN) Q.931 and *Signaling System 7* (SS7) protocols, an information element that identifies the number of an originating party

Calling Party Number (CPN)—A set of digits and related indicators that provide numbering information related to the calling party (the identification number of the party that is originating a call)

Cell Site—A transmitter-receiver tower operated by a wireless carrier (typically cellular or PCS) through which radio links are established between a wireless system and mobile and portable units

Central Office—A term used to describe local switches used by local telephone companies. The term is also synonymous with end office.

Centrex—A service for business customers that shifts the functions that are usually associated with a *Private Branch Exchange* (PBX) on a customer's premises to a central-office switching system

Channel Reuse—Refers to the practice of independently using radio channels that have the same radio frequency to cover different coverage areas

Coaxial Cable—A transmission line consisting of an inner conductor surrounded first by an insulating material and then by an outer conductor (either solid or braided). The mechanical dimensions of the cable determine its characteristic impedance.

Common Channel Signaling (CCS)—Also known as out-of-band signaling. Describes a scheme in which the content of the call is separated from the information that is used to set up the call (signaling information).

Competitive Access Providers (CAPs)—An alternative local exchange carrier that competes with the existing incumbent and dominant local exchange carrier(s)

Community-Access Television (CATV)—Also known as cable TV. Today, cable television is a popular form of sending entertainment programming into households via a physical transmission system (e.g, coaxial cable, DS1, etc.). Years ago, CATV enabled homes in remote areas to receive limited television programming when wireless transmission signals could not reach such areas.

Competitive Local Exchange Carrier (CLEC)—A description of the new competitive carriers (as encouraged by the Telecommunications Act of 1996) that are competing in the local loop marketplace

Conference Calling (CC)—A call routing service that provides a subscriber with the capability to establish a multi-connection call (i.e., a simultaneous communication between three or more parties or conferees)

Control Signals—Control signals are used for special auxiliary functions that are beyond a service provider's network. These signals communicate information that enable or disable certain types of calls. One example would be call barring.

Convergence—Addresses the technical and business aspects of integration of technology and business

Country Code—A one, two, or three-digit number that identifies a country or numbering plan to which international calls are routed. The first digit is always a world zone number. Additional digits define a specific geographic area (usually a specific country).

Customer Care System—A customer profile database system that is used to support customer complaints, to add new subscribers, to remove subscribers, and to maintain customer profile information

Customer Premises Equipment (CPE)—All telecommunications terminal equipment is located on the customer's premises, including telephone sets, PBXs, data terminals, and customer-owned coin-operated telephones.

Data Compression—A technique for encoding information so that fewer data bits of information are required to represent a given amount of data. Compression enables the transmission of more data over a given amount of time and circuit capacity and also reduces the amount of memory that is required for data storage.

Data Terminal Equipment (DTE)—In a data communications network, the data source, such as a computer, and the data sink, such as an optical storage device. (Also see Data Sink, Data Source, Network Channel Terminating Equipment, Channel Service Unit, and Data Service Unit.)

Data Warehouse—An information management service that stores, analyzes, and processes information that is derived from transaction systems

Database—A database is a collection of interrelated data that is stored in computer memory with a minimum of redundancy. Database information that is held in a computer-accessed memory usually is subdivided into pages, with each page being accessible to all users (unless it belongs to a closed user group).

De-multiplexing—A process that is applied to a multiplexed signal for recovering signals combined within it and for restoring these individual signals

Dedicated Circuits—A circuit that is designated for a specific use or customer

Dial-Line—A two-wire, line-side connection from an LEC end office. This connection is similar to the connections that are used for business and residential lines, and this connection can be used on a one-way or two-way basis. Dial-line connections enable the MSC to access any valid telephone number.

Dialing Parity—A company that is not an affiliate of a local phone company can provide phone services in such a manner that customers have the capability to route their calls automatically without the use of any access code.

Digital Signal—A signal that has a limited number of discrete states (usually two). In contrast, an analog signal varies continuously and has an infinite number of states.

Digital Signal 0 (DS0)—A 64 Kbps digital representation of voice

Digital Signal 1 (DS1)—Twenty-four voice channels packed into a 193-bit frame and transmitted at 1.544 Mbps. The unframed version, or payload, is 192 bits at a rate of 1.536 Mbps.

Digital Signal 2 (DS2)—Four T1 frames packed into a higher level frame transmitted at 6.312 Mbps

Digital Signal 3 (DS3)—Twenty-eight T1 frames packed into a level frame transmitted at 44.736 Mbps

Digital Subscriber Line (DSL)—A two-wire, full-duplex transmission system that transports user data between a customer's premises and a digital switching system or remote terminal at 144 kbit/s

Digital Television—A device that receives broadcasted, digitally formatted television signals. Such devices display programming with far greater resolution than current television sets.

Direct Distance Dialing (DDD)—A telephone service that lets a user dial a long-distance call without operator assistance. This feature is common today; however, many years ago, this capability was considered a special feature of telephone switches.

Direct Inward Dialing (DID)—A PBX or Centrex feature that enables the completion of an incoming call directly to an extension station without operator assistance. Direct inward dialing also is used in voice mail and radio paging systems.

Disconnect Signal—An on-hook signal that indicates the connection is being cleared. The signal that is responding to a disconnect signal but is applied in the direction opposite to the direction of propagation of the disconnect signal might also be considered a disconnect signal.

Distinctive Ringing—A feature that enables a subscriber's multiplicity of numbers (on a single line) to be identified by different ringing patterns

Do Not Disturb (DND)—A call processing feature that prevents a called subscriber from receiving calls. When this feature is active, no incoming calls will be offered to the subscriber. DND also blocks other alerting, such as the Call Forwarding-Unconditional abbreviated (or reminder) alerting and Message Waiting Notification alerting. DND makes the subscriber inaccessible for call delivery.

Domain Name—The unique name that identifies an Internet site

Downlink—The portion of a communication link used for transmission of signals from a satellite or airborne platform to a terrestrial terminal

Drop Wire—The wire from the telephone pole or underground street access to the connector block in the home or building.

Dual-Tone Multi-Frequency (DTMF)—A means of signaling that uses a simultaneous combination of one lower group of frequencies and one higher group of frequencies to represent each digit or character. These frequencies are tones.

E Carrier—A 10 MHz wireless carrier that is also known as a *Personal Communications Service* (PCS) carrier

E1—European basic multiplex rate, which packs 30 voice channels into a 256-bit frame and is transmitted at 2.048 Mbps

Earth Exploration Satellite Service—Services that facilitate observation of the Earth for various purposes, such as weather or geological

Electronic Mail (E-mail)—Messages, usually text, that are sent from one person to another via the computer

Emergency Services Access Point—An emergency services network element that is responsible for answering emergency calls

Encryption—A process of protecting voice or data information from being obtained by unauthorized users. Encryption involves the use of a data-processing algorithm (formula program) that uses one or more secret keys that both the sender and receiver of the information use to encrypt and decrypt the information. Without the encryption algorithm and key, unauthorized listeners cannot decode the message.

End Office (EO)—An EC switching system that terminates station loops and connects the loops to each other and to trunks

Engineered Capacity—The highest load level at which service objectives are met for a trunk group or for a switching system

Equal Access—An exchange carrier service that gives a customer an equal choice of trunk-side access to public switched interexchange carrier telephone networks in terms of such items as dialing plan and transmission quality. Another term for this feature is Feature Group D.

Erlang—Amount of voice connection time with reference to one hour. For example, a six-minute call is .1 Erlang.

Ethernet—A transmission protocol for packet-switched *Local-Area Networks* (LANs). Ethernet is a registered trademark of the Xerox Corporation.

Exchange—The typical shorthand term that is used to identify a telephone switching center

Exchange Access—The offering of access to telephone exchange services or facilities for the purpose of the origination or termination of telephone toll services

Exchange Carrier—A telephone company, generally regulated by a state regulatory body, that provides local (inter-LATA) telecommunications services

Facilities—The transmission parts (elements) of a service provider (sometimes used in more general terms to describe buildings and utilities)

Facsimile—Also known as a fax. This is a form of telegraphy supports the transmission of fixed images to some form of fixed media, such as paper.

Fax Mail (FxM)—A communications service that provides a subscriber with fax mail services, which include the capability to store faxes received and to forward faxes recorded, based on a number of subscriber set parameters

Fiber to the Home (FTTH)—A communications network where an optical fiber runs from the telephone switch to the subscriber's premises or home

Fiber to the Neighborhood (FTTN)—A communications network where an optical fiber forms a switch to a neighborhood of homes

Firewall—A term that refers to a physical and electronic method of protecting a computer from outside attack. This protection involves hardware and software.

Fixed Satellite Service (FSS)—A radio communication service that addresses communication between Earth stations at specified fixed points via one or more satellites (e.g., Intelsat)

Flat-Rate Pricing—Price-setting principles for a service provider that wishes to charge the same fee for calls, regardless of the number of calls made or the duration of each call. Usually, flat rates are single monthly or periodic charges.

Flexible Alerting (FA)—A call processing service that causes a call to a pilot directory number to branch the call into several legs to alert several termination addresses simultaneously. The mobile telephones in the group might be alerted by using distinctive alerting. Additional calls can be delivered to the FA pilot directory number at any time. The first leg to be answered is connected to the calling party, and the other call legs are abandoned.

Foreign Exchange (FX or FEX) Telecommunications trunk lines that connect directly to a foreign telephone company's switching system (exchange)

Four-Wire Circuits—A physical path in which four wires are represented to the terminal equipment. This path enables simultaneous transmission and reception. Two wires are used for transmitting in one direction, and the other two wires are used for transmitting in the other direction.

Fractional T1—Fractional T1 refers to data transmission speeds between 56 Kbps and 1.544 Mbps in a single, full T1.

Frame Relay—An access standard that is defined by the standards body called ITU. Frame relay services are telecommunications services that employ a form of packet switching that is analogous to a streamlined version of X.25 networks. The data packets are in the form of frames, and these frames are variable in length.

Frame Switching—Refers to devices that forward information frames based on the frame's layer 2 address. Frame switching can be done in two ways: via cut-through switching or via store and forward information switching.

Frequency Division Multiplexing (FDM)—A technique in which available transmission bandwidth of a facility is divided by frequency into narrower bands. Each band is used for a separate voice or data transmission channel. In this manner, multiple conversations can take place over a single transmission facility.

Full Duplex—Transferring of voice/data in both directions at the same time. This procedure becomes confusing in a TDMA system, because information is reconstructed to enable the transfer of voice information in both directions at the same time (although actual transmission does not occur simultaneously).

Gateway—A device or facility that enables information to be exchanged between two dissimilar computer systems or data networks. A gateway reformats data and protocols in such a way that the two systems or networks can communicate.

Global Title—In the *Signaling System 7* (SS7) protocol, an address, such as customer-dialed digits, that does not contain explicit information to enable routing in a signaling network, and therefore requires the signaling connection control part translation function

Global Title Translation (GTT)—In SS7, a procedure that translates a global title into a known address that enables message routing in the signaling network.

Grade of Service (GOS)—An estimate of customer satisfaction with a particular aspect of service; also refers to the probability that a call will fail due to the unavailability of links or circuits

Half Duplex—The capability to transfer voice or data information in either direction between communications devices, but not at the same time. The information is transmitted on one frequency and is received on another frequency.

Handoff—Handoff is the process of reassigning subscriber handsets to specific radio channels as the handsets move from cell site to cell site.

Hard Handoff—Hard handoff is a break-before-make form of call handoff between radio channels. In this scenario, the mobile handset temporarily (time measured in milliseconds) disconnects from the network as it changes channels. The radio protocols AMPS, TDMA, and GSM support only hard handoff.

High Data Rate Digital Subscriber Line (HDSL)—Modems on either end of one or more twisted-pair wires that deliver T1 or E1 speeds. At present, T1 requires two lines, and E1 requires three. (See SDSL for one-line HDSL.)

High-Level Data Link Control (HDLC)—A bit-oriented communications protocol in which control codes differ according to their bit positions and patterns

High-Usage Trunk Group—A transmission facility that is used only for routing large volumes of traffic to a single point or set of points

Hundred Call Seconds (CCS)—A measurement of telephone usage traffic that is used to express the average number of calls in progress or the average number of devices used. The first C in CCS represents the Roman numeral for hundred. There is 36CCS in one hour of telecommunications traffic.

Hybrid Fiber Coax—A system (usually CATV) where fiber is run to a distribution point that is close to the subscriber, then the signal is converted to run to the subscriber's premises over coaxial cable

Inband Signaling—A type of signaling in which the frequencies or time slots that are used to carry the signals are within the bandwidth of the information channel

Incumbent Local Exchange Carrier (ILEC)—A telephone service carrier that was operating in the local loop market prior to the Telecommunications Act of 1996 and the divestiture of the AT&T bell system (in other words, the local telephone company with which most people grew up)

Information Service—The offering of a capability for generating, acquiring, storing, transforming, processing, retrieving, utilizing, or making available information via telecommunications, including electronic publishing but not including any use of any such capability for the management, control, or operation of a telecommunications system or the management of a telecommunications service

Integrated Services Digital Network (ISDN)—A structured, all-digital telephone network system that was developed to replace (upgrade) existing analog telephone networks. The ISDN network supports advanced telecommunications services and defined universal standard interfaces that are used in wireless and wired communications systems.

Intelligent Network (IN)—A telecommunications network that is capable of providing advanced services through the use of centralized databases that can provide call processing and routing. IN systems are capable of providing enhanced services that can be used on wireless and wired networks.

Interconnection—The connection of telephone equipment or communications systems to the facilities of another network. The FCC regulates interconnection of systems to the public-switched telephone network. (Also see Bypass.)

Interexchange Carrier (IXC)—A carrier company in the United States, including Puerto Rico and the Virgin Islands, that is engaged in the provision of inter-LATA, interstate, and/or international telecommunications over its own transmission facilities or over facilities that are provided by other inter-exchange carriers

Interexchange Trunks—Transmission facilities that are used to support communications between switching centers (LEC switches) and the IXC

Inter-LATA service—Telecommunications between a point located in a local access and transport area and a point located outside such an area

International Carrier (INC)—A carrier that is authorized to provide interexchange communications services outside World Zone 1 by using the international dialing plan; however, the carrier has the option of providing service to World Zone 1 points outside the 48 contiguous United States.

International Gateway Facilities (IGFs)—Transmission facilities that are used by international gateway switches. An international gateway switch is used by the long-distance (interexchange) carrier to interface with international telecommunications networks.

International Routing Code—A three-digit code within the North American Numbering Plan, beginning with one, that classifies international calls as requiring either regular or special handling

International Telecommunications Union (ITU)—A European telecommunication standards body (counterpart of ANSI)

Internet—The major network running the Internet protocol across the United States and Canada. The Internet consists of more than 30,000 hosts and includes sites at universities, research laboratories, corporations, and nonprofit agencies.

Internet Service Provider (ISP) A vendor that provides access to the Internet and to the *World Wide Web* (WWW)

Intranet—A private network inside a company or organization that not only serves as the company information network but also provides access to the public Internet. The Intranet appears similar to another Internet server.

Intersatellite Service—Provides links between artificial Earth satellites

Inverse Multiplexing—The process of splitting a high-speed channel into multiple signals, transmitting the multiple signals over multiple facilities that are operating at a lower rate than the original signal (followed by a process of recombining the separately transmitted portions into the original signal at the original rate of speed)

Local Access and Transport Area (LATA)—As designated by the modification of final judgment, an area in which a local exchange carrier is permitted to provide service. The LATA contains one or more local exchange areas, usually areas with common social, economic, or other interests.

Local-Area Network (LAN)—A private network offering high-speed digital communications channels for the connection of computers and related equipment in a limited geographic area. LANs use fiber optic, coaxial, or twisted-pair cables or radio transceivers to transmit signals.

Local Exchange Carrier (LEC)—A company that provides telecommunications service within an LATA

Local Loop—A channel that connects customer equipment to the line-terminating equipment in a central office of the local exchange company. Typically, a loop is a two or four-wire cable circuit between a vertical main distributing frame appearance in a central office and the point of termination at a customer's premises. The cable from a frame to a cross-connect terminal is called a *fraeler*. The feeder also can be provided over digital loop or analog carrier systems. Digital systems can be served by either wire or fiber optic cable.

Message Waiting Notification (MWN)—A service that informs enrolled subscribers when a voice message is available for retrieval. MWN can use a pip tone, an MS indication, or an alert pip tone to inform a subscriber of an unretrieved voice message. MWN does not impact a subscriber's capability to originate calls or to receive calls.

Messaging Delivery Service (MDS)—A service that permits pending voice messages to be attempted for delivery to subscribers on a periodic basis until the subscriber acknowledges receipt of the messages

Mobile Access Hunting (MAH)—A call processing service that causes a call to a pilot directory number to search a list of termination addresses sequentially for one that is idle and can be alerted. If a particular termination address is busy, inactive, fails to respond to a paging request, or does not answer alerting before a timeout, then the next termination address in the list is tried. Only one termination address is alerted at a time. The mobile telephones in the group can be alerted by using distinctive alerting.

Mobile Identification Number (MIN)—The 10-digit number that represents a mobile telephone's (mobile station) identity

Mobile Satellite Services (MSS)—A communications service that provides communication between mobile Earth stations and one or more space stations or between mobile Earth stations via one or more space stations. Earth stations can be situated on-board ships, on-board aircraft, and on-board terrestrial vehicles. This service can also be used to detect and locate emergency signals from people who are in distress.

Mobile Switching Center (MSC)—The central switching system that is used for cellular and PCS networks. The MSC was formerly called the *Mobile Telephone Switching Office* (MTSO).

Mobile Telephone Switching Office (MTSO)—A cellular carrier switching system that includes switching equipment needed to interconnect mobile equipment with land telephone networks and associated data support equipment. (See also *Mobile Switching Center*, or MSC.)

Modem—The term is a contraction of modulator/demodulator and is a device or circuit that converts digital signals to and from analog signals for transmission over conventional analog telephone lines. The term *modem* also can refer to a device or circuit that converts analog signals from one frequency band to another. (Also see Smart Modem.)

Multi-Frequency (MF)—A type of inband address-signaling method in which decimal digits and auxiliary signals are each represented by selecting a pair of frequencies from the following group: 700, 900, 1100, 1300, 1500, and 1700 Hz. These audio frequencies are used to indicate telephone address digits, precedence, control signals (such as line-busy or trunk-busy signals), and other required signals.

Multiplexer—Electronic equipment that enables two or more communications signals to pass over a single communications circuit

Narrowband—A communications channel of restricted bandwidth, often resulting in degradation of the transmitted signal

N-Carrier—A predecessor of T-Carrier

National Number—The telephone number identifying a calling subscriber station within an area designated by a country code

Network (general)—A series of points interconnected by communications channels, often on a switched basis. Networks are either common to all users or privately leased by a customer for some specific application.

Network (antenna coupling)—A network that employs a radio frequency circulator, enabling two separate radio transmitters to use the same antenna at the same time

Network (balanced)—A network in which the series elements in both legs of the circuit are symmetrical with respect to ground

Network (bilateral)—A net

Network Element—A facility or the equipment that is used in the provision of a telecommunications service. The term includes subscriber numbers, databases, signaling systems, and information that is sufficient for billing and collection or that is used in the transmission, routing, or other provision of a telecommunications service.

Network Interconnection—The interconnecting of two more networks

Network Interface Card (NIC)—A circuit card or pack that connects a desktop computer to a larger network of computers. In telecommunications systems, network interface cards connect network devices to other network devices.

Network Operations Center (NOC)—A center that is responsible for the surveillance and control of telecommunications traffic flow in a service area

Noise (general)—Any random disturbance or unwanted signal in a communication system that tends to obscure the clarity of a signal in relation to its intended use

Noise (ambient)—The acoustic noise that is part of the environment in which a system transducer is located

Noise (atmospheric)—A component of sky noise arising from natural phenomena within the atmosphere, such as a lightning discharge

Noise (broadband)—A noise signal with significant spectral intensity over a band where the upper frequency is many times the strength of the information signal, thereby causing signal degradation.

North American Numbering Plan (NANP)—A telephone numbering system used in North America that uses 10-digit numbering. The number consists of a 3-digit area code, a 3-digit central office code, and a 4-digit line number.

Number Portability—The customer's current phone number can be transferred to other networks, carriers, or service providers.

Numbering Plan Area (NPA)—A three-digit code that designates one of the numbering plan areas in the North American Numbering Plan for direct-distance dialing. Originally, the format was *NO/IX*, where *N* is any digit two through nine and *X* is any digit. From 1995 on, the acceptable format is *NXX*.

Off-Hook—A signal used on lines and trunks to indicate in-use or request-for-service states

Offered Load—The total traffic load submitted to a group of servers, including any load that results from retries

On-Hook—A signal used to indicate that a line or trunk is not currently in use and is available for service. The signal is transmitted to a central office when the receiver is placed on its hook. Also, the state of being on hook

Open Systems Interconnection (OSI) Model—The OSI model is an internationally accepted framework of standards for telecommunications between different systems that are made by different vendors. The model organizes the telecommunications functions into seven different layers. This model enables engineers to isolate and classify telecommunications into discrete functions or activities.

Operations Administration and Maintenance (OA&M)—This term refers to managing a network or telecommunications system. Normally, one uses OA&M to reference a series of functions that include network diagnostics, network element management, alarm indications, and performance monitoring.

Operator Services—A term that is traditionally used by wireline telephone companies to refer to services that use a human operator or automated operator to render assistance to customers in the making of collect calls, third-party billed calls, credit-card calls, calling-card calls, and person-to-person calls

Optical Fiber—A threadlike filament of glass that is used to transmit digital voice, data, or video signals in the form of light pulses. Multi-mode step index fiber has relatively low capacity and seldom is used. Multi-mode graded index fiber is used for low-traffic routes. Single-mode step index fiber has a high capacity and is the most commonly used type.

Out-of-Band Signaling—A type of signaling in which the frequencies or timeslots that are used to carry the signals are outside the bandwidth of the information channel

Outside Plant—The part of a telephone system that is outside local exchange company buildings. Included are cables and supporting structures. Microwave towers, antennas, and cable system repeaters are not considered to be outside plant equipment.

Packet Data—Data transmission that breaks up or sends small packets of data by including the address and sequence number with each packet sent. Packets find their way through a packet-switching network that eventually routes them to their destination, where they are placed back in sequence by a *Packet Assembler/ Disassembler* (PAD).

Packet Switching—A mode of data transmission in which messages are broken into increments, or packets—each of which can be routed separately from a source and then can be reassembled in the proper order at the destination

Paging—A method of delivering a message, via a public communications system or radio signal, to a person whose exact whereabouts are unknown by the sender of the message. Users typically carry a small paging receiver that displays a numeric or alphanumeric message displayed on an electronic readout, or the message could be sent and received as a voice message or as other data.

Paging Audio Tone Signal—Tone only; no alphanumeric characters at all. The tone meant that the individual either called an operator or simply called a predetermined point. There are three types of tone-only protocols: two tone, three tone, and five tone.

Paging Message Service (PMS)—A service that permits paging messages to be attempted for delivery to subscribers (via the SMS)

on a periodic basis until the subscriber acknowledges receipts of the message. Messages can be up to 256 characters.

Password Call Acceptance (PCA)—A call screening feature that enables a subscriber to limit incoming calls to only those calling parties who can provide a valid PCA password (i.e., a series of digits)

Path Minimization (efficient routing of a mobile call)—Path minimization is the process of an efficient fixed network routing (the non-wireless portion of the call) of wireless calls tables in the wireless carrier switches

Permanent Virtual Circuit (PVC)—A virtual circuit that provides the equivalent of a dedicated private-line service over a packet-switching network between two data terminal devices. The path between the users is fixed; however, a PVC uses a fixed, logical channel to maintain a permanent association between the data terminal device stations.

Photonic Switches—Switches that use light to carry telecommunications signals within the switch. It is a next generation telecommunications switch.

Point of Interface (POI)—The physical location marking the point at which the local exchange carrier's service ends

Point of Presence—A physical location established by an interexchange carrier within an LATA for the purpose of gaining LATA access. The point of presence is usually a building that houses switching and/or transmission equipment, as well as the point of termination. (Also see Point of Termination.)

Pre-Subscription—A term that was created in 1984 to describe subscriber subscription to a pre-determined IXC. The term is a legacy word from the days of immediate AT&T post-divestiture.

Preferred Language (PL)—Provides a telephone service subscriber with the capability to specify the language for network services. This service enables the subscriber to specify service in English, Spanish, French, or Portuguese. Provisions will be made for additional languages in the future.

Primary Rate Interface (PRI)—In the *Integrated Services Digital Network* (ISDN), a channel that provides digital transmission capacity of up to 1.536 Mbit/s (1.984 Mbit/s in Europe) in each direction. The interface supports combinations of one 64-Kbit/s D channel and several Kbit/s B channels or H channel combinations.

Private Branch Exchange (PBX)—A private switching system serving an organization, business, company, or agency (usually located on a customer's premises)

Private Line—Often used to connect the wireless carrier switch to a cell site or to connect two cell sites. This connection can be two-wire or four-wire analog, DS−1, or DS−3 circuits.

Protocol (rules)—A precise set of rules and a syntax that govern the accurate transfer of information

Protocol (connection)—A procedure for connecting to a communications system to establish, carry out, and terminate communications

Protocol Conversion—The translation of the protocols of one system to those of another system to enable different types of equipment, such as data terminals and computers, to communicate

Provisioning—The operations that are necessary to respond to service orders, trunk orders, and special-service circuit orders and to provide the logical and physical resources necessary to fill those orders

Public Switched Telephone Network (PSTN)—An unrestricted dialing telephone network that is available for public use. The network is an integrated system of transmission and switching facilities, signaling processors, and associated operations support systems that are shared by customers. PSTN is also called a public network, public-switched network, or public telephone network.

Radio Common Carrier—A common carrier that is licensed by the *Federal Communications Commission* (FCC) to provide mobile telephone services using radio systems. (Also see Common Carrier.)

Radio Determination Satellite Service—A service that uses radio signals from satellites to determine the position and velocity of an object

Reconciliation—A clearinghouse function for multiple carriers. Interconnected carriers terminate calls in one another's network. Under such circumstances, the carriers are entitled to some form of compensation.

Regional Bell Operating Company (RBOC)—One of the seven U.S. telephone companies, also known as the baby Bells, that resulted from the breakup of AT&T

Ring Network—A data network of circular topology in which each node is connected to its neighbor to form an unbroken ring. (Also, a ring network in which one of the nodes often exercises central control is called a loop.)

Router—See Routing Switch.

Routing Switch (general)—An electronic device that connects a user-supplied signal (audio, video, and/or data) from any input to any user selected output. Inputs are called sources, and outputs are called destinations.

Routing Switch (network)—A device that forwards packets of a specific protocol type (such as *P*) from one logical network to another. These logical networks can be the same type or different types. A router receives physical layer signals from a network, performs data link and network layer protocol processing, then sends the data packet to its final destination.

Satellite (general)—A body that revolves around another body of greater mass and has a motion primarily determined by the force of attraction of the more massive body

Satellite (communications)—An orbiting space vehicle containing a set of transponders that receive signals from the ground and retransmit them to other ground-based receivers

Satellite System—A network of satellites and associated ground stations that transmit telephone, audio, video, and data signals between terrestrial points

Selective Call Acceptance—A CLASS feature that enables a customer to receive calls selectively from previously specified telephone numbers. All other calls are intercepted by a recorded denial announcement or are routed to an alternate directory number, depending on the subscriber's selection when the service is activated. CLASS is a service mark of Bellcore.

Selective Call Acceptance (SCA)—A call-screening service that enables a subscriber to receive incoming calls only from parties whose *Calling Party Numbers* (CPNs) are in an SCA screening list of specified CPNs

Server (telecommunications network)—The equipment or call carrying path that responds to a customer's attempt to use a network

Server (LAN)—A processor that serves users on a LAN, for example, by storing and managing data files or by connecting users to an external network

Service Access Code (SAC)—The three-digit codes in the NPA (N $^0/_1$ X) format, which are used as the first three digits of a 10-digit address in a North American Numbering Plan dialing sequence. Although NPA codes are normally used for the purpose of identifying specific geographical areas, certain NPA codes have been allocated to identify generic services or to provide access capability. These are known as SACs. The common trait, which is in contrast to an NPA code, is that SACs are non-geographic.

Service Objective—A statement of the quality of service that is to be provided to a customer

Service Provider—A generic name given to a company or organization that provides telecommunications service to customers (subscribers). (Also see Network Provider and Reseller.)

Short Message Service Point-to-Point (SMS-PP)—Provides bearer service mechanisms for delivering a short message as a packet of data between two service users, known as *Short Message Entities* (SMEs). SMEs are SMS endpoints that are capable of composing or disposing of a short message. One or both of the service users can be a mobile station. The data packets are transferred transparently between two service users. The network or destination application generates negative acknowledgments when it is unable to deliver the message as desired.

Signaling Link—A communication path that carries CCS messages between two adjacent signaling nodes

Signaling Point (SP)—In the SS7 protocol, a node in a signaling network that originates and receives signaling messages, transfers signaling messages from one signaling link to another, or both

Signaling System Number 7 (SS7)—An out-of-band CCS protocol standard that is designed to be used over a variety of digital telecommunication switching networks. SS7 is optimized to provide a reliable means for information transfer for call control, remote network management, and maintenance.

Signaling Transfer Point (STP)—In a CCS network, a packet switch that uses the SS7 protocol to connect signaling links to network switching systems and to other signaling transfer points

Service Control Point (SCP)—The database that contains the subscriber profiles. SCPs can also provide assistance with routing a call.

Service Switching Point (SSP)—The switch (or routing element) in the SS7 network. The switch would be a Class 5 central office in the wireline network (LEC). In the cellular or PCS world, the SSP would be the wireless switch/mobile switching center.

Soft Handoff—Soft handoff is the reverse of the hard handoff scenario. Soft handoff is a make-before-break form of call handoff between radio channels, whereby the mobile handset temporarily communicates with both the serving cell site and the targeted cell site (one or more cell sites can be targeted) before being directed to release all but the final target cell site radio channel. Currently, only CDMA supports soft handoff.

Space Operation Service—Concerned exclusively with the operation of spacecraft

Space Research Services—A research service where spacecraft are used for scientific or technical research

Star Network—A data network with a radial topology in which a central control node is the point to which all other nodes join

Supervisory Signals—Signals that are used to indicate or to control the status of operating states of circuits that establish a connection. A supervisory signal indicates that a particular state in a call has been reached and can signify the need for additional action.

Switched Circuits—Circuits that are used in the support of wireless 911

Switched Multi-Megabit Data Service (SMDS)—SMDS is a 1.544 Mbps public data service.

Switching (general)—The process of making and breaking (connecting and disconnecting) two or more electric circuits

Switching (telecommunications)—The process of connecting appropriate lines and trunks to form a communications path between two or more stations. Functions include transmission, reception, monitoring, routing, and testing.

Synchronous (general)—In step or in phase, as applied to two or more devices; a system in which all events occur in a predetermined timed sequence

Synchronous (data communications)—An operation that occurs at intervals directly related to a clock period. The bus protocol in such data transactions is controlled by a master clock and is completed within a fixed clock period.

Synchronous Optical Network (SONET)—A standard format for transporting a wide range of digital telecommunications services over optical fiber. SONET is characterized by standard line rates, optical interfaces, and signal formats.

T Carrier— A generic name given to describe any of the several analog transmission facilities that are used in the United States. T Carriers were first used in the Untied States during the early 1960s. The T Carrier first started as a single twisted pair of wires.

Tandem Switch—A switch that supports a network topology where connectivity between locations is attained by linking several locations through a single point. The tandem switch is similar to the traffic cop, because it directs traffic from several streets into other groups of streets.

Tariffs—Documents filed by a wireline telephone company with a state public utility commission or with the FCC. The tariff describes in detail the company's services and pricing of said services. The tariff essentially explains and justifies the pricing of the service.

Telecommunications—The transmission, between or among points specified by the user, of information of the user's choosing (including voice, data, images, graphics, and video), without change in the form or content of the information

Telecommunications Carrier—Any provider of telecommunications services. A telecommunications carrier shall be treated as a common carrier under this Act, only to the extent that the carrier is engaged in providing telecommunications services.

Telecommunications Service—The offering of telecommunications for a fee directly to the public, or to such classes of users as to be effectively available directly to the public, regardless of the facilities used

Telegraphy—A form of telecommunications that is concerned with the process of providing transmission and reproduction at a distance of text material or fixed images. The transmission of such information can be physical transmission facilities or over-the-air facilities that use some form of signaling protocol.

Television—The transmission and reception of visual images via electromagnetic waves

Terminal Equipment—Refers to the computers, telephones, and other data or voice devices at the end of a telephone line

Terminals—Devices that typically provide the interface between the telecommunications system and the user. Terminals can be fixed (stationary) or mobile (portable).

Three-Way Calling (3WC)—Provides the subscriber with the capability of adding a third party to an established two-party call, so that all three parties can communicate in a three-way call

Time Division Multiplexing (TDM)—A method for sending two or more signals over a common transmission path by assigning the path sequentially to each signal—each assignment being for a discrete time interval. All channels of a time-division multiplex system use the same portion of the transmission links' frequency spectrum, but not at the same time. Each channel is sampled in a regular sequence by a multiplexer.

Toll—Any message telecommunication charge for service provided beyond a local calling area

Toll Center—An old term (i.e., pre-AT&T divestiture) that is used to describe a Class 4 central office, which is also known as the tandem switching center

Toll Free—A term that is used to refer to free calling numbers. A company would buy an 800 number that would enable users to call them without any calling charges. Today, it also includes any calling number that users can call without incurring charges.

Traffic Engineering—Planning activity that determines the number and types of communications paths that are required between switching points and the call processing capacity of the switching equipment

Translation (general)—The conversion of information from one form to another

Translation (switching system)—The conversion of all or part of a telephone address destination code to routing instructions or routing digits

Translation (routing)—A short hand term used to describe the routing instructions and tables programmed into a telecommunications switch.

Transmission Control Protocol/Internet Protocol (TCP/IP)—The protocol that manages the bundling of outgoing data into packets, manages the transmission of packets on a network, and checks the bundles
for errors. This protocol is a networking protocol that supports communication across interconnected information networks (i.e., the Internet).

Trunk—A single transmission path connecting two switching systems. Trunks can be shared by many users but serve only one call at a time.

Trunk Group—A number of trunks that can be used interchangeably between two switching systems

Trunk Side Connection—Transmission facilities that connect switches to each other

Twisted Pair—A pair of insulated copper wires that are used in transmission circuits to provide bidirectional communications. The wires are twisted about one another to minimize electrical coupling with other circuits. Paired cable is made up of a few to several thousand twisted pairs.

Two-Wire Circuits—A transmission circuit comprised of two wires that support the sending and reception of information

Type 1—A connection that is a trunk-side connection to an end office. The end office uses a trunk-side signaling protocol in conjunction with a feature known as *Trunk with Line Treatment* (TWLT).

Type 1 with ISDN—An ANSI-standard ISDN line between the LEC end office and the MSC. The ISDN connection should be capable of providing connectivity to the PSTN that will support ISDN PRI (Primary Rate Interface) and ISDN BRI (Basic Rate Interface). The ISDN connection is a variation of the Type 1 connection.

Type 2A—A trunk-side connection to the LEC's access tandem. This connection enables the MSC to interface with the access tandem as if it were an LEC end office. This connection enables the wireless carrier's subscribers to obtain presubscription. The service provider will have access to any set of numbers within the LEC network.

Type 2B—A similar inter-connection type to the high-usage trunk groups established by the LEC for its own internal routing purposes. In the case of wireless carrier interconnection, the Type 2B should be used in conjunction with the Type 2A. When a Type 2B is used, the first choice of routing is through a Type 2B with overflow through the Type 2A.

Type 2C—A telephone system inter-connection that is intended to support interconnection to a public safety agency via an LEC E911 tandem or local tandem. This connection enables a wireless carrier to route calls through the PSTN to the *Public Safety Answering Point* (PSAP). This connection supports a limited capability to transport location coordinate information to the PSAP via the LEC.

Type 2D—A telephone system inter-connection that is intended to support interconnection to the LEC's operator service position. This connection enables the LEC to obtain *Automatic Number Identification* (ANI) information about the calling wireless subscriber in order to create a billing record.

Type S—A telephone system inter-connection that only carries control messages. The Type S is a SS7 signaling link from the wireless carrier to the LEC. The Type S supports call setup via the ISUP (ISDN User Part) portion of the SS7 signaling protocol and TCAP querying. The Type S is used in conjunction with the Type 2A, 2B, and 2D.

Universal Resource Locator (URL)—Used in the Internet. URLs refer to the Internet addresses that are used on the *World Wide Web* (WWW). The URL is a string expression that can represent any resource on the Internet.

Uplink—The Earth-to-satellite microwave link and related components, such as Earth station transmitting equipment. The satellite contains an uplink receiver. Various uplink components in the Earth station are involved processing and transmitting the signal to the satellite.

Validation—Validation is often confused with authentication. Where authentication essentially certifies the user as either real or fake and also as either good or bad, validation confirms the permission to complete the call. An example is as follows: Suppose a user calls another party using a mobile handset. The carrier certifies that the handset is the real handset and is not a clone that is being used by a criminal. Once the handset has been given the thumbs up, the carrier checks to see whether the call is even permitted under the user's billing plan. This next step is validation (being given the final green light to complete the call).

Very High Rate Digital Subscriber Line (VDSL)—A high-speed transmission technology that is intended to be used in the local loop. VDSL technology supports fiber optics to a neighborhood, while the last mile to the specific home still uses unshielded twisted pairs of wires.

Video—An electrical signal that carries TV picture information. (Also see Video Signal.)

Video on Demand (VOD)—A service that telephone companies were seeking to bring to their subscribers. VOD would enable a subscriber to request any video that they wished to see.

Virtual Circuits—A telecommunications link that appears to the user as a dedicated point-to-point circuit

Visitor Location Register (VLR)—The part of a wireless network (typically cellular or PCS) that holds the subscription and other information about visiting subscribers who are authorized to use the wireless network

Voice Mail (VM)—Provides the subscriber with voice mail services, including not only the basic voice recording functions but also time-of-day recording, time-of-day announcements, menu-driven voice recording functions, and time-of-day routing.

Wideband—The passing or processing of a wide range of frequencies. The meaning varies with the context. In an audio system, wideband might mean a band that is up to 20 kHz wide; in a TV system, however, the term might refer to a band that is many megahertz wide.

Wink—A telephone line signal that is a single supervisory pulse usually transmitted as an off-hook signal followed by an on-hook signal, where the off-hook signal is of a short, specified duration (compared to the on-hook signal)

Wireless Local Loop (WLL)—The providing of local telephone service via radio transmission

xDSL—A set of large-scale, high-bandwidth data technologies that can use standard twisted-pair copper wire to deliver high-speed digital services (up to 52 Mbps)

X.25—The X.25 protocol is a Data Link layer protocol. Specifically, X.25 uses the LAPB portion of this layer.

APPENDIX B

Network Interconnection Document Summary

Work on TIA's IS−93 began in 1992 when it became clear that technology development had elevated the wireless switch from the technical functionality of a PBX to that of a switching system equal to that of a telephone company switch. A number of wireless industry players had hoped to use IS−93 as a way of forcing the RBOCs into re-negotiating their interconnection agreements. The most significant difference between IS−93 and TR-NPL−000145 Issues 1 and 2 was the acceptance of a bi-directional signaling relationship between the cellular carriers and the PSTN. By acknowledging that the SS7 signaling occurred in two directions, the RBOCs could interact with the cellular carrier as a co-carrier; therefore, the cellular carriers could force acceptance of mutual compensation. GR−145-CORE was written (by Bellcore now known as Telcordia) in response to IS−93 Revision A. Bellcore's GR is a proprietary product sold by Bellcore. Some industry players believe the Bellcore document was funded by the owners of Bellcore in order to keep their existing interconnection agreements in place.

Why the documents were written is really not of any consequence. What is relevant is that both documents support the interconnection business.

Both documents were written to support technical and business perspectives and objectives of their respective interest groups: GR−145-CORE supports the objectives of the RBOCs, and IS−93 supports the objectives of the wireless industry (specifically the *Cellular Telecommunications Industry Association* (CTIA) and the TIA). IS−93 is intended to support the entire wireless community. IS−93 is a standard. GR−145-CORE is a company proprietary document sold for profit.

GR−145-CORE is service- and implementation-oriented. In other words, GR−145-CORE interconnection descriptions provide more than just technical signaling protocol and parameter information. GR−145-CORE addresses interconnection from the perspective of how the LEC is or is not capable of providing information. For example, GR−145-CORE describes how the LEC provides Feature Group A, B, C, and D support. Essentially, GR−145-CORE is an implementation document for LEC interconnection where the LEC is at the center of the interconnection view. GR−145-CORE also promotes Bellcore document products. TIA's IS−93 addresses the interconnection not only from a implementation perspective, but also from the perspective of a wireless carrier "wish list" where the wireless carrier is at the center of the interconnection view; the wireless carrier is assumed to be a network to which other carriers wish to interconnect.

Given its long history, Bellcore's GR−145-CORE is still the predominantly quoted industry specification. The documents are officially titled as follows:

- EIA/TIA IS−93 Revision A, September 1998, "Cellular Radio Telecommunications A_i-D_i Interfaces Standard"
- GR−145-CORE Issue 1, March 1996, "Compatibility Information for Interconnection of a Wireless Services Provider and a Local Exchange Carrier Network"

These documents are being updated constantly. IS-93 is now an ANSI standard known as ANSI-93.

Interface Types

GR−145-CORE focuses on 10 specific categories of interconnections. IS−93 focuses on 13 specific categories of interconnection. The categories of both documents are largely the same from a protocol perspective. GR−145-CORE supports the following categories:

- Direct *Wireless Service Provider* (WSP) connection through a LEC end office using *multifrequency* (MF) signaling, called Type 1
- Direct WSP connection through a LEC end office using the *Integrated Services Digital Network* (ISDN) protocol, also known as Type 1 with ISDN
- Direct WSP connection with a LEC tandem office using MF signaling called Type 2A
- Direct WSP connection with a LEC tandem office using SS7 signaling called Type 2A with SS7
- Direct WSP connection with a LEC *Common Channel Signaling* (CCS) *Signaling Transfer Point* (STP), called Type S
- Direct WSP connection with a specific LEC end office using MF signaling, called Type 2B
- Direct WSP connection with a specific LEC end office using SS7 signaling, called Type 2B with SS7

- Direct WSP connection with a LEC tandem office arranged for 911 emergency calls, called Type 2C
- Direct WSP connection with a LEC tandem office arranged for LEC operator assisted calls or directory service using MF signaling, called Type 2D
- Direct WSP connection with a LEC tandem office arranged for LEC operator assisted calls or directory service using SS7 signaling, called Type 2D with SS7

TIA's IS–93 supports the following categories:

- Trunk with Line Treatment using MF signaling, called *Point of Interface-T1* (POI-T1)
- General Trunk Access Signaling using MF signaling, called POI-T4
- General Trunk Access Signaling using SS7 signaling, called POI-T5 and POI-S5
- Direct Trunk Access Signaling using MF signaling, called POI-T6
- Direct Trunk Access Signaling using SS7 signaling, called POI-T7 and POI-S7
- Operator Services Access Signaling using MF signaling, called POI-T10
- Operator Services Access Signaling using SS7 signaling, called POI-T11 and POI-S11
- Call Management Features Signaling using MF signaling, called POI-T12
- Call Management Features Signaling using SS7 signaling, called POI-T13 and POI-S13
- Basic Signaling Transport using SS7 signaling, called POI-S14
- Global Title Signaling Transport using SS7 signaling, called POI-S15
- Cellular Nationwide Roaming Signaling using SS7 signaling, called POI-S16
- TCAP Applications using SS7 signaling, called POI-S17

As you have noticed, based on the numbering sequence, there are interfaces not described in the TIA document. The following are not fully supported in IS–93:

- Trunk with Line Treatment using ISDN-BRI signaling, called POI-T2
- Trunk with Line Treatment using ISDN-PRI signaling, called POI-T3
- Emergency Services Access Signaling, using MF signaling, called POI-T8

- Emergency Services Access Signaling, using SS7 signaling, called POI-T9 and POI-S9
- ISUP End-to-End Signaling using SS7 signaling, called POI-S18

The TIA acknowledges that these interfaces are not explicitly described in IS−93, but are recognized as legitimate interfaces that can be supported by the LEC. I will give details behind these interfaces and the lack of substantive industry support later in this chapter.

Comparisons

Bear in mind that IS−93 was written to effect changes on the business aspects of the interconnect environment between the wireless carrier and the LEC. Therefore, both documents can be overlaid with each other to see the similarities and the differences.

The comparison is important in order for you to gain an understanding of the interconnect perspectives. You will find after reading this comparative analysis just how similar the documents really are and see how interconnection can be applied by different service providers.

- POI-T is the designation used by TIA to represent the trunk point of interface. Bellcore shifted to this designation after TIA. The POI-T represents the voice part of the transmission facility between the service providers. In an SS7 environment, the POI-T only carries voice. In an MF environment, the POI-T carriers both voice and call control information.
- POI-S is the designation used by TIA to represent the SS7 point of interface or signaling point of interface. In an SS7 environment, the POI-S carries the call control information. Bellcore has also shifted to using this designation. The Type S connection is the equivalent of the POI-S connection.

The Bellcore document supports the following MF interfaces:

- Type 1 without ISDN
- Type 2A
- Type 2B
- Type 2C
- Type 2D

The TIA IS−93 equivalent MF interfaces are

- POI-T1
- POI-T4
- POI-T6
- POI-T8
- POI-T10

Bellcore GR−145-CORE	TIA IS−93 Revision 0
Type 1	POI-T1
Type 2A	POI-T4
Type 2B	POI-T6
Type 2C	POI-T8
Type 2D	POI-T10

SS7 Based Interface Descriptions

The Bellcore document supports the following SS7 interfaces:

- Type 1 with ISDN
- Type 2A with SS7
- *Type 2B with SS* Type 2C with SS7 (does not exist in GR−145-CORE but it is being implemented today in regions of the country)
- Type 2D with SS7

The TIA IS−93 equivalent SS7 interfaces are

- POI-T2 and POI-T3
- POI-T5 and POI-S5 (ISUP)
- POI-T7 and POI-S7 (ISUP)
- POI-T9 and POI-S9 (ISUP)
- POI-T11 and POI-S11 (ISUP)

GR−145-CORE	TIA IS−93 Revision 0
Type 1 with ISDN	POI-T2 and POI-T3
Type 2A with SS7	POI-T5 and POI-S5 (ISUP)
Type 2B with SS7	POI-T7 and POI-S7 (ISUP) POI-T9 and POI-S9 (ISUP)
Type 2D with SS7	POI-T11 and POI-S11 (ISUP)

IS−93 Revision 0 supports additional SS7-based interfaces and a MF interface that are not addressed by GR−145-CORE. Part of the reason for this difference is that these additional interfaces involve mobility issues not of concern to the fixed network RBOCs.

Multifrequency–Based Interface Descriptions

The following Bell system descriptions define *multifrequency* (MF) network interfaces. These are the de facto industry standards used to establish network interconnect and the associated interconnect business agreements.

GR−145-CORE Summary Descriptions of the Supported Interfaces

Type 1 The Type 1 connection is a trunk-side connection to an end office. The end office uses a trunk-side signaling protocol in conjunction with a feature known as *Trunk With Line Treatment* (TWLT). The TWLT feature allows the end office to combine some line-side and trunk-side features. TWLT enables the LEC to provide billing and IC presubscription. The wireless service provider will have access to numbers that reside in the interconnected LEC end office. This includes *NXX* codes within the LEC's territory as well as codes accessible through an interchange carrier's network, including international calls. Type 1 connections will also permit the mobile user to reach Directory Assistance, N11 codes (for example, 911, and so on), and Service Access Codes (for example, 700, 800, 900, and so on).

Type 2A The Type 2A connection is a trunk-side connection to the LEC's access tandem. This connection allows the MSC to interface with the access tandem as if it were a LEC end office. This connection enables the wireless carrier's subscribers to obtain presubscription. Wireless service providers will have access to any set of numbers within the LEC network. The Type 2A supports the following subscriber dialing patterns:

- 1+NPA+NXX+XXXX
- 101XXXX+NPA+XXXX
- 800
- 888
- 877
- 900

Note that prior to the Type 2D, the Type 1 was used in conjunction with the Type 2A to provide access to the LEC operator and N11 codes (411, 611, 911, and so on).

Type 2B The Type 2B connection is similar to the high usage trunk groups established by the LEC for its own internal routing purposes. In the case of wireless carrier interconnection, the Type 2B should be used in conjunction with the Type 2A. When a Type 2B is used, the first choice of routing is through a Type 2B with overflow through the Type 2A.

The Type 2B uses trunk-side signaling protocols, E&M supervision, and *multifrequency* (MF) address pulsing. The Type 2B restricts a subscriber to only those numbers served by the subtended LEC end office.

Type 2C The Type 2C connection is intended to support interconnection to a public safety agency via a LEC E911 tandem or local tandem. This connection enables a wireless carrier to route calls through the PSTN to the *Public Safety Answering Point* (PSAP). This connection supports a limited capability to transport location coordinate information to the PSAP via the LEC. The Type 2C is a fairly new type of interconnect that did not exist until IS−93 promoted the concept.

Prior to Type 2C, the LECs provided a MF-based interconnect to the LEC-owned 911 tandem. Note that the 911 tandem provides access to the *Public Safety Answering Point* (PSAP), which is typically owned and operated by local or state public safety officials. The difference between the Type 2C and the original MF based interconnection is that the Type 2C supports CAMA signaling, which is a MF-based format. The use of CAMA signaling is so some minimal level of location information, specifically cell site or cell face ID information, can be transmitted to the PSAP. Type 2C is adaptable to four different network configuration models known as Models A, B, C, and D.

Type 2D The Type 2D connection is intended to support interconnection to the LEC's operator service position. This connection enables the LEC to obtain *Automatic Number Identification* (ANI) information about the calling wireless subscriber in order to create a billing record.

The Type 2D connection is also a new connection. However, prior to Type 2D, the RBOCs provided the cellular carriers access to their respective operator positions in a variety of ways. One used a direct two-way MF trunk to the operator platform. Another used the Type 1 connection. Others even used the oldest and most basic type of interconnection: the dial line. Using a Type 1 or even a dial line would enable communication between the operator and the mobile subscriber, but would not allow the LEC to capture critical billing information. Needless to say, prior to many of the new interconnections, billing leakage had been an issue. However, given the tenor of the times (predivestiture), providing service was the priority, not profit-generation.

IS−93 Summary of Multifrequency (MF) Interfaces The following are the descriptions of MF based network interfaces as defined in the TIA's IS−93 standard.

POI-T1 Trunk With Line Treatment The POI-T1 is an inband MF trunk that possesses the feature *Trunk with Line Treatment* (TWLT). TWLT is an AT&T Bell Labs (pre-divestiture) feature. POI-T1 is the exact duplicate of the Type 1 trunk. POI-T1 supports the same signaling format and dialing patterns as does Type 1.

POI-T4 General Trunk Access The POI-T4 is an inband MF trunk that interconnects a wireless carrier's switch to the PSTN's tandem office. This interface exactly duplicates the Type 2A trunk. The same signaling format is supported. The same dialing patterns are supported.

The differences between POI-T4 and Type 2A is that POI-T4 was intended to allow the wireless carrier to interconnect to not only an LEC, but also an Interexchange Carrier tandem and an International Carrier tandem. The difference is one rooted in perspective, not in the technical protocol. The assumption the TIA made here is that the wireless carrier would interconnect to a variety of other service providers.

POI-T6 Direct Trunk Access The POI-T6 is an inband MF trunk that interconnects the wireless switch to the PSTN's end office. This interface exactly duplicates the Type 2B.

The differences between POI-T6 and the Type 2B is that POI-T6 was intended to allow the wireless carrier to interconnect not only to a LEC, but also an interexchange carrier switch and an international carrier switch in a fashion that restricted the wireless carrier to access subscribers residing in those other carriers' directly connected switch.

POI-T8 Emergency Services Access The POI-T8 is an inband MF trunk that interconnects the wireless switch to the PSTN's emergency services (911/E911) tandem switch. Note that typically the PSTN will connect via a two-way MF trunk to a PSAP position; how this is done varies from state to state.

POI-T8 is designed to support five different implementation models. Type 2C supports the same five POI-T8 models. The RBOCs Type 2C connection and the TIAs POI-T8 connection do not support location down to the level of latitude, longitude, and altitude. However, the TIA's POI-T9 and POI-S9 connections do.

The TIA has published a separate standard addressing emergency services under the TIA project number of PN−3581. The information in IS−93 Revision 0 represents early work (and thinking) on the provisioning of emergency services and should not be considered the final word on the subject. Network models are provided in an annex in IS−93.

POI-T10 Operator Services Access The POI-T10 is an inband MF operator trunk that interconnects the wireless switch to the PSTN's operator services position. Insufficient detail is provided by IS−93 to determine the level of compatibility with the Type 2D connection. However, when you reviews the TIA's IS−52 (Uniform Dialing Procedures), you can assume that the POI-T10 connection should be capable of supporting the same level of operator interaction as the Type 2D connection.

The Type 2D connection supports three signaling protocols: two of the protocols are MF-based, whereas the third is SS7-based. The MF-based protocols are called Interim Operator Services Signaling and Exchange Access Signaling. The Interim Operator Services Signaling is the most widely used protocol. Exchange Access Signaling supports communication between the LEC and the interexchange carrier. Note that LECs and wireless carriers must negotiate the use of Exchange Access Signaling. Again, given the dialing patterns supported by IS−52 and the fact that IS−93 is technically supportive of those dialing patterns, you can assume POI-T10 is capable of supporting both Interim Operator Services Signaling and Exchange Access Signaling.

GR−145-CORE Summary Descriptions of the CCS Supported Interfaces

The following descriptions define the common channel signaling based network interfaces in GR−145-CORE. These descriptions represent de facto network interconnect standards.

Type 1 with ISDN The *Integrated Services Digital Network* (ISDN) connection is a ANSI standard ISDN line between the LEC end office and the MSC. The ISDN connection should be capable of providing connectivity to the PSTN that will support ISDN PRI (Primary Rate Interface) and ISDN BRI (Basic Rate Interface). The ISDN connection is a variation of the Type 1 connection.

ISDN BRI provides two bearer channels (each 64 kb/sec) and a data channel (16 kb/sec). ISDN PRI provides 23 bearer channels and one data channel.

The Type 1 with ISDN provides the subscriber the same level of access to directory assistance, operator services, and service access codes as the Type 1 without ISDN. The Type 1 with ISDN can support data transmission and packet mode data transmission.

Type 2A with SS7 The Type 2A connection is a trunk-side connection to the LEC's access tandem. This connection allows the MSC to interface with the access tandem as if it were a LEC end office. This connection enables the wireless carrier's subscribers to obtain presubscription (1+ dialing). The Type 2A with SS7 requires the Type S connection. The Type 2A supports the following subscriber dialing patterns:

- 1+NPA+NXX+XXXX
- 101XXXX+NPA+XXXX
- 800
- 888
- 877
- 900

Note that prior to the Type 2D, the Type 1 was used in conjunction with the Type 2A to provide access to the LEC operator and N11 codes (411, 611, 911, and so on).

Type 2B with SS7 The Type 2B connection is similar to the high usage trunk groups established by the LEC for its own internal routing purposes. In the case of wireless carrier interconnection, the Type 2B should be used in conjunction with the Type 2A. When a Type 2B is used, the first choice of routing is through a Type 2B with overflow through the Type 2A. The Type 2B with SS7 requires the Type S connection. The Type 2B restricts a subscriber to only those numbers served by the subtended LEC end office.

Type 2C with SS7 The Type 2C connection is intended to support interconnection to a public safety agency via a LEC E911 tandem or local tandem. The Type 2C with SS7 is not supported by GR−145-CORE.

Type 2D with SS7 The Type 2D connection is intended to support interconnection to the LEC's operator service position. The Type 2D with SS7 connection is deployed on a limited basis today but requires more development work. The Type 2D with SS7 call setup is nearly the same as a basic SS7 call setup, with one exception: There is no answer message returned to the wireless service provider.

Type S The Type S connection is not a voice path connection. The Type S is a SS7 signaling link from the wireless carrier to the LEC. The Type S supports call setup via the ISUP (ISDN User Part) portion of the SS7 signaling protocol and TCAP querying. The signaling is described in detail in Bellcore documentation (GR−905-CORE and GR−1434-CORE). At a minimum, access links (A-links), cross links (C-links), and diagonal links (D-links) shall be supported. The fully associated links (F-Links) are not supported by all vendors. The Type S is used to support all SS7 versions of Type 2A, 2B, and 2D.

TIA IS−93 Revision "0" Summary Descriptions of the CCS Supported Interfaces The following descriptions define the common channel signaling based network interfaces developed by the wireless industry. The descriptions represent industry standards that are supported primarily by the wireless industry.

POI-T2 (ISDN BRI) and POI-T3 (ISDN PRI): Trunk With Line Treatment The POI-T2 interface uses ISDN *Basic Rate Interface* (BRI) signaling. The ISDN BRI protocol uses digital technology to provide services over a digital network. The BRI network interface provides two bearer channels plus one signaling channel (hence, 2B+D). This interface provides digital services such as 3.1 kHz audio, voice, data and packet mode transmission.

The POI-T3 interface uses ISDN *Primary Rate Interface* (PRI) signaling. The ISDN PRI protocol uses digital technology to provide services over a digital network. The network interface to the primary rate provides 23 bearer channels plus one signaling channel (hence, 23B+D). This interface provides digital services such as 3.1 kHz audio, voice, data, and packet mode transmission.

I had noted that IS−93 does not fully support these interfaces. ISDN over the wireless interface was not a supported product in the wireless industry. IS−93 indicates a need for further study on these interfaces. Theoretically, the interfaces can be supported, but more study is needed to determine what services should be provided.

POI-T5 and POI-S5 (ISUP): General Trunk Access The POI-T5 and POI-S5 interfaces use SS7 *Integrated Services Digital Network User Part* (ISUP) protocol signaling. The POI-S5 interface is used to control user traffic transferred across the POI-T5 interface. ISUP messages are used to establish and release the SS7 supported trunks and to provide supplementary ISDN services.

These interfaces are being implemented today between wireless carriers and the PSTN (RBOC). In some cases, the wireless carriers require only ISUP trunks without direct signaling links between the wireless carrier and the RBOC. This particular scenario describes the function of hub providers like Illuminet. Bell Atlantic is currently working towards becoming a regional hub provider covering the mid-Atlantic and North Eastern seaboard. Bell Atlantic still refers to this interconnection as Type 2A with SS7. Illuminet refers to this scenario as "signaling backbone services."

IS−93 positions this interconnection as a network function to enable the wireless carrier to provide SS7 supplementary services using this interconnection. GR−145-CORE positions the Type 2A in the same manner; however, the Bellcore document promotes specific Bellcore access products such as Bellcore's PCS Home Database Service as described by Bellcore's GR−1411-CORE and Bellcore's *PCS Access Service for Controllers* (PASC) as described by Bellcore's SR−3285.

POI-T7 and POI-S7 (ISUP): Direct Trunk Access

The POI-T7 and POI-S7 interfaces use SS7 ISUP protocol signaling. The POI-S7 interface is used to control user traffic transferred across the POI-T7 interface. ISUP messages are used to establish and release the SS7 supported trunks and to provide supplementary ISDN services.

GR−145-CORE and IS−93 Revision 0 position the interconnection for SS7 supplementary services. Again, GR−145-CORE promotes Bellcore products, whereas IS−93 refers to supplementary services as described by other standards.

POI-T9 and POI-S9 (ISUP): Emergency Services Access

The POI-T9 and POI-S9 interfaces use SS7 ISUP protocol signaling. The POI-S9 interface is used to control user traffic transferred across the POI-T9 interface. ISUP messages are used to establish and release the SS7 supported trunks and to provide supplementary ISDN services. The specification of the POI-T8, POI-T9, and POI-S9 interfaces is for further study. Refer to IS−93 Revision 0 Annex A—Emergency Services Models for the emergency services access models that were under study.

The difference between GR−145-CORE and IS−93 Revision 0 is that GR−145-CORE does not even attempt to address the location technology issue. It is likely GR−145-CORE will be updated by Bellcore to reflect location upon the completion of TIA's work. Location is largely a radio issue. The fixed network issue is one of information formatting and transmission of this information in a manner that is understandable by the PSAP.

POI-T11 and POI-S11 (ISUP): Operator Services Access

The POI-T11 and POI-S11 interfaces use SS7 ISUP protocol signaling. The POI-S11 interface is used to control user traffic transferred across the POI-T11 interface.

As was indicated, the Type 2D with SS7 call setup is nearly the same as a basic SS7 call setup, with one exception: there is no answer message returned to the wireless service provider. Given this fact, the implication is that the LECs' operator positions are limited in their ability to communicate call control information to the wireless switch. This further infers that although the POI-T11 and POI-S11 may be capable of carrying the answer message, the LEC is not capable of transmitting it; therefore, the interface may appear to be limited when it is not.

Other IS−93 Interfaces: Call Management Features

These network interfaces represent the TIA's work on network interfaces that support call related services.

POI-T12 (MF), POI-T13 and POI-S13 (SS7 ISUP): Call Management Features Call Management Features signaling allows a carrier network element to establish connections to a variety of features such as Calling Number Identification Presentation. Refer to EIA/TIA IS–53 and ANSI Standard T1.611 for the features supported by this interface. The key signaling information elements included in the address signaling sequence to obtain access to call management features are the following:

- Called number
- Calling number

Call Management Features signaling is provided via the following interface types:

- POI-T12 (MF)
- *POI-T13 and POI-S13 (ISUP)* The POI-T12 interface uses inband MF signaling.

The POI-T13 and POI-S13 interfaces use SS7 ISUP and SS7 *Transaction Capabilities Application Part* (TCAP) protocol signaling. The POI-S13 interface is used to control user traffic transferred across the POI-T13 interface or to provide transaction-oriented services required to complete a special call. ISUP messages are used to establish and release the SS7 supported trunks and to provide supplementary ISDN services. TCAP messages are used to perform network database queries and provide responses to those queries.

The POI-T12 interface is capable of supporting vertical service features like Call Forwarding and Call Waiting. POI-T12 is not capable of supporting Caller ID type features.

The significance in these interfaces is the wireless industry's recognition of providing supplementary services on a wholesale basis. Unlike Bellcore's GR–145-CORE, which acknowledges and even suggests ways to provide supplementary services on a wholesale basis, IS–93 defined a specific interface to perform that very function. In other words, IS–93 uniquely identified an interface to provide services to other carriers.

POI-S14: Basic Signaling Transport SS7 Basic Signaling Transport allows a carrier network element (an SS7 Signaling Point) to obtain basic message transport services via the SS7 signaling network (that is, to establish an SS7 Level 3 MTP logical network layer connection to route SS7 *message signal units* (MSUs) to destinations accessible within the interconnected SS7 network). SS7 Basic Signaling Transport is provided via the POI-S14 interface.

Basic Signaling Transport requires Levels 1, 2, and 3 of the SS7 MTP:

- Level 1 MTP defines the physical, electrical, and functional characteristics of a signaling link.
- Level 2 MTP defines the functions and procedures for the transfer of signaling messages over a signaling data link and ensures reliable transfer of signaling messages between two directly connected *Signaling Points* (SPs).
- Level 3 MTP defines the transport functions that are common to and independent of the operation of individual signaling links. These functions consist of Signaling Message Handling and Signaling Network Management. Signaling Network Management consists of Signaling Link Management, Signaling Route Management, and Signaling Traffic Management. The Level 3 MTP functions provide for the transfer of messages between any two *Signaling Points* (SPs) in an SS7 network.

This looks like a Type S connection, and it is. What is important to note here is that the POI-S14 is an interface specifically designed to support signaling as service. This interface enables a wireless carrier to sell signaling and not necessarily ISUP as a service. Unlike GR–145-CORE, IS–93 Revision 0 has explicitly identified this as a hub service. GR–145-CORE does not explicitly promote this type of interconnection. The RBOCS typically provides both signaling and ISUP.

POI-S15: Global Title Signaling Transport SS7 Global Title Signaling Transport allows a carrier network element (an SS7 Signaling Point) to obtain message transport services via the SS7 signaling network. The addition of the *Global Title Translation* (GTT) function allows the originating SP to send messages to a remote destination SP without providing the explicit signaling network point code address of the destination SP (provided that an intermediate Signaling Transfer Point is capable of translating the indicated Global Title type).

The Global Title is an address, such as dialed digits, which does not explicitly contain information that allows routing in the signaling network. The GTT function translates the Global Title into an address that can be routed through the signaling network. SS7 Global Title Signaling Transport is provided via the POI-S15 interface.

Global Title Signaling Transport requires the signaling information provided by Basic Signaling Transport and the SS7 *Signaling Connection Control Part* (SCCP). SCCP consists of connectionless control, connection-oriented control, management, and routing control functions. The routing control function of SCCP provides the GTT capability.

Again, IS−93 has uniquely identified an interface that will enable a wireless carrier to provide a specific signaling service. GR−145-CORE has not explicitly identified an interface for this service. However, given the way GR−145-CORE has changed within the last two years, the chances are that it will detail this interface.

POI-S16: Cellular Nationwide Roaming Cellular Nationwide Roaming signaling allows a cellular network element to exchange subscriber mobility information with another carrier network element. This type of signaling requires the signaling information provided by Basic and Global Title Signaling Transport.

The key signaling information included in the signaling sequence to obtain access to cellular nationwide roaming is a specific Global Title type that is translated at an intermediate *signaling transfer point* (STP) by the SCCP.

Cellular Nationwide Roaming signaling is provided via the POI-S16 interface. For TIA IS−41 Cellular Radio-Telecommunications Intersystem Operations, the following SCCP signaling criteria is used:

- SCCP Class 0 connectionless service
- The message types are *Unitdata* (UDT) and *Unitdata Service* (UDTS). The SCCP shall return a UDTS message when a received UDT message cannot be delivered to the specified destination and has the "return message on error" option set.
- Whether to set the "return message on error" or "discard message on error" option in the Protocol Class parameter of the UDT message is at the discretion of the implementation.
- The IS−41 Mobile Application Part is assigned six *Subsystem Numbers* (SSNs). Use of the following SSN values are recommended (although the use of other values is not prohibited):
  - 5 *Mobile Application Part* (MAP)
  - 6 *Home Location Register* (HLR)
  - 7 *Visitor Location Register* (VLR)
  - 8 *Mobile Switching Center* (MSC)
  - 9 *Equipment Identification Register* (EIR)
  - 10 *Authentication Center* (AC)

- In accordance with ANSI T1.112, an SSN shall be included in all messages even if message routing is based on *Global Title Translation* (GTT). An SSN value of zero should be used when the destination SSN is not specifically known by the source in order to allow the network to determine the actual SSN through GTT.
- Global Title Translation on *Mobile Identification Number* (MIN) can be used for communication with the HLR. Global Title Indicator Type 2 (0010 binary) is used. A translation type value of three is used for the "MIN to HLR" translation. The Global Title address Information field contains the 10-digit MIN. The encoding scheme is *Binary Coded Decimal* (BCD). Each address signal is coded as described in Section 3.4.2.3.1 of ANSI T1.112.
- Use of Signaling Point Codes, Global Titles, and Subsystem Numbers must meet ANSI T1.112 requirements so that any allowable combination of these addressing elements is supported. For example, as stated in T1.112.3, Section 3.4.1, "the address consists of any one or any combination of the following elements:
  - Signaling Point Code
  - Global Title (MIN to HLR for IS−41)
  - Subsystem Number where the referenced address is either the called party address or the calling party address fields in SCCP messages.

Needless to say, GR−145-CORE is not required to address roaming. However, as the industry begins to further deploy and integrate mobility and wireless local loop applications, roaming will eventually become an issue.

POI-S17: TCAP Applications TCAP Applications signaling allows a carrier network element (an SS7 Signaling Point) to exchange TCAP messages with another network SP for a variety of transaction-oriented operations. TCAP Applications signaling requires the signaling information provided by Basic and Global Title Signaling Transport along with the transaction capabilities of the SS7 TCAP. TCAP Applications signaling is provided via the POI-S17 interface. TCAP consists of transaction capabilities that manage remote operations via connectionless message transfer.

TCAP applications are only recently becoming a topic of interest to the RBOCs and other types of carriers. To TIA's credit, POI-S17 is an acknowledgment of the potential of wholesale services to other carriers. GR−145-CORE does not explicitly define an interface to support TCAP services.

POI-S18: ISUP End-to-End Signaling This interface requires further definition, work, and study. However, it is currently possible in the LEC network. The LECs currently provide ISUP message transport for the purpose of providing other service providers access to the LEC 800 database. At this time, there are no market drivers for such an interface.

Direct Inward Dialing and Dial Line Connections The Dial Line is the first interconnection type used between the wireless service provider and the LEC. The *Direct Inward Dialing* (DID) connection is the second oldest. Both are still in use today. However, IS−93 does not address these interconnections.

As the industry evolves, the need for interconnect standardization will grow. ANSI-93 and GR-145-CORE may or may not play a role in the evolution of the telecommunications industry. Regardless of whether or not current industry practices or standards are supporting the changing telecommunications industry, the industry must agree on a set of practices or standards. In the long run, common practices or industry standards should ensure the industry will keep itself from falling into a pit of chaos.

APPENDIX C

A Summary of the Telecommunications Act of 1996

In February 1996, the President of the United States signed into law the Telecommunications Act of 1996. The Telecommunications Act of 1996, also known as the Act, completely rewrote the rules of the Communications Act of 1934. The Act had the far-reaching effect of changing the rules for competition and regulation in most sectors of the telecommunications industry. The Act's provisions fall into five major areas:

- Telephone service
- Telecommunications equipment manufacturing
- Cable television
- Radio and television broadcasting
- The Internet and online computer services

The Act invalidated all state regulatory rules that restricted entry or limited competition in local and long distance wireline telephone service. The Act effectively superseded all of the MFJ's restrictions on AT&T, the *Baby Bell Operating Companies* (BOCs), and GTE. The Act enables AT&T long distance to enter the markets. The Act also allows GTE and the baby BOCs to enter the long distance business as long as they take certain steps as prescribed by the FCC. These steps are a "to do" checklist of that will ensure the FCC that barriers to entry into the local market have been removed. The checklist, also known as the Competitive Checklist, is summarized below:

- *Interconnection* The ILEC must allow other carriers to link networks to its network for the mutual exchange of traffic.
- *Access to Unbundled Network Elements* The telecommunications network is comprised of individual network elements. In order to provide "access" to an unbundled network element, the ILEC must provide a connection to the network element at any technically feasible point under rates, terms, and conditions that are just, reasonable, and nondiscriminatory.
- *Access to Poles, Ducts, Conduits, and Rights-of-Way* In order to serve customers, telephone company wires must be attached to, or pass through, poles, ducts, conduits, and rights-of-way. The ILEC must demonstrate that other carriers can obtain access to its telephone poles, ducts, conduits, and rights-of-way within reasonable time frames

and on reasonable terms and conditions, with a minimum of administrative costs, and in a manner that is consistent with fair and efficient practices.

- *Unbundled Local Loops* Local loops are the wires, poles, and conduits that the telephone company end office connects to the customer's home or business. The Act ensures nondiscriminatory access to unbundled local loops. This access ensures new entrants can provide quality telephone service promptly to new customers without constructing new loops to each customer's home or business.
- *Unbundled Local Transport* Nondiscriminatory access to the ILEC's transport facilities ensures that calls carried over competitors' lines are completed properly.
- *Unbundled Local Switching* A switch connects end user lines to each other and to trunks used for transporting calls. The ILEC is required to provide nondiscriminatory pricing for the switching of calls for competing carriers and non-competing carriers. As an illustration, this switching refers to switching calls from CLEC "A" to CLEC "B" or from CLEC "A" to the ILEC itself. Current pricing ranges from .0025 cents to .0045 cents for local switching and .0015 cents for tandem (regional) switching.
- *9-1-1 and E9-1-1 Services, Operator Services, and Directory Assistance* It is critical that the ILECs provide competing carriers with accurate and nondiscriminatory access to 9-1-1/E9-1-1 services so the competing carriers' customers are able to reach emergency assistance.
- *White Pages Directory Listings* These are the listings of customers' telephone numbers in a particular area. The ILEC shall provide competing carriers with access to all white page listings in the ILEC's operating area. Access shall include updated subscriber listings and permission to copy all white page listings.
- *Numbering Administration* The ILEC must demonstrate compliance with industry guidelines and FCC requirements to ensure that its competitors have the same access to new telephone numbers in a given area code as that provided to the ILEC.
- *Databases and Associated Signaling* New entrants must have the same access as the ILEC to these databases and signaling systems in order to have the same ability as the ILEC to transmit, route, complete, and bill for telephone calls.
- *Number Portability* Number portability enables customers to take their telephone number with them when they change local telephone companies. The ILEC shall ensure that all ILEC subscribers are

unimpeded in the transference of service and associated telephone number from the ILEC to the competing carrier. The ILEC will be facilitate the transfer of service and number to the competing carrier.

- *Local Dialing Parity* Local dialing parity ensures that end customers are not inconvenienced simply because they subscribe to a new entrant for local telephone service.
- *Reciprocal Compensation* The ILEC must compensate other carriers for the cost of transporting and terminating its local calls unless a mutually agreeable alternate arrangement is established.
- *Resale* The ILEC must offer other carriers all of its retail services at wholesale rates without unreasonable or discriminatory conditions or limitations so other carriers may resell those services to customers. This shall also include nondiscriminatory access to all Operational Support Systems.

The FCC effectively has the authorization to determine whether the baby BOCs and GTE have lowered the barriers to entry in the local market. The process has explicit parameters to measure the anticompetitive nature of the local telephone company. State regulatory commissions are required to implement the provisions of the Act.

On August 8, 1996, the *Federal Communications Commission* (FCC) released the First Report and Order, followed immediately by a Second Report and Order, in CC Docket No. 96–98 (collectively the Local Competition Orders), adopting rules to implement the local competition provisions of the Telecommunications Act of 1996. These decisions represent the first concrete national (and Federal government-supported) effort to fashion/create rules for the development of competitive local telephone markets. Given the recent entry of Bell Atlantic into the long distance market, the rules appear to have been successfully implemented.

All LECs, not just incumbent LECs, must make available all of their telecommunications services for resale to requesting telecommunications carriers. The FCC did not specify a detailed list of services that must be made available for resale; instead, it noted that state utility commissions, LECs, and resellers can determine the services that a LEC must provide at wholesale rates by examining that LEC's retail tariffs. LECs are neither required to resell exchange access services, nor are they required to resell services available at wholesale rates to parties who are not telecommunications carriers or who are purchasing the service for their own use.

LECs may not place restrictions and conditions on the use of resale services without making a showing (visible and documentable demonstration) to a state utility commission that such restrictions are not unreasonable.

The FCC, however, identified several exceptions to this rule. For example, retail services offered at promotional discounted prices by LECs for a period of 90 days or less need not be offered at the same discount to resellers, and restrictions prohibiting cross-class reselling of residential services to non-residential users are reasonable.

The Act enabled CLECs and long distance carriers to enter the local market in three different ways:

- Resale (also known as non-facilities based)
- Unbundled network elements
- Facilities

The reader should note that the Act also addresses:

- Availability
- Pricing methodology
- Provisioning

The following section addresses the three ways in which carriers can enter the local marketplace

Resale Based-Entry Into Local Competition

The following issues must be addressed in the resale market in order for the ILECs to open up their local markets to other carriers.

Availability The ILEC must be prepared to offer CLECs the technical and business capability to offer ILEC services at discounted prices to ILEC subscribers.

Pricing Methodology The ACT established national rules for use by state commissions in setting wholesale resale rates in order to promote "expeditious and efficient" entry into the local exchange market. The FCC pricing rules apply to incumbent LECs, but not to non-incumbent LEC's. The national rules have two components:

- Price setting methodology
- Default discount range for wholesale prices

Per the FCC, state commissions are to use the price setting methodology to establish discount rates within their individual states. If a state commission cannot establish a wholesale rate based on the FCC's methodology, the state should establish interim wholesale rates within the FCC default discount range.

Provisioning ILECs must ensure that resale service is at least equal in quality to that provided to its own subscriber by itself, any of its affiliates, or its end users. Resale services must be provisioned with the same timeliness as they are provisioned to the LEC's subsidiaries or end users.

Unbundling Based Entry into Local Competition: Unbundled Network Elements

The following issues must be addressed in the *Unbundled Network Element* (UNE) market in order for the ILEC to open up the local market.

Availability Incumbent LEC's must provide, to any requesting carrier, nondiscriminatory access to network elements on an unbundled basis at any "technically feasible" point. Incumbent LEC's must also provide access to the features, functions, and capabilities associated with unbundled network elements in a manner that will allow a requesting carrier to provide any service that can be offered through that element.

Incumbent LEC's must make available a minimum of seven unbundled network elements to CLECs:

- Local loops
- *Network interface devices* (NID)
- Local and tandem switches
- Interoffice transmission facilities
- Signaling and call-related database facilities
- Operations support systems and functions
- Operator and directory assistance facilities

The ILEC must provide both line and trunk side facilities when providing switching capabilities, plus the features, functions and capabilities of the switch, including all "vertical features," that the switch is able to provide.

When providing interoffice transmission facilities, an ILEC must, to the extent technically feasible, provide cross connects to the requesting carriers' equipment. In regard to offering the signaling and call-related database unbundled elements, ILEC's must offer requesting carriers access to

- Signaling networks—including access to both signaling links and signaling transfer points
- Call-related databases—databases for, among other things, billing, collection, transmission, and routing functions
- Service management systems—which allow competitors to create, modify, or update information in call-related databases

With respect to the operations support systems network element, ILEC's must provide access to its pre-ordering, ordering, provisioning, maintenance and repair, and billing functions by January 1, 1997. This is also known as Electronic Bonding. If a carrier requests an ILEC to unbundle the facilities and functionality's providing operator services and directory assistance, the ILEC must provide the requesting carrier with nondiscriminatory access at any technologically feasible point. To the extent technologically feasible, ILEC's must also offer customized routing that would route calls to a competitor's operator service or directory assistance platform.

Pricing Methodology The FCC developed one pricing methodology that applies to network interconnection, methods of obtaining interconnection, and unbundled elements. Under this methodology, an ILEC must base prices on forward-looking economic cost-based pricing. State regulatory commissions will approve the specific rates.

Incumbent LECs must base cost estimates on the use of the most efficient telecommunications technology currently available and are not permitted to consider embedded costs, retail costs, opportunity costs, or revenues to subsidize other services. Element rates shall be structured consistently with the manner in which the costs of providing the elements are incurred. For example, the cost of dedicated facilities shall be recovered through flat-rated charges. The FCC set specific price proxy ceilings for local loops in each state except Alaska for local switching and tandem switching.

Provisioning An ILEC must make unbundled elements available through both physical and virtual collocation. If an ILEC refuses to provide access to its unbundled elements, the ILEC is required to prove to the state commission that the requested method of access to its network elements is technically infeasible.

An ILEC must permit the collocation of any type of equipment used for interconnection or access to unbundled elements. It is not required to allow collocation of equipment to provide enhanced services. The FCC also provided detailed operational requirements with respect to collocation, including, among other requirements, that an ILEC must make space available on a first-come, first-serve basis and when planning renovations, must take into account projected demand for collocation of equipment. Further, when providing virtual collocation, an ILEC must service virtually collocated equipment at the same level as service is rendered to equipment used for its own use. When a telecommunications carrier purchases access to an unbundled network element, the ILEC retains the responsibility of maintaining, repairing, or replacing that element.

An ILEC must offer the same terms and conditions equally to all requesting carriers seeking collocation, and those terms and conditions must be no less favorable than the terms and conditions that the ILEC provides to itself. Incumbent LEC's are prohibited from "imposing limitations, restrictions, or requirements on requests for, or the sale or use of, any unbundled elements that would impair the ability of requesting carriers to offer telecommunications service in the manner they intend."

An ILEC shall provide unbundled network elements in such a way that a requesting carrier may combine them to provide a telecommunications service and, except upon request, may not separate elements that it currently combines. If requested and technically possible, the ILEC must combine requested network elements, even if they are not usually combined by the ILEC.

Facilities-Based Entry into Local Competition

The following issues must be addressed by the ILEC in order to open up its local markets.

Availability Telecommunications carriers seeking to enter the local telephone service market by deploying their own facility-based networks still require network "interconnection" with LECs. The FCC defines interconnection as the physical linking of two carriers' networks for the mutual exchange of traffic, independent of the transport and termination of traffic.

ILECs must offer network interconnection at any "technically feasible" point to requesting carriers providing exchange service or exchange access service. The ILECs must provide interconnection at the following six technically feasible points:

- Line side of a local switch
- Trunk side of a local switch
- Trunk interconnection points for a tandem switch

- Central office cross-connect points
- Out-of-band signaling facilities
- Points of access to unbundled

Pricing Methodology The FCC developed one pricing methodology that applies to interconnection, methods of obtaining interconnection, and unbundled elements. Under this methodology, an ILEC must base prices on forward-looking economic cost-based pricing. Once again, state commissions will approve the specific rates.

Incumbent LECs must base all cost estimates on the use of the most efficient telecommunications technology currently available and are not permitted to consider embedded costs, retail costs, opportunity costs, or revenues to subsidize other services. Network element rates shall be structured consistently with the manner in which the costs of providing the elements are incurred.

Provisioning ILECs must make interconnection or access to unbundled elements available through physical collocation, virtual collocation, and meet-point arrangements. If an ILEC refuses to provide a requested interconnection type, it must prove to the state commission that interconnection at that particular point is "technically infeasible."

An ILEC must permit the collocation of any type of equipment used for interconnection or access to unbundled elements. It is not required to allow collocation of equipment to provide enhanced services. The FCC provided detailed operational requirements with respect to collocation, including, among other requirements, that an ILEC must make space available on a first-come, first-serve basis and when planning renovations, it must take into account projected demand for collocation of equipment. Further, when providing virtual collocation the ILEC must service virtually collocated equipment at the same level as service is rendered to equipment used for its own use.

An ILEC must offer the same terms and conditions to all requesting carriers seeking collocation or meet point arrangements, and those terms and conditions must be no less favorable than the terms and conditions that the ILEC provides to itself. If a carrier requests interconnection at a quality that is superior to that which the ILEC provides itself, and a superior quality interconnection is technically feasible, the ILEC is obligated to comply with the request.

BIBLIOGRAPHY

Adamson, S., *Advanced Satellite Applications—Potential Markets* (1995), NJ, Booz Allen and Hamilton.

Alliance for Telecommunications Industry Solutions (ATIS), *Network Operations Forum* (NOF) documents (1990–1997).

American National Standards Institute—ANSI T1.110-118.

Azmak, Okan, member of technical staff, Lucent Technologies, New Jersey.

Bellcore, *BOC Notes on The LEC Networks* (1994), Issue 2, SR-TSV2002275, Piscataway, NJ.

Chu, Lawrence, consolidated work and teachings of L. Chu, New York Telephone, NYNEX, and Bell Atlantic-engineering and regulatory (1983–1999), retired. Chu currently is president of CCG, Inc.

Dahlbom, C.A., "Signaling Systems for Control of Telephone Switching" (1960), *Bell System Technical Journal*.

Eifinger, Charles P., consolidated teachings of Charles Eifinger, telecomm engineering consultant, NYNEX (1983–1999), retired.

Elbert, Bruce R., *Introduction to Satellite Communication*, (1987) ARTECH House.

Engineering and Operations in the Bell System (1982), AT&T Bell Laboratories, NJ.

Fix, Michael S., former network planner for Nextel Communications and NextWave Wireless. United States Air Force engineering and communications retired.

Grise, Don, telecommunications expert, formerly of Ameritech, director of interconnect for Bellcore, New Jersey.

Harte, Lawrence (1992), *Dual Mode Cellular*, Steiner Publishing.

Heldman, Robert, *Future Telecommunications* (1993), McGraw-Hill.

IEEE—802 series.

Joel, A.F., "What is Telecommunications Engineering?"

Lee, William (1989), *Mobile Cellular Telecommunications*, McGraw-Hill.

Lojko, Peter, interviews, 1996–1999, former Vice President for Engineering for American Mobile Satellite Corporation, former General Manager and Vice President of Operations for NextWave Wireless, and former Vice President of Wireline Business for Excel Switching.

Nickelson, R.L. (1984) Overview, "Space Communications."

Notes on the Network 1980, AT&T.

Ramler, J. (1984), *AIAA 10th CSSC*, Orlando.

Russell, Alicia, telecommunications expert, Austin, Texas.

Russell, Travis, Telecommunications Protocols, (1997) McGraw-Hill.

Schnipper, H. (1984), *SBS System*, IEEE Communications.

Shriver, Jube, October 3, 1996, "Satellite Firm Dealt Blow on Internet Plans," Los Angeles Times.

Smith, Lonnie, telecommunications expert, BellSouth Telecommunications interconnect (retired).

Telecommunication Transmission Engineering (1977), AT&T.

Telecommunications Act of 1996, S.652, One Hundred Fourth Congress of the United States of America.

Wilder, Floyd, *Guide to the TCP/IP Protocol Suite*, (1993), ARTECH House.

Wrobel, Leo (1990), *Disaster Recovery Planning*, ARTECH House.

Young, Harry, consolidated work of Harry Young, interconnection consultant (1992–1999).

INDEX

Symbols

A

B

C

D

I-K

L

M

N

R

T

U

V

W

X-Z

ABOUT THE AUTHOR

P.J. Louis is Vice President of Marketing and Product Management at TruePosition, Inc., a leader in wireless location systems. He is former director of network technology at NextWave Wireless, former senior managing engineer at Bell Communications Research, Inc., and former chief of staff and director of engineering for NYNEX. He has a B.S. in electrical engineering from Columbia University and a M.S. in technology management from Polytechnic University of New York. He is a registered engineer in the the state of New York. He is a senior member of the *Institute of Electrical and Electronics Engineers* (IEEE); having joined in 1979. He is a former officer of IEEE Communications Society–NY Section. He was a member of the editorial advisory board for Cellular Marketing magazine. He was a former chairman and vice chairman of several *Telecommunications Industry Association* (TIA) standards groups. He resides in New York state.